Climate Change and Agriculture

Climate Change and Agriculture

Perspectives, Sustainability and Resilience

Edited by

Noureddine Benkeblia
The University of the West Indies, Mona Campus, Kingston, Jamaica

WILEY

This edition first published 2023

Registered Offices
John Wiley & Sons, Inc., 111 River Street, Hoboken, NJ 07030, USA
John Wiley & Sons Ltd, The Atrium, Southern Gate, Chichester, West Sussex, PO19 8SQ, UK

Editorial Office
The Atrium, Southern Gate, Chichester, West Sussex, PO19 8SQ, UK

For details of our global editorial offices, customer services, and more information about Wiley products visit us at www.wiley.com.

Wiley also publishes its books in a variety of electronic formats and by print-on-demand. Some content that appears in standard print versions of this book may not be available in other formats.

Library of Congress Cataloging-in-Publication Data
Names: Benkeblia, Noureddine, editor.
Title: Climate change and agriculture : perspectives, sustainability and resilience / Noureddine Benkeblia.
Description: First edition. | Hoboken, NJ, USA : Wiley, 2023 | Includes bibliographical references and index.
Identifiers: LCCN 2022032131 (print) | LCCN 2022032132 (ebook) | ISBN 9781119789758 (hardback) | ISBN 9781119789765 (adobe pdf) | ISBN 9781119789772 (epub)
Subjects: LCSH: Crops and climate. | Agriculture–Environmental aspects. | Sustainable agriculture. | Climatic changes.
Classification: LCC S589.75 .C538 2023 (print) | LCC S589.75 (ebook) | DDC 630–dc23/eng/20220722
LC record available at https://lccn.loc.gov/2022032131
LC ebook record available at https://lccn.loc.gov/2022032132

Cover Design: Wiley
Cover Image: © David Holmes Geography/shutterstock

Set in 9.5/12.5pt STIXTwoText by Straive, Pondicherry, India

Noureddine Benarba
Sonia (Badra) Kheitmi
Françoise Varoquaux

List of Contributors

Kamruza Z. Ahmed
Climate and GIS Laboratory
Soils Department
Tocklai Tea Research Institute
Tea Research Association
Jorhat, Assam, India

Hiba Alatrash
Department of Field Crops
General Commission for Scientific Agricultural Research
Aleppo, Syria

Nujoud Alimad
Department of Plant Protection
Faculty of Agriculture
Damascus University
Damascus, Syria

Abdel Rahman Mohammad Said Al-Tawaha
Department of Biological Sciences
Al-Hussein Bin Talal University
Ma'an, Jordan

Abdel Razzaq Al-Tawaha
Department of Crop Science
Faculty of Agriculture
Universiti Putra Malaysia
Serdang, Selangor, Malaysia

Amanullah
Department of Agronomy
The University of Agriculture
Peshawar, Pakistan

Muhammad Arif
Department of Agronomy
The University of Agriculture
Peshawar, Pakistan

Azariah Babu
Tea Research Association
North Bengal Regional R & D Centre
Nagrakata, West Bengal, India

Bhaswatee Baroowa
Department of Environmental Science
Tezpur University
Tezpur, Assam, India

Humberto Alves Barbosa
Laboratório de Análise e Processamento de Imagens de Satélites (LAPIS)
Instituto de Ciências Atmosféricas
Universidade Federal de Alagoas
Alagoas, Brazil

Mina Bayanati
Department of Horticultural Science
Ferdowsi University of Mashhad
Mashhad, Iran

Sourajit Bayen
Department of Entomology
Tocklai Tea Research Institute
Tea Research Association
Jorhat, Assam, India

Goran Bekavac
Institute of Field and Vegetable Crops
Novi Sad, Serbia

Noureddine Benkeblia
Department of Life Sciences
The University of the West Indies
Kingston, Jamaica

Nilutpal Bhuyan
Department of Energy
Tezpur University
Tezpur, Assam, India

Mahyeddine Boumendjel
Laboratory of Biodiversity and Ecosystems Pollution
Faculty of Life and Nature Sciences
Chadli Bendjedid University
El-Tarf, Algeria

Sandra Cvejić
Institute of Field and Vegetable Crops
Novi Sad, Serbia

Javeid Ahmed Dar
SKUAST-K, Shalimar
Srinagar, Jammu and Kashmir, India

Catarina de Oliveira Buriti
National Semi-Arid Institute (INSA)
Ministry of Science
Technology, Innovations and Communications (MCTIC)
Campina Grande, Brazil

Rupanjali Deb Baruah
Climate and GIS Laboratory
Soils Department
Tocklai Tea Research Institute
Tea Research Association
Jorhat, Assam, India

Boško Dedić
Institute of Field and Vegetable Crops
Novi Sad, Serbia

Bhabesh Deka
Tea Research Association
North Bengal Regional R & D Centre
Nagrakata, West Bengal, India

Pratibha Deka
Department of Environmental Science
Tezpur University
Tezpur, Assam, India

Abhijit Dey
Department of Life Sciences
Presidency University
Kolkata, West Bengal, India

Rajarshi K. Gaur
Department of Biotechnology
Deen Dayal Upadhyaya Gorakhpur University
Gorakhpur, Uttar Pradesh, India

Jason Giovannettone
Dewberry, Fairfax, VA, USA

Nirmali Gogoi
Department of Environmental Science
Tezpur University
Tezpur, Assam, India

Mayuree Gohain
Department of Environmental Science
Tezpur University
Tezpur, Assam, India

Sonja Gvozdenac
Institute of Field and Vegetable Crops
Novi Sad, Serbia

Mohammad Ilyas
Ministry of Agriculture
Research Wing
Agriculture Research Institute
Mingora Swat, Pakistan

Imran
Department of Agronomy
The University of Agriculture
Peshawar, Pakistan;
Ministry of Agriculture
Agriculture Extension Wing
Merged Area
Khyber Pakhtunkhwa, Pakistan

Saher Islam
Department of Biochemistry and Biotechnology
Institute of Biochemistry and Biotechnology
University of Veterinary and Animal Sciences
Lahore, Pakistan

Yaman Jabbour
Department of Field Crops
General Commission for Scientific Agricultural Research
Aleppo, Syria

Siniša Jocić
Institute of Field and Vegetable Crops
Novi Sad, Serbia

Arun Karnwal
Department of Microbiology
School of Bioengineering and Biosciences
Lovely Professional University
Phagwara, Punjab, India

Rupam Kataki
Department of Energy
Tezpur University
Tezpur, Assam, India

Naila Kerkoub
Laboratory of Biodiversity and Ecosystems Pollution
Faculty of Life and Nature Sciences
Chadli Bendjedid University
El-Tarf, Algeria

Shah Khalid
Department of Agronomy
The University of Agriculture
Peshawar, Pakistan

Samia Khanum
Department of Botany
University of the Punjab
Lahore, Pakistan

Ankica Kondić -Špika
Institute of Field and Vegetable Crops
Novi Sad, Serbia

Zineb Labidi
Laboratory of Biodiversity and Ecosystems Pollution
Faculty of Life and Nature Sciences
Chadli Bendjedid University
El-Tarf, Algeria

Tumuluru Venkata Lakshmi Kumar
Atmospheric Science Research Laboratory
SRM University
Chennai, Tamilnadu, India

Tariq Mahmood
Department of Bioinformatics/Agriculture
Hazara University
Mansehra, Pakistan

Ana Marjanović Jeromela
Institute of Field and Vegetable Crops
Novi Sad, Serbia

Sanja Mikić
Institute of Field and Vegetable Crops
Novi Sad, Serbia

Dragana Miladinović
Institute of Field and Vegetable Crops
Novi Sad, Serbia

Milan Mirosavljević
Institute of Field and Vegetable Crops
Novi Sad, Serbia

Bojan Mitrović
Institute of Field and Vegetable Crops
Novi Sad, Serbia

Mousumi Mondal
Department of Agronomy
Bidhan Chandra Krishi Viswavidyalaya
Mohanpur
Nadia, West Bengal, India

Rachid Mrabet
Department of Environment and Natural Resources
Scientific Division
National Institute for Agricultural Research (INRA)
Rabat, Morocco

Hichem Nasri
Laboratory of Biodiversity and Ecosystems Pollution
Faculty of Life and Nature Sciences
Chadli Bendjedid University
El-Tarf, Algeria

Nidal Odat
Department of Medical Laboratories
Faculty of Science
Al-Balqa Applied University
Al Salt, Jordan

Ibrahim Ortas
Department of Soil Science and Plant Nutrition
Faculty of Agriculture
University of Çukurova
Adana, Turkey

Jelena Ovuka
Institute of Field and Vegetable Crops
Novi Sad, Serbia

Vineeta Pandey
Department of Biotechnology
Deen Dayal Upadhyaya Gorakhpur University
Gorakhpur, Uttar Pradesh, India

Franklin Paredes-Trejo
Department of Civil Engineering
Institute for Sustainable Agroindustry
University of the Western Plains
Ezequiel Zamora
San Carlos, Venezuela;
Laboratório de Análise e Processamento de Imagens de Satélites (LAPIS)
Instituto de Ciências Atmosféricas
Universidade Federal de Alagoas
Alagoas, Brazil

Sreyashi Paul
Department of Environmental Science
Tezpur University
Tezpur, Assam, India

José Prieto
Eumetsat (European Organisation for the Exploitation of Meteorological Satellites)
Darmstadt, Germany

Aleksandra Radanović
Institute of Field and Vegetable Crops
Novi Sad, Serbia

Somnath Roy
Department of Entomology
Tocklai Tea Research Institute
Tea Research Association
Jorhat, Assam, India

Jeyabalan Sangeetha
Department of Environmental Science
Central University of Kerala
Kasaragod, Kerala, India

Swapnil Ganesh Sanmukh
Institute for Bioengineering of Catalonia (IBEC)
The Institute of Science and Technology
Barcelona, Spain

Banothu Santhosh
Department of Crop Physiology
College of Agriculture
Rajendranagar, PJTSAU, Hyderabad, Telangana, India

Suman Sarkar
Tea Research Association
North Bengal Regional R & D Centre
Nagrakata, West Bengal, India

Eduard Torrents Serra
Institute for Bioengineering of Catalonia (IBEC)
The Institute of Science and Technology
Barcelona, Spain

Mohammad Shatnawi
Department of Biotechnology
Faculty of Agricultural Technology
Al Balqa Applied University
Salt, Jordan

Aarshi Srivastava
Department of Biotechnology
Deen Dayal Upadhyaya Gorakhpur University
Gorakhpur, Uttar Pradesh, India

Lee-Ann Steenkamp
Stellenbosch Business School
University of Stellenbosch
Cape Town, South Africa

Amreena Sultan
SKUAST-K, Shalimar
Srinagar, Jammu and Kashmir, India

Devarajan Thangadurai
Department of Botany
Karnatak University
Dharwad, Karnataka, India

Winnie Thebuho
Stellenbosch Business School
University of Stellenbosch
Cape Town, South Africa

Pratibha Vyas
Department of Microbiology
College of Basic Sciences and Humanity
Punjab Agricultural University
Ludhiana, Punjab, India

Miroslav Zorić
Institute of Field and Vegetable Crops
Novi Sad, Serbia

Preface

Food production is a key component of our survival. With the growing population, complex and increasingly globalized systems have developed in order to meet our quantitative and qualitative needs for food. On the other hand, agriculture is also a significant contributor to anthropogenic global warming and greenhouse gas (GHG) emissions. From the farm to the plate, food crops are grown, harvested, stored, processed, packaged, transported, and distributed, and this value chain releases greenhouse gases into the atmosphere, particularly farming, which releases significant amounts of methane and nitrous oxide, two powerful greenhouse gases. Therefore, it is urgent to transform the agricultural sector to respond to climate change by reducing agricultural emissions and sustainably increasing agricultural productivity, and preserving agroecosystems. The role of agriculture in climate mitigation should be seen much broadly than climate science alone can inform. The new approach of climate-smart agriculture (CSA), rooted in sustainable agriculture, will also contribute not only to achieving the reduction of GHG but also reducing the threat of hunger, supporting food security, and improving environmental management. Hence, both CSA and CA (conservation agriculture) can make important contributions in this area. With eighteen (18) chapters, this book covers the key topics on agriculture and climate change including novel and crop resilience, mitigation strategies, modern biological technologies, and climate-smart agriculture innovations. This book describes and analyses the sensitivity of the different agroecosystems and the diverse ways in which they might be affected by climate change. The different chapters describe the effects of climate change on agriculture and the impacts on crops, and offer broadened insights on cropping systems and the requirements to improve agricultural production under the adverse conditions of the future uncertain climate. A range of possible ways to adapt agriculture and agroecosystems in order to mitigate the disastrous consequences of climate change, particularly food security, are also presented. Agronomists, ecologists, climate change specialists, academia and graduate research students, and policy makers will certainly find the content of this book an appreciable addition to their collection.

Noureddine Benkeblia
The University of the West Indies, Kingston, Jamaica
November 2022

Foreword

No topic is more relevant in agriculture today than the unpredictable and potentially massive impacts of climate change. Among the swirling debates among politicians and vested commercial interests about long-term weather cycles and even total denial to calm the public and assure short-term profits, there is little disagreement among scientists that this phenomenon is real, and that it will have far-reaching effects on the global food supply and human wellbeing. The editor of *Climate Change and Agriculture*, Professor Noureddine Benkeblia from University of West Indies has assembled a series of well-focused chapters on this critical topic written by experts around the world. These deal with adapting crops and systems to changing conditions, with potential contributions of molecular methods to crop improvement, and with a number of specific components that need to be improved in order to meet the challenges ahead. The book is clearly written and meets a vital need at this time in history. It is recommended for those teaching courses in production and physiology, in pest management, and in providing adequate nutrients for crops that will likely be moving to new and less fertile soils. This new volume should be on the shelves of researchers and educators in agricultural science.

Charles Francis
Professor of Agronomy and Horticulture
University of Nebraska – Lincoln, Lincoln, NE, USA

and

Visiting Professor of Agroecology
Norwegian University of Life Sciences, Ås, Norway

1

Climate Change and Agriculture: State of the Art, Challenges, and Perspectives

Imran[1,2], Noureddine Benkeblia[3], Amanullah[1], and Abdel Rahman Mohammad Said Al-Tawaha[4]

[1] *Department of Agronomy, The University of Agriculture, Peshawar, Pakistan*
[2] *Ministry of Agriculture, Agriculture Extension Wing, Merged Area, Khyber Pakhtunkhwa, Pakistan*
[3] *Department of Life Sciences, The University of the West Indies, Kingston, Jamaica*
[4] *Department of Biological Sciences, Al- Hussein Bin Talal University, Ma'an, Jordan*

CONTENTS

1.1 Introduction

From the turn of the last century, the multiple effects of climate change, such as recurrent flooding, storms and hurricanes, global warming, sea level rising, and melting of ice, have been felt and observed in many regions of the world (Boettcher et al. 2021; Imran 2018; Imran et al. 2021a; Valone 2021). There is a general consensus among the scientific

Climate Change and Agriculture: Perspectives, Sustainability and Resilience,
First Edition. Edited by Noureddine Benkeblia.

communities that this change will likely impact not only agriculture production, food, and livelihood security but also the international economy and trade as well as human civilization (Brooks et al. 2009; Cottier and Shariff 2013; Imran et al. 2017). Over the last century, our planet has known a warming by 0.7 °C, and this warming is the result of human activities such as the use of fossil fuels and agriculture including livestock development. On another hand, learning about climate change and committing ourselves to protect our environment and mitigating this change is also one of the strategies that play a major role in addressing this challenge (Bentz 2020), and this human intervention will be a *sin qua non*-condition to mitigate these threatening changes. During the COP21 (https://www.un.org/sustainabledevelopment/blog/category/climate-change/cop21/) and Paris agreement where (i) a global framework was set out in order to avoid climate change by limiting global warming to well below 2 °C with a goal to limit it to 1.5 °C, (ii) a commitment to strengthen countries' ability to deal with the impacts of climate change and support them in their efforts, and the international community and most countries committed to take initiatives and put measures in place in order to limit greenhouse gas emissions, atmospheric carbon dioxide (CO_2) concentration is still increasing (Allen et al. 2018; Canadell et al. 2007; IPCC 2018; Ricke and Caldeira 2014). These increasing atmospheric carbon dioxide levels will be accompanied by temperature and precipitation changes (Figure 1.1) (Baker et al. 2018; Ricke and Caldeira 2014; Wigley et al. 1980), and even in the case emission are restricted, significant climate change is thought to be unavoidable (Ricke and Caldeira 2014).

Climate change affected lives across the world in drastic and unpredictable ways (Hardy 2003; Pecl et al. 2017). It is uncertain how climate change will affect agroecosystems and their ability to support productive farms or healthy natural ecosystems (Altieri et al. 2015; Imran 2017; Verchot et al. 2007). For example, researchers used digital techniques to predict how one vital soil characteristic, soil organic carbon, may be altered by climate change (Stevens et al. 2013; Stockmann et al. 2015; Taghizadeh-Mehrjardi et al. 2016). Research also established an association between the variation of soil organic carbon changes and soil types, land-use regimes, and current climate (Don et al. 2011; Gray and Bishop 2016; Hontoria et al. 1999; Ramesh et al. 2019; Smith 2008). For example, the projected average decline of soil organic carbon was less than one ton per hectare for sandy, low-fertility soils in dry conditions under cropping regimes. It was 15 times as much for clay-rich, fertile soils in wet conditions under native vegetation regimes. Predicting soil organic carbon changes is vital. It would allow us to better prepare for and adapt to altered soil conditions, and that would ultimately improve how we manage both agricultural and native ecosystems (ASA-CSSA-SSSA 2017; Gray and Bishop 2016).

1.2 Climate Change: A Global Perspective

Atmospheric pollution and thermal variation in biosphere harder the life of the living organism (plants, livestock, microbes). Fluctuation in environmental factors like floods and droughts making the life harder and harder for farmers to grow good quality and nutritious food (Hou et al. 2021; Ilyas et al. 2021). Farmers around the world depend on a stable climate to grow their crops and put food on their plates. But as climate change leads to more droughts, floods, and extreme weather, we see harvests wither or wash away

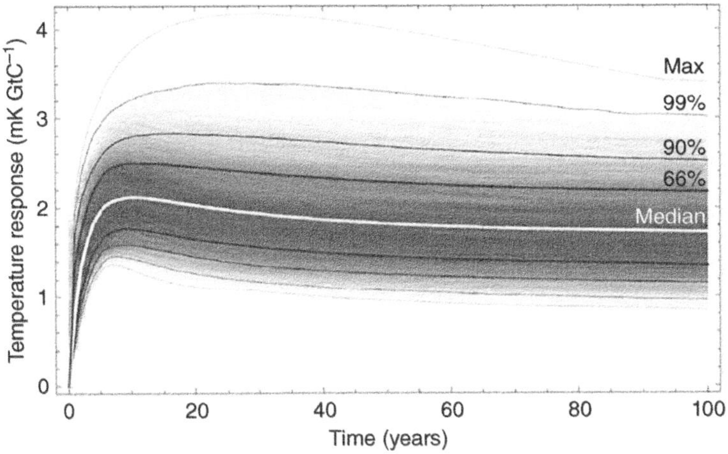

Figure 1.1 Temperature increases from an individual emission of carbon dioxide (CO_2). Time series of the marginal warming in mK (milliKelvin = 0.001 K) per GtC (=1015 g carbon) as projected by 6000 convolution-function simulations for the first 100 years after the emission. Maximum warming occurs a median of 10.1 years after the CO_2 emission event and has a median value of 2.2 mK GtC^{-1}. The colors represent the relative density of simulations in a given region of the plot. *Source:* Ricke and Caldeira 2014; published under Creative Commons CC BY 4.0 license.

(Abdul et al. 2016; Abrar et al. 2009; Ahmad et al. 2007). In 2011, during Mexico's worst-ever drought on record, more than two million acres of crops were lost. Climate change also makes soil less suitable for farming, as nutrients key to plant growth is stripped out by these same droughts and floods.

Some deniers will argue that carbon dioxide is 'plant food' and that more of it will help crops flourish, but this is a huge oversimplification of science (Farmer and Cook 2013; Pilkey et al. 2011). Plants also need water, but does that mean a flood is a good way to get it to them? Not. The same holds for carbon. Experiments show that some plants may do better in a world with more carbon dioxide, but many show signs of damage. Not convinced? Recent research has revealed that elevated levels of carbon dioxide (CO_2) in our atmosphere can have a 'junk food effect' on typically nutritious foods (Alamzeb et al. 2018). As increasing CO_2 speeds up photosynthesis, plants are producing more and more carbohydrates to keep up with their speed-up growth – at the expense of the minerals, vitamins, and proteins that take longer to build (Griffin and Seemann 1996; Kirschbaum 2011).

Climate refers to the average weather conditions in a place over many years (usually at least 30 years, to account for the range of natural variations from one year to the next). For example, the climate in Swat Pakistan (especially upper swat) is cold and snowy in the winter, while Peshawar Pakistan climate is hot and humid (Imran and Asad 2017; Imran and Khan 2015; Imran et al. 2016a,b). The climate in one area, like the Swat or Peshawar, is called a regional climate. The regional climate has great importance in respect of agricultural productivity, food security, and livelihood (Figure 1.2).

Recent studies have shown that plants and crops responded positively to regional climate and showed an ameliorating effect in terms of grain yield, plant height, thousand-grain weight, dry matter portioning, biological yield, oil yield, and quality of the crops (Baqa et al. 2015; Imran et al. 2015a,b; Iqbal et al. 2016; Khan 2015a,b; Naveed et al. 2016).

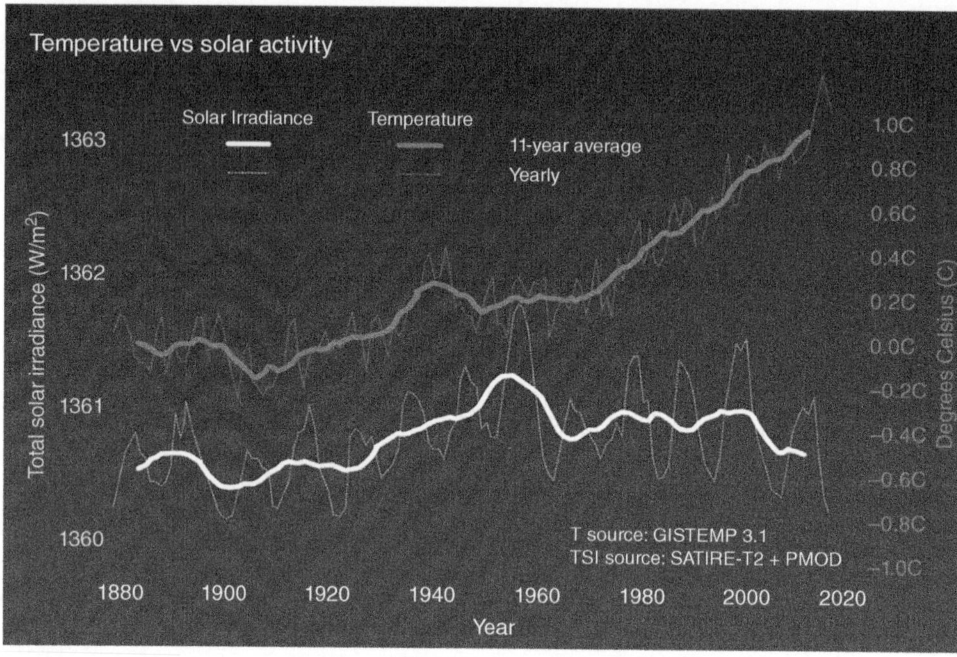

Figure 1.2 Increasing trend of solar radiation and temperature causing global warming.

A significant change in the Earth's climate is occurring slowly and gradually and influencing life on the planet earth. Climate can be defined as 'expected weather', however, when changes in the expected weather occur, it is known as climate changes which can be defined by the differences between average weather conditions at two separate times. Climate may change in different ways, over different time scales, and at different geographical scales (Clark 1985; Forchhammer and Post 2004). In recent times, scientists have become interested in global warming, due to mankind's impact on the climate system, through the enhancement of the natural greenhouse effect (Iqbal et al. 2016; Islam et al. 2015; Samreen et al. 2016). The earth is currently getting warmer because people are adding heat-trapping greenhouse gases to the atmosphere. The term 'global warming' refers to warmer temperatures, while 'climate change' refers to the broader set of changes that go along with warmer temperatures, including changes in weather patterns, the oceans, ice and snow, and ecosystems around the world. In their research, Anwar et al. (2016), Iqbal et al. (2016), and Samreen et al. (2016) reported that water vapour present in the atmosphere plays a vital role in GHGs trapping. The average climate around the world is called global climate. When scientists talk about global climate change, they are talking about a pattern of changes happening around the world over many years. One of the most important trends that scientists look for it is the average temperature of the Earth, which has been increasing for many years (Figures 1.3 and 1.4). The increase in temperature adversely affected all sectors of life but the most vulnerable is agriculture. Water shortage is enhancing day by day due to climate change in arid and semi-arid regions, which leads to demolish crop production practices (Imran and Asad 2015; Misra 2014; Naveed et al. 2016).

Figure 1.3 Increasing trend of the temperature and CO_2, accelerating climate change and global warming. *Source:* (a) Nirutft/Adobe Stock. (b) Dottedyeti/Adobe Stock. (c) Alexander/Adobe Stock.

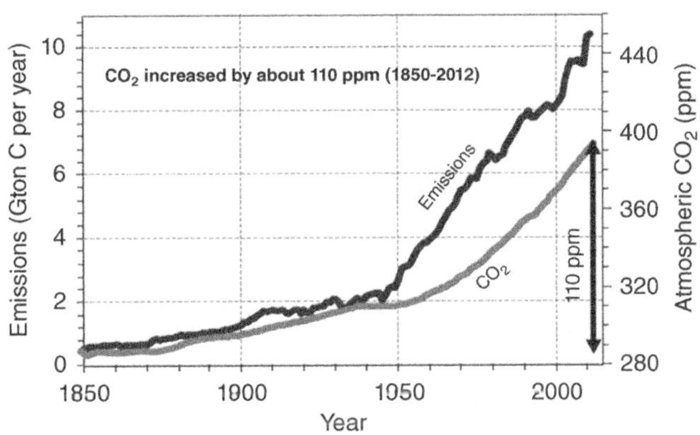

Figure 1.4 Average increasing trend and emission of CO_2 since 1850–2000.

1.3 Climate Change and Food Security

Food security has remained the foremost objective not only of the Government of Pakistan (Imran et al. 2020a,b,c) but also of the world (Duncan 2015). Therefore, the policymakers spend substantial time on designing sound food policy leading to food security (Figure 1.5). Various factors have an impact on food security including agronomic, institutional, and political factors, in addition to climatic factors (Imran et al. 2021c). Climate change is considered the most crucial factor influencing food security.

Food security includes different dimensions, namely production, distribution, and accessibility. Considering the food production dimension, the studies show a substantial impact of temperature and rainfall on food production (Mahmood et al. 2012; Janjua et al. 2010). Wheat is the staple food crop in Pakistan and many other countries as well and thus, the food security policy

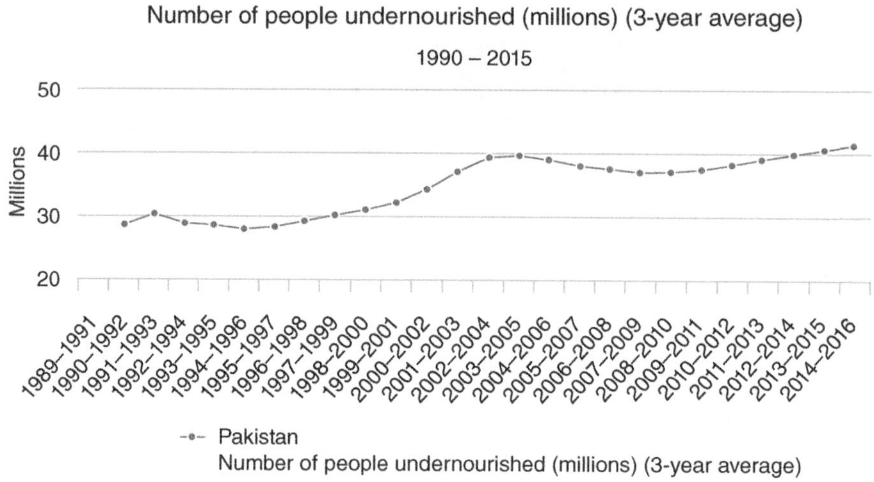

Figure 1.5 Food security and climate change a challenging issue showing an increasing trend of hunger. *Source:* Data from FAOStat 26 (2016).

mainly focuses on wheat production in the country. Farmers sow wheat in winter and harvest it in summer. Before the flowering stage, cold temperature accelerates the wheat growth process because high temperature can cause a delay in the growth of seedlings. Rainfall can damage wheat production at the time of harvesting, ultimately leading to a threat to food security in Pakistan (Janjua et al. 2010, 2014). In recent years, harsh weather conditions in Pakistan have affected adversely the supplies of agriculture production, particularly food crops, and it is expected that food production will be threatened in the coming years as well (GFSI 2012). Further, changes in temperature and rainfall during reproductive and vegetative stages are important in wheat production (Mitra and Bhatia 2008; Semenov 2009; Semenov et al. 2014). Some estimates show that wheat availability in the country has declined in the recent past (FAO 2013; GOP 2011). Per capita availability of wheat in Pakistan has an important position to achieve the objective of sustainable food security. Although pricing systems and regional trade strategies have important policy concerns (Hatfield and Dold 2018; Ortiz et al. 2008), climate change has a significant impact on wheat production. This demand determines the impact of climate change on wheat production in the country. Achieving the goal of food security, especially for poor populations, becomes difficult as a result of climate change (Gregory et al. 2005; Rosegrant and Cline 2003) coupled with other problems, such as rising population, shortage in water availability, and land degradation. Further, climate change is also linked to a decline in crop productivity (Cline 2007; Parry et al. 2005), thereby adversely affecting food security in Pakistan Adapting to climate change can reduce the negative impact on agricultural productivity (Di Falco and Chavas 2009). This paper focuses on wheat production in the presence of climate change in Pakistan. Small farms in Pakistan dominate as they account for more than 90% of total farms (GOP 2010) and food production is highly determined by these small farms. Punjab is the largest province of Pakistan with a contribution of around 80% in wheat production. Farmers in the non-irrigated region are more vulnerable to climate change compared to the irrigated region since farming depends on climatic factors (Al-Tawaha et al. 2020a; Alkahtani et al. 2020). Thus, it is imperative to consider both regions separately in determining the effects of changing temperature and rainfall on wheat production.

1.4 Ancestries of GHGs and Global Warming

Roughly one-third of the solar energy that reaches the top of Earth's atmosphere is reflected directly back to space (Figure 1.6). The remaining two-thirds are absorbed by the surface and, to a lesser extent, by the atmosphere. To balance the absorbed incoming energy, the Earth must, on average, radiate the same amount of energy back to space (Al-Tawaha et al. 2020c; Alwahibi et al. 2020).

Because the Earth is much colder than the sun, it radiates at much longer wavelengths, primarily in the infrared part of the spectrum. Much of this thermal radiation emitted by the land and ocean is absorbed by the atmosphere, including clouds, and reradiated back to Earth. This is called the greenhouse effect (Barnola et al. 1987).

It is wisely said that to solve a problem, first, you must know and understand the problem (Barnola et al. 1987). Climate change has happened in the past also and is a normal and natural process occurring cyclically due to changes in planetary movements. Research over the past century by scientific communities across the world has indicated that the alarming increase in the number of different greenhouse gases (GHGs), which include carbon dioxide, methane, nitrous oxide, and water vapour in the atmosphere, has buttressed the warming of the planet (Figures 1.5 and 1.6). It has also been discovered that there is a proportional change in the global temperature and GHGs released in the atmosphere which has caused the planet to warm by 1 °C in the last century and is continuing (Figure 1.7) (Ackerman and Stokes 2003; Khandekar et al. 2005).

Among the GHGs, the share of carbon dioxide in global warming is highest due to its higher proportional release (Figure 1.8). In the 1990s, the CO_2 levels today stand at 402 ppm (Ackerman and Stokes 2003; Barnett et al. 1999) and are further increasing at an alarming rate, and the last data showed a level of more than 420 ppm (Buis 2019). Figures 1.5 and 1.6 explains how the release of CO_2 by various anthropogenic activities determines its concentration levels in the atmosphere (Canadell et al. 2007; Harde 2019). The greenhouse effect caused by such gases is though important to keep the planet congenial for living, as in the absence of 'normal warming' earth will become a cold planet. The excess of anything is detrimental so is the case with GHGs (Figure 1.2).

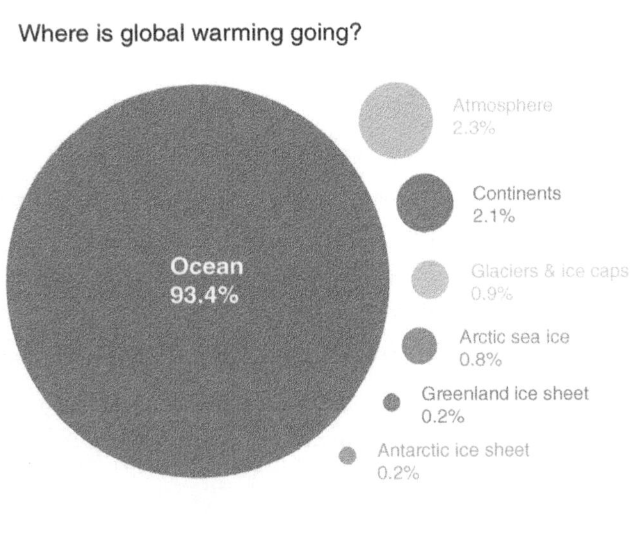

Figure 1.6 Impact of global warming and climate change on various components of the world.

Pakistan Meteorological Department

Projected impact of climate change			
	0 °C	1 °C →→→→→	→→→→→ +2 °C →→→→→
Food	Nil	Marginal crop eliminated Possible rise yield	Major crop eliminated Falling crop rise yield
Water	Nil	Small glaciers disappear Water shortage	Major glaciers disappear Significant water shortage
Risk	Nil	Extreme events (Floods, droughts, storms)	Frequent extreme events (Floods, droughts, storms)

Figure 1.7 Proportional change in the global temperature and GHGs released in the atmosphere and has caused the planet to warm by 1 °C.

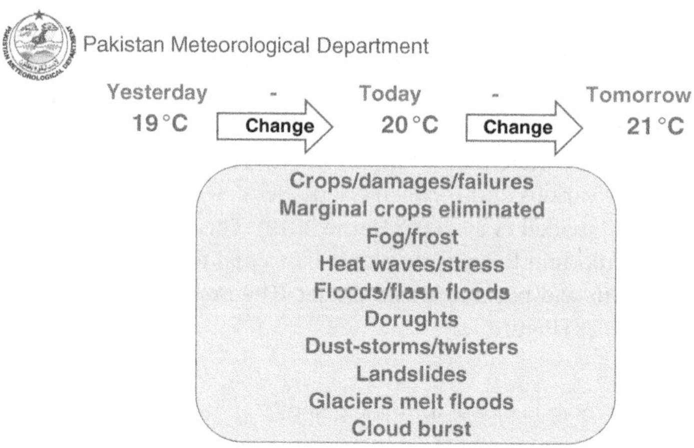

Figure 1.8 The greenhouse effect caused by such gases is though important to keep the planet congenial for living, as in the absence of 'normal warming' earth will become a cold planet.

1.5 Intensification of Global Health Issues with Climate Change

1.5.1 Healthy Life with Healthy Climate

Climate change impacts human health in countless ways, but four are worth emphasizing here.

a) *Temperatures*: As temperatures climb around the globe, we expect to see more heat-waves and ever more intense ones at that. Extreme heat can 'overpower the human

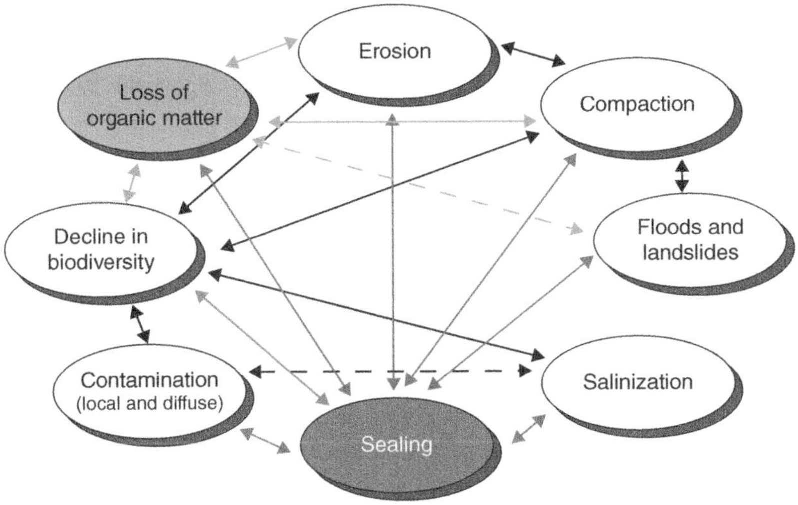

Figure 1.9 Increase in temperature and its devastating consequences.

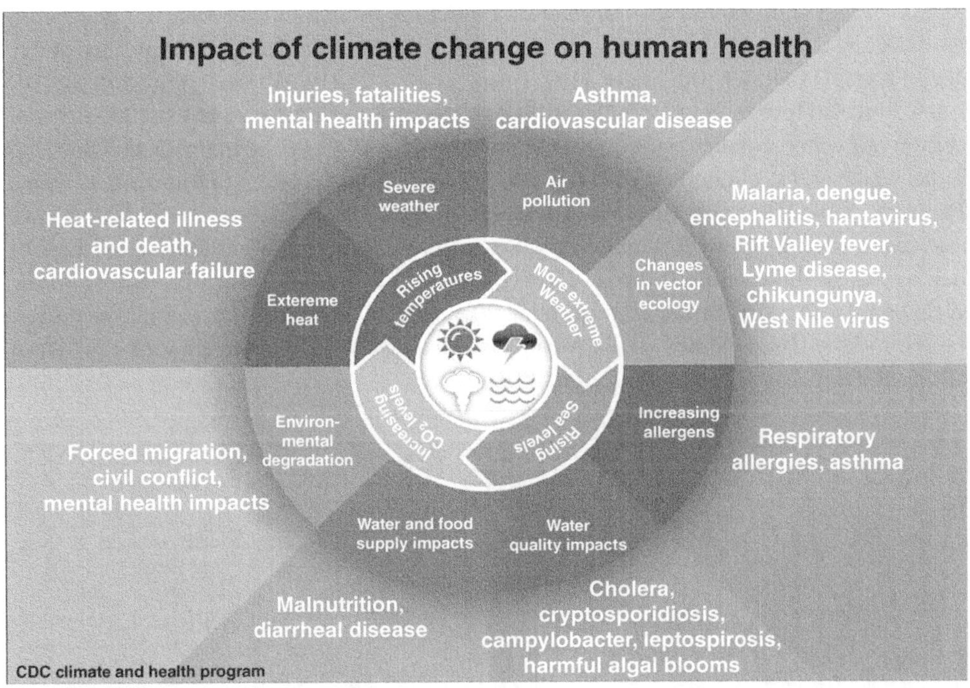

Figure 1.10 Life vulnerability towards climate change.

body' and cause dehydration, heatstroke, and major organs damage (Figure 1.10). And certain populations are more at risk from the impacts of heatwaves than others, including the elderly, children, and the poor (Alamzeb et al. 2018; Ali et al. 2015, 2019).

b) *Air pollution*: Pollution from burning fossil fuels is bad enough for the air we breathe, but many impacts of climate change also impact air quality. For instance, climate change has been linked to more wildfires. Wildfire smoke carries fine particles that 'can penetrate deep into your lungs.' Exposure has been linked to burning eyes, heart and lung diseases, and even death (Al-Tawaha et al. 2020b).

c) *Vector-borne diseases*: Vector-borne diseases are illnesses spread by insects or arachnids like mosquitoes, fleas, mites, and ticks. As our climate becomes warmer, some insects will see their geographic ranges grow – bringing the Lyme disease and West Nile or Zika viruses they carry along with them to new regions.

d) *Extreme weather*: While we go into more detail, climate change has been linked to many types of extreme weather, including hurricanes and floods. Not only can these extreme weather events have immediate fatal consequences, but they can lead to major injuries and the spread of waterborne illnesses such as wound infections, dermatitis, conjunctivitis, and ear, nose, and throat infections (Figures 1.9 and 1.10).

1.6 Climate Change Intensify Water Scarcity

Everyone deserves access to water for their survival. Climate change is making our water supply much less secure. Climate change poses a huge threat to something humans need above all else water (Frederick and Major 1997; Imran et al. 2021a,b). We need it for drinking, of course, but also for growing food. We are seeing the dire consequences climate change can have on our water supplies right now (Amanullah et al. 2015). While many factors have led to the water crisis, climate change has undoubtedly made a bad situation much worse (Figure 1.1). We know that increasing global temperatures can lead to droughts and have been experiencing record droughts for years in many regions getting only about half of their average annual rainfall since 2015 (Figures 1.1, 1.2, and 1.4). Many cities around the world could face a similar fate (Anwar et al. 2016, 2017). We already know that the largest reservoir in the United States – Lake Mead contained less than half the amount of water in 2015 as it did in 2000 (Karim and Imran 2019).

Figure 1.11 Tangible view of weather station for climatic data recording at Agriculture Research Institute, Mingora Swat.

Climate change significantly increasing water demand, limiting crop productivity, and reducing water availability in areas where irrigation is most needed or has a comparative advantage (Hanjra and Qureshi 2010; Imran et al. 2021a; Kang et al. 2009). Global atmospheric temperature is predicted to rise by approximately 4 °C by 2080, consistent with a doubling of atmospheric CO_2 concentration. Mean temperatures are expected to rise at a faster rate in the upper latitudes, with slower rates in equatorial regions (Baqa et al. 2015; Farooq et al. 2018). Mean temperature rise at altitude is expected to be higher than at sea level, resulting in an intensification of convective precipitation and acceleration of snowmelt and glacier retreat. In response to global warming, the hydrological cycle is expected to accelerate as rising temperatures increase the rate of evaporation from land and sea (Kundzewicz 2008; Ohmura and Wild 2002; Wu et al. 2010). Thus, rainfall is predicted to rise in the tropics and higher latitudes but decreases in the already dry semi-arid to arid mid-latitudes and the interior of large continents. Water-scarce areas of the world will generally become drier and hotter (Boettcher et al. 2021). Both rainfall and temperatures are predicted to become more variable, with a consequent higher incidence of droughts and floods, sometimes in the same place. Runoff patterns are harder to predict as they are governed by land use as well as uncertain changes in rainfall amounts and patterns (Hasanuzzaman 2019a,b; Hasanuzzaman and Fotopoulos 2019). Substantial reductions (up to -40%) in regional runoff have been modelled in south-eastern Australia and in other areas where annual potential evapotranspiration exceeds rainfall. Relatively small reductions in rainfall will translate into much larger reductions in a runoff, for example, a 5% fall precipitation in Morocco will result in a 25% reduction in runoff. In glacier-fed river systems, the timing of flows will change, although mean annual runoff may be less affected. As temperature rises, the efficiency of photosynthesis increases (Imran and Amanullah 2021; Tkemaladze and Makhashvili 2016).

1.7 Variation in Available Surface Water

Artificial irrigation provides approximately 40% of the world's food, including most of its agronomic and horticultural output, from an estimated 20% of agricultural land, or about 300 million ha worldwide (Imran and Amanullah 2021; Postel 1998). The green revolution technology of high inputs of nitrogen fertilizer, applied to responsive short-season varieties of rice and wheat, often required irrigation to realize its potential in Pakistan (Ibrahim and Khan 2017). During the same period, private investment in groundwater was stimulated by the availability of cheap pumps, power, and well construction methods not only did irrigated areas continue to grow, and access to groundwater services expanded. Consequently, aquifers are depleted in many parts of the world where they are most important in China, Pakistan, and the United States – sometimes facilitated by perverse incentives of subsidized energy and support prices for irrigated products (Boettcher et al. 2021). As the global population heads for more than nine billion people by 2050 (under medium growth projections), the world is rapidly becoming urbanized and wealthier. Food preferences are changing to reflect this, with declining trends in the consumption of staple carbohydrates, and an increase in demand for luxury products milk, meat, fruits, and vegetables – that are heavily reliant on irrigation in many parts of the world (Garnett 2014; Imran 2018). Future global food demand is expected to increase by some 70% by 2050 but will approximately double for developing countries. All other things being equal (that is a world without climate change), the amount of water withdrawn by irrigated agriculture will need to increase by 11% to match the demand for biomass production. Rice prices doubled and although commodity prices have fallen back since, the fundamentals (oil price, biofuel development, and continued rising food demand) are now expected to drive a period of high volatility in food prices. In the wake of this market turmoil, food security and agricultural livelihoods have regained importance in development planning, although some countries such as China seem ever more likely to balance further agricultural development and investment with imports (Imran 2018; Thrupp 2000). Pakistan has a large stock of under-performing canal irrigation infrastructure and a vibrant groundwater sector that is competitively depleting its lifeblood. Both create significant environmental externalities, which need to be managed (Imran et al. 2016a,b). Not only that, there are calls for water to be reserved to maintain environmental flows in rapidly developing river basins and restored to ecosystems in over-allocated ones (Ibrahim and Khan 2017; Opperman et al. 2019).

Water management in crop production tends to be concentrated on food crops where the timing and reliability of supply are critical. Water management (irrigation, drainage, and water conservation and control) achieves stability of crop production by maintaining soil conditions close to optimum for crop growth (Imran et al. 2020c; Hamdy et al. 2003; Stagnari et al. 2009). Irrigation allows the cultivation of crops when rainfall is erratic or insufficient, insures high-value, high-risk horticulture from failure, and has played a major role in achieving national and regional food security, as well as improving individual livelihoods (Fernald et al. 2012; Kahane et al. 2013). The extent and area of irrigation have grown massively in the twentieth century but have depleted surface and groundwater flows, often with severe consequences for aquatic ecosystems and those dependent on them (Burke and Moench 2000; FAO 2004; Emerton and Bos 2004). It is increasingly recognized, although rarely common practice, that greater net socio-economic benefit can be obtained from maintaining the integrity of managed ecosystems (Cai et al. 2001; FAO 2004). In the future, food security strategies will be more complex. Higher temperatures will increase water demand, and where rainfall declines, many will seek more irrigation to ensure food security and maintain livelihoods. At the same time, water supplies available for irrigation will become more variable and will decline in many parts of the world. New agricultural demands will be further tempered by the need to achieve better equity in access to reliable food supplies than in the past. As irrigation has been practised on only 20% of the world's cultivated land, there have been many, often the poorest, who have missed out on its benefits (Imran 2018; Imran et al. 2020c). The need to maintain viable aquatic ecosystems will place further stress on water resources, especially where the poorest are dependent on them for their livelihoods. Water allocations to agriculture may fall in many parts of the world owing to the combined impacts of climate change, environmental needs, and competition from higher-value economic sectors (Ali et al. 2019). There will be strong pressure to produce more with less water and to spread the benefits of all water use more widely and wisely. This task will be even more challenging because higher temperatures will reduce potential land and water productivity. Climatic variability in south-eastern Australia has had more profound impacts on water allocations and associated livelihoods in agriculture than even the most prudent farmers had anticipated and big changes lie ahead (Imran and Amanullah 2018).

Rainfed lands account for more than 80% of global crop area and 60% of global food output but are especially susceptible to the impacts of climate change, more so in the arid and semi-arid regions. In general, the productivity of rainfed agriculture in Pakistan is considerably lower than in more secure irrigated conditions, where the use of other factor inputs is generally higher. Many sophisticated adaptations have been developed by farmers to allow cropping in precarious arid and semi-arid conditions, including mixed and companion cropping, floodwater spreading, and runoff harvesting (Imran et al. 2019a,b). A substantial literature exists on the likely impacts of climate change on rainfed agriculture, with only a skimpy outline given here. Rainfed and pastoral agriculture dominate land use in many countries and are therefore a major determinant of hydrology and runoff in a river basin. Pakistan and especially Khyber Pakhtunkhwa agriculture is predominantly rainfed, with low and erratic rainfalls by small subsistence farms. There is a growing realization that the subsistence model of small-holder development cannot power rapid economic growth in Pakistan and that it is unlikely to make the required difference to the lives of the poorest (Schmidhuber et al. 2009). Various forms of foreign direct investment in agriculture (including land purchase) have generated much debate and apprehension over the past five to ten years, but the development of an indigenous commercially based agricultural sector is limited. The transaction costs of raising the intensity of farming are lower in commercial farming than in subsistence production, indicating a window of opportunity for production growth (Collier and Dercon 2009). One of the main debates on how to feed the world's growing population is the future balance between irrigated and rainfed agriculture. Higher productivity and relative security of irrigated agriculture are offset by the undesirable environmental consequences of diverting, storing, and consuming streamflow and groundwater (Iqbal et al. 2021; Russo et al. 2014). This debate will be further complicated by the impacts of climate change on agriculture. Rainfed agriculture is reliable and productive at higher latitudes and can be so in the humid tropics. However, in the mid-latitudes (sub-humid tropics, seasonally arid, semi-arid, and arid tropics), rainfed agriculture generally has low productivity and is prone to drought and crop failures (Hu et al. 1994; Sharifi 2012). Rainfed staples (such as wheat, maize, sorghum, and millet) are generally grown in conditions of deficient or sufficient rainfall. However, rice is naturally adapted to wet environments with excessive rainfall, requiring some measure of water control.

1.8 Climate Change and Soil Resources

Pakistan has made significant improvements in achieving food security in recent years and is close to being self-sufficient in terms of the food and dietary energy available to the population (Imran 2018). Soils are the support of primary production in terrestrial ecosystems (Woodwell and Whittaker 1968). Therefore, the agricultural production of food and fibre for the human population is strongly dependent on soil resources. At the ecological and human time scales, soil resources are non-renewable. Therefore, soil conservation is a critical factor to ensure increasing food demands. Soils can be a carbon (C) source and sink.

Therefore, soils contribute to the regulation of the carbon cycle and its consequent effects on climate change (Frank et al. 2015; Imran et al. 2021b,c). Changes in land use constitute the driving force that determines the soil's role as a source or as a sink of C. Tilling of land has led to a loss of organic carbon (which we will refer to hereinafter as OC) and an immediate increase in carbon emissions, whereas the reforestation of croplands causes an increase in carbon sequestration (Iqbal et al. 2016). However, sequestration of C under forestry or other non-agricultural land uses often only slowly replaces OC lost through cropping, and such differences in time scale between loss or change of soil properties due to man and their reversal are commonly several orders of magnitude different (Imran and Amanullah 2021). Furthermore, the properties of soils are sensitive to climate change. In mineral soils, the approximate relationship between organic matter and soil organic C is $1.724 \times \%\ OC = \%$ organic matter. A temperature increase would cause a higher decomposition rate of the OC (Imran et al. 2021d). Increased drought would have the opposite effect. The possible increase in plant productivity due to the fertilizing effect of an increase in atmospheric CO_2 would lead to an increase in OC inputs to the soil, especially with intensive cropping (assuming no hydrological or nutritional limitations). In natural ecosystems where productivity is limited by N shortage, increased atmospheric N deposition would increase OC inputs to the soil. On the contrary, decreased productivity due to the intensification of water stress leads to losses of soil OC (Amanullah et al. 2021). The foreseeable increase in the occurrence of forest fires would cause losses of OC and would increase the risk of erosion. Increased soil erosion causes the loss of upper soil horizons richer in OC. In areas in which forest fires are a recurring phenomenon, like in the Mediterranean basin, the production of highly stable forms of OC during the combustion of biomass (charcoal) can contribute to the stabilization of C in the medium term. All these processes do not exclude each other and some of the feedback may be positive, causing a multiplying effect (Gao et al. 2017; Khan 2017). OC is intimately linked with the natural fertility and productivity of soils as a source of macronutrients, especially N and P to the substrate of the soil's microbial activity humified carbon contributes greatly to the capacity to retain nutrients and pollutants (capacity for cationic and anionic exchange capacity). Humic substances of lower molecular weight (fulvic acids) improve the solubility of some essential micronutrients, and toxic metals (Khan 2017; Sarker et al. 2017; Wang et al. 2017). It is a critical factor in the development and maintenance of soil structure and in the stability of soil aggregates and, consequently, of the physical properties that depend on these factors: water infiltration capacity, water holding capacity for plants, aeration, compaction, and erodibility. Another process that would probably be affected by climate change is soil salinization. Projected increases in evapotranspiration and drought would raise the water table, saline intrusion, and accumulation of salts in the rooting soil depth in arid and semi-arid lands. These processes can be reduced with the application of appropriate farming techniques, tilling, irrigation control, and management of organic amendments, along with the reforestation of barren lands. In short, those measures promoting soil fertility would set off a multiplying effect on the ecosystems' capacity to fix atmospheric carbon in the long term. Of course, land use will almost certainly change as a consequence of climate change, opening opportunities for novel crops and varieties adapted to the new conditions, including the corresponding changes in cropping systems.

1.9 Relation of World Trade to Climate Change

There are no jobs, no production, and no exchange of goods on a dead planet. It may seem obvious, but it' is important to point out that the planet's economies depend on, well, the planet (Imran et al. 2020c). In 2017, a team of scientists and economists mapped out the potential economic damages different counties in the United States can expect as the climate continues to change. The researchers found 'that if warming continues at recent rates, it could shave 3–6% points off of the country's gross domestic product by century's end the warmer it gets, the bigger the hit to the economy.' Extreme weather linked to climate change also has huge economic repercussions. The year 2017 was the United States' costliest disaster year on record (Smith 2021). Between 2007 and 2017, the federal government spent an estimated US$ 350 billion responding to extreme weather and fires (Balch et al. 2018). On the other hand, climate solutions like renewable energy are fuelling our economy and creating good, well-paying jobs. In 2016, renewable energy employed nearly 10 million people around the globe. This sector has offered hope to laid off fossil fuel workers in places like Alberta – where (during the 2015 oil price crash) 'an estimated 100,000 Canadian oil workers were laid off, at some points causing Alberta's unemployment rate to hover around 10%.'

1.10 Anthropogenic Activities Ameliorate Climate Change

The human impact on climate during this era greatly exceeds that due to known changes in natural processes, such as solar changes and volcanic eruptions (Figure 1.2). However, human activities, primarily the burning of fossil fuels and clearing of forests, have greatly intensified the natural greenhouse effect, causing global warming. (Samreen et al. 2016; Shaw et al. 2010). Greenhouse gases and aerosols affect climate by altering incoming solar radiation and out-going infrared (thermal) radiation that are part of Earth's energy balance (Figure 1.6) (Lockwood 2012). Human activities contribute to climate change by causing changes in Earth's atmosphere in the amounts of greenhouse gases, aerosols (small particles), and cloudiness. The largest known contribution comes from the burning of fossil fuels, which releases carbon dioxide gas into the atmosphere. Water vapour is the most important greenhouse gas, and carbon dioxide (CO_2) is the second-most important one (Figure 1.6). Methane, nitrous oxide, ozone, and several other gases present in the atmosphere in small amounts also contribute to the greenhouse effect. In the humid equatorial regions, where there is so much water vapour in the air greenhouse effect is very large, adding a small additional amount of CO_2 or water vapour has only a small direct impact on downward infrared radiation. Changing the atmospheric abundance or properties of these gases and particles can lead to a warming or cooling of the climate system. Since the start of the industrial era (about 1750), the overall effect of human activities on climate has been a warming influence (Figure 1.2). The two most abundant gases in the atmosphere, nitrogen (comprising 78% of the dry atmosphere) and oxygen (comprising 21%), exert almost no greenhouse effect. Instead, the greenhouse effect comes from molecules that are more complex and much less common.

1.11 Relationship of Carbon and Water Cycle with Climate Change

1.11.1 Water and Carbon Cycle

Cycling of carbon and water is central to supporting life on earth and an understanding of these cycles underpins some of the most difficult international challenges of our times (Figure 1.1). Both these cycles are included in the core content elements of the specifications for A Level geography to be first taught in 2016. Whether we consider climate change, water security, or flood risk hazard an understanding of the physical process is central to the analysis of the geographical consequences of environmental change (Karim and Imran 2019). Both cycles are typically understood within the framework of a systems approach which is a central concept to much physical geographical inquiry. The concept of a global cycle integrates across scales. Systems theory allows us to conceptualize the main stores and pathways at a global scale. The systems framework also allows for more detailed (process detail) and local knowledge to be nested within the wider conceptual framework (Khan 2015a,b). Local studies on aspects of hydrology or carbon cycling can be understood as part of a broader attempt to understand in detail the nature of water and carbon cycling. Global environmental challenges frequently excite student interest in physical geography, but it can be difficult for students to see how they can conduct relevant investigations of fieldwork given the large scale and complexity of the issues. By embedding their knowledge within a systems framework, students can understand how measurements and understanding derived from their fieldwork and local studies contribute to the wider project of elucidating the cycling of water and carbon at the global scale (Ackerman and Stokes 2003; Adkins et al. 1998; Barnett et al. 1999; Marengo 2006).

This capacity to link scales means that the study of biogeochemical cycles is intrinsically geographical so that students can understand how processes operate in and impact upon particular places and how they are distributed in space.

1.11.2 The Water Cycle

Water is present in three phases on earth, liquid water, ice, and atmospheric moisture. On the global scale, there are stores of water in all three phases (Al-Tawaha et al. 2020c; Iqbal et al. 2021; Russo et al. 2014). Liquid water dominates with about 98% of the water in liquid form, predominantly in the oceans. Water is cycled between these stories by a range of key processes as identified in Figure 1.2. The mean residence times for various stores vary both with the size of the store (larger stores take longer to turn over) and with the rate of the processes which move water between stores (for example surface runoff is relatively rapid but groundwater flows are much slower so that groundwater residence times can be high).

1.12 Impacts of Climate Change on Water and Carbon Cycles

Climate change increases our risk of both heavy rains and extreme droughts. But why – and how? Science has shown that climate change touches every corner of our planet's ecosystem, and the water cycle is no exception. Because the processes involved are highly dependent on temperature, changes in one have consequences on the other. Specifically, as global temperatures have steadily increased at their fastest rates in millions of years, it's directly affected things like water vapour concentrations, clouds, precipitation patterns, and streamflow patterns, which are all related to the water cycle (Lal 2001; Trenberth et al. 2007). So how does climate change impact the water cycle? We've created an infographic below that illustrates what's going on, but we'll describe it here too. Put simply, water evaporates from the land and sea, which eventually returns to earth as rain and snow. Climate change intensifies this cycle because as air temperatures increase, more water evaporates into the air. Warmer air can hold more water vapour, which can lead to more intense rainstorms, causing major problems like extreme flooding in coastal communities around the world. But it doesn't end there. At the same time, some areas are experiencing stronger storms, others are experiencing more dry air and even drought. As we mentioned above, as temperatures rise, evaporation increases and soils dry out. Then when rain does come, much of the water runs off the hard ground into rivers and streams, and the soil remains dry.

1.12.1 Impacts on Carbon Cycle

Cycling of the element carbon is intimately associated with life on earth. Carbon is present in carbon-based molecules that are integral to all living creatures, like carbon dioxide and methane in the atmosphere, carbonate rocks in the lithosphere, and as organic molecules in soils and sediments that are derived from formerly living material (Imran et al. 2021c; Koçak and Ortaş 2021). Major carbon stores include the ocean, ocean sediments, soils, bedrock, vegetation, and the atmosphere (Ackerman and Stokes 2003; Adkins et al. 1998; Barnett et al. 1999). Atmospheric carbon has become a major policy focus because of the role of carbon dioxide and methane as greenhouse gasses. The magnitude of the major stores and how they are connected by key processes is illustrated in Figure 1.3 which is taken from the Intergovernmental Panel on Climate Change (IPCC). This diagram indicates significant anthropogenic perturbations of the carbon cycle since 1750 AD. About 90% of anthropogenic carbon release comes from the combustion of fossil fuels with the remainder driven by land-use change (Boettcher et al. 2021). The anthropogenic CO_2 released into the atmosphere about 24% is absorbed by the oceans and 26% is taken up by plants. Global CO_2 concentrations have increased from less than 320 ppm in 1960 to around 400 ppm at present (Figure 1.11) (see Earth System Research Laboratory: https://www.esrl.noaa.gov).

Carbon is a ubiquitous element on Earth. Most of the Earth's carbon is stored in rocks, but this carbon is essentially inert on the 100's to 1000's year timescales of interest to humans (Gavito et al. 2003; Khan et al. 2010). The rest of the carbon is stored as CO_2 (carbon dioxide) in the atmosphere (2%), as biomass in land plants and soils (5%), as fossil fuels in a variety of geologic reservoirs (8%), and as a collection of ions in the ocean (85%). These are the 'active' reservoirs of carbon of interest on this website. The earth is warmed by the sun, and this warmth is returned from Earth to the atmosphere in the form of heat radiation. Many gases in the atmosphere, including CO_2, absorb the Earth's heat energy and radiate in all directions. The energy radiated downward warms the surface and lower atmosphere (Boettcher et al. 2021; Koçak and Ortaş 2021; Sarker et al. 2017). Adding more CO_2 to the atmosphere means more heat radiation is captured by the atmosphere and radiated back to earth. Methane (CH_4) is another very important greenhouse gas that is part of the carbon cycle. This website addresses only CO_2.

Carbon is transferred between CO_2 and living or dead organic material by the very basic photosynthesis/respiration reaction (shown here in simplified form).

$$CO_2 + H_2O + \text{energy} \rightarrow CH_2O + O_2$$

When this reaction proceeds to the right, it is the fixation of carbon to organic matter by plants via photosynthesis; and when it proceeds to the left, it is respiration or combustion of that organic matter. Fossil fuels are the remnants of dead organic matter that lived millions of years ago.

1.12.2 Impacts on Global Carbon Cycle

The carbon cycle is a complex system of biological, chemical, and physical processes. A schematic from the IPCC AR4 report is shown here. The schematic shows the major reservoirs of carbon in gigatons of carbon, GtC (1 GtC = 1 PgC: Petagram of Carbon), and the major fluxes in GtC/year. Humans add CO_2 to the Atmosphere, Nature removes about half of it. In the 1990s, humans added 8.0×10^{15} g of carbon (10^{15} g of carbon = 1 PgC) to the atmosphere each year, primarily by burning fossil fuels (6.4 PgC/year) and clearing land in the tropics (1.6 PgC/year). The ocean took up 28% of this carbon, and the land absorbed 32%. Only 40% remained in the atmosphere to cause climate warming (Gavito et al. 2003). Natural processes are significantly damping the rate of carbon accumulation in the atmosphere. From 2000 to 2008, humans added 9.1 PgC to the atmosphere each year, 7.7 PgC/year from fossil fuels, and 1.4 PgC/year from land-use change. There is some evidence that a larger fraction of these recent emissions has remained in the atmosphere (45%) (Le Quéré et al. 2009). Future climate warming depends on both the CO_2 source from human emissions and the CO_2 sink from natural sinks in the ocean and the terrestrial biosphere.

1.13 Conclusion

The effect of climate change is not limited to water availability and uneven rainfall distribution only, but it also intensifies floods leads to dampen the soil, causes storms, and also increases the air temperature which increases the risk of drought and affects various agronomic crop yields and thus the food security and worsens the human living environments. Therefore, a well-coordinated program is necessary to create awareness among different sections of the society including the policymakers, general public, organizations, industrialists, and farmers to sustain food security and the environment in the days to come when climate change has to be experienced around the world.

1.14 Summary

Climate change is considered one of the most important factors affecting food security. Food safety includes various dimensions, namely production, distribution, and supply. Everyone deserves access to water for their survival and climate change is making our water supply much more violent. Excessive heat can cause dehydration, heatstroke, and major damage to organs. Carbon pollution and climate change mean that more floods and droughts make it harder and harder for growers to grow nutrient-rich foods or biofortified food. Climate change is a massive challenge worldwide. Climate change encourages through natural system manipulation such as deforestation, urbanization, and industrialization, living style of the people, and luxury in lifestyle on the cost of greenhouse gasses (GHG) emission. On the other hand, easiness in life changed the atmosphere of the earth, and aerosols (small particles) were created through anthropogenic activities. The average earth's temperature has been rising for many years. Greenhouse gases and aerosols affect the climate by changing incoming solar radiation and outgoing infrared radiation, which are part of the Earth's energy balance. Environmental fermentation or changing the properties of these gases and particles can cause heat or cooling in the climate system. A healthy life requires a healthy climate. Climate change affects human health in many ways. Pollution from fossil fuels is so bad for the air we breathe, but many factors accelerating climate change also disturb air quality.

References

Abdul, R., Khalil, S., Muhammad, A. et al. (2016). Response of sorghum (*Sorghum bicolor* L.) extract type, concentration and application time to weeds weight, grain and biomass yield of wheat. *Pure and Applied Biology* 5: 847–855.

Abrar, H., Athar, N., Imran, A., and Masood, A. (2009). Effect of weed competition periods on the growth and yield of black seed (*Nigella sativa* L.). *Pakistan Journal of Weed Science Research* 15: 71–81.

Ackerman, T. and Stokes, G. (2003). The atmospheric radiation measurement program. *Physics Today* 56: 38–44.

Adkins, J.F., Cheng, H., Boyle, E.A. et al. (1998). Deep-sea coral evidence for rapid change in ventilation of the deep North Atlantic 15,400 years ago. *Science* 280: 725–728.

Ahmad, M.A., Munir, I., Ali, W. et al. (2007). Assessment of genetic diversity of local and exotic *Brassica napus* germplasm. *Pakistan Journal of Biological Sciences: PJBS* 10: 2490–2494.

Alamzeb, M., Anwar, S., Iqbal, A. et al. (2018). Application of organic sources and nitrogen affect dry matter partitioning in wheat under tillage systems. *Pakistan Journal of Agricultural Research* 31: 106–115.

Ali, A., Hussain, I., Gul, H. et al. (2015). Effect of different doses of foliar fertilizer on yield and physiochemical characteristics of tomato (*Lycopersicon esculentum* Mill) cultivars under the agro-climatic condition of Peshawar. *International Journal of Biosciences* 7: 58–65.

Ali, I., Khan, A.A., Khan, A. et al. (2019). Humic acid and nitrogen levels optimizing productivity of green gram (*Vigna radiate* L.). *Russian Agricultural Sciences* 45: 43–47.

Alkahtani, J., Elshikh, M.S., Alwahibi, M.S. et al. (2020). Phosphorus and zinc fertilization improve productivity and profitability of rice cultivars under rice-wheat system. *Agronomy* 10: 1–15.

Allen, M.R., Dube, O.P., Solecki, W. et al. (2018). Framing and context. In: *Global Warming of 1.5 °C*. An IPCC Special Report on the impacts of global warming of 1.5 °C above pre-industrial levels and related global greenhouse gas emission pathways, in the context of strengthening the global response to the threat of climate change, sustainable development, and efforts to eradicate poverty (ed. V. Masson-Delmotte, P. Zhai, H.O. Pörtner, et al.). Geneva: IPCC. Available at: https://www.ipcc.ch/site/assets/uploads/sites/2/2019/05/SR15_Chapter1_Low_Res.pdf.

Al-Tawaha, A., Al-Tawaha, A., Sirajuddin, S. et al. (2020a). Ecology and adaptation of legumes crops: a review. *IOP Conference Series: Earth and Environmental Science* 492: 1–5.

Al-Tawaha, A.R.M., Amanullah, I., Al-Tawaha, A. et al. (2020b). Adapting crop management practices to climate change. *Advances in Environmental Biology* 14: 10–18.

Al-Tawaha, A.R.M., Jahan, N., Odat, N. et al. (2020c). Growth, yield and biochemical responses in barley to DAP and chitosan application under water stress. *Journal of Ecological Engineering* 21: 86–93.

Altieri, M.A., Nicholls, C.I., Henao, A., and Lana, M.A. (2015). Agroecology and the design of climate change-resilient farming systems. *Agronomy for Sustainable Development* 35: 869–890.

Alwahibi, M.S., Elshikh, M.S., Alkahtani, J. et al. (2020). Phosphorus and zinc fertilization improve zinc biofortification in grains and straw of coarse vs. fine rice genotypes. *Agronomy* 10: 1–5.

Amanullah, Khan, I., Jan, A. et al. (2015). Compost and nitrogen management influence productivity of spring maize (*Zea mays* L.) under deep and conventional tillage systems in semi-arid regions. *Communications in Soil Science and Plant Analysis* 46: 1566–1578.

Anwar, S., Iqbal, B., Khan, A.A. et al. (2016). Nitrogen and phosphorus fertilization of improved varieties for enhancing phenological traits of wheat. *Pure and Applied Biology* 5: 511–519.

Anwar, S., Ullah, W., Islam, M. et al. (2017). Effect of nitrogen rates and application times on growth and yield of maize (*Zea mays* L.). *Pure and Applied Biology (PAB)* 6: 908–916.

ASA-CSSA-SSSA (2017). Changing climate changes soils, American Society of Agronomy/Crop Science Society of America/Soil Science Society of America. *ScienceDaily*, 11 January 2017. Available at: www.sciencedaily.com/releases/2017/01/170111133851.htm.

Baker, H.S., Millar, R.J., Karoly, D.J. et al. (2018). Higher CO_2 concentrations increase extreme event risk in a 1.5 °C world. *Nature Climate Change* 8: 604–608.

Balch, J., Schoennagel, T., Williams, A. et al. (2018). Switching on the big burn of 2017. *Fire* 1: 1–9.

Baqa, S., Khan, A.Z., Khan, A.A. et al. (2015). Influence of farm yard manure and phosphorus application on yield and yield components of wheat. *Pure and Applied Biology* 4: 458–464.

Barnett, T.P., Hasselmann, K., Chelliah, M. et al. (1999). Detection and attribution of recent climate change: a status report. *Bulletin of the American Meteorological Society* 80: 2631–2660.

Barnola, J.M., Raynaud, D., Korotkevich, Y.S., and Lorius, C. (1987). Vostok ice core provides 160,000-year record of atmospheric CO_2. *Nature* 329: 408–414.

Bentz, J. (2020). Learning about climate change in, with and through art. *Climatic Change* 162: 1595–1612.

Boettcher, M., Brent, K., Buck, H.J. et al. (2021). Navigating potential hype and opportunity in governing marine carbon removal. *Frontiers in Climate* 3: 47.

Brooks, N., Grist, N., and Brown, K. (2009). Development futures in the context of climate change: challenging the present and learning from the past. *Development Policy Review* 22: 741–765.

Buis, A. (2019). *The Atmosphere: Getting a Handle on Carbon Dioxide*. NASA-Global Climate Change. Available at: https://climate.nasa.gov/news/2915/the-atmosphere-getting-a-handle-on-carbon-dioxide/.

Burke, J.J. and Moench, M. (2000). *Groundwater and Society: Resources, Tensions and Opportunities*. New York: United Nations.

Cai, X., Ringler, C., and Rosegrant, M.W. (2001). *Does Efficient Water Management Matter?: Physical and Economic Efficiency of Water Use in the River Basin, EPTD Discussion Paper # 72*. International Food Policy Research Institute (IFPRI). Available at: http://ebrary.ifpri.org/utils/getfile/collection/p15738coll2/id/70736/filename/70737.pdf.

Canadell, J.G., Le Quere, C., Raupach, M.R. et al. (2007). Contributions to accelerating atmospheric CO_2 growth from economic activity, carbon intensity, and efficiency of natural sinks. *Proceedings of the National Academy of Sciences USA* 104: 18866–18870.

Clark, W.C. (1985). Scales of climate impacts. *Climatic Change* 7: 5–27.

Cline, W.R. (2007). *Global Warming and Agriculture. Impact Estimates Per Country*. Washington, DC: Centre for Global Development, Peterson Institute for International Economics.

Collier, P. and Dercon, S. (2009). *African agriculture in 50 years. Smallholders in a rapidly changing world*? In: *Expert Meeting on How to Feed the World in 2050*. Rome: Economic and Social Development Department, Food and Agriculture Organization. Available at: http://ss.rrojasdatabank.info/ak983e00.pdf.

Cottier, T. and Shariff, N. (2013). International trade and climate change. In: *Research Handbook on Environment, Health and the WTO* (ed. G. Van Calster and D. Prévost), 413–447. Cheltenham, UK: Edward Elgar Publishing.

Di Falco, S. and Chavas, J.P. (2009). On crop biodiversity, risk exposure, and food security in the highlands of Ethiopia. *American Journal of Agricultural Economics* 91: 599–611.

Don, A., Schumacher, J., and Freibauer, A. (2011). Impact of tropical land-use change on soil organic carbon stocks – a meta-analysis. *Global Change Biology* 17 (4): 1658–1670.

Duncan, J. (2015). *Global Food Security Governance. Civil Society Engagement in the Reformed Committee on World Food Security*. London: Routledge.

Emerton, L. and Bos, E. (2004). *Value: Counting Ecosystems as an Economic Part of Water Infrastructure*. Gland, Switzerland: IUCN. Available at: https://portals.iucn.org/library/sites/library/files/documents/2004-046.pdf.

FAO (2004). *Economic Valuation of Water Resources in Agriculture: From a Sectoral to a Functional Perspective of Natural Resource Management*, FAO Water Report # 27. Rome: Food and Agriculture Organization.

FAO (2013). *Pakistan. Review of the Wheat Sector and Grain Storage Issues*. Rome: Food and Agriculture Organization. Available at: http://www.fao.org/3/i3251e/i3251e.pdf.

Farmer, G.T. and Cook, J. (2013). Understanding climate change denial. In: *Climate Change Science: A Modern Synthesis* (ed. G.T. Farmer and J. Cook), 445–466. Dordrecht: Springer.

Farooq, M., Bakhtiar, M., Khan, M.N. et al. (2018). Yield performance of selected varieties planted on different intervals as unirrigated wheat. *International Journal of Fauna and Biological Studies* 5: 95–99.

Fernald, A., Tidwell, V., Rivera, J. et al. (2012). Modeling sustainability of water, environment, livelihood, and culture in traditional irrigation communities and their linked watersheds. *Sustainability* 4: 2998–3022.

Forchhammer, M.C. and Post, E. (2004). Using large-scale climate indices in climate change ecology studies. *Population Ecology* 46: 1–12.

Frank, D., Reichstein, M., Bahn, M. et al. (2015). Effects of climate extremes on the terrestrial carbon cycle: concepts, processes and potential future impacts. *Global Change Biology* 21: 2861–2880.

Frederick, K.D. and Major, D.C. (1997). Climate change and water resources. *Climatic Change* 37: 7–23.

Gao, L., Wang, R., Shen, G. et al. (2017). Effects of biochar on nutrients and the microbial community structure of tobacco-planting soils. *Journal of Soil Science and Plant Nutrition* 17: 884–896.

Garnett, T. (2014). *What is a Sustainable Healthy Diet? A Discussion Paper*. Oxford, UK: Food Climate Research Network (FCRN). Available at: https://cgspace.cgiar.org/bitstream/handle/10568/35584/FCRN-sustainable-healthy-diet.pdf?sequence=1&isAllowed=y.

Gavito, M.E., Schweiger, P., and Jakobsen, I. (2003). P uptake by arbuscular mycorrhizal hyphae: effect of soil temperature and atmospheric CO_2 enrichment. *Global Change Biology* 9: 106–116.

GFSI (2012). *Global Food Security Index 2012. An Assessment of Food Affordability, Availability and Quality*, Report from the Economist Intelligence Unit. London: The Economist. https://impact.economist.com/sustainability/project/food-security-index/Downloads.

GOP (2010). *Agricultural Census 2010, Pakistan Report*. Pakistan Bureau of Statistics, Ministry of Finance, Government of Pakistan Available at: https://www.pbs.gov.pk/sites/default/files/aco/publications/agricultural_census2010/WRITE-UP%20AGRI.%20CENSUS%20 2010.pdf.

GOP (2011). *Agricultural Statistics of Pakistan*. Pakistan: Pakistan Bureau of Statistics, Ministry of Finance, Government of Pakistan.

Gray, J.M. and Bishop, T.F.A. (2016). Change in soil organic carbon stocks under 12 climate change projections over New South Wales. *Australia, Soil Science Society of America Journal* 80 (5): 1296–1307.

Gregory, P.J., Ingram, J.S.I., and Brklacich, M. (2005). Climate change and food security. *Philosophical Transactions of the Royal Society B* 360: 2139–2148.

Griffin, K.L. and Seemann, J.R. (1996). Plants, CO_2 and photosynthesis in the 21st century. *Chemistry & Biology* 3: 245–254.

Hamdy, A., Ragab, R., and Scarascia-Mugnozza, E. (2003). Coping with water scarcity: water saving and increasing water productivity. *Irrigation and Drainage* 52: 3–20.

Hanjra, M.A. and Qureshi, M.E. (2010). Global water crisis and future food security in an era of climate change. *Food Policy* 35: 365–377.

Harde, H. (2019). What humans contribute to atmospheric CO_2: comparison of carbon cycle models with observations. *Earth Sciences* 8: 139–158.

Hardy, J.T. (2003). *Climate Change: Causes, Effects and Solutions*. Chichester: Wiley.

Hasanuzzaman, M. (2019a). *Agronomic crops*. In: *Management Practices*, vol. *2*. Singapore: Springer.

Hasanuzzaman, M. (2019b). *Agronomic Crops*. Singapore: Springer.

Hasanuzzaman, M. and Fotopoulos, V. (2019). *Priming and Pretreatment of Seeds and Seedlings. Implication in plant stress tolerance and enhancing productivity in crop plants*. Singapore: Springer.

Hatfield, J.L. and Dold, D. (2018). Agroclimatology and wheat production: coping with climate change. *Frontiers in Plant Science* 9: 1–5.

Hontoria, C., Saa, A., and Rodríguez-Murillo, J.C. (1999). Relationships between soil organic carbon and site characteristics in peninsular Spain. *Soil Science Society of America Journal* 63 (3): 614–621.

Hou, L., Zhang, X., Feng, G. et al. (2021). Arbuscular mycorrhizal enhancement of phosphorus uptake and yields of maize under high planting density in the black soil region of China. *Scientific Reports* 11: 1–11.

Hu, Y., Ma, Y., Sun, Y., and Guo, G. (1994). Effect of B application on the agronomic traits, yields and oil contents of a double-row rape (*Brassica napus* L.) cultivar. *Oil Crops (China)* 16: 43–46.

Ibrahim, M. and Khan, A. (2017). Phenology and maize crop stand in response to mulching and nitrogen management. *Sarhad Journal of Agriculture* 33: 426–434.

Ilyas, M., Ayub, G., Imran, I. et al. (2021). Calcium and boron effect production and quality of autumn potato crop under chilling temperature. *Communications in Soil Science and Plant Analysis* 52: 369–382.

Imran (2017). Climate change is a real fact confronting to agricultural productivity. *International Journal of Environmental Sciences and Natural Resources* 3: 65–73.

Imran (2018). Climate change is threat toward agronomy (base of food, fiber system), and food security. *Food and Nutrition* 3: 1–8.

Imran and Amanullah (2018). Global impact of climate change on water, soil resources and threat towards food security: evidence from Pakistan. *Advances in Plants & Agriculture Research* 8: 350–355.

Imran and Amanullah (2021). Phosphorus and boron application Optimizing biofortification of phosphorus and productivity of french bean (*Phaseolus vulgaris* L.). *Communication in Soil Science and Plant Analysis* 52: 2876–2883.

Imran and Asad, A.K. (2015). Phenological characteristics of *Brassica napus* L. as influenced by biochar application and shoot cutting duration (days). *Civil and Environmental Research* 7: 104–107.

Imran and Asad, A.K. (2017). Grain yield and phenology of maize cultivars influenced by various phosphorus sources. *Journal of Food Science and Quality Management* 37: 74–78.

Imran and Khan, A.A. (2015). Influence of compost application and seed rates on production potential of late sown maize on high elevation in Swat, Pakistan. *Journal of Environmental Science* 5: 36–40.

Imran, Asad, A.K., and Fayaz, A. (2015a). Phenology of various rice genotypes as affected by different transplanting dates under cold climatic region of Khyber Pakhtunkhwa-Pakistan. *Journal of Environmental and Earth Science* 5: 119–122.

Imran, Khan, A.A., Ahmad, F., and Ullah, I. (2015b). Influence of hydrated calcium sulphate ($CaSO_4{\cdot}2H_2O$) and nitrogen levels on water infiltration rate and maize varieties productivity in rainfed area of Swat, Pakistan. *Chemistry and Materials Research* 7: 15–20.

Imran, Khan, A.A., Khan, U.I., and Naveed, S. (2016a). Weeds density and late sown maize productivity influenced by compost application and seed rates under temperate environment. *Pakistan Journal of Weed Science Research* 22: 169–181.

Imran, Khan, A.A., Inamullah et al. (2016b). Yield and yield attributes of rapeseed cultivars as influence by sulfur level under Swat valley conditions. *Pure and Applied Biology (PAB)* 4: 296–301.

Imran, Bari, A., Roshan, A. et al. (2017). Traditional rice farming accelerate CH_4 and N_2O emissions functioning as a stronger contributors of climate change. *Agricultural Research & Technology* 9: 89–92.

Imran, A., Bari, A., and Ali, R. (2019a). Peach organic sources, phosphorous and beneficial microbes application enhance productivity and profitability of soybean. *Soybean Research* 16: 39–48.

Imran, Amanullah, Abdul, B. et al. (2019b). Climatic variability and agronomic cropping pattern. In: *Agronomic Crops* (ed. M. Hasanuzzaman), 33–44. Singapore: Springer.

Imran, A., Arif, M., Shah, Z., and Bari, A. (2020a). Soil application of trichoderma and peach (*Prunus persica* L.) residues possesses biocontrol potential for weeds and enhances growth and profitability of soybean (*Glycine max*). *Sarhad Journal of Agriculture* 36: 10–20.

Imran, A., Mahmood, T., Sajid, M. et al. (2020b). Improvement of rice quality via biofortification of selenium, iron, and zinc and its starring role in human health. In: *Rice Research for Quality Improvement: Genomics and Genetic Engineering* (ed. A. Roychoudhury), 699–710. Singapore: Springer.

Imran, Amanullah, and Al-Tawaha, A.R.M. (2020c). The productivity of subsequent wheat enhanced with residual carbon sources and phosphorus under improved irrigation system. *Communications in Soil Science and Plant Analysis* 51: 1306–1314.

Imran, Amanullah, Al-Tawaha, A.R.M.S. et al. (2021a). Micronutrient biofortification in rice for better quality. In: *Rice Research for Quality Improvement: Genomics and Genetic Engineering* (ed. A. Roychoudhury), 639–653. Singapore: Springer.

Imran, A., Khalid, S., Arif, M. et al. (2021b). The application of biochar for the mitigation of abiotic stress-induced damage. In: *Sustainable Soil and Land Management and Climate Change* (ed. S. Fahad, O. Sönmez, S. Saud, et al.), 45–56. Boca Raton, FL: CRC Press.

Imran, Amanullah, and Al-Tawaha, A.R. (2021c). Carbon sources application increase wheat yield and soil fertility. *Communication in Soil Science and Plant Analysis* 52: 695–703.

Imran, Amanullah, and Al Tawaha, A.R.M. (2021d). Management of nano-black carbon, phosphorous and bio fertilizer improve soil organic carbon and ensilage biomass of soybean and maize. *Communication in Soil Science and Plant Analysis* 52: 2837–2851.

IPCC (2018). *Global Warming of 1.5 °C*. An IPCC Special Report on the impacts of global warming of 1.5 °C above pre-industrial levels and related global greenhouse gas emission pathways, in the context of strengthening the global response to the threat of climate change, sustainable development, and efforts to eradicate poverty (ed. V. Masson-Delmotte, P. Zhai, H.O. Pörtner, et al.). Geneva: IPCC. Available at: https://www.ipcc.ch/site/assets/uploads/sites/2/2019/06/SR15_Full_Report_High_Res.pdf.

Iqbal, B., Bashir, A., Inam, U. et al. (2016). Effect of phosphorus, sulphur and different irrigation levels on phenological traits of Triticale. *Pure and Applied Biology* 5 (2): 303–310.

Iqbal, M.T., Ahmed, I.A.M., Isik, M. et al. (2021). Role of mycorrhizae inoculations on nutrient uptake in rice grown under aerobic and anaerobic water management. *Journal of Plant Nutrition* 44: 550–568.

Islam, M., Shazma, A., and Saqib, B. (2015). Growth and yield components of wheat varieties as affected by dual purpose practices. *Pure and Applied Biology* 4: 491–496.

Janjua, P.Z., Samad, G., Khan, N.U., and Nasir, M. (2010). Impact of climate change on wheat production: a case study of Pakistan. *The Pakistan Development Review* 49: 799–822.

Janjua, P.Z., Samad, G., and Khan, N.U. (2014). Climate change and wheat production in Pakistan: an autoregressive distributed lag approach. *NJAS – Wageningen Journal of Life Sciences* 68: 13–19.

Kahane, R., Hodgkin, T., Jaenicke, H. et al. (2013). Agrobiodiversity for food security, health and income. *Agronomy for Sustainable Development* 33: 671–693.

Kang, Y., Khan, S., and Ma, X. (2009). Climate change impacts on crop yield, crop water productivity and food security – a review. *Progress in Natural Science* 9: 1665–1674.

Karim, S.A. and Imran (2019). Reduction in substrate moisture content reduce final yield of wheat. *Biomedical Journal of Scientific & Technical Research* 15: 11150–11154.

Khan, A. (2015a). Biochar application and shoot cutting duration (days) influenced growth, yield and yield contributing parameters of *Brassica napus* L. *Journal of Biology, Agriculture and Healthcare* 5: 1–6.

Khan, A. (2015b). Phenological characteristics of *Brassica napus* L. as influenced by biochar application and shoot cutting duration (days). *Civil and Environmental Research* 7: 104–107.

Khan, A.A. (2017). Canola yield and quality enhanced with sulphur fertilization. *Russian Agricultural Sciences* 43: 113–119.

Khan, Ali, H.G., Ahmad, A., and Siraj, eA. (2010). Impact of rising prices of fertilizers on crops production in Pakistan. *Global Journal of Management and Business Research* 10: 54–61.

Khandekar, M.L., Murty, T.S., and Chittibabu, P. (2005). The global warming debate: a review of the state of science. *Pure and Applied Geophysics* 162: 1557–1586.

Kirschbaum, M.U.F. (2011). Does enhanced photosynthesis enhance growth? Lessons learned from CO_2 enrichment studies. *Plant Physiology* 155: 117–124.

Koçak, B. and Ortaş, İ. (2021). Short-term eucalyptus and phragmites biochar's efficiency in mineralization of soil carbon. *Journal of Soil Science and Plant Nutrition* 21: 3346–3353.

Kundzewicz, Z.W. (2008). Climate change impacts on the hydrological cycle. *Ecohydrology & Hydrobiology* 8: 195–203.

Lal, R. (2001). Potential of desertification control to sequester carbon and mitigate the greenhouse effect. *Climatic Change* 51: 35–72.

Le Quéré, C., Raupach, M.R., Canadell, J.G. et al. (2009). Trends in the sources and sinks of carbon dioxide. *Nature Geoscience* 2: 831–836.

Lockwood, J. (2012). Atmospheric moisture. In: *An Introduction to Physical Geography and the Environment* (ed. J. Holden), 875p. Harlow: Pearson.

Mahmood, N., Ahmad, B., Hassan, S., and Bakhsh, K. (2012). Impact of temperature and precipitation on rice productivity in rice wheat cropping system of Punjab province. *The Journal of Animal & Plant Sciences* 22: 993–997.

Marengo, J. (2006). On the hydrological cycle of the Amazon basin: a historical review and current state of the art. *Revista Brasileira de Meteorologia* 21: 1–19.

Misra, A.K. (2014). Climate change and challenges of water and food security. *International Journal of Sustainable Built Environment* 3: 153–165.

Mitra, R. and Bhatia, C.R. (2008). Bioenergetic cost of heat tolerance in wheat crop. *Current Science* 94: 1049–1053.

Naveed, S., Muhammad, I., Inayat, K. et al. (2016). Anthelmintic, antilice, insecticidal, cytotoxic and phytotoxic potential of ethanolic extracts of two wild medicinal plants *Iphiona grantioides* and *Pluchea arguta*. *Woulfenia Journal* 23: 13–25.

Ohmura, A. and Wild, M. (2002). Is the hydrological cycle accelerating? *Science* 298: 1345–1346.

Opperman, J.J., Kendy, E., and Barrios, E. (2019). Securing environmental flows through system reoperation and management: lessons from case studies of implementation. *Frontiers in Environmental Science* 7: 104.

Ortiz, R., Sayre, K.D., Govaerts, B. et al. (2008). Climate change: can wheat beat the heat? *Agriculture, Ecosystems & Environment* 126: 46–58.

Parry, M., Rosenzweig, C., and Livermore, M. (2005). Climate change, global food supply and risk of hunger. *Philosophical Transactions of the Royal Society B* 360: 2125–2138.

Pecl, G.T., Araújo, M.B., Bell, J.D. et al. (2017). Biodiversity redistribution under climate change: impacts on ecosystems and human well-being. *Science* 355: 1–9.

Pilkey, O.H., Pilkey, K.C., and Fraser, M. (2011). *Doubts, Uncertainties, and Qualms. Global Climate Change: A Primer*, vol. 3, 29–41. New York, USA: Duke University Press.

Postel, S.L. (1998). Water for food production: will there be enough in 2025. *Bioscience* 48: 629–637.

Ramesh, T., Bolan, N.S., Kirkham, M.B. et al. (2019). Chapter 1 - Soil organic carbon dynamics: impact of land use changes and management practices: a review. *Advances in Agronomy* 156: 1–107.

Ricke, K.L. and Caldeira, K. (2014). Maximum warming occurs about one decade after a carbon dioxide emission. *Environmental Research Letters* 9: 1–8.

Rosegrant, M.W. and Cline, S.A. (2003). Global food security: challenges and policies. *Science* 302: 1917–1919.

Russo, T., Alfredo, K., and Fisher, J. (2014). Sustainable water management in urban, agricultural, and natural systems. *Water* **6**: 3934–3956.

Samreen, U., Ibrar, M., Lalbadshah et al. (2016). Ethnobotanical study of subtropical hills of Darazinda, Takht-e-Suleman range F.R D.I. Khan, Pakistan. *Pure and Applied Biology* 5: 149–164.

Sarker, J.R., Singh, B.P., He, X. et al. (2017). Tillage and nitrogen fertilization enhanced belowground carbon allocation and plant nitrogen uptake in a semi-arid canola crop–soil system. *Scientific Reports* 7: 1–13.

Schmidhuber, J., Bruinsma, J., and Boedeker, G. (2009). *Capital Requirements for Agriculture in Developing Countries to 2050, Expert Meeting on How to Feed the World in 2050*. Rome: Economic and Social Development Department, Food and Agriculture Organization. Available at: https://rrojasdatabank.info/ak974e00.pdf.

Semenov, M.A. (2009). Impacts of climate change on wheat in England and Wales. *Journal of the Royal Society Interface* 6: 343–350.

Semenov, M.A., Stratonovitch, P., Alghabari, F., and Gooding, M.J. (2014). Adapting wheat in Europe for climate change. *Journal of Cereal Science* 59: 245–256.

Sharifi, R. (2012). Study of yield, yield attribute and dry matter accumulation of canola (*Brassica napus* L.) cultivars in relation to sulfur fertilizer. *International Journal of Agriculture and Crop Sciences* 4: 409–415.

Shaw, E.M., Beven, K.J., Chappell, N., and Lamb, R. (2010). *Hydrology in Practice*. London: CRC Press.

Smith, P. (2008). Land use change and soil organic carbon dynamics. *Nutrient Cycling and Agroecosystems* 81: 169–178.

Smith, A.B. (2021). *2010-2019: A Landmark Decade of U.S. Billion-Dollar Weather and Climate Disasters*. Colorado, USA: NOAA's National Centers for Environmental Information. Available at: https://www.climate.gov/news-features/blogs/beyond-data/2010-2019-landmark-decade-us-billion-dollar-weather-and-climate.

Stagnari, F., Ramazzotti, S., and Pisante, M. (2009). Conservation agriculture: a different approach for crop production through sustainable soil and water management: a review. In: *Organic Farming, Pest Control and Remediation of Soil Pollutants. Sustainable Agriculture Reviews*, vol. 1 (ed. E. Lichtfouse), 55–83. Dordrecht: Springer.

Stevens, A., Nocita, M., Tóth, G. et al. (2013). Prediction of soil organic carbon at the European scale by visible and near infrared reflectance spectroscopy. *PLOS ONE* 8: e66409.

Stockmann, U., Padarian, J., McBratney, A. et al. (2015). Global soil organic carbon assessment. *Global Food Security* 6: 9–16.

Taghizadeh-Mehrjardi, R., Nabiollahi, K., and Kerry, R. (2016). Digital mapping of soil organic carbon at multiple depths using different data mining techniques in Baneh region, Iran. *Geoderma* 266: 98–110.

Thrupp, L.A. (2000). Linking agricultural biodiversity and food security: the valuable role of agrobiodiversity for sustainable agriculture. *International Affairs* 76: 265–281.

Tkemaladze, G.S. and Makhashvili, K.A. (2016). Climate changes and photosynthesis. *Annals of Agrarian Science* 14: 119–126.

Trenberth, K.E., Smith, L., Qian, T. et al. (2007). Estimates of the global water budget and its annual cycle using observational and model data. *Journal of Hydrometeorology* 8: 758–769.

Valone, T.F. (2021). Linear global temperature correlation to carbon dioxide level, sea level, and innovative solutions to a projected 6 °C warming by 2100. *Journal of Geoscience and Environment Protection* 09: 85–135.

Verchot, L.V., Van Noordwijk, M., Kandji, S. et al. (2007). Climate change: linking adaptation and mitigation through agroforestry. *Mitigation and Adaptation Strategies for Global Change* 12: 901–918.

Wang, Y., Ghouri, F., Shahid, M.Q. et al. (2017). The genetic diversity and population structure of wild soybean evaluated by chloroplast and nuclear gene sequences. *Biochemical Systematics and Ecology* 71: 170–178.

Wigley, T., Jones, P., and Kelly, P. (1980). Scenario for a warm, high-CO_2 world. *Nature* 283: 17–21.

Woodwell, G.M. and Whittaker, R.H. (1968). Primary production in terrestrial ecosystems. *American Zoologist* 8: 19–30.

Wu, P., Wood, R., Ridley, J., and Lowe, J. (2010). Temporary acceleration of the hydrological cycle in response to a CO_2 rampdown. *Geophysical Research Letters* 37: 1–5.

2

Climate-Smart Plants Combat Climate Change and Liability for Food Security

Imran[1,2], Amanullah[1], and Abdel Rahman Mohammad Said Al-Tawaha[3]

[1] *Department of Agronomy, The University of Agriculture, Peshawar, Pakistan*
[2] *Ministry of Agriculture, Agriculture Extension Wing, Merged Area, Khyber Pakhtunkhwa, Pakistan*
[3] *Department of Biological Sciences, Al- Hussein Bin Talal University, Ma'an, Jordan*

CONTENTS

Climate Change and Agriculture: Perspectives, Sustainability and Resilience, First Edition. Edited by Noureddine Benkeblia.

2.1 Introduction

Climate-smart plants depend on growing period (Imran et al. 2021a). These are grown and categorized as winter season crops such as chickpeas, lentils, *Lathyrus*, broad bean, and vetch (Imran et al. 2021b). Another category of summer season crops is soybeans and peas (Imran and Amanullah 2021; Imran 2021; Imran et al. 2021c). Food legumes are widely grown throughout the world, with a long cultivation history and in a diversity of cropping systems (Imran et al. 2016a; Ali et al. 2019). Legumes are important food crops that can play a major role in achieving food security, nutrition, and human health, contributing to making agriculture more sustainable and helping mitigate and adapt to climate change (CC) (Imran et al. 2016b). With the advantages of short growing periods and biological nitrogen fixation (BNF), food legumes play an important role in cropping systems (Anwar et al. 2016; Samreen et al. 2016). Legumes can be integrated into rotations with cereals, tuber, and vegetable crops. Such systems provide a quality food harvest after non-leguminous crops and improve soil fertility on farms. According to FAO (1994), legumes are crop plants belonging to the *Leguminosae* family (commonly known as the pea family) that produce edible seeds and are used for human and animal consumption. However, only harvested for dry grain are classified as legumes (Naveed et al. 2016; Baqa et al. 2015). Legume species used for oil extraction (e.g. soybean (*Glycine max* (L.), groundnut (*Arachis hypogaea* L.)) and sowing purposes (e.g. clover (different species belonging to the genus *Trifolium* L.), and alfalfa (*Medicago sativa* L.)) are not considered legumes. Likewise, legume species are not considered as legumes when they are used as vegetables (e.g. green peas (*Pisum sativum* L.) and green beans (*Phaseolus vulgaris* L.)). Consequently, when common bean (*Phaseolus vulgaris* L.) is harvested for dry grain, it is considered legumes; but when the same species are harvested unripe (known as green beans), it is not treated as a legume. Major constraints in legumes production are diseases and insect pests along with minimum attention of the farmers (Ali et al. 2019; Anwar et al. 2016; Samreen et al. 2016). Common insects and pests of mung bean are pods borer, whitefly, pods sucking bug, caterpillar, Jassids, and aphids. Pigeon pea common diseases are rust leaf spot, Fusarium wilt, and occasionally sterility mosaic virus may also attack (Imran et al. 2021d). This disease may spread and be transmitted by Eriophyid mite. The affected plant is totally or partially sterile and could not produce seed. Common insect pests of pigeon pea may be moth and pod fly (Imran et al. 2016c). These diseases can reduce yield from 10 to 30%, depending upon the severity, pathogen, and weather conditions (Aise et al. 2011; Imran et al. 2016a). Cowpea is also susceptible to various diseases and insect pests. Common diseases of cowpea are aphid-born mosaic virus, Fusarium wilt, leaf smut, and scab. General insect pests of cowpea might be beetles, aphids, pod borers, caterpillars, and whiteflies which seriously damage the crop (Wani et al. 2012; Naveed et al. 2016; Baqa et al. 2015). Certain insects can also provide access to disease organisms and transmit them directly to plants. Therefore, understanding the relationships between the insect and the crop will enable farmers to manage pests much better. Major insect damage to soybean crops is stem fly, whitefly, green stink bug, cutworm, and larvae. Insects attack all parts of the soybean plant and feed throughout the growing season. A new practice to control insect attacks is based on knowledge of the economic injury levels of the consequential insects (Anchal et al. 1997; Akpalu et al. 2014). The economic injury level is the population of insects that is capable of producing an amount of economic damage which is at least equal to the cost of controlling the insects. Wise monitoring of major insects is required to effectively make a decision relative to insecticide application (Ankomah et al. 1996; Imran and Amanullah 2018; Imran et al. 2015a, 2019; Imran et al. 2021c).

In this context to obtain farm productivity of legumes under harsh condition and hostile environments to increase food security and overcome hunger, intercropping or relay intercropping of food with other crops is another good choice. As ideal intercrops, food can increase crop yields and economic benefits of intercropping systems, and much more innovation will be needed in the future for adaptation to CC and increasing demand for food for the overwhelming population.

2.2 Climate Change, Adaptability, and Ecology of Legumes Crop

Pulses depend on the growing period, firstly, winter season crops such as chickpeas, lentils, *Lathyrus*, broad bean, and vetch, secondly, summer season types, such as soybeans and peas. They have many mark's in our globe they supply human with food, a feed also provide the soil with nitrogen fixation through bacterial nodes and increase the CO_2 concentration of the ambiance (Mousavi-Derazmahalleh et al. 2019; Imran et al. 2016b). Fabaceae, syn. Leguminosae is the third-largest family of angiosperms, including over 750 genera and 19000 species ranging from small herbs to large trees. Studying the pulses species under stress conditions helps in the choice for pasture breeding and soil preservation programs, mainly in arid and semi-arid regions. A study was undertaken to create biogeographical manners in the pulses flora of southern Africa to assist the selection of species with agricultural potency (Trytsman et al. 2016; Imran and Amanullah 2018; Imran et al. 2015b, 2019; Imran et al. 2021a).

Pulses crops acting an important part in human nourishment, mammal feeding, and soil fecundity improvement; it plays a vast task in decreasing paucity and contributes to the persistence of economic progress in the global. Embrace new varieties of pulses that have latent endure under stress situation aid in expanding the planting zones which guide to get better the country's revenue (Amanullah et al. 2021; Imran et al. 2020a). Leaves from pulses trees and shrubs can be a precious supplement for nourishment livestock, particularly through the desiccated time when other nurture raw materials are insufficient or their alimentary value is low. The agronomic assessment at two agro-ecologically discrete regions in eastern DRC verified that diverse forage trees and shrub species were tailored distinctly (Imran et al. 2020b; Ilyas et al. 2020). The species with top forage superiority, *L. diversifolia* (especially accession ILRI 15551) and *L. leucocephala* execute preferably with advanced soil productiveness (Katunga 2014; Imran and Amanullah 2018; Imran et al. 2015c, 2019; Imran et al., 2021b).

Legumes are generally developed rain-fed and are showing distinct kinds of drought zones from terminal drought to sporadic drought (Purushothaman et al. 2013). Acclimation technologies are endeavours to reinstate the synchrony or to generate new management between ambience and plant communities. Both chickpeas and pigeon peas are important grain legumes developed for their protein-rich seeds used in human utilization, for their capacity to renovate and conserve soil fecundity by nitrogen fixation, and for their convenience to robust very well into different cropping manners (Upadhyaya et al. 2010). Heat stress through venereal progress can issue noteworthy yield loss. Chickpea (*Cicer arietinum* L.) is an important food legume and heat stress influences chickpea ontogeny over a range of milieu (Devasirvatham et al. 2012; Imran and Amanullah 2018; Imran et al. 2015d, 2019; Imran et al., 2021e). Four accessions of *C. reticulatum* and one accession of *C. pinnatifidum* were found to be an impedance to water deficiency and heat stress (up to 41.8 °C) (Canci and Toker 2009).

In chickpea, total root biomass in precocious development stages has been revealed to appreciably participate in seed products under incurable water deficit in middle and south India (Kashiwagi 2015). Deeper root coordination and superior root desiccated mass could be allied with enhanced prevention of soil water restriction in drought-tolerant legumes than in drought-sensitive ones (Abdelhamid et al. 2010; Ali et al. 2019; Imran and Khan 2015a; Imran et al. 2014, 2015e). Reproduction programs are ongoing to promote cultivars tailored to the spacious range of latitudes and with disease resistance (Siddique et al. 2012).

Conformation strategies that are debated include prior seeding of pulse crops, use of coldness pulses, crop succession within crop rotation, and modulations to the microclimate such as forthright seeding into standing straw (Cutforth 2007; Ali et al. 2019; Imran and Khan 2015b; Imran et al. 2014, 2015f).

Prospect pulses genotypes will require to preserve or smooth to be greater under the expected CCs. The most significant way to enhance the granule yield of cool period food legumes per unit area under stress niches will be the enlargement of cultivars with conflict to numerous abiotic stresses in the circumstance of endurable incorporated pulses output, insect and pest executive, nutrients organization, and irrigation use if accessible (Toker and Yadav 2010). Faba bean has a lot of proper imputes for sustainable recent cropping patterns but these outcomes propose that yield will be inadequate by anticipated CC, presuppose the expansion of heat-tolerant cultivars, or enhanced flexibility by other mechanisms such as earlier flowering periods (Bishop et al. 2016).

Faba bean yield will turn into ever more uneven with anticipated CC and underline the requirement to recover heat stress resilience of faba bean floral advance and anthesis to endorse future (Bishop et al. 2016; Ali et al. 2019; Imran and Khan 2015c; Imran et al. 2014, 2015g). Faba bean yield was minimized by heat stress within the hotness range recognized to contrive yield decreasing in other output species, and as in other species (Bishop et al. 2016). Replacement of imported soybean food by locally created pulses could have a convenient effect, mainly for pulses produced and used on-farm. Modest is identified about the natural adjustment of faba bean germplasm, or about the regions and impacts of genes that affect traits associated with drought adjustment, chiefly stomatal morphology and role as key manners for gas reciprocity between plant and ambience (Khazaei 2014). Katerji (2011) drought at flowering, early podding, and grain-filling phases decrease both grain and straw yields.

Promoting new varieties that can be tolerated or conflict with environmental invasion mainly CC will be sharing in raising the prime creation of pulses during minimizing the stomata conductance, CO_2 improvement, and universal warming tolerates. Enhancing the leaf area index also takes part in a large role in adjusting the legumes which offer to the CO_2 raising. On the other hand, the nitrogen fixation outfitted by legume and C:N ratio can be enhanced.

To enhance crop execution, it is necessary to be aware of how crop species react and adjust to water deficit circumstances (Khazaei 2014; Ali et al. 2019; Imran and Khan 2015d; Imran et al. 2014, 2015h). Due to variability in quantity, interval, strength, and allocation of rainfall overall regions' global enhancement of new varieties are required to evade drought incidents. Adjustment based on the genetic makeup and phenotype elasticity, species with high flexibility are more adapted to stress environments and can be survived for long period and generate high yield ingredients. Drought is a very imperative ecological cause restraining the output of pulses crops as well as other field crops globally (Imran 2021). CC patterns estimate better variableness in rainfall patterns and expand periods of winter and spring drought on continents will affect areas of pulses planting. The planting required for varieties showing prior drought bearing and water use efficiency is fabricated in the enhancing stressed within pulses plant improving programs on the selection of varieties preferable to tolerate expanded times of water deficit (Imran et al. 2020b; Ilyas et al. 2020).

Modern molecular markers help in limiting quantitative trait loci that participated in the selection of the favourable traits used in breeding programs. Amelioration that demanded pulses depended on conductors of water deficit, for example chlorophyll fluorescence, stomatal density, and root density. Genetic resource gathering and classification can have an important part in distinguishing pulses genotypes adapted to water deficit-prone environments. In legumes, the influences of water deficit or other stresses on nitrogen fixation require to be quantified alongside the influences on the plant itself (Abberton et al. 2008).

Molecular analysis stressed that many genes are concerned in plant acclimation to all stresses that are exposed by pulses as well other crops. Direct gene transfer can be shared in enhancing the yield and yield ingredients during the genetic stability of legume crops under various ecology microclimates. Molecular biology has a wide range of capacity enforcements in pasture pulses breeding, including marker-assisted and genomics-assisted selection and the recognition of quantitative trait loci and electing genes for necessitating characters (Nichols et al. 2012).

On the other hand, classical breeding programs have an immense role in the acclimatization of pulses based on the introgression of genes. Nichols et al. (2012) pointed out that new perennial pulses have recently been progressed to defeat the soil acidity and waterlogging output oblige of Lucerne and white clover and to minimize groundwater recharge and the dispersal of arid land salinity. Several studies have been undertaken on the adaptation. Adequate drought tolerance to combat the moisture stress and expanded droughts typical of Mediterranean-like. For any of these selected species to be prosperous on a marketable level, numerous other traits need to be considered for their improvement. As a minimum, they would be demanded to have extensive soil and climatic adaptation to help endorse a bulky seed market (Real et al. 2011). Real et al. (2011) and Imran et al. (2018b) stated that the three-year assessment of six regions in southern Australia recognized a set of 10 preference perennial forage legumes (*B. bituminosa* var. *albomarginata*; *C. australasicum*; *D. hirsutum*; *K. prostrata*; *L. bainesii*; *L. pedunculatus*; *L. cytisoides*; *M. sativa* subsp. *sativa*; *M. Sativa* subsp. *caerulea*; and *M. sativa* subsp. *falcata*) with appropriate adaptation, rendering, and insistence. Voisin et al. (2013) reported that even though the properties of pulses concerning the nitrogen cycle may raise the elasticity of output systems exposed to CC, their involvement in extenuating CC earns to be better determined at global levels. The ecology and adaption point of pulses are dependent on the emergent cropping systems and rotation to decrease the using pesticides and mineral N fertilizer and enhance the yield stability. Himanen (2016) concluded that intercropping can avail as one adjustment strategy to reinforce the adaptive ability of Finnish farms. Promoting varieties with the ameliorative time of flowering, disease resistance, and yield will share in the financial system and beneficial business. Biodiversity farming and building of strong soils are assured, flourishing strategies to acclimatize to the augment in droughts presaged under CC.

Abd El-Moneim and Cocks (1992) reported that the two selections, 347 and 311 of *Lathyrus* spp., generated from Syria and Turkey, respectively, joined both high pasture and seed yields with extensive adaptation and steadiness and could be deemed the most broadly adapted lines. Various entities brightened up to have abroad adaptation, in provisions of both yield and firmness (Abd El Moneim et al. 1988). Collaborative adaptation arrangements by farmers and researchers can construct approximately rationally, environmentally, and economically resonance adaptation events as well as motivate the precedence building of farm adaptive ability (Himanen 2016). CC adaptations exemplary link to exploit technological regenerations variegation strategies changed organizations and the construction of public networks (Darnhofer 2010). Participating farmers and pastoral stakeholders in limiting and boosting adaptation strategies for agriculture to expose CC, with affirmation on the part of intercropping in structuring the adaptive strength of farms to have highly executions of climate prompt manners. Because CC has a great part in food output for present and next generation that predictable to turnoff more than 3 billion suffering from starvation, the next congress and strategic planning issued by governorates should take into consideration all factors that have direct influences on growth and allocation of the field crops focusing on the pulses crops which contributing highly percentage of food nutritive value.

2.3 Role of Legumes in Body Health Maintenance

Legumes (mung bean, mash bean, pigeon pea, cowpea, chickpea, field pea, lentil, and soybean) are hailed as the most protective bean (Aziz et al. 2016; Imran 2018a; Imran and Khan 2015a; Imran et al. 2015i, 2017a, 2018; Imran et al. 2016b). It has the highest protein content amongst plant products. As animal protein contains all the essential amino acids, lacking in legumes protein, 'Soy protein' refers to the protein found in soybeans. Soy is often used to replace the animal proteins in an individual's diet. Soybean belongs to the family Leguminacea but is mostly known as an oilseed. Here it should be needed to know about soybean as a leguminous crop instead of oilseed (Devi et al. 2012). Soybean is the only vegetable food that contains all eight essential amino acids. Soybeans are processed into various soya products namely soy flour, soy milk, cottage cheese-like tofu, and fermented products like tempeh and miso.

2.3.1 Legumes in Diet Can Help in Cardiovascular Health Stimulation (Legumes and Soybean)

Soybean oil is a rich source of polyunsaturated fatty acids, including the two essential fatty acids called linoleic acid and linolenic acid. Substituting saturated fat sources with unsaturated fats may help lower your cholesterol levels (Devi et al. 2013). Elevated serum cholesterol levels lead to atherosclerosis and an increased risk for heart disease. These two essential fatty acids are precursors to hormones that regulate smooth muscle contractions and blood pressure. Because your heart is a muscle, this health benefit is significant. Soybeans are a source of soluble fibre, which can help lower cholesterol levels and potentially reduce the risk of heart disease (Gaur et al. 2012).

2.3.2 Legumes and Recovery of Digestive Health and Prevent Colon Cancer

Mostly legumes and especially soybeans are a source of insoluble fibre, which adds bulk to stools and helps waste pass quickly through the digestive tract (Jukanti et al. 2012). Adding soybeans to your diet may prevent colon cancer, diverticular disease, constipation, and haemorrhoids. Soybeans have a high concentration of calcium, which is a mineral crucial to bone health (Imran 2018b,c; Imran and Khan 2015e; Imran et al. 2015j, 2016c, 2017b, 2018; Siddique et al. 2012). Isoflavones says isoflavones can interfere as well as activate the activity of your oestrogen. Soybeans contain isoflavones, which are substances that have a chemical structure very similar to oestrogen. During menopause, when oestrogen levels drop, isoflavones bind to the oestrogen receptors in cells and ease the symptoms of menopause, such as hot flashes. The isoflavones in soybeans increase bone density in women and offer protection against osteoporosis (Figure 2.1).

2.4 Climate Change and Legume Productivity and Profitability

Legumes production technology may be almost similar and undistinguishable throughout the world in this advanced era of the internet and make the world global village but here are a few available technologies that can increase the productivity and profitability of legumes (Sinclair and Vadez 2012).

Figure 2.1 Tangible view of soybean experimental plots at ARI Agriculture Research Institute Mingora Swat with bio fertilization of Trichoderma and PSB.

2.4.1 Short-Duration, High-Yielding Varieties

Matching legume's crop maturity duration to the available cropping window, including soil moisture availability, is a major strategy to avoid drought stress. Hence, emphasis in crop improvement programs has been to develop high-yielding, short-duration cultivars that escape terminal drought (Imran et al. 2015k). These short-duration varieties provide opportunities for the inclusion of a given crop or variety in the cropping systems with a narrow cropping window or new production niches.

The key factors for significant increase might be achieved with the following technique.

1) Introduction of high yielding, short duration, Fusarium wilt resistant varieties
2) High adoption of improved cultivars and production technologies mechanized field operations and effective management of pod-borer
3) Availability of grain storage facilities to farmers at the local level at an affordable cost.

2.4.2 Improved Varieties with Drought Tolerance

Wider dissemination of drought-tolerant material would provide sustenance to the livelihoods of farmers who are more vulnerable to shocks of crop failure. The drought-tolerant varieties can provide cost-effective long-term solutions against the adverse effects of drought. Returns to investment in breeding for drought tolerance are likely to be higher compared to those in other drought management strategies (Imran et al. 2015l). On the other hand, even though the potential economic benefits of drought-tolerance breeding research are attractive, farmers may not benefit from it if appropriate institutional arrangements are not in place for the multiplication and distribution of seeds of improved varieties. This is more so in the case of large-seeded legumes whose seed requirement is very high (Varshney et al. 2010; Imran 2018c; Imran and Khan 2015c,e; Imran et al. 2016a,c).

2.5 New Niches

Legumes crops have a great diversity of maturity durations that enable their cultivation in many niches and different production systems to increase production (Table 2.1). A few examples are given below, but there are many more in other crops and niches that can be exploited successfully.

The main cropping systems involving major food could be the following.

2.5.1 Faba Bean

a) Faba bean – rice – barley intercropping with green manure – cotton.
b) Faba bean – transplanting cotton – faba bean intercropping with green manure – maize intercropping with soybean, sweet potatoes, or rice.
c) Faba bean intercropping with green manure – corn multiple cropping with rice – barley – cotton.
d) Faba bean intercropping with green manure – cotton – faba bean or barley – transplanting cotton.
e) Faba bean intercropping with green manure – maize intercropping soybean – barley intercropping with green manure – cotton.
f) Faba bean intercropping with green manure – maize intercropping with cotton – Faba bean intercropping with green manure – maize intercropping soybean or sweet potatoes.

Table 2.1 Several species of legumes are taxonomically similar.

Subfamily: Papilionoideae	
Tribe	**Species**
Genisteae	*Lupinus albus* L.
	Lupinus luteus L.
	Lupinus angustifolius L.
	Lupinus mutabilis Sweet
Indigofereae	*Cyamopsis etragonoloba* (L.) Taub.
Phaseoleae	*Canavalia ensiformis* (L.) DC.
	Mucuna pruriens (L.) DC.
	Cajanus cajan (L.) Huth
	Psophocarpus tetragonolobus (L.) DC.
	Sphenostylis stenocarpa (Hochst. ex A. Rich.) Harms
	Lablab purpureus (L.) Sweet
	Vigna angularis (Willd.) Ohwi and H. Ohashi
	Vigna radiata (L.) R. Wilczek
	Vigna mungo (L.) Hepper
	Vigna umbellata (Thunb.) Ohwi and H. Ohashi
	Vigna aconitifolia (Jacq.) Maréchal
	Vigna unguiculata (L.) Walp.
	Vigna subterranea (L.) Verdc.
	Phaseolus vulgaris L.
	Phaseolus lunatus L.
	Phaseolus coccineus L.
	Phaseolus acutifolius A. Gray
Cicereae	*Cicer arietinum* L.
Fabeae	*Vicia faba* L.
	Vicia sativa L.
	Lens culinaris Medik.
	Pisum sativum L.

The main biennial cropping patterns, generally with two crops a year, are as follows:

a) Faba bean – corn relay intercropping with sweet potatoes – wheat (canola) – peanut.
b) Wheat – cotton – faba bean – corn relay intercropping with sweet potatoes.
c) Wheat (barley) – corn relay intercropping with sweet potatoes – faba bean – peanut.
d) Faba bean – early rice – late rice – canola – semi-late rice.
e) Wheat – semi-late rice – faba bean – semi-late rice – barley.

2.5.2 Pea

a) The first year: peas – early rice – late rice.
b) The second year: barley (wheat) – early rice – late rice.
c) The third year: canola – early rice – late rice.

2.5.3 Intercropping and Relay Intercropping

a) Pea – maize – maize.
b) Pea – canola – wheat.
c) Pea – wheat – potato.
d) Pea – barley – maize.

2.5.4 Mung Bean

a) Wheat (canola) – mung bean – wheat – maize (millet).
b) Wheat – mung bean – wheat – mung bean (sweet potato).
c) Wheat – mung bean – cotton – wheat – mung bean (millet) (three crops in two years).
d) Wheat – mung bean – sweet potato – maize – wheat – mung bean (Five crops in three years).

2.5.5 Adzuki Bean

a) Adzuki bean – maize (sorghum) – millet.
b) Adzuki bean – maize – sorghum.
c) Adzuki bean – spring wheat – maize (flax) – maize.
d) Adzuki bean – wheat – wheat.
e) Adzuki bean – millet – maize – wheat.
f) Winter wheat (barley) – adzuki bean – winter wheat (barley) – adzuki.
g) Winter wheat (barley) – adzuki bean – cotton (maize, sorghum, or millet).
h) Winter wheat (barley) – summer maize intercropping with adzuki bean – spring maize intercropped with soybean.
i) Spring wheat – millet (broomcorn millet) – adzuki bean.
j) Winter wheat (barley) – maize intercropped with adzuki bean – winter wheat (barley) - maize intercropped with adzuki.
k) Winter wheat (barley) – sunflower intercropped with adzuki bean – winter wheat (barley) – sunflower intercropped with adzuki.

2.6 Advance Legumes Production Technology

2.6.1 Seed Inoculum

Mungbean (green gram) inoculation with various beneficial microbes (phosphate solubilizing bacteria (PSB), Trichoderma, Rhizobium, *Bradyrhizobium japonicum*, Azospirillum, Azotobacter, etc.) have been tested and noticed the highest green forage, dry matter, and crude protein (Figure 2.1). It has been reported that the highest number of plants m^{-2}, plant height (cm), number of leaves $plant^{-1}$, pods length (cm), numbers of seeds pod^{-1}, number of pods plants-[1], thousand seed weight, seed yield, and harvest index was noted with the inoculation of biofertilizer (PSB) (Sinclair and Vadez 2012; Imran 2018d,e; Imran and Khan 2015b,e; Imran et al. 2015b,g). The nitrogen-fixing bacteria (Rhizobia) that live on legumes roots in nodules are not native to most soils. The best way to introduce these bacteria is to inoculate the seed. Once, introduced, the Rhizobia population remains active in the soil for a long time. In the presence of the appropriate inoculant of *Rhizobium japonicum*, more nodules are formed on the roots of the legumes plant which can fix atmospheric nitrogen from the air that is almost as effective as nitrogen applied as fertilizer to promote the growth and development of the plant (Imran 2018d; Imran and Khan 2017, 2015b).

2.6.2 Method of Inoculation

Inoculum is a black powder containing nitrogen-fixing bacteria that are mixed with ground peat or some similar carrier and applied on seed just before planting time. Seeds are moistened with concentrated sugar solution; inoculant is applied at the rate of 1250 gm per 100 kg seeds and then mixed thoroughly to have a uniform coating of inoculum on the seeds. This process should be done in a shady place (Imran et al. 2016a,c). The use of fungicide in case of a seed treatment may interfere with inoculated seed and with symbioses of the Rhizobium system. Thus, compatible fungicides (i.e. Benlate and Dithane Z-78 (Zineb)) with no toxicity to Rhizobia should be used. Treat seed immediately before planting and use inoculum dose a little higher than recommended.

2.6.3 Method of Sowing

The method of sowing should meet three objectives (i) adequate and uniform depth of seed placement, (ii) adequate seed numbers and uniform distribution, and (iii) sufficient soil-seed contact for germination. Under existing conditions, legumes seed must be planted with a single row cotton drill or tractor drill because of their consistency in producing good stands (Imran et al. 2016c).

2.6.4 Time of Sowing (*Kharif* Legumes/*Rabi* Legumes)

Variation in yield loss due to delay in the planting is the result of variation in weather and its influence on disease, plant height, flower abortion, amount of vegetative growth, etc. (Imran et al. 2015h, 2016b). Therefore, the planting date is considered to be one of the important factors for maximum crop yield. Late planting may result in a significant decrease in seed yield (Figures 2.1 and 2.2).

2.6.5 Seed Rate

The optimum seed for *Kharif* or *Rabi* legumes at a depth of 3 to 5 cm with 30 to 45 cm row spacing or optimum space for each legume crop gives the optimum population of 120 000 to 130 000 plants per acre. As many as 20 to 24 plants per meter of the row are generally

Figure 2.2 Tangible view of mungbean (green gram) experimental plots at Agriculture Research Institute Mingora Swat with bio fertilization of Trichoderma and PSB.

satisfactory. However, within a wide range of plant populations for a particular row spacing, legumes yield does not vary significantly. However, low populations result in low poding height and excessive branches but good lodging resistance. Conversely, high plant populations result in increased lodging, high poding height, and less branching. Generally, legumes in narrow rows are higher-yielding because they capture more of the sun's energy which drives the photosynthetic machinery of the plant. However, wide rows are used if the varieties are tall and bushy type.

2.6.6 Irrigation

The number of irrigations varies with climatic conditions, management practices, and length of the growing season. Moisture stress during flowering, pod filling, and seed development stages reduces yield. Usually, four to six irrigations are required for *Kharif* legumes and two to three irrigations for *Rabi* crop depending upon the rains.

Therefore, irrigation must be given at the following stages:

- Three weeks after germination
- Initiation of flowering
- Pod filling stage
- Seed development stage.

2.6.7 Weed Management

Weeds are complete with legumes crops for nutrients, moisture, and light and thus, reduce yield. The most effective measure for developing weed control in legumes varies; depending on types of weeds, degree of weed infestation, soil type, weather patterns, crop rotation, tillage methods, row spacing, and pieces of equipment available. A good weed control program should include a combination of preventive, cultural, mechanical, and chemical practices.

2.6.8 Preventive Measures

All these measures taken to prevent the introduction and spread of weeds include the use of weed-free crop seeds.

- Certified seed of legumes
- Clean planting/harvesting equipment
- Weed-free soil, seed, and farmyard manure/organic residues.

2.6.8.1 Cultural Control

2.6.8.1.1 Crop Rotation

Crop rotation reduces the weed populations of certain weeds common to a particular crop. It results in improved crop yield, quality, and improved soil conditions, and reduces the chance of plant disease and insect infestation.

2.6.8.1.2 Tillage Practices

Deep and dry plugging gives substantial control of perennial weeds. By the tillage method, weeds are buried with soil, thrown over, and with disruption of the ultimate relationship between the weed, crop, and soil.

2.6.8.2 Mechanical (Physical) Control

Hand weeding is the most common practice and used employed by the farmers, two weedings are recommenced during the growing period for each leguminous crop. Mechanical weeding by rotary weeder controls the annual weeds effectively and economically.

2.6.8.3 Chemical Control

The experimental results have shown that proper use of pre-emergence herbicides, i.e. Pendimethalin 8(Stomp), Trifluralin (Treflan), and Oxadiazon (Ronstar) can be applied after planting before the germination of legumes crop. These have been found very effective in controlling a wide range of grasses and broadleaf weeds.

2.6.8.4 Integrated Weed Control

This involves the use of two or more above-mentioned weed control techniques. The combined use of more than one weed control technique is advantageous because one technique rarely achieves complete eradication of weeds.

2.7 Diseases and Insect Pest of Legumes and Their Control

About 35 diseases of economic importance are known to affect leguminous crops throughout the world. Mostly on *Kharif* legumes (mungbean), fungal diseases of leaf spot, root rot, and powdery mildews may attack in wet and warm weather while bacterial leaf blight and yellow mosaic virus are also serious diseases in Asian countries like Pakistan and India. Common insects and pests of mung bean are pods borer, whitefly, pods sucking bug, caterpillar, Jassids, and aphids. Pigeon pea common diseases are rust leaf spot, Fusarium wilt, and occasionally sterility mosaic virus may also attack (Imran et al. 2016a). This disease may spread and be transmitted by Eriophyid mite. The affected plant is totally or partially sterile and could not produce seed. Common insect pests of pigeon pea may be moth and pod fly. These diseases can reduce yield from 10 to 30%, depending upon the severity, pathogen, and weather conditions.

Cowpea is also susceptible to various diseases and insect pests. Common diseases of cowpea are aphid-born mosaic virus, Fusarium wilt, leaf smut, and scab. General insect pests of cowpea might be beetles, aphids, pod borers, caterpillars, and whiteflies which seriously damage the crop. Certain insects can also provide access to disease organisms and transmit them directly to plants. Therefore, understanding the relationships between the insect and the crop will enable farmers to manage pests much better. Major insect damage to soybean crops is stem fly, white fly; green stink bug, cutworm, and larvae. Insects attack all parts of the soybean plant and feed throughout thc growing season. A new practice to control insect attacks is based on knowledge of the economic injury levels of the consequential insects (Imran et al. 2016b). The economic injury level is the population of insects that is capable of producing an amount of economic damage which is at least equal to the cost of controlling the insects. Wise monitoring of major insects is required to effectively make a decision relative to insecticide application.

In Pakistan soybean, major diseases have been observed and these are anthracnose, charcoal rot, purple seed stain, pod, and stem blight, and bacterial blight/pustule. Careful diagnosis is very important for the disease control strategies, because more than one pathogen or a complex may be responsible for the final loss in yield and seed quality. Thus, the collection of samples is essential throughout the growing season (Imran et al. 2016c).

Major *Rabi* legumes such as chickpea, lentil, and field pea might produce low yield and profitability due to various insect pests and disease attacks. A common disease of chickpea is, blight disease while the most dangerous enemy of chickpea, root rot, and wilting. Insects of chickpea are aphids, gram caterpillar, leaf folder, and pods borer attack and reduce the yield from 30 to 70% relating to the severity (Imran et al. 2020a, 2021e). Reduction in yield could be overcome by planting disease and insect pest-resistant cultivars, proper seed treatment, and crop rotation. Like chickpea, lentil crop is also susceptible to the attack of aphids, gram caterpillar, leaf folder, pods borer, and weevils. They can be checked by crop rotation and appropriate pesticides (Imran et al. 2016a). The most common disease of lentils is, leaf rust, downy mildew, blight, and Fusarium wilt whereas sometimes root rot or root wilting may occur. A decrease in yield could be overcome by planting disease and insect pest-resistant cultivars, proper seed treatment, and crop rotation. Another *Rabi* crop, filed pea/pea attack by a fungus such as powdery mildew, downy mildew, pods and leaf spots, grey mould, Fusarium wilt, and root rot attack may seriously damage the crop. The insect of pea includes pea moth, pea aphids, and weevil. The damage of the crop and yield reduction can be controlled by the planting of disease-resistant cultivars, a fungicide treatment, and crop rotation (Imran et al. 2016b).

2.8 Disease Management Program Should Include the Following Methods

2.8.1 Preventive Measures

- Avoid planting in wet and poorly drained soils to reduce the chances of the development of soil-borne diseases (Root rot).
- Harvest seed soybeans as soon as they are mature.
- Keep the crop free from weeds because they may be the hosts for any diseases.
- Plant quality and healthy seed, free of mechanical damage.
- Seeds should be stored at 8 to 10% moisture at 15 °C temperature to have more seed viability.

2.8.2 Cultural Control

- Crop rotation with non-leguminous crop.
- Plant early before the soil temperature rises.
- Plough down crop residues.
- Reduce plant population, increase row width and avoid high fertility to prevent a closed canopy, improve aeration and increase drying in the canopy.

2.8.3 Chemical Control

2.8.3.1 Seed Treatment

To protect from soil-borne diseases, seed should be treated before planting with one of the available fungicides; Captan, Dithane M-45, Benlate, and Tecto at the rate of 1.5–2 gm per 1 kg seed (Imran et al. 2020).

2.8.3.2 Foliar Spray

At the appearance of disease symptoms, after fielding survey foliar spray with one of the systemic fungicides, Dithane M-45 at the rate of 1 kg ha^{-1}, Benlate, and Tecto at the rate of 120–150 gm ha^{-1} dissolved in 250 l of water should be done after 10 to 15 days interval. This process should be repeated 2–3 times depending upon the severity of the disease (Imran et al. 2016a,b; Ali et al. 2019; Naveed et al. 2016; Baqa et al. 2015; Anwar et al. 2016; Samreen et al. 2016).

2.8.3.3 Integrated Disease Control

- Use integrated control; combines high tolerate cultivar, good drainage, complete tillage, seed treatment, and rotation with cereals.
- Use multi-race resistance varieties or least susceptible cultivars and avoid tall, Viney cultivars that may lodge.

2.8.4 BioControl

Many insect predators, parasites, and pathogens occur in legume fields that help in keeping the population of pest species below economic levels. However, experience has shown us that beneficial insects and pathogens do not do a complete job and that chemical control becomes necessary.

2.8.5 Cultural Control

Early planted legumes tend to receive the majority of the overwhelming adult of flies and bugs, while late planted will alleviate some of the problems with these two insects and cutworms. Therefore, it is still more economical to plant early for high yields and control any potential insect problems with insecticides than it is to plant late for insect control (Imran et al. 2016b,c).

2.8.6 Chemical Control

- For cutworm and termite apply powder of BHC at the rate of 7 kg ha^{-1} or Dieldrin 20 EC at the rate of 5–7 l ha^{-1} mixed with irrigation water.
- For other insects especially flies, thrips and larvae spray; Dimecron 100% at the rate of 600 ml ha^{-1} or Methyl-Parathion 50% at the rate of 800 to 1200 ml ha^{-1} Somicidin 20 EC at the rate of 400 to 600 ml ha^{-1} dissolved in 250 l of water if the attack is severe spray two times with an interval of 8–10 days.

2.9 Harvesting and Threshing

Various legumes crop matures in 80 to 120 days depending upon growing season and the variety that was planted. Senescence is the decline in chemical activity associated with the ageing of plants and maturation is only the loss of water from plants or seeds which are physiologically mature. Seeds are physiologically mature when they are no longer synthesizing food. Physiological loss of chlorophyll and acceleration of senescence is characteristics of dry, dehiscent fruits (soybean pods). Ethylene and abscisic acid play an important role in the abscission and dehiscence of the pod and often capsule dry fruits (soybean). Oil and storage protein have reached their maximum dry weight. At this stage seed moisture is 45 to 55%, pods and stems are yellow, and leaves are yellow or have dropped. Delay in harvesting not only reduces seed quality but also reduces harvesting efficiency and increases shattering losses. As soon as the pods are dry enough to open easily, harvest them, thresh after drying within 7–10 days, and threshed seed must be cleaned before storage or marketing.

2.10 Yield

The number of pods per plant is a function of spacing and intercepted light while leaf N is a principal factor in determining legumes seed yield (Imran et al. 2020a; Imran et al. 2021e). The average farmer's yield ranges from 1500 to 2500 $kg\,ha^{-1}$ of soybean, 1200 to 1800 of mungbean, and mash bean varies with seed size and crop variety. At Agricultural Research Institutes under high-level management practices, the yields range from 2500 to 3500 $kg\,ha^{-1}$ in the case of the soybean crop.

2.11 Storage

Well-dried seed should be stored at about 8 to 10% moisture content and 15 °C in tropical regions. To maintain dried seed at a low moisture level, two practices are feasible. These are:

- To grow legumes for seed in an area where the relative humidity is low.
- To use moisture-proof containers for seed storage.

For long-term benefit and effectiveness, air-conditioned storage in tropical and subtropical areas are more suitable. Adequately air-conditioned storage should be maintained at a temperature of 20 to 22 °C or less and relative humidity of 60% or less when the storage period is of 89 months duration. Poor quality legume seeds will deteriorate quickly in storage than high-quality seeds (Imran et al. 2020b). Also, the practice of 'Carrying over' legume seeds should be discouraged because this crop does not store well and the quality of seed quickly diminishes during the second overwintering period.

2.12 Marketing

Beside Government Agencies private agencies should be directed to procure the entire production from the farmers at a high cost and effective rate.

2.13 Advantages of Legume Planting

- Has a short duration season and farmers could utilize rice, cotton, and rainfed fallow areas.
- Well, fit in the existing cropping system without clashing with major crops.
- Economics often dictates crop sequence, but where choices are available, legumes should follow crops other than legumes like cereals that make better use of the nitrogen provided by legume crops.
- Legume also provides good quality protein which is high in one of the limiting amino acids lysines and is useful as a supplement to other cereals.

2.14 Conclusions

Agronomic management of legumes under CC stress conditions a quick alternative need of Nanotechnology and biotechnology application should be needed to develop new cultivars for the new environments. The successful and critical solution may be to develop resilient legume varieties, demonstrating great potential in difficult conditions, or to combat CC to enhance food security. The efforts in this admiration to breed lentils with vertical nodulation and nodule clusters are associated with high nitrogen fixation. Legume varieties with mature leaf concentration and pea varieties with waxy leaf surfaces help to combat heat stress. The progress of additional efforts to screen germplasm for heat tolerance and disease resistance in chickpeas, lentils, mung beans, and soybean. Almost all crops of human interest are having with advanced production technologies beyond their routine traditional production methods. Legume crops can contribute to alleviating hunger as well as malnutrition, especially in poorer countries. Fortunately, legume crops need no special attention unlike rice-wheat-potato and most oil seeds. Also, they can adapt to a wide range of soil and climate and demand very little nutrient and water management. Legumes can fix atmospheric nitrogen also. Legumes production could be a great option economically, nutritionally, and ecologically. Some trends seem to be the ability to fix legumes to make good use of the elevated CO_2 to increase growth by utilizing excess C for N fixation and nutrient uptake. This will require varieties and treatments that allow conversion of this growth to the grain. The reduction of some moderate, but not severe, stresses (e.g. drought, early N, P, and K) under high CO_2 but a need to manage the increased demand put on these factors by the greater growth. There is a potential need to rework the phenology of varieties for regions as temperature and water profiles change.

2.15 Summary

Improving the production capacity of legumes, so these climate-smart plants can play a more important role in environmental security, which can be relied upon cover production gaps that can lead to rapid CC. Against the environment. Enhancing the manufacturing capability of legumes, so these local weather brilliant crops can play an extra essential task in environmental security, which might be relied on cowl manufacturing gaps that may result in speedy local weather change. In opposition to the environment. CC is an increased risk of pests and diseases, which is a significant obstacle to the production of different crops. Particularly small landholder farmers can struggle to control it. Local weather varies (drought stress, water stress, rising temperatures, rising CO_2, and different greenhouse gases (GHGs)) is an escalated threat of pests and diseases, which is a big impediment to various crop production. In particularly limited landholder farmers can wrestle to regulate it. Successful and critical solutions may be to develop flexible fruit and crop varieties, demonstrate great potential in grim environments, or counter CC to enhance food security. Pulses production with vertical nodulation and nodule clusters associated is fixed with soil health with nitrogen fixation. Pea varieties with solid leaf concentration and pea types with waxy leaf surfaces that help withstand heat stress. Development of additional efforts for screening germs for heat tolerance and disease prevention in peppers, lentils, peanuts, and soybeans. Almost all crops of human interest have advanced production technologies, beyond the traditional methods of production. Fruit crops can help reduce hunger as well as nutritional deficiencies, especially in poorer countries. Fortunately, fruit crops do not need any special attention unlike rice, wheat, potatoes, and most oil seeds. Pulses can accommodate a wide range of soil and climate and demand very little nutrition and water management. Production of legumes can be an economically, nutritionally, and environmentally sound option under CC. Propitious and important options could also exhibit immense potential under local weather alter to boost food security. Improvement of extra efforts for screening germs for warmth tolerance and illness prevention in peppers, lentils, peanuts, and soybeans. Just about all crops of human curiosity possess superior Manufacturing technologies, past the normal strategies of production. Fruit crops may also help cut back starvation in addition to dietary deficiencies, mostly in poorer countries. Happily, fruit crops don't want any particular consideration, not like rice, wheat, potatoes, and more importantly oil seeds. Pulses can accommodate a big selection of soil and local weather and demand little or no vitamin and water management. Legumes may also fix atmospheric nitrogen. Manufacturing of legumes may be an economically, nutritionally, and environmentally sound choice beneath altering climate.

References

Abberton, M.T., MacDuff, J.H., Marshall, A.H., and Mike, W. (2008). The genetic improvement of forage grasses and legumes to enhance adaptation of grasslands to climate change. FAO, May 2008.

Abd El Moneim, A.M., Cocks, P.S., and Swedan, Y. (1988). Yield stability of selected forage vetches (*Vicia* spp.) under rainfed conditions in west Asia. *The Journal of Agricultural Science* 111 (2): 295–301.

Abd El-Moneim, A.M. and Cocks, P.S. (1992). Adaptation and yield stability of selected lines of *Lathyrus* spp. under rainfed conditions in West Asia. *Euphyica* 66 (1–2): 69–89.

Abdelhamid, M., Shokr, M., and Bekheta, M. (2010). Growth, root characteristics, and leaf nutrients accumulation of four faba bean (*Vicia faba* L.) cultivars differing in their broomrape tolerance and the soil properties in relation to salinity. *Communications in Soil Science and Plant Analysis* 41 (22): 2713–2728.

Aise, D., Erdal, S., Hasanand, A., and Ahment, M. (2011). Effects of different water, phosphorus and magnesium doses on the quality and yield factors of soybean in Harran plain conditions. *International Journal of Physical Sciences* 6 (6): 1484–1495.

Akpalu, M.M., Siewobr, H., Oppong, D., and Akpalu, S.E. (2014). Phosphorus application and Rhizobia inoculation on growth and yield of soybean (*Glycine max* L. Merrill). *American Journal of Experimental Agriculture* 4 (6): 674–685.

Ali, I., Khan, A.A., Imran et al. (2019). Humic acids and nitrogen levels optimizing productivity of green gram (*Vigna radiate* L.). *Russian Agricultural Sciences* 45 (1): 43–47.

Amanullah, Khalid, S., Khalil, F. et al. (2021). Growth and dry matter partitioning response in cereal-legume intercropping under full and limited irrigation regimes. *Scientific Reports* https://doi.org/10.1038/s41598-021-92022-4.

Anchal, D., Kharwara, P.C., Rana, S.S., and Dass, A. (1997). Response of gram varieties to sowing dates and phosphorus levels under on farm condition. *Himachal Journal of Agricultural Research* 23: 112–115.

Ankomah, A.B., Zapata, F., and Hardarson, G. (1996). Yield, nodulation and N_2 fixation by cowpea cultivars at different phosphorus levels. *Biology and Fertility of Soils* 22: 10–15.

Anwar, S., Israeel, Iqbal, B. et al. (2016). Nitrogen and phosphorus fertilization of improved varieties for enhancing phenological traits of wheat. *Pure and Applied Biology* 5 (3): 511–519. https://doi.org/10.19045/bspab.2016.50065.

Aziz, A.L.A., Ahiabor, B.D.K., Opoku, A., and Abaidoo, R.C. (2016). Contributions of Rhizobium inoculants and phosphorus fertilizer to biological nitrogen fixation, growth and grain yield of three soybean varieties on a fluvic luvisol. *American Journal of Experimental Agriculture* 10 (2): 1–11.

Baqa, S., Khan, A.Z., Inamullah et al. (2015). Influence of farm yard manure and phosphorus application on yield and yield components of wheat. *Pure and Applied Biology* 4 (4): 37–45.

Bishop, K.A., Leakey, A.D.B., and Ainsworth, E.A. (2016). How seasonal temperature or water inputs affect the relative response of C3 crops to elevated [CO_2]: a global analysis of open top chamber and free air CO_2 enrichment studies. *Food and Energy Security* 3 (1): 33–45. https://doi.org/10.1002/fes3.44.

Canci, H. and Toker, C. (2009). Evaluation of annual wild Cicer species for drought and heat resistance under field conditions. *Genetic Resources and Crop Evolution* 56: 1. https://doi.org/10.1007/s10722-008-9335-9.

Cutforth, H.W. (2007). Adaptation of pulse crops to the changing climate of the northern great plains. *Agronomy Journal* 99: 1684–1699.

Darnhofer, L. (2010). Adaptiveness to enhance the sustainability of farming systems. A review. *Agronomy for Sustainable Development* 30: 545–555.

Devasirvatham, V., Tan, D.K.Y., Gaur, P.M. et al. (2012). High temperature tolerance in chickpea and its implications for plant improvement. *Crop & Pasture Science* 63 (5): 419–428.

Devi, K.N., Singh, L.N.K., Devi, T.S. et al. (2012). Response of soybean to sources and levels of phosphorus. *The Journal of Agricultural Science* 4 (6): 44–53.

Devi, K.N., Singh, T.B., Athokpam, H.S. et al. (2013). Influence of inorganic, biological and organic manures on nodulation and yield of soybean and soil properties. *The Australian Journal of Science* 7 (9): 1407–1415.

FAO (1994). Definition and classification commodities, 4. Pulses and derived products. Also available at http://www.fao.org/es/faodef/fdef04e.htm

Gaur, P.M., Jukanti, A.K., Srinivasan, S., and Gowda, C.L.L. (2012). Chickpea (*Cicer arietinum* L.). In: *Breeding of Field Crops* (ed. D.N. Bharadwaj), 165–194. Jodhpur, India: Agrobios (India).

Himanen, S.J. (2016). Engaging farmers in climate change adaptation planning: assessing intercropping as a means to support farm adaptive capacity. *Agriculture* 6 (34): https://doi.org/10.3390/agriculture6030034.

Ilyas, M., Ayub, G., Imran et al. (2020). Calcium and boron effect on production and quality of autumn potato under chilling temperature. *Communications in Soil Science and Plant Analysis* https://doi.org/10.1080/00103624.2020.1854286.

Imran (2018a). Climate change is threat toward agronomy (base of food, fiber system), and food security. *Food & Nutrition Journal* FDNJ-160. https://doi.org/10.29011/2575-7091.100060.

Imran (2018b). Phosphorous fertilization influenced weeds attributes and phenological characteristics of mungbean cultivars (*Vigna radiata* L.). *Russian Agricultural Sciences* 44 (3): 229–234.

Imran (2018c). Physiological and morphological traits of agronomic crops influenced by climate change. *Modern Concepts & Developments in Agronomy* 1 (4): MCDA.000524.

Imran (2018d). Ecological environmental variability influence growth and yield potential of rice under Northern Climatic Scenario. *Russian Agricultural Sciences* 44 (1): 18–24.

Imran (2018e). Organic matter amendments improve soil health, productivity and profitability of maize and soybean. *Annals of Reviews and Research* 1 (3): 555564.

Imran (2021). The bioavailability of phosphorus in composite vs. hybrid maize differ with phosphorus and boron fertilization. *Phosphorus, Sulfur, and Silicon and the Related Elements*. https://doi.org/10.1080/10426507.2021.1920588.

Imran and Amanullah (2018). Review. Global impact of climate change on water, soil resources and threat towards food security: evidence from Pakistan. *Advances in Plants & Agriculture Research* 8 (5): 350–355. https://doi.org/10.15406/apar.2018.08.00349.

Imran and Amanullah (2021). Phosphorus biofortification and uptake in maize enhanced with integrated phosphorus management. *Phosphorus, Sulfur, and Silicon and the Related Elements* https://doi.org/10.1080/10426507.2021.2022677.

Imran and Khan, A.A. (2015a). Grain yield and phenology of maize cultivars influenced by various phosphorus sources. *Food Science and Quality Management*. ISSN 2224-6088 (Paper); ISSN 2225-0557.

Imran and Khan, A.A. (2015b). Biochar application and shoot cutting duration (days) influenced growth, yield and yield contributing parameters of *Brassica napus* L. *Journal of Biology, Agriculture and Healthcare. care.*. ISSN 2224-3208 (Paper); ISSN 2225-093X (Online).

Imran and Khan, A.A. (2015c). Influence of compost application and seed rates on production potential of late sown maize on high elevation in Swat-Pakistan. *Journal of Environment and Earth Science*. ISSN 2224-3216 (Paper); ISSN 2225-0948.

Imran and Khan, A.A. (2015d). Phenological characteristics of *Brassica napus* L. as influenced by biochar application and shoot cutting duration (days). *Civil and Environmental Research*. ISSN 2224-5790 (Paper); ISSN 2225-0514 (Online).

Imran and Khan, A.A. (2015e). Effect of transplanting dates on yield and yield components of various rice genotypes in hilly area cold climatic region of Khyber Pakhtunkhwa-Pakistan. *Journal of Biology, Agriculture and Healthcare*. ISSN 2224-3208 (Paper); ISSN 2225-093x (Online).

Imran and Khan, A.A. (2017). Canola yield and quality enhanced with sulphur fertilization. *Russian Agricultural Sciences* 43: 113. https://doi.org/10.3103/S1068367417020100.

Imran, Khan, A.A., Irfanullah, and Ahmad, F. (2014). Production potential of rapeseed (*Brassica napus* L.) as influenced by different nitrogen levels and decapitation stress under the rainfed agro-climatic condition of Swat Pakistan. *Journal of Global Innovations in Agricultural and Social Sciences*. ISSN: 2311-3839 (Online); ISSN: 2312-5225 (Print).

Imran, Khan, A.A., and Ahmad, F. (2015a). Phenology of various rice genotypes as affected by different transplanting dates under cold climatic region of Khyber Pakhtunkhwa-Pakistan. *Journal of Environment and Earth Science*. ISSN 2224-3216 (Paper); ISSN 2225-0948.

Imran, Khan, A.A., Ahmad, F., and Irfanullah (2015b). Influence of hydrated calcium sulphate ($CaSO_4 \cdot 2H_2O$) and nitrogen levels on water infiltration rate and maize varieties productivity in rainfed area of Swat, Pakistan. *Chemical and Environmental Sciences.* ISSN 2224-3224 (Print); ISSN 2225-0956 (Online).

Imran, Khan, A.A., Ahmad, F., and Irfanullah (2015c). Nitrogen levels, tillage practices and irrigation timing influenced yield, yield components and oil contents of Canola. *Civil and Environmental Research.* ISSN 2224-5790 (Paper); ISSN 2225-0514 (Online).

Imran, Khan, A.A., Ullah, I. et al. (2015d). Yield and yield attributes of rapeseed cultivars as influence by sulfur level under Swat valley conditions. *Pure and Applied Biology* 4 (3): 296–301.

Imran, Khan, A.A., Inamullah, and Luqman (2015e). Weeding stages and their effect on yield and yield components of rice in upper Swat, Pakistan. *Pakistan Journal of Weed Science Research* 21 (4): 555–563.

Imran, Khan, A.A., Zada, H. et al. (2015f). Graine yield and yield components of wheat cultivar "Siran 2010" as affected by phosphorous levels under rain fed condition. *Journal of Natural Sciences Research.* ISSN 2224-3186 (Paper); ISSN 2225-0921 (Online).

Imran, Khan, A.A., Akhtar, K. et al. (2015g). Rice seedling characteristics of various genotypes influenced by different sowing dates in Swat-Pakistan. *Journal of Environment and Earth Science.* ISSN 2224-3216 (Paper); ISSN 2225-0948.

Imran, Maula, F., Uzair, M., and Zada, H. (2015h). Farmers income enhancement through off-season vegetables production under natural environment in Swat-Pakistan. *Journal of Environment and Earth Science.* ISSN 2224-3216 (Paper); ISSN 2225-0948.

Imran, Hussain, I., Khattak, I. et al. (2015i). Roots nodulation, yield and yield contributing parameters of mungbean cultivars as influenced by different phosphorous level in Swat-Pakistan. *Pure and Applied Biology* 4 (4): 557–567.

Imran, Khattak, I., Hussain, I. et al. (2015j). Growth and yield of maize hybrids as effected by different sowing dates in Swat Pakistan. *Pure and Applied Biology* 4 (4).

Imran, Uzair, M., Maula, F. et al. (2015k). Introduction and promotion of off-season vegetables production under natural environment in hilly area of upper Swat-Pakistan. *Journal of Biology, Agriculture and Healthcare.*

Imran, Naveed, S., Khan, A.A., and Khattak, I. (2015l). Impact of phosphorus levels and seed rates on growth and yield of late sown maize on high elevation in Swat, Pakistan. *Pakistan Journal of Agricultural Research* 28 (4): 406–413.

Imran, Khan, A.A., Khan, I.U., and Naveed, S. (2016a). Weeds density and late sown maize productivity influenced by compost application and seed rates under temperate environment. *Pakistan Journal of Weed Science Research* 22 (1): 169–181.

Imran, Khan, A.A., Inam, I., and Ahmad, F. (2016b). Yield and yield attributes of mungbean (*Vigna radiata* L.) cultivars as affected by phosphorous levels under different tillage systems. *Cogent Food & Agriculture* 2: 1151982.

Imran, Zada, H., Naveed, S. et al. (2016c). Variable rates of phosphorous application influenced phenological traits of green gram (*Vignaradiata* L.). *Journal of Agricultural Research* 1 (3).

Imran, Bari, A., Ali, R. et al. (2017a). Traditional rice farming accelerate CH4 & N_2O emissions functioning as a stronger contributors of climate change. *International Journal of Environmental Sciences & Natural Resources* https://doi.org/10.19080/ARTOAJ.2017.09.555765.

Imran, Jamal, N., Alam, A., and Khan, A.A. (2017b). Grain yield, yield attributes of wheat and soil physio-chemical characteristics influenced by biochar, compost and inorganic fertilizer application. *Agricultural Research & Technology: Open Access Journal* https://doi.org/10.19080/ARTOAJ.2017.10.555795.

Imran, Amanullah, Bari, A., and Ali, R. (2018). Peach sources, phosphorous and beneficial microbes enhance productivity of soybean. *Soybean Research* 16 (2): 39–48.

Imran, Amanullah, Khan, A.A. et al. (2019). Production statistics and modern technology of maize cultivation in Khyber Pakhtunkhwa Pakistan. *Plant Science Archives* 2 (2): 1–12.

Imran, Amanullah, Arif, M. et al. (2020a). Soil application of *Trichoderma* and Peach (*Prunus persica* L.) residues possesses biocontrol potential for weeds and enhances growth and profitability of soybean (*Glycine max*). *Sarhad Journal of Agriculture* 36 (1): 10–20. https://doi.org/10.17582/journal.sja/2020/36.1.10.20.

Imran, Amanullah, and Al-Tawaha, A.R. (2020b). The productivity of subsequent wheat enhanced with residual carbon sources and phosphorus under improved irrigation system. *Communications in Soil Science and Plant Analysis* https://doi.org/10.1080/00103624.2020.1763387.

Imran, Amanullah, Khan, A.A. et al. (2020c). Adequate fertilization, application method and sowing techniques improve maize yield and related traits. *Communications in Soil Science and Plant Analysis* (36)-1-10-20.

Imran, Amanullah, Arif, M. et al. (2020d). Integration of Peach (*Prunus persica* L.) residues, beneficial microbes and phosphorus enhance phenology, growth and yield of soybean. *Russian Agricultural Sciences* 46 (3): 223–230. ISSN 1068-3674. © Allerton Press, Inc.

Imran, Amanullah, and Al Tawaha, A.R.M. (2021a). Management of nano-black carbon, phosphorous and bio fertilizer improve soil organic carbon and ensilage biomass of soybean and maize. *Communications in Soil Science and Plant Analysis*. https://doi.org/10.1080/00103624.2021.1966439.

Imran, Amanullah, and Altawaha, A.R. (2021b). Carbon assimilation and dry matter partitioning in soybean ameliorates with the integration of nano-black carbon, along with beneficial microbes and phosphorus fertilization. *Journal of Plant Nutrition*.

Imran, Amanullah, Khan, A.A. et al. (2021c). Adequate fertilization, application method and sowing techniques improve maize yield and related traits. *Communications in Soil Science and Plant Analysis* https://doi.org/10.1080/00103624.2021.1925688.

Imran, Amanullah, and Al-Tawaha, A.R. (2021d). Carbon sources application increase wheat yield and soil fertility. *Communications in Soil Science and Plant Analysis* https://doi.org/10.1080/00103624.2020.1865397.

Imran, Amanullah, Khan, A.A. et al. (2021e). Adequate fertilization, application method and sowing techniques improve maize yield and related traits. *Communications in Soil Science and Plant Analysis* https://doi.org/10.1080/00103624.2021.1925688.

Jukanti, A.K., Gaur, P.M., Gowda, C.L.L., and Chibbar, R.N. (2012). Nutritional quality and health benefits of chickpea (*Cicer arietinum* L.): a review. *British Journal of Nutrition* 108: S11–S26.

Kashiwagi, J. (2015). Scope for improvement of yield under drought through the root traits in chickpea (*Cicer arietinum* L.). *Field Crops Research* 175: 74–54.

Katerji, N. (2011). Faba bean productivity in saline–drought conditions. *European Journal of Agronomy* 35 (1): 2–12.

Katunga, M.M.D. (2014). Agro-ecological adaptation and participatory evaluation of multipurpose tree and shrub legumes in mid altitudes of Sud-Kivu, D. R. Congo. *American Journal of Plant Sciences* 5: 2031–2039.

Khazaei, H. (2014). Leaf traits associated with drought adaptation in faba bean (*Vicia faba* L.). Doctoral thesis. Helsinki, p. 59.

Mousavi-Derazmahalleh, M., Bayer, P.E., Hane, J.K. et al. (2019). Adapting legume crops to climate change using genomic approaches. *Plant, Cell & Environment* 42: 6–19.

Naveed, S., Ibrar, M., Khattak, I. et al. (2016). Anthelmintic, antilice, insecticidal, cytotoxic and phytotoxic potential of ethanolic extracts of two wild medicinal plants *Iphiona grantioides* and *Plucheaarguta*. *Wulfenia Journal Austria* 23 (11): 13–25.

Nichols, P.G.H., Revell, C., Humphries, A.W. et al. (2012). Temperate pasture legumes in Australia - their history, current use and future prospects. *Crop & Pasture Science* 63: 691–725.

Purushothaman, R., Zaman-Allah, M., Mallikarjuna, N. et al. (2013). Root anatomical traits and their possible contribution to drought tolerance in grain legumes. *Plant Production Science* 16 (1): 1–8.

Real, D., Li, G., Clark, S.G. et al. (2011). Evaluation of perennial forage legumes and herbs in six Mediterranean environments. *Chilean Journal of Agricultural Research* 71 (3): 357–369.

Samreen, U., Ibrar, M., Lalbadshah et al. (2016). Ethnobotanical study of subtropical hills of Darazinda, Takht-e-Suleman range F.R. D.I. Khan, Pakistan. *Pure and Applied Biology* 5 (1): 149–164. https://doi.org/10.19045/bspab.2016.50020.

Siddique, K.H.M., Johansen, C., Turner, N.C. et al. (2012). Innovations in agronomy for food legumes. A review. *Agronomy for Sustainable Development* 32: 45–64.

Sinclair, T.R. and Vadez, V. (2012). The future of grain legumes in cropping systems. *Crop & Pasture Science* https://doi.org/10.1071/CP12128.

Toker, C. and Yadav, S.S. (2010). Legumes cultivars for stress environments. In: *Climate Change and Management of Cool Season Grain Legume Crops* (ed. S. Yadav and R. Redden), 351–376. Dordrecht: Springer.

Trytsman, M., Westfall, R.H., Breytenbach, P.J.J. et al. (2016). Diversity and biogeographical patterns of legumes (*Leguminosae*) indigenous to southern Africa. *PhytoKeys* 70: 53–96.

Upadhyaya, H.D., Kashiwagi, J., Varshney, R.K. et al. (2010). Phenotyping chickpeas and pigeon peas for adaptation to drought. In: *Drought Phenotyping in Crops: From Theory to Practice*, 347–355. Generation Challenge Programme.

Varshney, R.K., Thudi, M., May, G.D., and Jackson, S.A. (2010). Legume genomics and breeding. *Plant Breeding Reviews* 33: 257–304.

Voisin, A.-S., Guéguen, J., Huyghe, C. et al. (2013). Legumes for feed, food, biomaterials and bioenergy in Europe: a review. In: *Agronomy for Sustainable Development*, 1–20. Springer Verlag/EDP Sciences/INRA.

Wani, S.P., Sarvesh, K.V., Krishnappa, K. et al. (2012). *Bhoochetana: Mission to Boost Productivity of Rainfed Agriculture Through Science-Led Interventions in Karnataka*. Patancheru, Andhra Pradesh, India: International Crops Research Institute for the Semi-Arid Tropics, BOE 058; pp. 84. ISBN: 978-92-9066-548-9.

3

Adapting Crops to Climate Change

Abdel Rahman Mohammad Said Al-Tawaha[1], Samia Khanum[2], Noureddine Benkeblia[3], Amanullah[4], Imran[4,5], Shah Khalid[4], Abdel Razzaq Al-Tawaha[6], Mousumi Mondal[7], Nidal Odat[8], Abhijit Dey[9], Nujoud Alimad[10], Devarajan Thangadurai[11], Jeyabalan Sangeetha[12], Saher Islam[13], and Mohammad Shatnawi[14]

[1] Department of Biological Sciences, Al-Hussein Bin Talal University, Ma'an, Jordan
[2] Department of Botany, University of the Punjab, Lahore, Pakistan
[3] Department of Life Sciences, The University of the West Indies, Kingston, Jamaica
[4] Department of Agronomy, The University of Agriculture, Peshawar, Pakistan
[5] Ministry of Agriculture, Agriculture Extension Wing, Merged Area, Khyber Pakhtunkhwa, Pakistan
[6] Department of Crop Science, Faculty of Agriculture, Universiti Putra Malaysia, Serdang, Selangor, Malaysia
[7] Department of Agronomy, Bidhan Chandra Krishi Viswavidyalaya Mohanpur, Nadia, West Bengal, India
[8] Department of Medical Laboratories, Faculty of Science, Al-Balqa Applied University, Al Salt, Jordan
[9] Department of Life Sciences, Presidency University, Kolkata, West Bengal, India
[10] Department of Plant Protection, Faculty of Agriculture, Damascus University, Damascus, Syria
[11] Department of Botany, Karnatak University, Dharwad, Karnataka, India
[12] Department of Environmental Science, Central University of Kerala, Kasaragod, Kerala, India
[13] Department of Biochemistry and Biotechnology, Institute of Biochemistry and Biotechnology, University of Veterinary and Animal Sciences, Lahore, Pakistan
[14] Department of Biotechnology, Faculty of Agricultural Technology, Al Balqa Applied University, Salt, Jordan

CONTENTS

Climate Change and Agriculture: Perspectives, Sustainability and Resilience, First Edition. Edited by Noureddine Benkeblia.

3.1 Introduction

The diversity in genetics and trait of crops has been reduced because of domestication, intensification of agricultural production by green revolution, and native crops displacement with major crops (Pingali 2012). In spite of decrease in diversity of crops, there is an increase in major staple crops in the past century (Fischer et al. 2009). This is because of conventional breeding with simplification and intensification of production system. It is also due to the selection of edible yield and adaptations against factors affecting yield of crops such as pathogens, pests, and optimization of husbandry practices of crop with increased input of fertilizers, pesticides, herbicides, and mechanization in order to decrease the environmental flux impact (Abu-Darwish et al. 2009; Al-Rifaee et al. 2004; Supanjani et al. 2005; Sulpanjani et al. 2005; Tawaha and Turk 2004). Selection in this ideal environment leads to loss of some important genes for the efficiency and adaptation against stresses (Brown 2003). This results in three changes such as selection criteria based on efficiency and adaptation characteristics of the yield, making sure of the presence of efficient and stress-tolerant genes and their exploitability in a variety of breeding germplasms, and increase the use of minor crops with good nutritional values, environmental sustainability, and needing low inputs. Global climate change is a major threat to the productivity of crops. Decrease in productivity leads to increased food prices and global food insecurity (FAO 2014). If this continues, it will worsen the existing situation and lead to famine or social unrest. Climate change will continue to affect agriculture unless resilience of crops is not improved (Tawaha and Turk 2002a). Major decrease in the production of cereals by 2020 is estimated to be 9% for maize, 11% for rice, and 14% for wheat (Hisas 2011). The most common environmental changes are global warming, change in rainfall patterns, and other extreme weather events, and all these changes result in increased pathogens and pest attacks. Increase in CO_2 level may be the reason for the reduction in nutritional value of different crops and increase in the toxicity of the crops due to the change in the chemical composition of tissues (Dwivedi et al. 2013). In majority of the world, due to climate change, there are significant gaps in yield caused by poor and unsustainable cropland management (Abu Obaid et al. 2018; Al-Tawaha and Seguin 2006; Al-Tawaha et al. 2006, 2007, 2010, 2018a,b,c; Al-Tawaha and Odat, 2010). Use of genetic engineering and molecular breeding techniques can help in the improvement of resilience of the crops to the change in climate, which can ensure food security. Genetic engineering is the manipulation of already existing genes or transfer of useful genes to generate required phenotypes. On the other hand, breeding is improving the germplasm by introducing novel alleles. As genetic engineering is controversial in some countries, breeding is the acceptable method for the improvement of crops.

3.2 Potential of Genomics-Assisted Breeding in Producing Climate-Resilient Crops

There are many tools offered by genomics to improve food yield, stability, and quality by breeding strategies. Crossbreeding use of DNA markers in MAS (marker-aided selection) is successful for crop improvement. Advancement in genomics helps in better understanding of species and gene level diversity in crop. It also can help DNA markers to accelerate the pace of genetic improvement (Muthamilarasan et al. 2013, 2014). Breeding strategies to generate climate-ready new cultivars (Varshney et al. 2005) required details about stresses that can affect the productivity and yield of the crops under certain climatic conditions. Multi-environment testing can help in monitoring the impact of stress on crop productivity. Searching morphological and physiological traits in available germplasm, which can help plants to survive in unfavorable conditions such sudden change in climate, is the need of the hour for plant breeders. Crop physiology can help in this matter by identifying ideotypes. Biometric and phenotypic analyses help in the identification of responses of different genotypes under different physiological and weather conditions. These findings can help in genomics-based breeding techniques like selection of germplasm for improved cultivars productions. Quantitative trait locus (QTL) analysis and genetic mapping by association mapping populations have increased the dissection of agricultural traits and can allow the MAS, QTL, genomic selection, and AM studies in the name of breeding programs (Kulwal et al. 2011). Advancement in next-generation sequencing (NGS) helps in generating a variety of genetic markers such as InDels and single nucleotide polymorphism (SNP), and then novel genes of any required trait can be discovered by using genotyping-by-sequencing (GBS) techniques. Genome-wide association study (GWAS) can help in the identification of traits of interest by doing statistical association between variant traits and DNA polymorphism. The precise location of genes or alleles is achieved by increasing mapping resolution when NGS is combined with GWAS (Liu et al. 2013; Ma et al. 2012; Varshney et al. 2014). New genes that have been evolved, shuffled, and selected by nature in different environments result in diversity in wild species over the period of evolution. On the other hand, selection and domestication of crops is recent by humans. Genetic diversity has been reduced in many crops because of domestication and breeding. Recently, breeders try to produce crops with increased yield by mixing relative alleles. The reshuffling is limited because of limited recombination within species in crops with large genomes that results in continuous inheritance of block genes. It also causes fixed linkage block and linkage drag having no required combination alleles. These limitations allow breeders to access only the combinations of novel genes from elite parents. Through genomics the diversity of alleles can be increased by mining gene pool of crop wild relatives (CWRs). Rapid identification and selection of gene of interest also can be performed by genomics tools. Through all these recent techniques the security of food can be improved. Thus food production is increased by enhancement of germplasm of different crops with the help of genomics. The advancement in genomics can help in identification and transfer of gene of interest from allied gene pool into elite crops, which will help in meeting the global food requirement challenge.

3.3 Genomics of Climate-Resilient Crops

This section covers the challenges, breeding needs, and priority for climate resilience in many leading crops. Genomics tools are now widely available and used all over the world.

3.3.1 Cereals

Cereal is a staple crop but global warming and sudden climate changes have affected its production and also have a socioeconomic impact especially in underdeveloped countries. To overcome this challenge and ensure global food security, it is the need of the hour to crossbreed with omics strategies by dissecting and manipulation of genetic makeup of crop against abiotic stresses (Langridge and Fleury 2011). In order to understand the adaptive response to abiotic stresses, many genomics studies have been carried out since past decade and the most common way is QTL analysis for identification of genetic loci that regulate these responses in cereals. It consists of root architecture (Lynch et al. 2014; Uga et al. 2013; Wasson et al. 2012), drought-adaptive traits (Serraj et al. 2009; Tuberosa 2012), accumulation of carbohydrates that are water soluble (Landi et al. 2005; Rebetzke et al. 2008; Salem et al. 2007; Snape et al. 2007), concentration of abscisic acid (Rebetzke et al. 2008, Rehman et al. 2011), stay-green (Borrell et al. 2014; Yang et al. 2007), temperature of canopy (Lopes et al. 2014), and discrimination of carbon isotope (113C) (Pinto et al. 2010). The increase in temperature is linked to global warming, which results in disturbed accumulation of starch, leaf senescence, and reduced yields especially when coupled with drought stress. In wheat about 27% and 17% reduction of yield due to drought and heat stress correspondingly (Pinto et al. 2010). The QTL yield is linked also with other traits and supports the crop performance dissection in abiotic stress situation into genetic and physiological components leading to a strategic approach of breeding (Reynolds and Tuberosa 2008). Wang et al. (2012) and Paliwal et al. (2012) explain the QTL with concurrent effects in both drought and heat situation. Through various studies it is proved that warming has adverse impact on the yield of rice (Welch et al. 2010). High temperature decreases the rice yield by two possible ways: (i) high temperature when combined with high humidity results in spikelet sterility and (ii) high temperature at night reduced accumulation of assimilates (Wassmann and Jagadish 2009). Global warming is also the reason for flooding and disturbed rainfall patterns. It is the need of the hour to produce crops that can stand adverse climatic conditions such as waterlogging or submergence. Among cereals rice is very prone to submergence stress and causes US$1 billion loss in Asia (Mackill et al. 2012). In rice there is QTL for variability for surviving ability in prolonged submergence condition (Xu et al. 2006). After its identification with the help of marker-assisted backcrossing (MABC) technique, we can generate rice cultivars that can tolerate submergence stress by only three backcross generations. DNA markers can form for the production of six cultivars to meet the challenge in flood prone areas (Bailey-Serres et al. 2010). This study shows the importance of MAS for the introgression of important QTL into elite materials. In maize, QTL has been identified that leads to adventitious root formation, which can help in absorption of moisture when water content is low (Mano et al. 2005a,b). Climate Change and Salinity impact adversely on agriculture because of global change in climate. In durum wheat genes are identified to tolerate the salt accumulation and salinity (James et al. 2006). When barley is crossed with wild barley it results in the identification of QTL that reduce the accumulation of Na^+ in shoots when grow in saline soil (Shavrukov et al. 2010). In rice, many QTL have been identified that have the ability to tolerate salinity (Wang et al. 2012).

3.3.2 Oilseeds and Pulses

These are considered as major crops because they are rich in proteins and oil content. Biotic and abiotic stresses are the main reason for the reduction in their production. Soybeans have the highest protein content and second highest oil content, yet studies are required to increase tolerance in soybean against difference stresses. Phaseolus beans are enriched in vitamins, proteins, and minerals (Gepts and Aragão 2008). Heat and drought stress tolerance in *Phaseolus vulgaris* is under specific genetic control (Asfaw et al. 2012; Schneider et al. 1997). Lines selection for improved drought adaptation is successful (Beebe et al. 2008; Singh 2007; Urrea et al. 2009). Both biotic and abiotic stresses are the major threat to the yield of pigeon pea and chickpea. To overcome this challenge, research to develop genomic resources at large scale has been carried out at national and international level (Varshney et al. 2013a). These resources can be used for breeding programs. The intraspecific and interspecific genetic map of chickpea has been developed (Gaur et al. 2011; Gujaria et al. 2011; Hiremath et al. 2012; Thudi et al. 2011), and genetic region related to abiotic (Rehman et al. 2011; Vadez et al. 2012), biotic (Anbessa et al. 2009; Anuradha et al. 2011; Kottapalli et al. 2009), and agronomic characteristics (Bajaj et al., 2014, 2015; Cobos et al. 2009; Das et al. 2015; Kujur et al. 2015a, 2015b; Rehman et al. 2011) has been identified. About 3000 simple sequence repeat (SSR) markers (Bohra et al. 2011; Dutta et al. 2011; Saxena et al. 2010), SNP (Saxena 2008) transcript reads (Dubey et al. 2011; Dutta et al. 2011), ESTs (Raju et al. 2010), and transcriptome assemblies (Dubey et al. 2011; Kudapa et al. 2012) are developed in pigeon pea to improve yield of crop by using them in genomics-assisted breeding for crop improvement. Genome sequence of *kabuli* and *desi* chickpeas and pigeon pea has been published recently (Jiang et al. 2013; Varshney et al. 2012, 2013b). The sequence data of these peas has been used for the improvement of crop and food security in arid and semiarid regions. Other oilseed crops like peanuts are also studied on the basis of genomics-assisted breeding. Genomic resources haves been developed on large scale to aid in molecular breeding of peanut and the QTL for stress adaptation has been identified for oil quality (Gautami et al. 2012; Varshney et al. 2009; Sarvamangala et al. 2011) and resistance against late leaf spot and rust (Khedikar et al. 2010; Sujay et al. 2012).

3.3.3 Millets

These are small grain gramineous crops. Their most important features are efficiency of water use, good nutrient content, ability to grow in nutrient-poor soils, and adaptability to ecological stresses. The most important millet crops are foxtail, pearl, proso, barnyard, kodo, and finger millet, which are cultivated worldwide. Among them foxtail millet is C4 crop and is used as a model for the study of neology of millets and biofuel grasses (Lata et al. 2013; Muthamilarasan and Prasad 2015). The genome of foxtail millet has been sequenced in China and the United States (Bennetzen et al. 2012; Zhang et al. 2012). Foxtail millet is very rich in sources of alleles, QTL and genes improving cereals, and biofuel grasses genetically. Therefore, genomic resources are developed on large scale that also include SSRs (Pandey et al. 2013, Zhang et al. 2014b), eSSRs (Kumari et al. 2013), intron length polymorphism (Muthamilarasan et al. 2014), transposable element-based markers

(Yadav et al. 2015), and miRNA-based markers (Yadav et al. 2014). The open access online database is designed also for foxtail millet miRNA database (Khan et al. 2014), foxtail millet marker database (Suresh et al. 2013), and foxtail millet transposable elements-based marker database (Yadav et al. 2015). Moreover, the use of these markers in association mapping, population genetics, genomics-assisted breeding, and comparative genomics also is demonstrated (Muthamilarasan and Prasad 2015). Three major QTL in pearl millet has been identified for grain yield but with low interaction has been identified for flowering (Bidinger et al. 2007). Studies about millets are at an initial level as compared to other crops. Millets serve as an important source of alleles, novel genes, and QTL for tolerance against stresses, so there is a need for them to be identified and characterized.

3.3.4 Forest and Fruit Tree Crops

Clones of Pinus, Populus, Eucalyptus, and Abies can be used for afforestation because of the production of wood. It is important to develop clones of this important trees of forest. Many researches have been carried out for the DNA genotyping and marker identification of forest trees. Successful test of whole-exome capture using in-solution target enrichment (Neves et al. 2011) and genome complexity reduction DArT (Alves-Freitas et al. 2011) for identification of genome-wide marker has been carried out, which is required for GS in Pinus. Conifer genome project has been started based on the importance of genome sequencing for genetic marker development. Gene catalog and genetic resource of *Picea tidea* and *Picea glauca* has been developed. The aim of GENOAK project is to develop a reference genome for pedunculate oak. The genome of *Eucalyptus grandis* is interpreted and helps in agro-forestry by using hardwood trees. This will help industries based on hardwood products and eucalyptus production feedstock for cellulose biofuels. This will increase the breeding of forest tree to meet the challenge of adapted population, which withstands the changes in climate. Fruit trees are as important as cereal crops, and to maintain the nutrients source for daily intake, climate-resilient clones of trees are essential to be maintained. Bioinformatics in combination with genome-based breeding and omic resources is essential for the breeding of perennial fruits and their adaptation to changes in climate. Annotation projects and genome sequencing consist of perennial fruits such as grape (Velasco et al. 2007), apple (Velasco et al. 2010), banana (D'Hont et al. 2012), peach (Ahmad et al. 2011), cocoa (Argout et al. 2011), and sweet orange (Xu et al. 2013). Advancement in genome sequencing, high resolution gene mapping, and accurate phenotyping increase the development of variations in allele and functional alleles, which is linked with trait of interest. But very limited research has been carried out in such genome-based breeding. Genomic and genetic techniques help in identification of genome regions with the trait of interest.

3.4 Genomics-Assisted Breeding Strategies for Climate-Resilient Traits

Genomics approaches help in sequence-based breeding that further leads to dissection and cloning of the trait of interest that controls stresses resulting in increase of genetic diversity resources (Edwards et al. 2013; Tuberosa et al. 2011). GS is the major approach to increase the yield of cereal crops under abiotic stresses especially during drought. However, for QTL the major approach is MAS, while use of QTL cloning is accelerated by using high output, sequencing (Edwards and Wang 2012; Imelfort et al. 2009), precise phenotyping (Araus and Cairns 2014), and identification of suitable candidate genes by 'omics' profiling (Gupta et al. 2013). QTL cloning leads to chances of genetic engineering of genes with adaptation traits against abiotic stresses (Salvi et al. 2007). Advanced genomics technologies, alleviating the adverse impact on productivity of crop due to climate change is still an overwhelming task. To ensure fruit security, there is a need of integrated approaches so breeders can openly select crops having more adaptive traits for climate change.

3.5 Flowering Time and Drought Adaptation

Flowering is influenced by the temperature like day length control the transition of flowers by interlinked with genetic pathways. Global warming will result in increase in ambient temperature with no change in photoperiod. Increase in temperature increased the development and growth with shortening the life cycles, decreasing photosynthesis (Reynolds et al. 2010), lowering yield, and shortening reproductive phase in annual plants (Ainsworth and Ort 2010). High temperature at the time of sensitive developmental will increase the reproductive tissue damage. Detailed information about genetic control of flowering, epistatic interactions, allelic variants, and phenotypic variation at various growth conditions is required for the identification of climate change breeding targets. Germplasm resources are increased with accurate near isogenic lines (NILs) (Bentley et al. 2011, 2013), and multifounder populations are developed for wheat (Mackay et al. 2014) and other important crops to support further research on climate-smart crops. Initial studies should explain the impact of flowering on crop yield under various environmental stresses. Identification of how much loci will impact the stage of flowering will help in shortening or lengthening the flowering stage, which can help survival and adaptation under various environmental stresses. Many researches are under way for the improvement of responses in plant to drought and efficiency of water use. Molecular breeding and identification and use of QTL in breeding help in the development of cultivars with drought tolerance. To determine the role of roots to reduce the negative impact of drought, many studies have been carried out in maize. QTL of root traits was explained in different maize populations (Hund et al. 2011, Ruta et al. 2010; Tuberosa et al. 2011). In these studies it is showed that some QTL impact on yield of grain under drought stress (Landi et al. 2005). It is suggested to select maize hybrids with drought stress tolerance (Cooper et al. 2014). Yield of QTL is linked with many other components of traits leading to dissection performance of crop in abiotic stress into genetic and physiological components to support breeding (Reynolds and Tuberosa 2008). To map adaptation of drought stress in rice about 15 populations used for studies and 4 regions are identified for under stress yield components and also for drought components of tolerance (Kamoshita et al. 2008). The first region was linked with traits of grain yield of drought resistance, traits of plant type (Zhang et al. 2001a), QTL of stability of cell membrane (Tripathy et al. 2000), traits of root (Robin et al. 2003), and osmotic adjustment (Lilley et al. 1996). Second genomic region is enriched with QTL of root trait (Boopathi et al. 2005; Hemamalini et al. 2000; Kamoshita et al. 2002; Nguyen et al. 2004; Zheng et al. 2000; Zhang et al. 2001b) under both watered and drought situations. Third region consists of QTL of water status, yield of grain, stability of cell membrane, water loss other than stomata, osmotic adjustment, and thick and deep root traits (Boopathi et al. 2005, Hemamalini et al. 2000, Nguyen et al. 2004, Zhang et al. 2001b, Zheng et al. 2000). The fourth region for drought was categorized by QTL of root traits, stability of cell membrane, leaf rolling, plant water status, leaf drying, per panicle number of grains, biomass, relative fertility of spikelet, and delay in time of flowering (Babu et al. 2003; Boopathi et al. 2005; Courtois et al. 2003; Gomez et al. 2005; Hemamalini et al. 2000; Jearakongman 2005; Kamoshita et al. 2002; Lafitte et al. 2004; Lanceras et al. 2004; Li et al. 2005; Nguyen et al. 2004; Price et al. 2002a,b; Robin et al. 2003; Tripathy et al. 2000; Xu et al. 2005; Yue et al. 2006; Zhang et al. 2001a; Zheng et al. 2003). All these four regions are linked with drought tolerance, reported through various studies, and suggested to be part of marker-assisted breeding program for tolerance against drought in rice.

3.6 Tolerance for Heat and Cold Stress

Tolerance for freezing temperature is the key reason for the survival of plants during winter. It also helps to withstand other stresses such as wind, desiccation, heaving, ice encasement, low light, winter pathogens, snow cover, and fluctuating temperatures. The same genetic responses are observed during freezing temperature, drought salinity such as maintaining integrity of cell membrane, and cellular water retention to develop resistance against desiccation (Yue et al. 2006). Cold acclimation with time also develops the adaptation against drought stress (Seki et al. 2002). The most important traits like winter hardiness (Thomas and James 1993), ABA production induced by drought and cold stress (Laurie et al. 1995), osmotic stress adaptation (Galiba et al. 1993), frost tolerance, and vernalization response (Galiba et al. 1995; Hayes et al. 1998) have been located on five chromosomes of Triticeae group. This chromosome is considered as the major region for genome research in many cereals and grasses. In areas with high temperature favorable for plant growth combined with less rainfall in winter results in sudden winter drought situation requiring new breeding strategies for stress tolerance. The expression of cold-regulated genes is regulated by C-repeat building factor genes, which are most commonly found in eudicots and monocots. There is also cis acting CRT/DRE element recognized by CBF transcription factors on COR genes (Stockinger et al. 1997). In barley about 20 CBFT genes are identified (Francia et al. 2007; Skinner et al. 2006). The CBF genes vary among cultivars of wheat and barley (Francia et al. 2007; Knox et al. 2010). The co-segregation of frost tolerance loci of wheat and barley, role in cold acclimation (Stockinger et al. 1997), and interaction of CBF gene with frost tolerance (Vagujfalvi et al. 2003) are the reasons to make barley and wheat a strong candidate of QTL for frost tolerance. Prevailing rise in temperature will reduce the yield in cereals and legumes (Quarrie et al. 1997). Temperature is increasing both during day and night (Hall and Ziska 2000), so heat-resistant cultivars should have high yield in both high and cold temperatures. Many heat-resistant cultivars of various crops have been developed by public breeding programs. Commercial breeding companies are still struggling in it, but they developed successful heat-resistant commercial cultivars. Before that there is no focus on heat-resistant cultivars breeding.

3.7 Salinity and Submergence Tolerance

In cereal the main issue is salinity and submergence stress because the soil is affected by irrigation canal seepage and rainfall just after irrigation causes wetting. Genetic diversity of waterlogging tolerance has been reported in wheat, maize, barley, and oats (Kerr 1986). Diverse mechanism of tolerance related to morphology, phenology, metabolism, nutrition balances, anoxia tolerance, damage, and recovery have also been reported (Setter and Waters 2003). At the time of germination and early growth of rice seedling, tolerance of flooding is required for both rainfed and irrigated regions because in such soil waterlogging is enough to cause reduction in crop due to hypoxia sensitivity (Ismail et al. 2009). Substantial variation in genetics in crops has been reported with the ability to grow normally in flood areas. Tolerance genotype of seedlings growing under hypoxia condition has ability to catabolize starch into simple sugar to generate energy for the normal growth of embryo (Miro and Ismail 2013; Septiningsih et al. 2013). Different QTL have been identified from different rice races – two of them are chromosome 9 and chromosome 7 with major effects (Angaji et al. 2010; Septiningsih et al. 2013). These QTL are used for cloning and MAB for the production of high yield cultivars of rice and have the ability to grow better in flood areas. The most important QTL SUB1 on chromosome 9 responsible for flood tolerance is identified. This is responsible for encoding ERF to stop ethylene-mediated response when grown in submergence condition, which also limits the unnecessary elongation and chlorophyll degradation because both of these are responsible for starvation of carbohydrates. These varieties have the ability to survive for 4–18 days in submergence with maximum yield as compared to normal cultivars with no negative impact on other important traits (Bailey-Serres et al. 2010; Ismail et al. 2013; Mackill et al. 2012; Singh et al. 2009). These studies on rice crop may facilitate the identification and production of climate-resilient cultivars for other crops for different stresses. Increase in salt concentration because of unnecessary irrigation of poor quality drainage of water result in increased salt levels in the soil (Ismail et al. 2007). Many studies revealed complexity and various responses because of high salt concentrations in various plant species (Munns and Tester 2008). Many different major and small loci are linked with salinity tolerance. Many QTL were identified in rice and many other crops. Many genes are also identified while studying responses of salt stress through functional genomics. In recent studies NGS is used to target important QTL for cloning, development of InDel and SNP markers for the manipulation of loci in MAB. The considerable genetic diversity of tolerance of salt stress and the responses to tolerate increase in salt levels in soils and water facilitate the production of cultivars of cereals having tolerance against salt stress. Though it required a large investment to develop such cultivars that have the ability to withstand salt stress and many other suddenly worsening climatic conditions.

3.8 Tools of Genetic Engineering to Target Mutagenesis

Breeders use mutagenesis technique to produce cultivars with improved yield through genetic variation. This strategy also is used for production of cultivars with improved adaptability. X-rays of the most common mutagens were used because of their availability

(Muller 1927). Later on gamma rays were used for point and chromosomal mutations (Devreux and Scarascia Mugnozza 1964). Recently, chemical mutagenesis is most common because of it is easy to handle, there is no requirement of specialized equipment, and it results in high frequency mutation. Chemical mutagens have the ability to induce SNPs instead of chromosomal mutation as compared to radiation. EMS is a chemical mutagen used for the induction of random mutation of genome, and it helps in the isolation of DNA from mutated genome by using TILLING technique and also facilitates screening of M2 population with the help of molecular strategies. The identification of single mutation can be possible by mismatching base cleavages formed by the means of frequent melting and reannealing of products of PCR, which are amplified from group of alleles related to a specific gene (Caldwell et al. 2004; McCallum et al. 2000). NGS has ability to identify mutation at genome level. The identification of promotor and other regulatory mutations can be done by RNA-seq, microarray, and qPCR. When the mutant allele is identified in a gene of interest by backcrossing, we can identify that mutation linked with specific phenotype of stress resistance. This is considered as reverse genetic technique to link a mutant allele with its phenotype. On the other hand, TILLING is also helpful in forward genetic method with phenotypic screening of adaptation to different stresses followed by characterization of phenotypes with resequencing by using combination of complete genome, microarrays, and linkage maps for detailed explanation of changes in gene expression because of SNPs than original lines. Many other molecular resources and tools are available now for genetic engineering and also for reverse genetic tests to manipulate the regulatory mechanism and genetic building blocks for the production of climate change resilient cultivars with improved yield and also helps in the correction of mutations (Hsu et al. 2014). Targeted genetic engineering is an alternative technique to classic breeding for the production of improved crops with maximum food production (Belhaj et al. 2013; Osakabe and Osakabe 2015). To edit genome four types of nucleases are found: zinc finger nucleases (ZFNs) (Townsend et al. 2009), engineered homing meganucleases/endonucleases (EMNs) (Silva et al. 2011), CRISPR (Cong et al. 2013; Mali et al. 2013), and transcription activator-like effector nucleases (TALENs) (Cermak et al. 2011). SSN with the use of DSB assist accurate genome engineering. This mechanism initiates DNA repairing mechanism by homology direct repair or non-homologous end joining. HDR will help in direct correction at certain location of genome while NHEJ leads to site-directed mutagenesis. Another difference is HDR is only active in late G2 phase while NHEJ is active throughout the cell cycle. These are the reasons why NHEJ is considered as the main DSB repair mechanism. The illustration of genome editing is clearly explained in different crops by Belhaj et al. (2013) and Osakabe and Osakabe (2015). The targeted modification of genome that consists of mutation, insertion, replacements, and then rearrangements of chromosome in various crops is generated by SSN effect. Every method has pros and cons based on cost, target efficiency, construction ease, and specificity (Chen and Gao 2014; Gao 2015). The most important advantage of ZFNs is no production of transgenic because viral vectors are used for the expression of nuclease without integrating into genome, but they also have disadvantages such as limited target sites and difficulty in designing the experiment. CRISPR is now considered as the important tool for editing the genome (Gaj et al. 2013) and emerged as a substitute of classical plant breeding for production of improved crop. CRISPR has many advantages such as direct construct design, assembly, and high rate of mutation than ZFNs and TALENs. In order to identify another target, it required modification of 20 nucleotides of gRNA. CRISPR is successfully applied in *Nicotiana benthamiana* (Jiang et al. 2013; Li et al. 2013; Nekrasov et al. 2013), *Solanum lycopersicum* (Brooks et al. 2014), *Arabidopsis thaliana* (Feng et al. 2013, 2014; Jiang et al. 2013; Li et al. 2013; Mao et al. 2013), *Oryza sativa* (Feng et al. 2013; Jiang et al. 2013; Mao et al. 2013; Miao et al. 2013; Shan et al. 2013; Xie and Yang 2013; Zhang et al. 2014a), *Triticum aestivum* (Wang et al. 2014), *Sorghum bicolor* (Jiang et al. 2013), and *Citrus sinensis* (Jia and Wang 2014). The CRISPR–Cas9 technology is used to edit the genes responsible for adaptation traits. gRNAs are used to target genes that encode response to stress conditions to increase disease resistance in rice (Xie and Yang 2013). The transgenic wheat has mutation in TaMLO-A1 allele, which is responsible to show resistance against powdery mildew (Wang et al. 2014). OsSWEET14 and OsSWEET11 are the genes of susceptibility to bacterial blight and are also targeted regions for mutation (Jiang et al. 2013). CRISPR technique is useful for increased breeding of crops with targeted genes of mutation regions but having no foreign DNA. Genome engineering combined with targeted mutagenesis through genes editing assists in producing elite cultivars of crop with durable climate resilience.

3.9 Conclusion

Change in global climate is a serious threat to the production of agricultural crops, which leads to food and nutritional security challenge all over the world. Advancement in technology like molecular breeding and trans-gene technologies can develop cultivars with modified genotype to survive in severe climate change. Genomic-mediated breeding programs are very important for the production of climate-resilient crops. Many model organisms have been identified just like green foxtail, foxtail millet, and Brachypodium to translate the traits that are needed to decode for introgression in crops. Advancement in DNA sequencing, genomics tools, and CWR sequencing will help in the identification of important genes and regions with key regulatory stress tolerance leading to the production of new cultivars with resistance. It is difficult to identify the impact of change in climate on resistance of crop, and it also varies among different crops and relative environment but genomic-mediated breeding programs will help in the reduction of the adverse effects of climate change on future crops.

3.10 Summary

Legumes and cereals are grown in more than 120 countries across all the continents. Both legume and cereal are great worth crops and play a dynamic role in crop diversity in dry areas. Most of the legume and cereal crops are grown under rainfed agroecosystem with poor level of input use, which has resulted in low production. Crop growth and productivity is affected by many biotic and abiotic factors. On the other hand, physiological approaches of plants including grain legumes provide general outlines of plant responses and finding stress tolerance-related traits or elite cultivars. During the past decade, several physiological variables have been evaluated and recommended to understand the influence of environmental stress on grain legumes such as nutrient uptake and translocation, shoot architecture, leaf gas exchange (photosynthesis, transpiration, and stomatal conductance), chlorophyll content and fluorescence, yield, seed quality, root dynamic, and its interaction with soil biota.

References

Abu Obaid, A.M., Melnyk, A.V., Onichko, V.I. et al. (2018). Evaluation of six sunflower cultivars for forage productivity under salinity conditions. *Advances in Environmental Biology* 12: 13–15.

Abu-Darwish, M.S., Abu Dieyeh, Z.H., Mufeed, B. et al. (2009). Trace element contents and essential oil yields from wild thyme plant (*Thymusserpyllum* L.) grown at different natural-variable environments, Jordan. *Journal of Food, Agriculture and Environment* 7: 920–924.

Ahmad, R., Parfitt, D.E., Fass, J. et al. (2011). Whole genome sequencing of peach (*Prunus persica* L.) for SNP identification and selection. *BMC Genomics* 12: 569. https://doi.org/10.1186/1471-2164-12-569.

Ainsworth, E.A. and Ort, D.R. (2010). How do we improve crop production in a warming world? *Plant Physiology* 154: 526–530. https://doi.org/10.1104/pp.110.161349.

Al-Rifaee, M., Turk, M.A., and Tawaha, A.M. (2004). Effect of seed size and plant population density on yield and yield components of local faba bean (*Viciafaba* L.) major. *International Journal of Agriculture and Biology* 6: 294–299.

Al-Tawaha, A.R.M. and Odat, N. (2010). Use of sorghum and maize allelopathic properties to inhibit germination and growth of wild barley (*Hordeum spontaneum*). *NotulaeBotanicae, HortiAgrobotanici Cluj-Napoca* 38: 124–127.

Al-Tawaha, A.M. and Seguin, P. (2006). Effects of seeding date, row spacing, and weeds on soybean seed isoflavone concentration. *Canadian Journal of Plant Science* 86: 1079–1082.

Al-Tawaha, A.M., Seguin, P., Smith, D.L., and Beaulieu, C. (2006). Foliar application of elicitors alters isoflavone concentrations and other seed characteristics of field-grown soybean. *Canadian Journal of Plant Science* 86: 677–684.

Al-Tawaha, A.M., Seguin, P., Smith, D.L. et al. (2007). Effects of irrigation on isoflavone concentrations of soybean grown in southwestern Québec. *Journal of Agronomy and Crop Science* 193: 238–246.

Al-Tawaha, A.M., Yadav, S.S., Turk, M. et al. (2010). *Crop Production and Management Technologies for Drought Prone Environments, Chapter in Book. Climate Change and Drought Management in Cool Season Grain Legume Crops*. Springer.

Al-Tawaha, A.M., Al-Karaki, G., Al Tawaha, A.M. et al. (2018a). Effects water flow rate on quantity and quality of lettuce (*Lactuca Sativa* L.) in nutrient film technique (Nft) under hydroponics conditions. *Bulgarian Journal of Agricultural Science* 24: 791–798.

Al-Tawaha, A.R., Turk, M.A., Al-Tawaha, A.R. et al. (2018b). Using chitosan to improve growth of maize cultivars under salinity conditions. *Bulgarian Journal of Agricultural Science* 24: 437–442.

Al-Tawaha, A.M., Al-Tawaha, A.M., Aludatt, M.H. et al. (2018c). Effects of soil type and rainwater harvesting treatments in the growth, productivity and morphological trains of barley plants cultivated in semi-arid environment. *Australian Journal of Crop Science* 12: 975–979.

Alves-Freitas, D., Kilian, A., and Grattapaglia, D. (2011). Development of DArT (Diversity Arrays Technology) for high-throughput genotyping of *Pinus taeda* and closely related species. *BMC Proceedings* 5: 22. https://doi.org/10.1186/1753-6561-5-S7-P22.

Anbessa, Y., Taran, B., Warkentin, T.D. et al. (2009). Genetic analyses and conservation of QTL for ascochyta blight resistance in chickpea (*Cicer arietinum* L.). *Theoretical and Applied Genetics* 119: 757–765. https://doi.org/10.1007/s00122-009-1086-2.

Angaji, S., Septiningsih, E.M., Mackill, D.J., and Ismail, A.M. (2010). QTLs associated with tolerance of flooding during germination in rice (*Oryza sativa* L.). *Euphytica* 172: 159–168. https://doi.org/10.1007/s10681-009-0014-5.

Anuradha, C., Gaur, P.M., Pande, S. et al. (2011). Mapping QTL for resistance to botrytis grey mould in chickpea. *Euphytica* 182: 1–9. https://doi.org/10.1007/s10681-011-0394-1.

Araus, J.L. and Cairns, J.E. (2014). Field high-throughput phenotyping: the new crop breeding frontier. *Trends in Plant Science* 19: 52–61. https://doi.org/10.1016/j.tplants.2013.09.008.

Argout, X., Salse, J., Aury, J.M. et al. (2011). The genome of Theobroma cacao. *Nature Genetics* 43: 101–108. https://doi.org/10.1038/ng.736.

Asfaw, A., Blair, M.W., and Struik, P.C. (2012). Multienvironment quantitative trait loci analysis for photosynthate acquisition, accumulation, and remobilization traits in common bean under drought stress. *G3 (Bethesda)* 2: 579–595. https://doi.org/10.1534/g3.112.002303.

Babu, R.C., Sarkarung, S., Wade, L.J. et al. (2003). Genetic analysis of drought resistance in rice by molecular markers: association between secondary traits and field performance. *Crop Science* 43: 1457–1469. https://doi.org/10.2135/cropsci2003.1457.

Bailey-Serres, J., Fukao, T., Ronald, P. et al. (2010). Submergence tolerant rice: Sub1's journey from landrace to modern cultivar. *Rice* 3: 138–147. https://doi.org/10.1007/s12284-010-9048-5.

Bajaj, D., Saxena, M.S., Kujur, A. et al. (2014). Genome-wide conserved non-coding microsatellite (CNMS) marker-based integrative genetical genomics for quantitative dissection of seed weight in chickpea. *Journal of Experimental Botany* 66: 1271–1290. https://doi.org/10.1093/jxb/eru478.

Bajaj, D., Upadhyaya, H.D., Khan, Y. et al. (2015). A combinatorial approach of comprehensive QTL-based comparative genome mapping and transcript profiling identified a seed weight-regulating candidate gene in chickpea. *Scientific Reports* 5: 9264. https://doi.org/10.1038/srep09264.

Beebe, S., Rao, I.M., Cajiao, C., and Grajales, M. (2008). Selection for drought resistance in common bean also improves yield in phosphorus limited and favorable environments. *Crop Science* 48: 582–592. https://doi.org/10.2135/cropsci2007.07.0404.

Belhaj, K., Chaparro-Garcia, A., Kamoun, S., and Nekrasov, V. (2013). Plant genome editing made easy: targeted mutagenesis in model and crop plants using the CRISPR/Cas system. *Plant Methods* 9: 39. https://doi.org/10.1186/1746-4811-9-39.

Bennetzen, J.L., Schmutz, J., Wang, H. et al. (2012). Reference genome sequence of the model plant Setaria. *Nature Biotechnology* 30: 555–561. https://doi.org/10.1038/nbt.2196.

Bentley, A.R., Turner, A.S., Gosman, N. et al. (2011). Frequency of photoperiod-insensitive Ppd-A1a alleles in tetraploid, hexaploid and synthetic hexaploid wheat germplasm, *Plant. Breed* 130: 10–15. https://doi.org/10.1111/j.1439-0523.2010.01802.x.

Bentley, A.R., Horsnell, R., Werner, C.P. et al. (2013). Short, natural, and extended photoperiod response in BC2F4 lines of bread wheat with different Photoperiod-1 (Ppd-1) alleles. *Journal of Experimental Botany* 64: 1783–1793. https://doi.org/10.1093/jxb/ert038.

Bidinger, F.R., Napolean, T., and Hash, C.T. (2007). Quantitative trait loci for grain yield in pearl millet under variable post flowering moisture conditions. *Crop Science* 47: 969–980. https://doi.org/10.2135/cropsci2006.07.0465.

Bohra, A., Dubey, A., Saxena, R.K. et al. (2011). Analysis of BAC-end sequences (BESs) and development of BES-SSR markers for genetic mapping and hybrid purity assessment in pigeonpea (*Cajanus* spp.). *BMC Plant Biology* 11: 56. https://doi.org/10.1186/1471-2229-11-56.

Boopathi, N.M., Chezhain, P., Jeyaprakash, P. et al. (2005). QTL mapping of drought-resistance traits using indica rice (*Oryza sativa* L.) lines adapted to target population of environment. *Proceedings of the 2nd International Conference on Integrated Approaches to Sustain and Improve Plant Production under Drought Stress (Rome).*

Borrell, A.K., Mullet, J.E., George-Jaeggli, B. et al. (2014). Drought adaptation of stay-green sorghum is associated with canopy development, leaf anatomy, root growth, and water uptake. *Journal of Experimental Botany* 65: 6251–6263. https://doi.org/10.1093/jxb/eru232.

Brooks, C., Nekrasov, V., Lippman, Z.B., and Van Eck, J. (2014). Efficient Gene editing in tomato in the first generation using the clustered regularly interspaced short palindromic repeats/CRISPR-Associated9 System. *Plant Physiology* 166: 1292–1297. https://doi.org/10.1104/pp.114.247577.

Brown, J.K.M. (2003). A cost of disease resistance: paradigm or peculiarity? *Trends in Genetics* 19: 667–671. https://doi.org/10.1016/j.tig.2003.10.008.

Caldwell, D.G., McCallum, N., Shaw, P. et al. (2004). A structured mutant population for forward and reverse genetics in Barley (*Hordeum vulgare* L.). *The Plant Journal* 40: 143–150. https://doi.org/10.1111/j.1365-313X.2004.02190.x.

Cermak, T., Doyle, E.L., Christian, M. et al. (2011). Efficient design and assembly of custom TALEN and other TAL effector-based constructs for DNA targeting. *Nucleic Acids Research* 39: e82. https://doi.org/10.1093/nar/gkr218.

Chen, K. and Gao, C. (2014). Targeted genome modification technologies and their applications in crop improvements. *Plant Cell Reports* 33: 575–583. https://doi.org/10.1007/s00299-013-1539-6.

Cobos, M.J., Winter, P., Kharrat, M. et al. (2009). Genetic analysis of agronomic traits in a wide cross of chickpea. *Field Crops Research* 111: 130–136. https://doi.org/10.1016/j.fcr.2008.11.006.

Cong, L., Ran, F.A., Cox, D. et al. (2013). Multiplex genome engineering using CRISPR/Cas systems. *Science* 339: 819–823. https://doi.org/10.1126/science.1231143.

Cooper, M., Gho, C., Leafgren, R. et al. (2014). Breeding drought-tolerant maize hybrids for the us corn-belt: discovery to product. *Journal of Experimental Botany* 65: 6191–6204. https://doi.org/10.1093/jxb/eru064.

Courtois, B., Shen, L., Petalcorin, W. et al. (2003). Locating QTLs controlling constitutive root traits in the rice population IAC 165 × Co39. *Euphytica* 134: 335–345. https://doi.org/10.1023/B:EUPH.0000004987.88718.d6.

Das, S., Upadhyaya, H.D., Bajaj, D. et al. (2015). Deploying QTL-seq for rapid delineation of a potential candidate gene underlying major trait-associated QTL in chickpea. *DNA Research* 22: 193–193. https://doi.org/10.1093/dnares/dsv004.

Devreux, M. and Scarascia Mugnozza, G.T. (1964). Effects of gamma radiation of the gametes, zygote and proembryo in *Nicotiana tabacum* L. *Radiation Botany* 4: 373–386. https://doi.org/10.1016/S0033-7560(64)80004-1.

D'Hont, A., Denoeud, F., Aury, J.M. et al. (2012). The banana (*Musa acuminata*) genome and the evolution of monocotyledonous plants. *Nature* 21: 213–217. https://doi.org/10.1038/nature11241.

Dubey, A., Farmer, A., Schlueter, J. et al. (2011). Defining the transcriptome assembly and its use for genome dynamics and transcriptome profiling studies in pigeonpea (*Cajanus cajan* L.). *DNA Research* 18: 153–164.

Dutta, S., Kumawat, G., Singh, B. et al. (2011). Development of genic-SSR markers by deep transcriptome sequencing in pigeonpea [*Cajanus cajan* (L.) Millspaugh]. *BMC Plant Biology* 11: 17. https://doi.org/10.1186/1471-2229-11-17.

Dwivedi, S.L., Sahrawat, K.L., Upadhyaya, H.D., and Ortiz, R. (2013). Food, nutrition and agrobiodiversity under global climate change. *Advances in Agronomy* 120: 1–128. https://doi.org/10.1016/B978-0-12-407686-0.00001-4.

Edwards, D. and Wang, X. (2012). Genome sequencing initiatives. In: *Genetics, Genomics and Breeding of Oilseed Brassicas* (ed. D. Edwards, J. Batley, I.A.P. Parkin and C. Kole), 152–157. Enfield, NH: Science Publishers.

Edwards, D., Batley, J., and Snowdon, R.J. (2013). Accessing complex crop genomes with next-generation sequencing. *Theoretical and Applied Genetics* 126: 1–11. https://doi.org/10.1007/s00122-012-1964-x.

FAO (2014). *The State of Food Insecurity in the World 2014. Strengthening the Enabling Environment for Food Security and Nutrition*. Rome: FAO.

Feng, Z., Zhang, B., Ding, W. et al. (2013). Efficient genome editing in plants using a CRISPR/Cas system. *Cell Research* 23: 1229–1232. https://doi.org/10.1038/cr.2013.114.

Feng, Z.Y., Mao, Y.F., Xu, N.F. et al. (2014). Multi-generation analysis reveals the inheritance, specificity and patterns of CRISPR/Cas induced gene modifications in Arabidopsis. *Proceedings of the National Academy of Sciences United States of America* 111: 4632–4637. https://doi.org/10.1073/pnas.1400822111.

Fischer, R.A., Byerlee, D., and Edmeades, G.O. (2009). *Can Technology Deliver on the Yield Challenge to 2050?* Rome: Food and Agriculture Organization of the United Nations.

Francia, E., Pecchioni, N., Barabaschi, D. et al. (2007). Fine mapping of a HvCBF gene cluster at the frost resistance locus Fr-H2 in barley. *Theoretical and Applied Genetics* 115: 1083–1091. https://doi.org/10.1007/s00122-007-0634-x.

Gaj, T., Gersbach, C.A., and Barbas, C.F. (2013). ZFN, TALEN, and CRISPR/Casbased methods for genome engineering. *Trends in Biotechnology* 31: 397–395. https://doi.org/10.1016/j.tibtech.2013.04.004.

Galiba, G., Tuberosa, R., Kocsy, G., and Sutka, J. (1993). Involvement of chromosomes 5A and 5D in cold induced abscisic acid accumulation and in frost tolerance of wheat calli. *Plant Breeding* 110: 237–242. https://doi.org/10.1111/j.1439-0523.1993.tb00583.x.

Galiba, G., Quarrie, S.A., Sutka, J. et al. (1995). RFLP mapping of the vernalization (VRN1) and Frost-resistance (FR1) genes on chromosome 5A of wheat. *Theoretical and Applied Genetics* 90: 1174–1179. https://doi.org/10.1007/BF00222940.

Gao, C. (2015). Developing genome editing technologies for crop improvement. *Meeting on Plant Genome Engineering, Plant & Animal Genome XXIII (San Diego, CA)*. Available online at: https://pag.confex.com/pag/xxiii/webprogram/Paper14087.html (accessed February 15, 2015).

Gaur, R., Sethy, N.K., Choudhary, S. et al. (2011). Advancing the STMS genomic resources for defining new locations on the intraspecific genetic linkage map of chickpea (*Cicer arietinum* L.). *BMC Genomics* 12: 117. https://doi.org/10.1186/1471-2164-12-117.

Gautami, B., Foncéka, D., Pandey, M.K. et al. (2012). An international reference consensus genetic map with 897 marker loci based on 11 mapping populations for tetraploid groundnut (*Arachis hypogaea* L.). *PLoS ONE* 7: e41213. https://doi.org/10.1371/journal.pone.0041213.

Gepts, P. and Aragão, F. (2008). Genomics of Phaseolus beans, a major source of dietary protein and micronutrients in the tropics. In: *Genomics of Tropical Crop Plants* (ed. P. Moore and R. Ming), 113–143. Berlin: Springer.

Gomez, M., Babu, R.C., Shanmugasundaram, P. et al. (2005). QTL mapping and marker assisted selection for drought tolerance in rice (*Oryza sativa* L.). *Proceedings of the 2nd International Conference on Integrated Approaches to Sustain and Improve Plant Production under Drought Stress (Rome).*

Gujaria, N., Kumar, A., Dauthal, P. et al. (2011). Development and use of genic molecular markers (GMMS) for construction of a transcript map of chickpea (*Cicer arietinum* L.). *Theoretical and Applied Genetics* 122: 1577–1589. https://doi.org/10.1007/s00122-011-1556-1.

Gupta, B., Sengupta, A., Saha, J., and Gupta, K. (2013). Plant abiotic stress: 'Omics' approach. *Journal of Plant Biochemistry & Physiology* 1: e108. https://doi.org/10.4172/2329-9029.1000e108.

Hall, A.E. and Ziska, L.H. (2000). Crop breeding strategies for the 21st century. In: *Climate Change and Global Crop Productivity* (ed. K.R. Reddy and H.F. Hodges), 407–423. New York, NY: CAB International.

Hayes, P.M., Blake, T., Chen, T.H. et al. (1998). Quantitative trait loci on barley (*Hordeum vulgare* L.) chromosome 7 associated with components of winter hardiness. *Genome* 36: 66–71.

Hemamalini, G.S., Shashidhar, H.E., and Hittalmani, S. (2000). Molecular marker assisted tagging of morphological and physiological traits under two contrasting moisture regimens at peak vegetative stage in rice (*Oryza sativa* L). *Euphytica* 112: 69–78. https://doi.org/10.1023/A:1003854224905.

Hiremath, P.J., Kumar, A., Penmetsa, R.V. et al. (2012). Large-scale development of cost-effective SNP marker assays for diversity assessment and genetic mapping in chickpea and comparative mapping in legumes. *Plant Biotechnology Journal* 10: 716–732. https://doi.org/10.1111/j.1467-7652.2012.00710.x.

Hisas, S. (2011). *The Food Gap. The Impacts of Climate Change in Food Production: A 2020 Perspective*. Alexandria, VA: Universal Ecological Fund.

Hsu, P.D., Lander, E.S., and Zhang, F. (2014). Development and applications of CRISPR-Cas9 for Genome Engineering. *Cell* 157: 1262–1278. https://doi.org/10.1016/j.cell.2014.05.010.

Hund, A., Reimer, R., and Messmer, R. (2011). A consensus map of QTLs controlling the root length of maize. *Plant and Soil* 344: 143–158. https://doi.org/10.1007/s11104-011-0735-9.

Imelfort, M., Batley, J., Grimmond, S., and Edwards, D. (2009). Genome sequencing approaches and successes. *Methods in Molecular Biology* 513: 345–358. https://doi.org/10.1007/978-1-59745-427-8_18.

Ismail, A.M., Heuer, S., Thomson, M.J., and Wissuwa, M. (2007). Genetic and genomic approaches to develop rice germplasm for problem soils. *Plant Molecular Biology* 65: 547–570. https://doi.org/10.1007/s11103-007-9215-2.

Ismail, A.M., Ella, E.S., Vergara, G.V., and Mackill, D.J. (2009). Mechanisms associated with tolerance to flooding during germination and early seedling growth in rice (*Oryza sativa*). *Annals of Botany* 103: 197–199. https://doi.org/10.1093/aob/mcn211.

Ismail, A.M., Singh, U.S., Singh, S. et al. (2013). The contribution of submergence-tolerant (Sub1) rice varieties to food security in flood-prone rainfed lowland areas in Asia. *Field Crops Research* 152: 83–93. https://doi.org/10.1016/j.fcr.2013.01.007.

James, R.A., Davenport, R.J., and Munns, R. (2006). Physiological characterization of two genes for Na^+ exclusion in durum wheat, nax1 and nax2. *Plant Physiology* 142: 1537–1547. https://doi.org/10.1104/pp.106.086538.

Jearakongman, S. (2005). Validation and discovery of quantitative trait loci for drought tolerance in backcross introgression lines in rice (*Oryza sativa* L.) Cultivar IR64. Ph.D. Thesis, Kasetsart University.

Jia, H. and Wang, N. (2014). Targeted genome editing of sweet orange using Cas9/sgRNA. *PLoS ONE* 9: e93806. https://doi.org/10.1371/journal.pone.0093806.

Jiang, W., Zhou, H., Bi, H. et al. (2013). Demonstration of CRISPR/Cas9/sgRNA-mediated targeted gene modification in Arabidopsis, tobacco, sorghum and rice. *Nucleic Acids Research* 41: e188. https://doi.org/10.1093/nar/gkt780.

Kamoshita, A., Wade, L., Ali, M. et al. (2002). Mapping QTLs for root morphology of a rice population adapted to rainfed lowland conditions. *Theoretical and Applied Genetics* 104: 880–893. https://doi.org/10.1007/s00122-001-0837-5.

Kamoshita, A., Chandra Babu, R., Boopathi, N.M., and Fukai, S. (2008). Phenotypic and genotypic analysis of drought-resistance traits for development of rice cultivars adapted to rainfed environments. *Field Crops Research* 109: 1–23. https://doi.org/10.1016/j.fcr.2008.06.010.

Kerr, R.A. (1986). Greenhouse warming still coming. *Science* 232: 573–575. https://doi.org/10.1126/science.232.4750.573.

Khan, Y., Suresh, B.V., Muthamilarasan, M. et al. (2014). Comprehensive genome-wide identification and expression profiling of foxtail millet [*Setaria italica* (L.)] miRNAs in response to abiotic stress and development of miRNA database. *Plant Cell, Tissue and Organ Culture* 118: 279–292. https://doi.org/10.1007/s11240-014-0480-x.

Khedikar, Y.P., Gowda, M.V., and Sarvamangala, C. (2010). A QTL study on late leaf spot and rust revealed one major QTL for molecular breeding for rust resistance in groundnut (*Arachis hypogaea* L.). *Theoretical and Applied Genetics* 121: 971–984. https://doi.org/10.1007/s00122-010-1366-x.

Knox, A.K., Dhillon, T., Cheng, H. et al. (2010). CBF gene copy number variation at frost resistance-2 is associated with levels of freezing tolerance in temperate-climate cereals. *Theoretical and Applied Genetics* 121: 21–35. https://doi.org/10.1007/s00122-010-1288-7.

Kottapalli, P., Gaur, P.M., Katiyar, S.K. et al. (2009). Mapping and validation of QTLs for resistance to an Indian isolate of ascochyta blight pathogen in chickpea. *Euphytica* 165: 79–88. https://doi.org/10.1007/s10681-008-9762-x.

Kudapa, H., Bharti, A.K., Cannon, S.B. et al. (2012). A comprehensive transcriptome assembly of pigeonpea (*Cajanus cajan* L.) using sanger and second-generation sequencing platforms. *Molecular Plant* 5: 1020–1028. https://doi.org/10.1093/mp/ssr111.

Kujur, A., Upadhyaya, H.D., Shree, T. et al. (2015a). Ultra-high density intra-specific genetic linkage maps accelerate identification of functionally relevant molecular tags governing important agronomic traits in chickpea. *Scientific Reports* 5: 9468. https://doi.org/10.1038/srep09468.

Kujur, A., Bajaj, D., Upadhyaya, H.D. et al. (2015b). A genome-wide SNP scan accelerates trait-regulatory genomic loci identification in chickpea. *Scientific Reports* 5: 11166. https://doi.org/10.1038/srep11166.

Kulwal, P.L., Thudi, M., and Varshney, R.K. (2011). Genomics interventions in crop breeding for sustainable agriculture. In: *Encyclopedia of Sustainability Science and Technology* (ed. R.A. Meyers), 2527–2540. New York, NY: Springer.

Kumari, K., Muthamilarasan, M., Misra, G. et al. (2013). Development of eSSR-markers in *Setaria italica* and their applicability in studying genetic diversity, cross-transferability and comparative mapping in millet and non-millet species. *PLoS ONE* 8: e67742. https://doi.org/10.1371/journal.pone.0067742.

Lafitte, H.R., Price, A.H., and Courtois, B. (2004). Yield response to water deficit in an upland rice mapping population: associations among traits and genetic markers. *Theoretical and Applied Genetics* 109: 1237–1246. https://doi.org/10.1007/s00122-004-1731-8.

Lanceras, J.C., Pantuwan, G., Jongdee, B., and Toojinda, T. (2004). Quantitative trait loci associated with drought tolerance at reproductive stage in rice. *Plant Physiology* 135: 384–399. https://doi.org/10.1104/pp.103.035527.

Landi, P., Sanguineti, M.C., and Salvi, S. (2005). Validation and characterization of a major QTL affecting leaf ABA concentration in maize. *Molecular Breeding* 15: 291–293. https://doi.org/10.1007/s11032-004-7604-7.

Langridge, P. and Fleury, D. (2011). Making the most of 'omics' for crop breeding. *Trends in Biotechnology* 29: 33–40. https://doi.org/10.1016/j.tibtech.2010.09.006.

Lata, C., Gupta, S., and Prasad, M. (2013). Foxtail millet: a model crop for genetic and genomic studies in bioenergy grasses. *Critical Reviews in Biotechnology* 33: 328–343. https://doi.org/10.3109/07388551.2012.716809.

Laurie, D.A., Pratchett, N., Snape, J.W., and Bezant, J.H. (1995). RFLP mapping of five major genes and eight quantitative trait loci controlling flowering time in a winter × spring barley (*Hordeum vulgare* L.) cross. *Genome* 38: 575–585. https://doi.org/10.1139/g95-074.

Li, Z., Mu, P., Li, C. et al. (2005). QTL mapping of root traits in a doubled haploid population from a cross between upland and lowland japonica rice in three environments. *Theoretical and Applied Genetics* 110: 1244–1252. https://doi.org/10.1007/s00122-005-1958-z.

Li, J., Aach, J., Norville, J.E. et al. (2013). Multiplex and homologous recombination-mediated genome editing in *Arabidopsis* and *Nicotiana benthamiana* using guide RNA and Cas9. *Nature Biotechnology* 31: 688–691. https://doi.org/10.1038/nbt.2654.

Lilley, J.M., Ludlow, M.M., McCouch, S.R., and O'Toole, J.C. (1996). Locating QTL for osmotic adjustment and dehydration tolerance in rice. *Journal of Experimental Botany* 47: 1427–1436.

Liu, S., Wang, X., Wang, H. et al. (2013). Genome wide analysis of ZmDREB genes and their association with natural variation in drought tolerance at seedling stage of *Zea mays* L. *PLoS Genetics* 9: e1003790. https://doi.org/10.1371/journal.pgen.1003790.

Lopes, M.S., Rebetzke, G.J., and Reynolds, M. (2014). Integration of phenotyping and genetic platforms for a better understanding of wheat performance under drought. *Journal of Experimental Botany* 65: 6167–6177. https://doi.org/10.1093/jxb/eru384.

Lynch, J.P., Chimungu, J.G., and Brown, K.M. (2014). Root anatomical phenes associated with water acquisition from drying 1 soil: targets for crop improvement. *Journal of Experimental Botany* 65: 6155–6166. https://doi.org/10.1093/jxb/eru162.

Ma, Y., Qin, F., and Tran, L.S. (2012). Contribution of genomics to gene discovery in plant abiotic stress responses. *Molecular Plant* 5: 1176–1178. https://doi.org/10.1093/mp/sss085.

Mackay, I.J., Bansept-Basler, P., Barber, T. et al. (2014). An eight-parent multiparent advanced generation inter-cross population for winter-sown wheat: creation, properties and validation. *G3 (Bethesda)* 4: 1603–1610. https://doi.org/10.1534/g3.114.012963.

Mackill, D.J., Ismail, A.M., Singh, U.S. et al. (2012). Development and rapid adoption of submergence-tolerant (Sub1) rice varieties. *Advances in Agronomy* 115: 299–252. https://doi.org/10.1016/B978-0-12-394276-0.00006-8.

Mali, P., Yang, L., Esvelt, K.M. et al. (2013). RNA-guided human genome engineering via Cas9. *Science* 339: 823–826. https://doi.org/10.1126/science.1232033.

Mano, Y., Muraki, M., Fujimori, M. et al. (2005a). Identification of QTL controlling adventitious root formation during flooding conditions in teosinte (*Zea mays* ssp. huehuetenangensis) seedlings. *Euphytica* 142: 33–42. https://doi.org/10.1007/s10681-005-0449-2.

Mano, Y., Omori, F., Muraki, M., and Takamizo, T. (2005b). QTL mapping of adventitious root formation under flooding conditions in tropical maize (*Zea mays* L.) seedlings. *Breeding Science* 55: 343–347. https://doi.org/10.1270/jsbbs.55.343.

Mao, Y., Zhang, H., Xu, N. et al. (2013). Application of the CRISPR-Cas system for efficient genome engineering in plants. *Molecular Plant* 6: 2008–2011. https://doi.org/10.1093/mp/sst121.

McCallum, C.M., Comai, L., Greene, E.A., and Henikoff, S. (2000). Targeted screening for induced mutations. *Nature Biotechnology* 18: 455–457. https://doi.org/10.1038/74542.

Miao, J., Guo, D., Zhang, J. et al. (2013). Targeted mutagenesis in rice using CRISPR-Cas system. *Cell Research* 23: 1233–1236. https://doi.org/10.1038/cr.2013.123.

Miro, B. and Ismail, A.M. (2013). Tolerance of anaerobic conditions caused by flooding during germination and early growth in rice (*Oryza sativa* L). *Frontiers in Plant Science* 4: 269. https://doi.org/10.3389/fpls.2013.00269.

Muller, H.J. (1927). Artificial transmutation of the gene. *Science* 66: 84–87. https://doi.org/10.1126/science.66.1699.84.

Munns, R. and Tester, M. (2008). Mechanisms of salinity tolerance. *Annual Review of Plant Biology* 59: 651–681. https://doi.org/10.1146/annurev.arplant.59.032607.092911.

Muthamilarasan, M. and Prasad, M. (2015). Advances in Setaria genomics for genetic improvement of cereals and bioenergy grasses. *Theoretical and Applied Genetics* 128: 1–14. https://doi.org/10.1007/s00122-014-2399-3.

Muthamilarasan, M., Theriappan, P., and Prasad, M. (2013). Recent advances in crop genomics for ensuring food security. *Current Science* 105: 155–158.

Muthamilarasan, M., Venkata Suresh, B., Pandey, G. et al. (2014). Development of 5123 intron-length polymorphic markers for large-scale genotyping applications in foxtail millet. *DNA Research* 21: 41–52. https://doi.org/10.1093/dnares/dst039.

Nekrasov, V., Staskawicz, B., Weigel, D. et al. (2013). Targeted mutagenesis in the model plant *Nicotiana benthamiana* using Cas9 RNA-guided endonuclease. *Nature Biotechnology* 31: 691–693. https://doi.org/10.1038/nbt.2655.

Neves, L., Davis, J., Barbazuk, B., and Kirst, M. (2011). Targeted sequencing in the loblolly pine (*Pinus taeda*) megagenome by exome capture. *BMC Proceedings* 7: O48. https://doi.org/10.1186/1753-6561-5-S7-O48.

Nguyen, T.T.T., Klueva, N., Chamareck, V. et al. (2004). Saturation mapping of QTL regions and identification of putative candidate genes for drought tolerance in rice. *Molecular Genetics and Genomics* 272: 35–46. https://doi.org/10.1007/s00438-004-1025-5.

Osakabe, Y. and Osakabe, K. (2015). Genome editing with engineered nucleases in plants. *Plant & Cell Physiology* 56: 389–400. https://doi.org/10.1093/pcp/pcu170.

Paliwal, R., Röder, M.S., Kumar, U. et al. (2012). QTL mapping of terminal heat tolerance in hexaploid wheat (*T. aestivum* L.). *Theoretical and Applied Genetics* 125: 561–575. https://doi.org/10.1007/s00122-012-1853-3.

Pandey, G., Misra, G., Kumari, K. et al. (2013). Genome-wide development and use of microsatellite markers for largescale genotyping applications in foxtail millet [*Setaria italica* (L.)]. *DNA Research* 20: 197–197. https://doi.org/10.1093/dnares/dst002.

Pingali, P.L. (2012). Green revolution: impact, limits and the path ahead. *Proceedings of the National Academy of Sciences United States of America* 109: 12302–11230. https://doi.org/10.1073/pnas.0912953109.

Pinto, R.S., Reynolds, M.P., Mathews, K.L. et al. (2010). Heat and drought adaptive QTL in a wheat population designed to minimize confounding agronomic effects. *Theoretical and Applied Genetics* 121: 1001–1021. https://doi.org/10.1007/s00122-010-1351-4.

Price, A.H., Cairns, J.E., Horton, P. et al. (2002a). Linking drought-resistance mechanisms to drought avoidance in upland rice using a QTL approach: progress and new opportunities to integrate stomatal and mesophyll responses. *Journal of Experimental Botany* 53: 989–984. https://doi.org/10.1093/jexbot/53.371.989.

Price, A.H., Townend, J., Jones, M.P. et al. (2002b). Mapping QTLs associated with drought avoidance in upland rice grown in the Philippines and West Africa. *Plant Molecular Biology* 48: 683–695. https://doi.org/10.1023/A:1014805625790.

Quarrie, S.A., Laurie, D.A., Zhu, J. et al. (1997). QTL analysis to study the association between leaf size and abscisic acid accumulation in droughted rice leaves and comparisons across cereals. *Plant Molecular Biology* 35: 155–165. https://doi.org/10.1023/A:1005864202924.

Raju, N., Gnanesh, B., Lekha, P. et al. (2010). The first set of est resource for gene discovery and marker development in pigeonpea (*Cajanus cajan* L.). *BMC Plant Biology* 10: 45. https://doi.org/10.1186/1471-2229-10-45.

Rebetzke, G.J., Herwaarden, A.F., Jenkins, C. et al. (2008). Quantitative trait loci for water-soluble carbohydrates and associations with agronomic traits in wheat. *Australian Journal of Agricultural Research* 59: 891–805. https://doi.org/10.1071/AR08067.

Rehman, A.U., Malhotra, R.S., Bett, K. et al. (2011). Mapping QTL associated with traits affecting grain yield in chickpea (*Cicer arietinum* L.) under terminal drought stress. *Crop Science* 51: 450–463. https://doi.org/10.2135/cropsci2010.03.0129.

Reynolds, M. and Tuberosa, R. (2008). Translational research impacting on crop productivity in drought-prone environments. *Current Opinion in Plant Biology* 11: 171–179. https://doi.org/10.1016/j.pbi.2008.02.005.

Reynolds, M.P., Hays, D., and Chapman, S. (2010). Breeding for adaptation to heat and drought stress. In: *Climate Change and Crop Production* (ed. M.P. Reynolds), 71–91. CABI: Oxfordshire.

Robin, S., Pathan, M.S., Courtois, B. et al. (2003). Mapping osmotic adjustment in an advanced back-cross inbred population of rice. *Theoretical and Applied Genetics* 107: 1288–1296. https://doi.org/10.1007/s00122-003-1360-7.

Ruta, N., Liedgens, M., Fracheboud, Y. et al. (2010). QTLs for the elongation of axile and lateral roots of maize in response to low water potential. *Theoretical and Applied Genetics* 120: 621–631. https://doi.org/10.1007/s00122-009-1180-5.

Salem, K.F.M., Roder, M.S., and Borner, A. (2007). Identification and mapping quantitative trait loci for stem reserve mobilisation in wheat (*Triticum aestivum* L.). *Cereal Research Communications* 35: 1367–1374. https://doi.org/10.1556/CRC.35.2007.3.1.

Salvi, S., Sponza, G., Morgante, M. et al. (2007). Conserved noncoding genomic sequences associated with a flowering-time quantitative trait locus in maize. *Proceedings of the National Academy of Sciences United States of America* 104: 11376–11381. https://doi.org/10.1073/pnas.0704145104.

Sarvamangala, C., Gowda, M.V.C., and Varshney, R.K. (2011). Identification of quantitative trait loci for protein content oil content and oil quality for groundnut (*Arachis hypogaea* L.). *Field Crops Research* 122: 49–59. https://doi.org/10.1016/j.fcr.2011.02.010.

Saxena, K. (2008). Genetic improvement of pigeonpea-a review. *Tropical Plant Biology* 1: 159–178. https://doi.org/10.1007/s12042-008-9014-1.

Saxena, R., Prathima, C., Saxena, K. et al. (2010). Novel SSR markers for polymorphism detection in pigeonpea (*Cajanus* spp.). *Plant Breeding* 129: 142–148. https://doi.org/10.1111/j.1439-0523.2009.01680.x.

Schneider, K.A., Brothers, M.E., and Kelly, J.D. (1997). Marker-assisted selection to improve drought tolerance in common bean. *Crop Science* 37: 51–60. https://doi.org/10.2135/cropsci1997.0011183X003700010008x.

Seki, M., Narusaka, M., Ishida, J. et al. (2002). Monitoring the expression profiles of 7000 Arabidopsis genes under drought, cold and high-salinity stresses using a full-length cDNA microarray. *The Plant Journal* 31: 279–292. https://doi.org/10.1046/j.1365-313X.2002.01359.x.

Septiningsih, E.M., Ignacio, J.C., Sendon, P.M. et al. (2013). QTL mapping and confirmation for tolerance of anaerobic conditions during germination derived from the rice landrace Ma-Zhan Red. *Theoretical and Applied Genetics* 126: 1357–1366. https://doi.org/10.1007/s00122-013-2057-1.

Serraj, R., Kumar, A., McNally, K.L. et al. (2009). Improvement of drought resistance in rice. *Advances in Agronomy* 103: 41–99. https://doi.org/10.1016/S0065-2113(09)03002-8.

Setter, T.M. and Waters, I. (2003). Review of prospects for germplasm improvement for waterlogging tolerance in wheat, barley and oats. *Plant and Soil* 253: 1–34. https://doi.org/10.1023/A:1024573305997.

Shan, Q., Wang, Y., Li, J. et al. (2013). Targeted genome modification of crop plants using a CRISPR-Cas system. *Nature Biotechnology* 31: 686–688. https://doi.org/10.1038/nbt.2650.

Shavrukov, Y., Gupta, N.K., Miyazaki, J. et al. (2010). Hvnax3-a locus controlling shoot sodium exclusion derived from wild barley (*Hordeum vulgare* ssp *spontaneum*). *Functional & Integrative Genomics* 10: 277–291. https://doi.org/10.1007/s10142-009-0153-8.

Silva, G., Poirot, L., Galetto, R. et al. (2011). Meganucleases and other tools for targeted genome engineering: perspectives and challenges for gene therapy. *Current Gene Therapy* 11: 11–27. https://doi.org/10.2174/156652311794520111.

Singh, S.P. (2007). Drought resistance in the race durango dry bean landraces and cultivars. *Agronomy Journal* 99: 1219–1225. https://doi.org/10.2134/agronj2006.0301.

Singh, S., Mackill, D.J., and Ismail, A.M. (2009). Responses of Sub1 rice introgression lines to submergence in the field: yield and grain quality. *Field Crops Research* 113: 12–23. https://doi.org/10.1016/j.fcr.2009.04.003.

Skinner, J.S., Szucs, P., Zitzewitz, J.V. et al. (2006). Mapping of barley homologs to genes that regulate low temperature tolerance in Arabidopsis. *Theoretical and Applied Genetics* 112: 832–842. https://doi.org/10.1007/s00122-005-0185-y.

Snape, J.W., Foulkes, J., Simmonds, J. et al. (2007). Dissecting gene × environmental effects on wheat yields via QTL and physiological analysis. *Euphytica* 154: 401–408. https://doi.org/10.1007/s10681-006-9208-2.

Stockinger, E.J., Gilmour, S.J., and Thomashow, M.F. (1997). *Arabidopsis thaliana* CBF1 encodes an AP2 domain-containing transcriptional activator that binds to the C-repeat/DRE, a cis-acting DNA regulatory element that stimulates transcription in response to low temperature and water deficit. *Proceedings of the National Academy of Sciences United States of America* 94: 1035–1040. https://doi.org/10.1073/pnas.94.3.1035.

Sujay, V., Gowda, M.V., Pandey, M.K. et al. (2012). Quantitative trait locus analysis and construction of consensus genetic map for foliar disease resistance based on two recombinant inbred line populations in cultivated groundnut (*Arachis hypogaea* L.). *Molecular Breeding* 30: 773–788. https://doi.org/10.1007/s11032-011-9661-z.

Sulpanjani, A., Yang, M.S., Tawaha, A.R.M. et al. (2005). Effect of Magnesium application on yield, mineral contents and active components of *Chrysanthemum coronarium* L. under Hydroponics Conditions. *Bioscience Research* 2: 73–79.

Supanjani, Tawaha, A.R.M., Yang, M.S. et al. (2005). Role of calcium in yield and medicinal quality of *Chrysanthemum coronarium* L. *Journal of Agronomy* 4: 188–192.

Suresh, B.V., Muthamilarasan, M., Misra, G., and Prasad, M. (2013). FmMDb: a versatile database of foxtail millet markers for millets and bioenergy grasses research. *PLoS ONE* 8: e71418. https://doi.org/10.1371/journal.pone.0071418.

Tawaha, A.M. and Turk, M.A. (2002a). Lentil (*Lens culinaris* Medic.) productivity as influenced by rate and method of phosphate placement in a Mediterranean environment. *Acta Agronomica Hungarica* 50: 197–101.

Tawaha, A.M. and Turk, M.A. (2004). Field pea seeding management for semi-arid Mediterranean conditions. *Journal of Agronomy and Crop Science* 190: 86–92.

Thomas, H. and James, A.R. (1993). Freezing tolerance and solute changes in contrasting genotypes of *Lolium perenne* L. acclimated to cold and drought. *Annals of Botany* 72: 249–254. https://doi.org/10.1006/anbo.1993.1105.

Thudi, M., Bohra, A., Nayak, S.N. et al. (2011). Novel SSR markers from bac-end sequences, dart arrays and a comprehensive genetic map with 1, 291 marker loci for chickpea (*Cicer arietinum* L.). *PLoS ONE* 6: e27275. https://doi.org/10.1371/journal.pone.0027275.

Townsend, J.A., Wright, D.A., Winfrey, R.J. et al. (2009). High-frequency modification of plant genes using engineered zinc-finger nucleases. *Nature* 459: 442–445. https://doi.org/10.1038/nature07845.

Tripathy, J.N., Zhang, J., Robin, S. et al. (2000). QTLs for cell-membrane stability mapped in rice (*Oryza sativa* L.) under drought stress. *Theoretical and Applied Genetics* 100: 1197–1202. https://doi.org/10.1007/s001220051424.

Tuberosa, R. (2012). Phenotyping for drought tolerance of crops in the genomics era. *Frontiers in Physiology* 3: 347. https://doi.org/10.3389/fphys.2012.00347.

Tuberosa, R., Salvi, S., Sanguineti, M.C. et al. (2011). Genomics of root architecture and functions in maize. In: *Root Genomics* (ed. A.C. de Oliveira and R. Varshney), 179–104. Berlin: Springer.

Uga, Y., Sugimoto, K., Ogawa, S. et al. (2013). Control of root system architecture by deeper rooting 1 increases rice yield under drought conditions. *Nature Genetics* 45: 1097–1002. https://doi.org/10.1038/ng.2725.

Urrea, C.A., Yonts, C.D., Lyon, D.J., and Koehler, A.E. (2009). Selection for drought tolerance in dry bean derived from the Mesoamerican gene pool in western Nebraska. *Crop Science* 49: 2005–2010. https://doi.org/10.2135/cropsci2008.12.0694.

Vadez, V., Krishnamurthy, L., Thudi, M. et al. (2012). Assessment of ICCV 2 × JG 62 chickpea progenies shows sensitivity of reproduction to salt stress and reveals QTL for seed yield and yield components. *Molecular Breeding* 30: 9–21. https://doi.org/10.1007/s11032-011-9594-6.

Vagujfalvi, A., Galiba, G., Cattivelli, L., and Dubcovsky, J. (2003). The cold regulated transcriptional activator Cbf3 is linked to the frost-tolerance locus Fr-A2 on wheat chromosome 5A. *Molecular Genetics and Genomics* 269: 60–67. https://doi.org/10.1007/s00438-003-0806-6.

Varshney, R.K., Graner, A., and Sorrells, M.E. (2005). Genomics-assisted breeding for crop improvement. *Trends in Plant Science* 10: 621–630. https://doi.org/10.1016/j.tplants.2005.10.004.

Varshney, R.K., Penmetsa, R.V., Dutta, S. et al. (2009). Pigeonpea genomics initiative (PGI): an international effort to improve crop productivity of pigeonpea (*Cajanus cajan* L.). *Molecular Breeding* 26: 393–398. https://doi.org/10.1007/s11032-009-9327-2.

Varshney, R.K., Chen, W., Li, Y. et al. (2012). Draft genome sequence of pigeonpea (*Cajanus cajan*), an orphan legume crop of resource-poor farmers. *Nature Biotechnology* 30: 83–89. https://doi.org/10.1038/nbt.2022.

Varshney, R.K., Mohan, S.M., Gaur, P.M. et al. (2013a). Achievements and prospects of genomics-assisted breeding in three legume crops of the semi-arid tropics. *Biotechnology Advances* 31: 1120–1134. https://doi.org/10.1016/j.biotechadv.2013.01.001.

Varshney, R.K., Song, C., Saxena, R.K. et al. (2013b). Draft genome sequence of kabuli chickpea (*Cicer arietinum*): genetic structure and breeding constraints for crop improvement. *Nature Biotechnology* 31: 240–246. https://doi.org/10.1038/nbt.2491.

Varshney, R.K., Terauchi, R., and McCouch, S.R. (2014). Harvesting the promising fruits of genomics: applying genome sequencing technologies to crop breeding. *PLoS Biology* 12: e1001883. https://doi.org/10.1371/journal.pbio.1001883.

Velasco, R., Zharkikh, A., Troggio, M. et al. (2007). A high quality draft consensus sequence of the genome of a heterozygous grapevine variety. *PLoS ONE* 2: e1326. https://doi.org/10.1371/journal.pone.0001326.

Velasco, R., Zharkikh, A., Affourtit, J. et al. (2010). The genome of the domesticated apple (Malus x domestica Borkh.). *Nature Genetics* 42: 833–839. https://doi.org/10.1038/ng.654.

Wang, H., Penmetsa, R.V., Yuan, M. et al. (2012). Development and characterization of BAC-end sequence derived SSRs, and their incorporation into a new higher density genetic map for cultivated peanut (*Arachis hypogaea* L.). *BMC Plant Biology* 12: 10. https://doi.org/10.1186/1471-2229-12-10.

Wang, Y., Cheng, X., Shan, Q. et al. (2014). Simultaneous editing of three homoeoalleles in hexaploid bread wheat confers heritable resistance to powdery mildew. *Nature Biotechnology* 32: 947–951. https://doi.org/10.1038/nbt.2969.

Wassmann, R. and Jagadish, S. (2009). Regional vulnerability of climate change impacts on Asian rice production and scope for adaptation. *Advances in Agronomy* 102: 91–133. https://doi.org/10.1016/S0065-2113(09)01003-7.

Wasson, A.P., Richards, R.A., Chatrath, R. et al. (2012). Traits and selection strategies to improve root systems and water uptake in water-limited wheat crops. *Journal of Experimental Botany* 63: 3485–3498. https://doi.org/10.1093/jxb/ers111.

Welch, J.R., Vincent, J.R., Auffhammer, M. et al. (2010). Rice yields in tropical/subtropical Asia exhibit large but opposing sensitivities to minimum and maximum temperatures. *Proceedings of the National Academy of Sciences United States of America* 107: 14562–14567. https://doi.org/10.1073/pnas.1001222107.

Xie, K. and Yang, Y. (2013). RNA-guided genome editing in plants using a CRISPR-Cas system. *Molecular Plant* 6: 1975–1980. https://doi.org/10.1093/mp/sst119.

Xu, J.L., Lafitte, H.R., Gao, Y.M. et al. (2005). QTLs for drought escape and tolerance identified in a set of random introgression lines of rice. *Theoretical and Applied Genetics* 111: 1642–1650. https://doi.org/10.1007/s00122-005-0099-8.

Xu, K., Xu, X., Fukao, T. et al. (2006). Sub1A is an ethylene-response-factor-like gene that confers submergence tolerance to rice. *Nature* 442: 705–708. https://doi.org/10.1038/nature04920.

Xu, Q., Chen, L.L., Ruan, X. et al. (2013). The draft genome of sweet orange (*Citrus sinensis*). *Nature Genetics* 45: U59–U92. https://doi.org/10.1038/ng.2472.

Yadav, C.B., Muthamilarasan, M., Pandey, G. et al. (2014). Development of novel microRNA-based genetic markers in foxtail millet for genotyping applications in related grass species. *Molecular Breeding* 34: 2219–2224. https://doi.org/10.1007/s11032-014-0137-9.

Yadav, C.B., Bonthala, V.S., Muthamilarasan, M. et al. (2015). Genome-wide development of transposable elements-based markers in foxtail millet and construction of an integrated database. *DNA Research* 22: 79–90. https://doi.org/10.1093/dnares/dsu039.

Yang, J., Zhang, J., Liu, K. et al. (2007). Abscisic acid and ethylene interact in rice spikelets in response to water stress during meiosis. *Journal of Plant Growth Regulation* 26: 318–328. https://doi.org/10.1007/s00344-007-9013-8.

Yue, B., Xue, W., Xiong, L. et al. (2006). Genetic basis of drought resistance at reproductive stage in rice: separation of drought tolerance from drought avoidance. *Genetics* 172: 1213–1228. https://doi.org/10.1534/genetics.105.045062.

Zhang, J., Nguyen, B.D., Sarkarung, S. et al. (2001a). Locating genomic regions associated with components of drought resistance in rice: comparative mapping within and across species. *Theoretical and Applied Genetics* 103: 19–29. https://doi.org/10.1007/s001220000534.

Zhang, W.P., Shen, X.Y., Wu, P. et al. (2001b). QTLs and epistasis for seminal root length under a different water supply in rice (*Oryza sativa* L.). *Theoretical and Applied Genetics* 103: 118–123. https://doi.org/10.1007/s001220100561.

Zhang, G., Liu, X., Quan, Z. et al. (2012). Genome sequence of foxtail millet (*Setaria italica*) provides insights into grass evolution and biofuel potential. *Nature Biotechnology* 30: 549–554. https://doi.org/10.1038/nbt.2195.

Zhang, H., Zhang, J., Wei, P. et al. (2014a). The CRISPR/Cas9 system produces specific and homozygous targeted gene editing in rice in one generation. *Plant Biotechnology Journal* 12: 797–797. https://doi.org/10.1111/pbi.12200.

Zhang, S., Tang, C., Zhao, Q. et al. (2014b). Development of highly polymorphic simple sequence repeat markers using genome-wide microsatellite variant analysis in Foxtail millet [*Setaria italica* (L.) P. Beauv]. *BMC Genomics* 2: 15. https://doi.org/10.1186/1471-2164-15-78.

Zheng, H.G., Babu, R.C., Pathan, M.S. et al. (2000). Quantitative trait loci for root-penetration ability and root thickness in rice: comparison of genetic backgrounds. *Genome* 43: 53–61. https://doi.org/10.1139/g99-065.

Zheng, B.S., Yang, L., Zhang, W.P. et al. (2003). Mapping QTLs and candidate genes for rice root traits under different water supply conditions and comparative analysis across three populations. *Theoretical and Applied Genetics* 107: 1505–1515. https://doi.org/10.1007/s00122-003-1390-1.

4

Role of Biotechnology in Climate-Resilient Agriculture

Abdel Rahman Mohammad Said Al-Tawaha[1], Rachid Mrabet[2], Mina Bayanati[3], Banothu Santhosh[4], Noureddine Benkeblia[5], Imran[6,7], Amanullah[6], Shah Khalid[6], Abdel Razzaq Al-Tawaha[8], Hiba Alatrash[9], Yaman Jabbour[9], Abhijit Dey[10], Devarajan Thangadurai[11], Jeyabalan Sangeetha[12], and Saher Islam[13]

[1] *Department of Biological Sciences, Al-Hussein Bin Talal University, Ma'an, Jordan*
[2] *Department of Environment and Natural Resources, Scientific Division, National Institute for Agricultural Research (INRA), Rabat, Morocco*
[3] *Department of Horticultural Science, Ferdowsi University of Mashhad, Mashhad, Iran*
[4] *Department of Crop Physiology, College of Agriculture, Rajendranagar, PJTSAU, Hyderabad, Telangana, India*
[5] *Department of Life Sciences, The University of the West Indies, Kingston, Jamaica*
[6] *Department of Agronomy, The University of Agriculture, Peshawar, Pakistan*
[7] *Ministry of Agriculture, Agriculture Extension Wing, Merged Area, Khyber Pakhtunkhwa, Pakistan*
[8] *Department of Crop Science, Faculty of Agriculture, Universiti Putra Malaysia, Serdang, Selangor, Malaysia*
[9] *Department of Field Crops, General Commission for Scientific Agricultural Research, Aleppo, Syria*
[10] *Department of Life Sciences, Presidency University, Kolkata, West Bengal, India*
[11] *Department of Botany, Karnatak University, Dharwad, Karnataka, India*
[12] *Department of Environmental Science, Central University of Kerala, Kasaragod, Kerala, India*
[13] *Department of Biochemistry and Biotechnology, Institute of Biochemistry and Biotechnology, University of Veterinary and Animal Sciences, Lahore, Pakistan*

CONTENTS

4.1 Crop Yield Sensitivity to Climatic Variability as the Basis for Creating Climate Resilient Agriculture

Climate change has already had an impact on food security as a result of warming, changing precipitation patterns, and the increased frequency of some extreme events (IPCC 2019). Specifically, crop growth, development, gas exchange water relations, and yield formation

Climate Change and Agriculture: Perspectives, Sustainability and Resilience,
First Edition. Edited by Noureddine Benkeblia.

processes are differentially affected by climate variables (temperature, rainfall, drought, heat) and patterns (Abu-Darwish et al. 2009; Sulpanjani et al. 2005; Supanjani et al. 2005; Tawaha and Turk 2004; Turk et al. 2004). Which complicate extreme weather impact on crop production, is that magnitude of warming and drought are driven by complex interactions among precipitation, temperature, vapour pressure, wind intensity, and solar radiation (Porter et al. 2014). Increasing the concentration of atmospheric carbon dioxide has a positive effect on the capacity for photosynthesis and water-use efficiency. These effects vary quite considerably among C3 and C4 type crops. It is widely accepted that C02 fertilization have a positive stimulating effect on crop yields of wheat, rice, and soybeans. Understanding crop yield sensitivity to climate variability and change has been studied by several authors using diverse research approaches, but combining modelling, crop statistics time series, and climate generated data are the most commonly used (Turk and Tawaha 2002a,b). In this chapter, the focus will on four crops, i.e. wheat, corn, rice, and soybeans, that provide 75% of all calories consumed by humans (Lobell et al. 2011). Droughts are estimated to have caused a cereal (maize, rice, and wheat) loss of 1820 million Mg globally over the last four decades (Lesk et al. 2016). In the same period, yield growth per year in these crops is significantly lower than the annual growth rate of 2.4% required for doubling cultivation by 2050 (Ray et al. 2012, 2013). In other terms, according to Pradhan et al. (2015), crop yields have either stagnated or collapsed over the period of 1961–2008 across 24 to 39% of their cultivated areas globally. Climate change effects on agriculture are felt everywhere (Tawaha and Turk 2002a; Turk and Tawaha 2001). Corn yields, for example, have steadily increased in the United States and China over the last 50 years and show no signs of slowing. In the case of wheat, average yields in Western Europe have tripled since 1960 but have been largely flat and volatile since the mid-1990s, potentially indicating a slowing of yield growth (Lin and Huybers 2012). China doubled average rice yields between 1960 and 1980 but has struggled to maintain those rates since then, with yield trends also flattening since the mid-1990s.

Food security and stability are impacted by climate variability which accounts for up to 60% of yield variability in major parts of the world (Frieler et al. 2017). According to Ray et al. (2015), climate variability influenced crop yields on 78, 52, 75, and 67% of corn, rice, wheat, and soybean harvesting regions. Negative impacts of climate trends on crop yields have been more common than positive ones. Crop productivity impacts are expected to be negative in low latitude and tropical regions, even at low levels of warming, but somewhat positive in high-latitude regions (FAO 2015). The effects are expected to be more mixed in the mid to high latitudes, especially at lower levels of warming. Global warming may increase yields and extend the growing season and acreage of agricultural crops in high latitudes (such as Russia, northern Europe, Canada, and South America), though yields may be low in some regions due to poor soil fertility and water shortages (FAO 2015). As summarized by Ray et al. (2015), many of the most productive regions such as the Midwestern United States and the China corn belt for corn and the Western Europe and Australia for wheat are highly sensitive to climate variability. Porter et al. (2014) reported from an analysis of 66 international studies that yields of corn and wheat begin to decline with 1 to 2 °C of local warming in the tropics. They also found that temperate corn and tropical rice yields are less clearly affected at these temperatures, but significantly affected with warming of 3 to 5 °C. The authors also concluded that future impacts are expected to be consistent with the trajectory of past impacts, with the majority of locations experiencing negative impacts while some locations benefit (Tawaha et al. 2003; Turk et al. 2003a,b,c). At the global level, Iizumi et al. (2018) used a counterfactual analysis to determine that between 1981 and 2010, climate change reduced global mean corn, wheat, and soybean yields by 4.1, 1.8, and 4.5%, respectively, relative to preindustrial climate. Matiu et al. (2017) found that the combined effects of drought and high temperature significantly decreased yields of corn, soybean, and wheat by 11.6, 12.4, and 9.2%, respectively. They also reported that rice was least affected by climate variability both globally and by top country producers.

In Australia, Hochman et al. (2017) and Innes et al. (2015) stressed an important decline (27%) in wheat yields due to reduced rainfall and rising daily maximum temperatures. The same findings were reported by Gupta et al. (2017) and Jha and Tripathi (2017) for India on wheat. In India and China, Ray et al. (2015) detected that on average 32 and 31% of wheat yield variability was due to climate variability respectively. Overall, negative impacts of drought and temperature on agricultural yields were assessed in Asia except in Northern China for both wheat and rice.

According to the Intergovernmental Panel on Climate Change's Fifth Assessment Report (AR5), under more optimistic scenarios, climate change could reduce food crop yields in parts of Africa by 10 to 20%, a significant drop for already vulnerable populations and regions. Across Africa, hotspot climate change continent, yields of staple crops have significantly decreased due to rainfall decline and drought severity. In South Africa, 50% of corn yield variability can be explained by climate variability as found by Ray et al. (2015). According to these authors, overall precipitation variability in sub-Saharan Africa is more critical due to the dominance of corn rainfed cropping systems. In Europe, long-term temperature and precipitation trends are reducing continent-wide wheat and barley yields by 2.5 and 3.8%, respectively, and having slightly increased corn productivity (Moore and Lobell 2015). In France and Italy, 41 to 49% of year-to-year corn yield variability can be explained by climate variability. In Western Europe, climate variability explained from 31 to 51% of wheat yield changes; however, in Eastern Europe, the climate variability explained up to 66% of wheat yield variability (Ray et al. 2015). For soybeans, climate variability explained 43, 36, and 26–34% for Argentina, the United States, and consortium countries of Brazil, India, and China (Ray et al. 2015). In terms of climate-yield relationships, other weather factors are important, mainly, heat stress, dry and wet spells, solar radiation (cloud cover), wind speed, and ozone exposure. In addition, altitudinal effects, cultivars, and agricultural management (i.e. irrigation, fertilization) are important factors explaining crop responses to climate variability and change (Asseng et al. 2014; Sakurai et al. 2012). In fact, climate change impacts are worsened or intensified by inappropriate seeding and harvest dates, increased infestation of pests and diseases, declining viability of crop varieties, and advanced degradation of soil quality and health. In addition to climate change, substantial yield gaps caused by imperfect and unsustainable cropland management exist in most parts of the world (Abu Obaid et al. 2018; Al-Tawaha and Seguin 2006; Al-Tawaha et al. 2006, 2007, 2010, 2018a,b,c,d; Al-Tawaha and Odat 2010; Ray et al. 2012). In other words, adaptation measures such as shifting dates, changing cultivars, and in-season crop management could moderate or regulate adverse climate change impacts on crops depending on locality features and weather fluctuations (Asseng et al. 2013). Studies by Innes et al. (2015)

demonstrated a reinforced relationship between observed climate variables and crop yields that indicate future expected warming will have more severe impacts on crop production. Hence, interactions of environmental, technological, and management changes need a holistic understanding of competing effects in order to modulate crop sensitivity to climate change and propose more resilient crop and food systems. Innovation process should follow a transdisciplinary approach in order to integrate systemic changes in climate variability.

In 2017, world agricultural land (Croplands plus Permanent Meadows and Pastures) occupied 4813 Mha, an increase of 4% (198 Mha) since the 1970s; an increase in the Croplands area has been primarily responsible for this increase. In face of reduced availability of land for agriculture, worldwide investment in research and development was directed to intensify production and increase efficiency of resource uses using an array of advanced technologies and systems (Tripathi et al. 2017). With the ambition of limiting climate change to below 2 °C or even 1.5 °C land-based mitigation options have moved into the limelight of discussions (IPCC 2018; Minx et al. 2018; Smith et al. 2019). Land-based mitigation can deliver about 30% of global mitigation between 2020 and 2050 (Griscom et al. 2017; IPCC 2019; Roe et al. 2019). Knowledge and technologies generated from conservation agriculture, agroecology, and regenerative agriculture are growing fast and are promoted in both developed and developing countries in order to bridge yield gaps, mitigate climate change, bring resilience and broad-based productivity, and seeking more sustainable food systems (Garnett et al. 2013; Giller et al. 2015, 2017; Kirkegaard et al. 2014). However, linking technical, economic, and political feasibility is still not sufficiently addressed and has not yet achieved a substantial deviation in climate change trends. In other terms, substantial gaps still exist between the upscaling and diffusion of such systems and barriers to their implementation should be rooted in the decision-making process and incentivized by policies (Wezel et al. 2014). International organizations and research institutions have harnessed this momentum and governments must respond to these demands. Over the next 30 years, food systems will face an exceptional convergence of challenges and emerging tensions. In other terms, stringent and substantial technological advancement is an obligatory condition for agricultural development worldwide because unabated climate warming associated with population growth and changing dietary patterns will exacerbate many of the food insecurity issues and negate any pathway for future improvement. In addition, the availability and productivity of land, water, and energy will be altered. Hence, it is critically important that farmers get access to technologies that provide sustainable and healthy food and feed, secure water availability, reduce energy dependence, mitigate advanced soil degradation, and protect biodiversity. On the Americas and Australia's large farms, climate-smart agriculture employs modern science, especially advanced high-yielding genotypes, digital, precision, and highly sophisticated machines, state-of-the-art agrochemicals, and biotechnology. In fact, Giller et al. (2017) stated that while agroecology is fundamentally about harnessing biology and biological processes, which can mitigate climate change impact, then it could be argued that genetic engineering is one of the most powerful (potential) tools of agroecology. *Small farmers in* less developed countries should also explore technological advances in order to acquire sustainable intensification options (De Pinto and Ulimwengu 2017; Rosegrant et al. 2014).

4.2 Role of Biotechnology for Breeding Climate Resilient Varieties of Field Crops

Classical breeding including introduction, mass selection, pure line selection, recurrent selection, hybrid breeding, mutation breeding, and recombination breeding (Kumar et al. 2020) have been playing an essential role in improving crops varieties not just for high production but also for biotic stresses resistance (diseases and pests), and abiotic stresses tolerance (drought, low, and high temperatures) (Kumar et al. 2020), through phenotypic selections of the important agronomical traits (Wortmann et al. 1998). There are many factors that determine the method of breeding such as the crop, the aim of the breeding program, the volume of the breeding program, germplasm availability, and knowing of the trait and its correlation with other traits under the selection (Kumar et al. 2020). In general, breeding cycle takes from 7 to 10 years (Kumar et al. 2020), and so it is not an easy process, this is especially true when enhancing crop varieties for environmental stresses, because stresses effects on crops are multiple and mutable, particularly since crops are exposed to several stresses at the same time (Zhao et al. 2017).

However, by 2050, the percentage of the population will rise significantly to more than nine billion (Tester and Langridge 2010), and with the current climate change which directly affects all forms of life and intensifies current problems, for instance, floods and droughts, cold and heat stresses, extreme weather events, pests and pathogens, and incidence of weeds, besides to the emergence of many new challenges depending upon geographical areas (Irigoyen et al. 2014; Shabani et al. 2014; Shiferaw et al. 2011; Wassman et al. 2009), which strongly impact on crop productivity and major cropping systems (Ackerly et al. 2010). So, it requires well knowledge of physiological, genetic, and molecular mechanisms, new approaches, and policies with the new technologies applications, and tools, so as to face the growing need for food and to face the climate change challenges (Taranto et al. 2018). Classical breeding needs to be integrated with biotechnological methodologies including **genetic engineering** and **genome editing** and **molecular breeding** to produce climate-resilient crops (Munawar et al. 2020).

4.2.1 Genetic Engineering

Genetic engineering, also called (transgenic), is the approach of transferring desirable genes from any source directly to plants so as to produce desirable traits for instance resistance to biotic (pathogens and pests) stresses, tolerance to abiotic (drought, salinity, low and high temperatures) stresses, herbicide tolerance, and enhanced nutrition in plants (ISAAA 2018; Kumar et al. 2020; Newman and Newman 2008). Since 1980, *Agrobacterium* has been used to develop transgenic plants for plant breeding (Bevan et al. 1983). Genetic engineering techniques have the potential to solve unsolved issues, for example scientists were able to give the solution of vitamin A lack in the poor of developing countries, by genetically engineered the pro-vitamin A pathway into the rice endosperm (Potrykus 2001). Flavr Savr was the first transgenic tomato cultivar that was approved for commercial cultivation in 1994 (Singh and Singh 2015). Several transgenic barley cultivars have been released with favourable traits, for example (drought, low, and high temperatures) tolerance (Ramireddy et al. 2018; Soltész et al. 2013), (Stem rust, Leaf rust) resistance (Golegaonkar et al. 2009; Horvath et al. 2003), besides high production.

In the same context, by using genetic engineering, the scientists were able to isolate the gene coding for toxic protein called Bt gene from *Bacillus Thurenginesis* bacterium and engineered into tomato, tobacco, cotton, maize, soybean, sugarcane plants, to give insect-resistant genetically engineered (IRGE) (ISAAA 2018). Cultivation of (IRGE) crops has limited the spread of some crop worms such as

Pod borer (*Helicoverpa armigera*) in chickpea (Eapen et al. 1987), bollworm (*Helicoverpa armigera*) in cotton (Wu et al. 2008), and earworm (*Helicoverpa zea*) in corn (Dively et al. 2018), this resulted in a reduction in the use of chemical insecticide (Brookes and Barfoot 2018; Li et al. 2020, NASEM (National Academies of Sciences, Engineering, and Medicine) 2016).

4.2.2 Genome Editing

Genome editing (GE) tools are recorded among the biggest set of the new plant breeding techniques (NPBT) (HLG-SAM 2017). GE technologies which including clustered regularly interspaced short palindromic repeats-associated CAS9 protein (CRISPR). transcription activator-like effector nucleases (TALEN) (Munawar et al. 2020; Varshney et al. 2018), opening a new horizon for specific changes in cultivation crops for defence against biotic (diseases, pests), abiotic (drought, heat) stresses, accelerating crop breeding process, and finally increases crop production with low production cost (Courtier-Orgogozo et al. 2017). These technologies enable changes in targeted DNA sequences by removing, replacing, or adding of one or more bases (Zhang et al. 2019), which allow producing novel favourable alleles to accelerate the improvement of producing novel varieties this means the ability to release novel varieties in a one-generation without the need to select for desirable groups of alleles from large populations (Scheben and Edward 2017; Varshney et al. 2018) also expanding the genepool of favourable alleles (Varshney et al. 2018). CRISPR-Cas9 method is reported to be the most widely used system for GE in model crops (Wang et al. 2014). It was utilized to edit *Lotus japonicus* genes which is related to symbiotic nitrogen fixation (Wang et al. 2016). Also, it was applied successfully by Nekrasov et al. to improve resistance against powdery mildew in tomato plants by targeting the (*SlMLO*1) gene (Zheng et al. 2016). Moreover, it has also been used for enhancing several important traits in bread wheat (*Triticum aestivum*) such as (plant high, kernel length and weight, low-gluten wheat, and Male fertility) by targeting (*TaGASR7, TaGW2, TaDEP1*, α-gliadin, and *TaMs45*) genes, respectively (Sánchez-León et al. 2018; Singh et al. 2018; Zhang et al. 2016), and to improve the resistance against (powdery mildew) by targeting (*TaEDR1*) gene (Zong et al. 2017). TALEN was first utilized on rice crop to improve resistance against major forms of bacterial blight disease by targeting (*OsSWEET14*) gene, which is responsible for the susceptibility to bacterial blight disease in rice (Li et al. 2012). Also, by applying TALEN's technique, it was possible to give resistance against powdery mildew disease in the hexaploid wheat (*Triticum aestivum* L.), by editing three Homoalleles of the susceptible (*MLO*) gene (Wang et al. 2014). On the contrary, the progress in the applications of (GE) technologies associated to abiotic stresses remains delimited to Arabidopsis species, because abiotic stresses are usually controlled by multiple genetic mechanisms, which necessitate synchronous tuning of several genes. Lately, some good results have been recorded in soybean and coco against (drought, salinity) tolerance (Cardi et al. 2017; Curtin et al. 2018).

4.2.3 Molecular Breeding

Approaches that target specific regions in plant's DNA to optimize them for specified aims are identified as molecular breeding (Acquaah 2012). Also called as marker-assisted breeding (Lodhi et al. 2020), including many markers such as restriction fragment length polymorphisms (RFLPs), by Botstein et al. (1980). Simple sequence repeats (SSRs), by Litt and Luty (1989) and by Edwards et al. (1991). Amplified fragment length polymorphisms (AFLPs), by Keygene (Vos et al. 1995). Diversity arrays technology (DArT), by Jaccoud et al. (2001), these tools help in the understanding of genomic structure, to locate desirable genes, and to make the breeding process easier (Acquaah 2012; Jiang 2013). Several markers, such as DArT and SSR, are used in barleys to improve some morphological traits (Islamovic et al. 2013), also in chickpea to improve abiotic stresses tolerance (drought, salinity, high, and low temperature) (Jha et al. 2019; Mugabe et al. 2019; Rehman et al. 2011; Sivasakthi et al. 2018; Vadez et al. 2012). Marker-assisted selection (MAS) means selecting and removing of undesirable plants in early phase of the breeding process (Xu and Crouch 2008). Leading to accelerate gain from selection by focusing on fewer promising plants (Collard and Mackill 2008). MAS tools have been successfully utilized to solve the difficulties related with selections based on phenotype (Xu and Crouch 2008), raise the selection gain (Brumlop and Finckh 2011), shorten the time required to develop varieties (Kellya et al. 2003), and provided the possibility to precisely group target alleles into a single genotype (Kumar et al. 2020). By using MAS techniques, it was possible to give resistance against phomopsis stem blight and anthracnose in lupin, and give resistance against fusarium wilt in chickpea (Mannur et al. 2019). Genetic linkage map which including quantitative trait locus (QTL), mapping, and genome-wide association study (GWAS) are valuable approaches which can be used for identifying (QTL) (Zhang et al. 2018). QTL can be controlled by few or many loci with large or minor effects on the phenotype. Anyway, the transfer of QTLs and the analysis of their genetic basis are possible through crossbreeding programs, by making a cross leading to the development of mapping populations (Xu 2010).

So, any genetic studies have to use one or more genetic populations as a plant material, for example recombinant inbred lines (RIL), backcross (BC), nested association mapping (NAM), which is resulted from a multiple-cross sharing one common parent, and these genetic populations can be categorized according to their genetic structure and level of segregation:

a) Temporary populations (or Sergeant) such as F2 generation or BC, in this population the plants or lines will continue in segregating after self-pollination till they become homozygous.
b) Permanent populations: the lines in this kind of population are homozygous so they will not segregate and will not change even through self-pollination for example RIL and Double Haploid (Xu 2010).

Many studies involving mapping of QTL and GWAS techniques have been reported for a lot of crops to identify QTLs for many of important agronomical traits like grain yield and phenology traits in durum wheat (Soriano et al. 2017), quality in cotton crop (Fan et al. 2018), and under drought stress for wheat (Gupta et al. 2017; Zaïm et al. 2020), heat stress in durum wheat (Sukumaran et al. 2018), and for biotic stress like rust disease in wheat (Kthiri et al. 2019; Limbalkar et al. 2019).

Shi et al. (2017) used RIL population in bread wheat, using 90 k wheat iSelect SNP (single nucleotide polymorphism) array to detect QTLs for number of Kernel per spike trait under five environments, by GWAS analysis of data they identified 11 QTLs for number of kernel per spike trait on 2A, 3A, 4A, 1B, 2B, 3B, 4B, 6B, 7B, and 2D chromosomes, three of them were identified in multiple environments. Börner et al. (2002) and Kirigwi et al. (2007) were able to identify QTLs on 4A and 7D chromosomes in hexaploid wheat which controlling number of kernel per spike trait, by using RIL populations. In the same way (Huang et al. 2004; Narasimhamoorthy et al. 2006), utilized BC populations and identified QTLs in 2A, 6A, 7A, 3B, 1D, 3D, and 7D chromosomes for the same trait (Liu et al. 2020). Utilizing doubled haploid population in bread wheat to detect QTLs for thousand kernel weight, using the wheat 55-k (SNP) genotyping assay, 9 QTLs were identified 1A, 5A, 6A 4B, 2D, 4D, 5D, and 6D chromosomes, the QTLs *qTgw.nwipb-4DS* and *qTgw.nwipb-6AL* were stable under different environments 2015–2017, with explaining a phenotyping variance ranging from 15.31 to 32.43% and 21.34 to 29.46%, and mapped within genetic distances of 2.609 and 5.256 cM separately, so they can be used as a markers in (MAS) for high yielding in wheat.

Wang et al. (2011) used 27 F2:3 populations, with 179 polymorphic SSR and EST-SSR markers, they identified a 76 QTLs for most of the important traits, which are located on 1A, 5A, 4A, 7A, 3B, 4B, 5B, 7B, and 2D. The use of marker system in order to detect QTL related to the important traits under abiotic stress (drought, salinity, high, and low temperature) has been informed in several researches (Hamwieh et al. 2013; Mwadzingeni et al. 2017; Paul et al. 2018; Roselló et al. 2019; Vadez et al. 2012). Zaïm et al. (2020) used four RILs durum wheat populations to determine a QTL under drought conditions for grain yield and 1000 kernel weight traits, with using 3202 high-confidence polymorphic markers, the genotypes were genotyped by sequencing, Six QTLs were detected to be related with grain yield on 4A, 7A, 5B, 2B, and 7B chromosomes and explained 4.3 to 13.4% of the observed phenotypic variance.

4.3 Conclusion

Climate change, overexploitation and depletion of natural resources, collateral damage from agricultural industrialization, and rampant overpopulation all have an effect on the quality of agricultural production and jeopardize food self-sufficiency. In order to improve the performance of crops against environmental stress, the role of biotechnology in agriculture that is resilient to climate is very important: activities accumulating positive alleles contributing to stress tolerance.

Eapen, S., Köhler, F., Gerdemann, M., and Schieder, O. (1987). Cultivar dependence of transformation rates in moth bean after co-cultivation of protoplasts with *Agrobacterium tumefaciens*. *Theoretical and Applied Genetics* 75: 207–210.

Edwards, A., Civitello, A., Hammond, H.A., and Caskey, C.T. (1991). DNA typing and genetic-mapping with trimeric and tetrameric tandem repeats. *American Journal of Human Genetics* 49: 746–756.

Fan, L., Wang, L., Wang, X. et al. (2018). A high-density genetic map of extra-long staple cotton (*Gossypium barbadense*) constructed using genotyping-bysequencing based single nucleotide polymorphic markers and identification of fiber traits-related QTL in a recombinant inbred line population. *BMC Genomics* 19: 489.

FAO (2015). Climate change and food systems: global assessments and implications for food security and trade. Food Agriculture Organization of the United Nations (FAO).

Frieler, K., Schauberger, B., Arneth, A. et al. (2017). Understanding the weather signal in national crop-yield variability. *Earth's Future* 5: 605–616. https://doi.org/10.1002/2016EF000525.

Garnett, T., Appleby, M.C., Balmford, A. et al. (2013). Sustainable intensification in agriculture: premises and policies. *Science* 341: 33–34.

Giller, K.E., Andersson, J.A., Corbeels, M. et al. (2015). Beyond conservation agriculture. *Frontiers in Plant Science* 6: 870.

Giller, K.E., Andersson, J.A., Sumberg, J., and Thompson, J. (2017). A golden age for agronomy? In: *Agronomy for Development* (ed. J. Sumberg), 150–160. London: Earthscan.

Golegaonkar, P.G., Karaoglu, H., and Park, R.F. (2009). Molecular mapping of leaf rust resistance gene Rph14 in *Hordeum vulgare*. *Theoretical and Applied Genetics* 119: 1281–1288.

Griscom, B.W., Adams, J., and Ell, P.W. (2017). Natural climate solutions. *Proceedings of the National Academy of Sciences of the United States of America* 114: 11645–11650.

Gupta, P.K., Balyan, H.S., and Gahlaut, V. (2017). QTL analysis for drought tolerance in wheat: present status and future possibilities. *Agronomy* 7: 5.

Hamwieh, A., Imtiaz, M., and Malhotra, R.S. (2013). Multi-environment QTL analyses for drought-related traits in a recombinant inbred population of chickpea (*Cicer arietinum* L.). *Theoretical and Applied Genetics* 126: 1025–1038.

HLG-SAM (2017). New Techniques in Agricultural Biotechnology, Explanatory note 02. Available online: https://ec.europa.eu/research/sam/pdf/topics/explanatory_note_new_techniques_agricultural_biotechnology.pdf#view=fit&pagemode=none (accessed 23 November 2018).

Hochman, Z., Gobbett, D.L., and Horan, H. (2017). Climate trends account for stalled wheat yields in Australia since 1990. *Global Change Biology* 23: 2071–2081.

Horvath, H., Rostoks, N., Brueggeman, R. et al. (2003). Genetically engineered stem rust resistance in barley using the Rpg1 gene. *Proceedings of the National Academy of Sciences* 100: 364–369.

Huang, X.Q., Kempf, H., Canal, M.W., and Röder, M.S. (2004). Advanced backcross QTL analysis in progenies derived from a cross between a German elite winter wheat variety and a synthetic wheat (*Triticum aestivum* L.). *Theoretical and Applied Genetics* 109: 933–943. https://doi.org/10.1007/s00122-004-1708-7.

Innes, P.J., Tan, D.K.Y., Van Ogtrop, F., and Amthor, J.S. (2015). Effects of high-temperature 6 episodes on wheat yields in New South Wales, Australia. *Agricultural and Forest Meteorology* 208: 95–107.

IPCC (2018). Global warming of 1.5°C. Summary for Policy Makers. World Meteorological Organization, United Nations Environment Program, and Intergovernmental Panel on Climate Change, Switzerland.

IPCC (2019). Special Report on Climate Change, Desertification, Land Degradation, Sustainable Land Management, Food Security, and Greenhouse Gas Fluxes in Terrestrial Ecosystems. U.N. Environmental Programme. https://ar5-syr.ipcc.ch/ipcc/ipcc/resources/pdf/IPCC_SynthesisReport.pdf.

Iizumi, T. et al. (2018). Crop production losses associated with anthropogenic climate change for 1981–2010 compared with preindustrial levels. *International Journal of Climatology*, 38(14): 5405–5417.

Irigoyen, J.J., Goicoechea, N., Antolín, M.C. et al. (2014). Growth, photosynthetic acclimation and yield quality in legumes under climate change simulations: an updated survey. *Plant Science* 226: 22–29. https://doi.org/10.1016/j.plantsci.2014.05.008.

ISAAA (2018). *Brief 54: Global Status of Commercialized Biotech/GM Crops: 2018*. ISAAA Briefs 54. Ithaca, NY: ISAAA.

Islamovic, E., Obert, D.E., Oliver, R.E. et al. (2013). A new genetic linkage map of barley (*Hordeum vulgare* L.) facilitates genetic dissection of height and spike length and angle. *Field Crops Research* 154: 91–99.

Jaccoud, D., Peng, K., Feinstein, D., and Kilian, A. (2001). Diversity arrays: a solid state technology for sequence information independent genotyping. *Nucleic Acids Research* 29: E25.

Jha, B., and Tripathi, A. (2017). How susceptible Is India's food basket to climate change?. *Social Change*, 47(1): 11–27.

Jha, U.C., Kole, P.C., and Singh, N.P. (2019). QTL mapping for heat stress tolerance in chickpea (*Cicer arietinum* L.). *Legume Research* https://doi.org/10.18805/LR-4121.

Jiang, G.L. (2013). Molecular markers and marker-assisted breeding in plants. In: *Plant Breeding from Laboratories to Fields* (ed. S.B. Andersen), 45–83. Rijeka: InTech.

Kellya, J.D., Gepts, P., Miklas, P.N., and Coyne, D.P. (2003). Tagging and mapping of genes and QTL and molecular marker-assisted selection for traits of economic importance in bean and cowpea. *Field Crops Research* 82: 135–154.

Kirigwi, F.M., Ginkel, M.V., Brown-Guedira, G. et al. (2007). Markers associated with a QTL for grain yield in wheat under drought. *Molecular Breeding* 20: 401–413. https://doi.org/10.1007/s11032-007-9100-3.

Kirkegaard, J.A., Conyers, M.K., Hunt, J.R. et al. (2014). Sense and nonsense in conservation agriculture: principles, pragmatism and productivity in Australian mixed farming systems. *Agriculture Ecosystems & Environment* 187: 133–145.

Kthiri, D., Loladze, A., N'Diaye, A. et al. (2019). Mapping of genetic loci conferring resistance to leaf rust from three globally resistant durum wheat sources. *Frontiers in Plant Science* 10: 1247.

Kumar, S., Gupta, P., Choukri, H., and Siddique, K.H.M. (2020). Efficient breeding of pulse crops. In: *Accelerated Plant Breeding*, vol. 3 (ed. S.S. Gosal and S.H. Wani), 1–30. Springer Nature Switzerland AG.

Lesk, C., Rowhani, P., and Ramankutty, N. (2016). Influence of extreme weather disasters on global crop production. *Nature* 529: 84–87.

Li, T., Liu, B., Spalding, M.H. et al. (2012). High-efficiency TALEN-based gene editing produces disease-resistant rice. *Nature Biotechnology* 30: 390.

Li, Y., Hallerman, E.M., Wu, K., and Peng, Y. (2020). Insect-resistant genetically engineered crops in China: development, application, and prospects for use. *Annual Review of Entomology* 65: 273–292.

Limbalkar, O.M., Sharma, J.B., Jha, S.K. et al. (2019). Mapping of genes for leaf and stem rust resistance in bread wheat genotype selection 212. *Indian Journal of Genetics and Plant Breeding* 79: 1–8.

Lin, M. and Huybers, P. (2012). Reckoning wheat yield trends. *Environmental Research Letters* 7: 024016.

Litt, M. and Luty, J.A. (1989). A hypervariable microsatellite revealed by in vitro amplification of a dinucleotide repeat within the cardiac-muscle actin gene. *American Journal of Human Genetics* 44: 397–301.

Liu, T., Wu, L., Gan, X. et al. (2020). Mapping quantitative trait loci for 1000-grain weight in a double haploid population of common wheat. *International Journal of Molecular Sciences* 21: 3960.

Lobell, D.B., Schlenker, W., and Costa-Roberts, J. (2011). Climate trends and global crop production since 1980. *Science* 333: 616–620.

Lodhi, S.S., Maryam, S.H., Rafique, K.H. et al. (2020). Overview of the prospective strategies for conservation of genomic diversity in wheat landraces. In: *Climate Change and Food Security with Emphasis on Wheat* (ed. M. Ozturk and A. Gul), 293–309. Academic Press Elsevier.

Mannur, D.M., Babbar, A., Thudi, M. et al. (2019). Super Annigeri 1 and improved JG 74: two Fusarium wilt-resistant introgression lines developed using marker-assisted backcrossing approach in chickpea (*Cicer arietinum* L.). *Molecular Breeding* 39: 2. https://doi.org/10.1007/s11032-018-0908-9.

Matiu, M., Ankerst, D.P., and Menzel, A. (2017). Interactions between temperature and drought in global and regional crop yield variability during 1961–2014. *PLoS One* 12: e0178339.

Minx, J.C., Lamb, W.F., Callaghan, M.W. et al. (2018). Negative emissions - Part 1: Research landscape and synthesis. *Environmental Research Letters* 13: https://doi.org/10.1088/1748-9326/aabf9b.

Moore, F.C. and Lobell, D.B. (2015). The fingerprint of climate trends on European crop yields. *Proceedings of the National Academy of Sciences* 112: 2670–2675; 201409606.

Mugabe, D., Coyne, C.J., Piaskowski, J. et al. (2019). Quantitative trait loci for cold tolerance in chickpea. *Crop Science* 59: 573–582.

Munawar, S., ul Qamar, M.T., Mustafa, G. et al. (2020). Role of biotechnology in climate resilient agriculture. In: *Environment, Climate, Plant and Vegetation Growth* (ed. S. Fahad et al.). Springer Nature Switzerland AG https://doi.org/10.1007/s11032-018-0908-9.

Mwadzingeni, L., Shimelis, H., Jasper, D. et al. (2017). Genomewide association analysis of agronomic traits in wheat under drought-stressed and non-stressed conditions. *PLoS One* 12: e0171692.

Narasimhamoorthy, B., Gill, B.S., Fritz, A.K. et al. (2006). Advanced backcross QTL analysis of a hard winter wheat × synthetic wheat population. *Theoretical and Applied Genetics* 112: 787–796.

NASEM (National Academies of Sciences, Engineering, and Medicine) (2016). *Genetically Engineered Crops: Experiences and Prospects*. Washington, DC: The National Academies Press.

Newman, R.K. and Newman, C.W. (2008). *Barley for Food and Health: Science, Technology, and Products*. Hoboken, NJ: Wiley.

Paul, P.J., Samineni, S., Thudi, M. et al. (2018). Molecular mapping of QTLs for heat tolerance in chickpea. *International Journal of Molecular Sciences* 19: 2166.

Porter, J.R., Xie, L., Challinor, A.J. et al. (2014). Food security and food production systems. In: *Climate Change 2014: Impacts, Adaptation, and Vulnerability Part A: Global and Sectoral Aspects Contribution of Working Group II to the Fifth Assessment Report of the Intergovernmental Panel on Climate Change* (ed. V.R. Field Barros, D.J. Dokken, K.J. Mach, et al.), 485–433. Cambridge, UK/New York, NY, USA: Cambridge University Press.

Potrykus, I. (2001). Golden rice and beyond. *Plant Physiology* 125: 1157–1161.

Pradhan, P., Fischer, G., van Velthuizen, H. et al. (2015). Closing yield gaps: how sustainable can we be? *PLoS One* 10: e0129487.

Ramireddy, E., Hosseini, S.A., Eggert, K. et al. (2018). Root engineering in barley: increasing cytokinin degradation produces a larger root system, mineral enrichment in the shoot and improved drought tolerance. *Plant Physiology* 177: 1078–1095.

Ray, D., Ramankutty, N., Muellerm, N. et al. (2012). Recent patterns of crop yield growth and stagnation. *Nature Communications* 3: 7.

Ray, D.K., Gerber, J.S., MacDonald, G.K., and West, P.C. (2015). Climate variation explains a third of global crop yield variability. *Nature Communications* 6: 5989. https://doi.org/10.1038/ncomms6989.

Ray, D.K., Mueller, N.D., West, P.C., and Foley, J.A. (2013). Yield trends are insufficient to double global crop production by 2050. *PLoS One* 8 (6): e66428. https://doi.org/10.1371/journal.pone.0066428. PMID: 23840465.

Rehman, A.U., Malhotra, R.S., Bett, K. et al. (2011). Mapping QTL associated with traits affecting grain yield in chickpea (*Cicer arietinum* L.) under terminal drought stress. *Crop Science* 51: 450–463.

Roe, S., Streck, C., Obersteiner, O. et al. (2019). Contribution of the land sector to a 1.5°C 34 World. *Nature Climate Change* 9: 817–828.

Rosegrant, M.W., Koo, J., Cenacchi, N. et al. (2014). *Food Security in a World of Natural Resource Scarcity: The Role of Agricultural Technologies*. Washington, DC: International Food Policy Research Institute.

Roselló, M., Royo, C., Sanchez-Garcia, M., and Soriano, J.M. (2019). Genetic dissection of the seminal root system architecture in Mediterranean durum wheat landraces by genome-wide association study. *Agronomy* 9: 364.

Sakurai, G., Lizumi, T., and Yokozawa, M. (2012). Varying temporal and spatial effects of climate on maize and soybean affect yield prediction. *Climate Research* 49: 143–154.

Sánchez-León, S., Gil-Humanes, J., Ozuna, C.V. et al. (2018). Low-gluten, nontransgenic wheat engineered with CRISPR/Cas9. *Plant Biotechnology Journal* 16: 902–910.

Scheben, A. and Edward, D. (2017). Genome editors take on crops. *Science* 355: 1122–1123.

Shabani, F., Kumar, L., and Esaeili, A. (2014). Future distributions of *Fusarium oxysporum* f. spp. in European, middle eastern and north African agricultural regions under climate change. *Agriculture Ecosystems & Environment* 197: 96–05.

Shi, W., Hao, C., Zhang, Y. et al. (2017). A combined association mapping and linkage analysis of Kernel Number per spike in common wheat (*Triticum aestivum* L.). *Frontiers in Plant Science* 8: 1412.

Shiferaw, B., Prasanna, B.M., Hellin, J., and Bänziger, M. (2011). Crops that feed the world. Past successes and future challenges to the role played by maize in global food security. *Food Security* 3: 307–327.

Singh, B.D. and Singh, A.K. (2015). *Marker-Assisted Plant Breeding: Principles and Practices*. New Delhi: Springer.

Singh, M., Kumar, M., Albertsen, M.C. et al. (2018). Concurrent modifications in the three homeologs of Ms45 gene with CRISPR-Cas9 lead to rapid generation of male sterile bread wheat (*Triticum aestivum* L.). *Plant Molecular Biology* 97: 371–383.

Sivasakthi, K., Thudi, M., Tharanya, M. et al. (2018). Plant vigour QTLs co-map with an earlier reported QTL hotspot for drought tolerance while water saving QTLs map in other regions of the chickpea genome. *BMC Plant Biology* 18: 29.

Smith, P., Calvin, K., Nkem, J. et al. (2019). Which practices co-deliver food security, climate change mitigation and adaptation, and combat land-degradation and desertification? *Global Change Biology* https://doi.org/10.1111/gcb.14878.

Soltész, A., Smedley, M., Vashegyi, I. et al. (2013). Transgenic barley lines prove the involvement of TaCBF14 and TaCBF15 in the cold acclimation process and in frost tolerance. *Journal of Experimental Botany* 64: 1849–1862.

Soriano, J.M., Malosetti, M., Roselló, M. et al. (2017). Dissecting the old Mediterranean durum wheat genetic architecture for phenology.; biomass and yield formation by association mapping and QTL meta-analysis. *PLoS One* 12: e0178290.

Sulpanjani, A., Yang, M.S., Tawaha, A.R.M. et al. (2005). Effect of magnesium application on yield, mineral contents and active components of *Chrysanthemum coronarium* L. under hydroponics conditions. *Bioscience Research* 2: 73–79.

Supanjani, A., Tawaha, A.M., Yang, M.S., and Lee, K.D. (2005). Role of calcium in yield and medicinal quality of *Chrysanthemum coronarium* L. *Journal of Agronomy* 4: 188–192.

Sukumaran, S., Reynolds, M.P., and Sansaloni, C. (2018). Genome-wide association analyses identify QTL hotspots for yield and component traits in durum wheat grown under yield potential, drought, and heat stress environments. *Frontiers in Plant Science* 9: 81.

Tawaha, A.M. and Turk, M.A. (2002a). Lentil (*Lens culinaris* Medic.) productivity as influenced by rate and method of phosphate placement in a Mediterranean environment. *Acta Agronomica Hungarica* 50: 197–101.

Tawaha, A.M., Singh, V.P., and Turk, M.A. (2003). A review on growth, yield components and yield of barley as influenced by genotypes, herbicides and fertilizer application. *Research on Crop* 4: 1–9.

Tawaha, A.M. and Turk, M.A. (2004). Field pea seeding management for semi-arid Mediterranean conditions. *Journal of Agronomy and Crop Science* 190: 86–92.

Taranto, F., Nicolia, A., Pavan, S. et al. (2018). Biotechnological and digital revolution for climate-smart plant breeding. *Agronomy* 8: 277.

Tester, M. and Langridge, P. (2010). Breeding technologies to increase crop production in a changing world. *Science* 327: 818–822.

Tripathi, V., Edrisi, S.A., Chen, B. et al. (2017). Bio-technological advances for restoring degraded land for sustainable development. *Trends in Biotechnology* 35: 847–859.

Turk, M.A. and Tawaha, A.M. (2001). Common vetch (*Vicia sativa* L.) productivity as influenced by rate and method of phosphate fertilization in a Mediterranean environment. *Agricultura Mediterranea* 131: 108–111.

Turk, M.A. and Tawaha, A.M. (2002a). Onion (*Allium cepa* L.) as influenced by rate and method of phosphorus placement. *Crop Research* 23: 105–107.

Turk, M.A. and Tawaha, A.M. (2002b). Impact of seeding rate, seeding date and method of phosphorous application in faba bean (*Vicia faba* L. minor) in the absence of moisture stress. *Biotechnology, Agronomy, Society and Environment* 6: 171–178.

Turk, M.A., Tawaha, A.M., and Shatnawi, M. (2003a). Lentil (*Lens culinaris* Medik) response to plant density, sowing date, phosphorus fertilization and Ethephon application in the absence of moisture stress. *Journal of Agronomy and Crop Science* 189: 1–6.

Turk, M.A., Tawaha, A.M., and Samara, N. (2003b). Effects of seeding rate and date, and phosphorus application on growth and yield of Narbon vetch (*Vicia narbonensis*). *Agronomie* 23: 1–4.

Turk, M.A., Shatnawi, M.K., and Tawaha, A.M. (2003c). Inhibitory effects of aqueous extracts of black mustard on germination and growth of Alfalfa. *Weed Biology and Management* 3: 37–40.

Turk, M.A., Tawaha, A.M., and Lee, K.D. (2004). Seed germination and seedling growth of three lentil cultivars under moisture stress. *Asian Journal of Plant Sciences* 3: 394–397.

Vadez, V., Krishnamurthy, L., Thudi, M. et al. (2012). Assessment of ICCV 2 9 JG 62 chickpea progenies shows sensitivity of reproduction to salt stress and reveals QTL for seed yield and yield components. *Molecular Breeding* 30: 9–21.

Varshney, R.K., Thudi, M., Pandey, M.K. et al. (2018). Accelerating genetic gains in legumes for the development of prosperous smallholder agriculture: integrating genomics, phenotyping, systems modelling and agronomy. *Journal of Experimental Botany* 69: 3293–3212.

Vos, P., Hogers, R., Bleeker, M. et al. (1995). AFLP – a new technique for DNA-fingerprinting. *Nucleic Acids Research* 23: 4407–4414.

Wang, H., La Russa, M., and Qi, L.S. (2016). CRISPR/Cas9 in genome editing and beyond. *Annual Review of Biochemistry* 85: 227–264.

Wang, J., Liu, W., Wang, H. et al. (2011). QTL mapping of yield-related traits in the wheat germplasm 3228. *Euphytica* 177: 277–292.

Wang, Y., Cheng, X., Shan, Q. et al. (2014). Simultaneous editing of three homoeoalleles in hexaploid bread wheat confers heritable resistance to powdery mildew. *Nature Biotechnology* 32: 947–951.

Wassman, R., Jagadish, S.V.K., Sumfleth, K. et al. (2009). Regional vulnerability of climate change impacts on Asian rice production and scope for adaptation. *Advances in Agronomy* 102: 91–133.

Wezel, A., Casagrande, M., Colette, F. et al. (2014). Agroecological practices for sustainable agriculture. A review. *Agronomy for Sustainable Development* 34: 1–20.

Wortmann, C.S., Kiluby, R.A., Eledu, C.A., and Arron, D.J. (1998). *Atlas of Common Bean (Phaseolus vulgaris L.) Production in Africa*, 133. Cali, CO: CIAT.

Wu, K.M., Lu, Y.H., Feng, H.Q. et al. (2008). Suppression of cotton bollworm in multiple crops in China in areas with Bt toxin–containing cotton. *Science* 321: 1676–1678.

Xu, Y. (2010). *Molecular Plant Breeding*. Wallingford: CAB International.

Xu, Y. and Crouch, J.H. (2008). Marker-assisted selection in plant breeding: from publications to practice. *Crop Science* 48: 391–307.

Zaïm, M., Kabbaj, H., Kehel, Z.A. et al. (2020). Combining QTL analysis and genomic predictions for four durum wheat populations under drought conditions. *Frontiers in Genetics* 11: 316.

Zhang, J., Long, Y., Wang, L. et al. (2018). Consensus genetic linkage map construction and QTL mapping for plant height-related traits in linseed flax (*Linumusitatissimum* L.). *BMC Plant Biology* 18: 160. https://doi.org/10.1186/s12870-018-1366-6.

Zhang, Y., Liang, Z., Zong, Y. et al. (2016). Efficient and transgene-free genome editing in wheat through transient expression of CRISPR/Cas9 DNA or RNA. *Nature Communications* 7: 12617. https://www.nature.com/articles/ncomms12617.

Zhang, Y., Malzahn, A.A., Sretenovic, S., and Qi, Y. (2019). The emerging and uncultivated potential of CRISPR technology in plant science. *Nature Plants* 5: 778–794.

Zhao, C., Liu, B., Piao, S. et al. (2017). Temperature increase reduces global yields of major crops in four independent estimates. *Proceedings of the National Academy of Sciences of the United States of America* 114: 9326–9331.

Zheng, Z., Appiano, M., Pavan, S. et al. (2016). Genome-wide study of the tomato SlMLO gene family and its functional characterization in response to the powdery mildew fungus Oidium neolycopersici. *Frontiers in Plant Science* 7: 380. https://doi.org/10.3389/fpls.2016.00380.

Zong, Y., Wang, Y., Li, C. et al. (2017). Precise base editing in rice, wheat and maize with a Cas9-cytidine deaminase fusion. *Nature Nanotechnology* 35: 438–440.

5

Breeding Crops for Tolerance to Salinity, Heat, and Drought

Abdel Rahman Mohammad Said Al-Tawaha[1], *Nidal Odat*[2], *Noureddine Benkeblia*[3], *Naila Kerkoub*[4], *Zineb Labidi*[4], *Mahyeddine Boumendjel*[4], *Hichem Nasri*[4], *Imran*[5,6], *Amanullah*[5], *Shah Khalid*[5], *Abdel Razzaq Al-Tawaha*[7], *Mina Bayanati*[8], *Hiba Alatrash*[9], *Abhijit Dey*[10], *Devarajan Thangadurai*[11], *Jeyabalan Sangeetha*[12], *and Saher Islam*[13]

[1] *Department of Biological Sciences, Al-Hussein Bin Talal University, Ma'an, Jordan*
[2] *Department of Medical Laboratories, Faculty of Science, Al-Balqa Applied University, Al Salt, Jordan*
[3] *Department of Life Sciences, The University of the West Indies, Kingston, Jamaica*
[4] *Laboratory of Biodiversity and Ecosystems Pollution, Faculty of Life and Nature Sciences, Chadli Bendjedid University, El-Tarf, Algeria*
[5] *Department of Agronomy, The University of Agriculture, Peshawar, Pakistan*
[6] *Ministry of Agriculture, Agriculture Extension Wing, Merged Area, Khyber Pakhtunkhwa, Pakistan*
[7] *Department of Crop Science, Faculty of Agriculture, Universiti Putra Malaysia, Serdang, Selangor, Malaysia*
[8] *Department of Horticultural Science, Ferdowsi University of Mashhad, Mashhad, Iran*
[9] *Department of Field Crops, General Commission for Scientific Agricultural Research, Aleppo, Syria*
[10] *Department of Life Sciences, Presidency University, Kolkata, West Bengal, India*
[11] *Department of Botany, Karnatak University, Dharwad, Karnataka, India*
[12] *Department of Environmental Science, Central University of Kerala, Kasaragod, Kerala, India*
[13] *Department of Biochemistry and Biotechnology, Institute of Biochemistry and Biotechnology, University of Veterinary and Animal Sciences, Lahore, Pakistan*

CONTENTS

Climate Change and Agriculture: Perspectives, Sustainability and Resilience, First Edition. Edited by Noureddine Benkeblia.

5.1 Genetic Engineered Plants for Stress Tolerance

Plants in their natural and agricultural habitat encountered a plethora of abiotic stress such as drought, salinity, and mineral toxicity that limit their production and sustainability (Odat 2018a,b; Turk and Tawaha 2002a,b). Abiotic stress causes ionic toxicity, induces water stress and nutrient imbalance, which lead to adverse effects on plant growth, development, and establishment (Turk and Tawaha 2001; Tawaha and Turk 2002). Moreover, abiotic stress reduces the metabolic activities of important enzymes, decreases nutrient uptake, and leads to nutritional disorders which result in a reduction in productivity (Odat 2020). The development of abiotic tolerant crops offers a cost-effective management strategy against abiotic stress (Tawaha et al. 2003; Turk et al. 2003a,b). The prevailing genetic variety in germplasm could be used effectively for the development of abiotic stress tolerant genotypes (Odat et al. 2015). Conventional strategies for breeding crop with improved stress tolerance were hitherto faced with small achievements due to the difficulty of breeding for tolerance phenotypes in conventional breeding programs. Currently, there is an increasing interest in using the emerging technologies of genomics and genetic engineering as means to identify key genetic loci underlining stress tolerance and as a tool to select for allelic variation in the wild and germplasm gene pools.

Advances in molecular biology and genetic engineering and their utilization in agriculture have now been largely acknowledged and proven to be highly profitable for the development of tolerant crop plants (Tester and Langridge 2010). During the last 50 years, there has been a notable advance in the field of genetics and genetic engineering that enabled the scientific community to better understand how plants face and respond to abiotic stresses (Apse and Blumwald 2002). These developments in the tools and the methodology in genetic technology have been successful in attaining important information for ameliorate stress conditions and contribute to producing tolerant variants of crops (Nongpiur et al. 2016). It is now become possible to determine the location of genetic loci that are involved in salinity stress response, determine the molecular pathways, and ultimately use data for establishing salinity tolerance varieties of existing crops. The utilization of advanced technology in genetic engineering such as genome editing has largely improved our knowledge to generate stress-tolerant crops.

The traditional ways of genetic technology such as overexpression of genes which is involved in regulating abiotic stresses typically done with certain limitations (Cabello et al. 2014). The response of crop plant to abiotic stress are highly coordinated and involve multiple genes, polygenic, that are part of varied signalling, developmental, and biochemical pathways and this is further complicated by the high level of interconnection among these processes. Identifying genes involved in stress conditions and controlling their gene expression are enabled us to choose favourable phenotypes that are of variable agriculturally importance. For example, crop biomass can increase and the harmful effects can be eliminated by establishing gene variants of increased uptake of cell resources necessary for increased growth and establishment by the overexpressed target genes (Zhu 2001). Additionally, genetically produced genetic variants of crops involved a single change in gene expression of transgene which would target to positively regulate stress tolerance to a particular stress factor such as salinity or drought. However, in the field of crops, two or more abiotic stresses can occur either simultaneously or one after the other, thus tolerance to just one becomes limited (Mittler and Blumwald 2010). Therefore, to develop new transgenics that are suitable for field conditions scientists are required to invent approaches. For polygenic and pleiotropic effects, researchers currently use stress induce promotors as a viable strategy (Checker et al. 2012). Another approach involves ubiquitination in which scientists engineer protein post-translation modification process (Lyzenga and Stone 2012). By using this methodology, scientists showed that alternation of the protein during the period of post-translation by the expression of, for example, a monoubiquitin has contributed not only to an increase in the ability of plant to adapt to various abiotic stresses but also enhance crop biomass and establishment (Guo et al. 2008). Additionally, scientists are now capable to interfere with gene expression by altering and manipulating various RNA molecules namely MicroRNA and miRNA. The expression of miRNA target sequences is a further promising method for ameliorating the stress response of crop plants (Ebert et al. 2004).

Genome-editing has revolutionized agriculture as it provides a new avenue for generating crop plants tolerant to abiotic stress with traits such as improved resistance to salinity, drought, efficiency, and herbicide tolerance. Genome editing is the targeted mutagenesis; researchers introduce a given alteration at known sites in the genome. The technology is done by utilizing DNA cleavage chemicals and DNA repair pathways (Orellana et al. 2010). The chemicals are usually engineered endonucleases which digest target DNA at scientist specified sequences and these called double strand breaks are re-joined either by what is called non-homologous end joining (NHEJ) or homologous recombination (HR) (Carroll 2014). Custom targeted mutations of DNA, such as deletions and insertions, are obtained through the manipulation of these DNA repair mechanisms or through repair. Various genome-editing platforms such as the zinc finger nucleases (ZFNs), the transcription activator-like effector nucleases (TALENs), the meganucleases, and the CRISPR/Cas RNA-guided nucleases are among the most recent approaches in crop improvement (Curtin et al. 2012). ZFNs are chimeric or fusion protein which is produced by fusing the cleavage gene of the *Fok*I restriction enzyme with a customized gene or domain of Cys2-His2 Zinc finger (DNA binding). Since the DNA cleavage of *Fok*I require dimerization *Fok*I is necessary (Bitinaite et al. 1998). Therefore, a single ZFN molecule is established as such that it harbours a few nucleotide sequences of about a 5–6 bp spacer region and site-specific division at this spacer sequence which is obtained through heterodimerization of ZFNs. Each zinc finger region is consisting of 3–6 single fingers, and each finger is designed to target a 3 bp nucleotide sequence (Urnov et al. 2010). Thus, a ZFN heterodimer consists of about 2 DNA-binding regions containing 4 different fingers each, and each finger targets a 3 bp sequence that will identify a DNA region of about 24 bp which can exist only once in any organism's DNA (Urnov et al. 2010). This makes ZFNs highly site-specific and recognizes precisely the targeted DNA (Nongpiur et al. 2016). Similarly, TALENs are also great tools which consist of typically fusion polypeptides with specific binding properties that are commonly integrated into the *Fok*I restriction endonuclease. In this case, the DNA-binding region contains conserved amino acid repeats which are equivalent to transcription activator-like effectors (TALEs) extracted from bacterial plant pathogens, typically *Xanthomonas* spp. (Boch and Bonas 2010). When infection with this pathogen occurs, the TALEs are produced into the target cells and bind to the host genome, therefore, enhance the host gene activation. On the other hand, several restriction enzymes or endonucleases

A high saline concentration in the soil is first of all perceived by the plant as a significant decrease in water availability thereby imposing osmotic stress conditions. When the osmotic adjustment is insufficient, the water tends to leave the cells, causing a water deficit and consequently the loss of turgidity. On the other hand, the entry of salt into the plant generally causes an ionic imbalance, which is reflected in deficiencies or excess in certain elements. Depending on the ionic composition of the solute, ionic toxicity or nutritional deficiencies may occur either by the predominance of specific ions or the competitive effects between cations and anions (Karan and Subudhi 2012; Munns and James 2003; Parihar et al. 2015). All the above-mentioned events are reported to be among the key reasons for the impairing of plant growth and development under salt stress. In general, the effects of salinity on plant morphological traits are mainly manifested by the decrease in the growth of the vegetative apparatus, characterized by low branching, small organ diameters, reduced number of nodes, reduction in the number of leaves and stem length, and consequently an increase in the root-shoot ratio (Acosta-motos et al. 2017).

A stressful environment that affects growth affects photosynthesis. Indeed, many studies reported that under severe salt conditions, the photosynthesis is inhibited (Parihar et al. 2015). The inhibition of this process can be due to several factors namely: dehydration of cell membranes, which reduces their CO_2 permeability; the toxicity of the ions forming the salt; the reduction of CO_2 supply due to the hydro-active closure of the stomata; decreased in chlorophyll and pigments contents, increased leaf senescence induced by salinity; the change in enzyme activity caused by the change in their conformation in the cytoplasmic structure (Hernandez et al. 1995; Parihar et al. 2015).

The oxidative stress is the other well-established consequence of salinity. The generated ROS can react with different biomolecules, leading to the dysfunction of membranes and cell death (Parihar et al. 2015). The salinity condition is also associated with the qualitative and quantitative changes in protein synthesis. Indeed, several studies have shown that sodium chloride reduces protein synthesis and increases protein hydrolysis in cultivated plants such as peas, grapes, and beans (Tremblin and Coudret 1986). Agastian et al. (2000) reported that soluble proteins increase at low salinity levels and decrease at high salt concentration thresholds in blackberries. The accumulation of amino acids and carbohydrates has also been described as a response to the salt stress (He et al. 2016; Mansour 2000; Murakeözy et al. 2003).

5.3.3 Heat

While heat is a necessary component for the healthy growth of plants, overheating can cause negative reactions that will harm the proper development and, ultimately, the quality of crop yield. Heat stress or overheating is defined by Waqas (2019) as 'a condition when the intensity of temperature increases above the threshold level of the plant tolerance which causes irreversible damage to them'. Wahid et al. (2007) for their part, described it as a complex combination of several factors including severity, exposure time, and rate of increase in temperature. They also reported that the heat stress is usually set up when the temperature is 10 to 15 °C above ambient. According to the Intergovernmental Panel on Climate Change (IPCC 2013), the global average temperature is predicted to increase by

+1 °C compared to the current values and by +3 °C by 2100. This alarming situation has led researchers to intensify studies in order to understand how this parameter impacted the growth and development of plants.

Many studies have reported that high temperature can cause negative changes in different plant phenological development (Jha et al. 2014). These abnormalities can be noticed from the early growing stages by the delay, reduction, or inhibition of seed germination and seedling emergence (Akter and Islam 2017; Siddiqui et al. 2015). At the developmental stage, plants are particularly sensitive to high temperatures during flowering and reproductive phases (Bita and Gerats 2013). This can result in the reduction of the size and number of floral organs, leading to a significant loss in crop productivity (Röth et al. 2015). Additionally, the elevated temperature is likely to accelerate and shorten the flowering process, thereby, limiting the accumulation of the required resources for the successful production of gametes and fruit setting (Röth et al. 2015). The adverse effects of high temperature can also be observed on root and shoot development. Indeed, elevated in soil or air temperatures may strongly decrease the number, biomass, and elongation of the roots (Fahad et al. 2017). Regarding the shoot part, the major visual symptoms of long and severe heat stress include growth inhibition, reduction in internode length as well as rolling, thinning, scorching, and senescent leaves (Fahad et al. 2017; Wahid et al. 2007).

At the physiological and biochemical levels, high temperatures are at the origin of different alterations (Ashraf and Harris 2013). However, the severity of these alterations highly depends on the species and the interactions between several factors such as nutrient availability, soil water status, and solar radiation. The photosynthesis is among vital processes which can severely be disturbed by heat. In general, the photosynthesis in higher plants stops when the temperature reaches 40–45 °C (Badger et al. 1982). This phenomenon may be due to the destruction of chloroplast thylakoid membranes, the inhibition of the electron transport chain (PSII) as well as the deactivation of enzymes (such as RubisCo) involved in photosynthesis (Allakhverdiev et al. 2008). Stomata, through which the plant sweats and absorbs the CO_2 needed for photosynthesis, tighten in high heat to prevent water loss. This mechanism also impairs the input of CO_2 and therefore the photosynthesis reaction, limiting the growth of plants (Ref). Additionally, it was reported that heat stress induces a reduction in the chlorophyll content, either by blocking its biosynthesis or by accelerating its degradation (Dutta et al. 2009). Damages in the ATP generating system and carbon assimilation processes have also been established under intense heat exposure (Allakhverdiev et al. 2008). Moreover, it has been shown that temperature above threshold levels markedly diminishes plant respiration, probably by altering the mitochondrial respiratory chain complexes (Akter and Islam 2017).

Numerous studies have confirmed that elevated temperature induces excessive ROS production in different cell compartments (Akter and Islam 2017; Bita and Gerats 2013). The accumulation of these toxic products can cause serious injuries, including increases in lipid peroxidation and membrane thermo-instability (Wahid et al. 2007). It also appears that high heat causes the rupture of hydrogen bonds and an increase in hydrophilic forces at the membrane level, resulting in a fluid physical state (Los and Murata 2004). Phytohormones, which are frequently associated with the regulation of plant growth and development, can widely be impacted by heat. According to Obata and Fernie 2012, heat affects phytohormones production in different ways depending on species and phenological stage. ABA, ET, and AUX are among the most well-documented phytohormones during heat stress, an increase or decrease of these latter can negatively influence various plant processes (Wu et al. 2016). Similarly, changes in the amounts of primary and secondary metabolites have been noticed when plants suffer from heat stress. The accumulation of primary metabolites concerns mainly soluble sugars, proline, and glycine betaine (Bita and Gerats 2013). Regarding secondary metabolites, several studies have shown the accumulation of phenolic compounds under the influence of heat stress (Wahid 2007). This accumulation is can be explained by the increase in phenylalanine ammonia lyase (PAL), the first enzyme in the biosynthesis pathway of phenolic compounds, and a decrease in polyphenol oxidase activity (Rivero et al. 2001; Röth et al. 2015).

5.4 Conclusion

Drought, salinity, temperature, and saline/acidity stress result in primary and secondary stress and loss of productivity, and alterations in physiological, biochemical, and molecular processes in plants. Salinity and drought stress both reduce crop yields by affecting plant growth and limiting land use. Abiotic stresses such as salinity, heat, and drought play a critical role in determining the physiological and metabolic activities of plants. Average yields in all major crops are a small fraction of record levels, and drought and soil salinity are the primary factors limiting yields. As a result, crop productivity must be increased by improving crop adaptation. On the other hand, success in developing drought and salt-tolerant crops is contingent on the efficient and combined use of genetic engineering and traditional breeding tools.

5.5 Summary

Various abiotic stress conditions such as drought, salinity, and heat have been to show a considerable alteration in physiological, biochemical, and molecular processes in agriculturally important plants due to primary and secondary stresses that cause a significant decrease in biomass and crop productivity. Salinity and drought limit crop yield by directly affecting plant growth pathways and restricting the use land of utilization. These abiotic stress factors have a fundamental role in a plant's life due to their effects on the physiological and metabolic activities of different crop plants. Agriculture of today is facing an unprecedented threat due to these stress factors. The world population is predicted to hit over 9 billion by 2050 which indicates that the need to feed this population is urgent. There is not only limited availability of resource lands for agriculture today but also the agricultural system worldwide is threatened by climate change which makes stress conditions even greater contributors in limiting effort for bringing new land for agricultural resources and cultivation. In this chapter, we discuss the following issues: (i) The tools of genetic engineering that have potential use in producing stress tolerance genotypes; (ii) the application of GS for crop enhancement and improvement; and (iii) the major impact of abiotic stresses (salinity, heat, and drought) on plants' morpho-biochemical and physiological processes.

References

Abera, T., Feyisa, D., Yusuf, H. et al. (2005). Grain yield of maize as affected by biogass slurry and NP fertilizer rate at Bako, Western Oromiya, *Ethopia. Bioscience Research* 2: 31–38.

Abu Obaid, A.M., Melnyk, A.V., and Onichko, V.I. (2018). Evaluation of six sunflower cultivars for forage productivity under salinity conditions. *Advances in Environmental Biology* 2: 13–15.

Abu-Darwish, M.S., Abu Dieyeh, Z.H., Batarseh, I. et al. (2009). trace element contents and essential oil yields from Wild thyme plant (*Thymusserpyllum* L.) grown at different natural-variable environments, Jordan. *Journal of Food, Agriculture and Environment* 7: 920–924.

Acosta-Motos, J.R., Hernández, J.A., Álvarez, S. et al. (2017). Long-term resistance mechanisms and irrigation critical threshold showed by *Eugenia myrtifolia* plants in response to saline reclaimed water and relief capacity. *Plant Physiology and Biochemistry* 111: 244–256.

Agastian, P., Kingsley, S., and Vivekanandan, M. (2000). Effect of salinity on photosynthesis and biochemical characteristics in mulberry genotypes. *Photosynthetica* 38: 287–290.

Ahmad, S., Wahid, A., Rasul, R.E., and Wahid, A. (2005). Comparative morphological and physiological responses of green gram genotypes to salinity applied at different growth stages. *Botanical Bulletin of Academia Sinica* 46: 135–142.

Akter, N. and Islam, M.R. (2017). Heat stress effects and management in wheat. *Agronomy for Sustainable Development* 37: 1–27.

Allakhverdiev, S.I., Kreslavski, V.D., Klimov, V.V. et al. (2008). Heat stress: an overview of molecular responses in photosynthesis. *Photosynthesis Research* 98: 541–550.

Allen, C.D., Macalady, A.K., Chenchouni, H. et al. (2010). A global overview of drought and heat-induced tree mortality reveals emerging climate change risks for forests. *Forest Ecology and Management* 259: 660–684.

Al-Rifaee, M., Turk, M.A., and Tawaha, A.M. (2004). Effect of seed size and plant population density on yield and yield components of local faba bean (*Viciafaba* L. major). *International Journal of Agriculture and Biology* 6: 294–299.

Al-Tawaha, A.R.M. and Odat, N. (2010). Use of sorghum and maize allelopathic properties to inhibit germination and growth of wild barley (*Hordeum spontaneum*). *Notulae Botanicae Horti Agrobotanici Cluj-Napoc* 38: 124–127.

Al-Tawaha, A.M. and Seguin, P. (2006). Effects of seeding date, row spacing, and weeds on soybean seed isoflavone concentration. *Canadian Journal of Plant Science* 86: 1079–1082.

Al-Tawaha, A.M., Seguin, P., Smith, D.L., and Beaulieu, C. (2006). Foliar application of elicitors alters isoflavone concentrations and other seed characteristics of field-grown soybean. *Canadian Journal of Plant Science* 86: 677–684.

Al-Tawaha, A.M., Seguin, P., Smith, D.L., and Bonnell, B. (2007). Effects of irrigation on isoflavone concentrations of soybean grown in southwestern Québec. *Journal of Agronomy and Crop Science* 193: 238–246.

Al-Tawaha, A.R., Al-Karaki, G., Al-Tawaha, A.R. et al. (2018a). Effects water flow rate on quantity and quality of lettuce (*Lactuca sativa* L.) in nutrient film technique (NFT) under hydroponics conditions. *Bulgarian Journal of Agricultural Science* 24: 791–798.

Al-Tawaha, A.R., Turk, M.A., Al-Tawaha, A.R., and Aludatt, M.H. (2018b). Using chitosan to improve growth of maize cultivars under salinity conditions. *Bulgarian Journal of Agricultural Science* 24: 437–442.

Al-Tawaha, A.R., Al-Tawaha, A.R., Alu'datt, M. et al. (2018c). Effects of soil type and rainwater harvesting treatments in the growth, productivity and morphological trains of barley plants cultivated in semi-arid environment. *Australian Journal of Crop Science* 12: 975–979.

Apse, M.P. and Blumwald, E. (2002). Engineering salt tolerance in plants. *Current Opinion in Biotechnology* 13: 146–150.

Arruda, M.P., Lipka, A.E., Brown, P.J. et al. (2016). Comparing genomic selection and marker-assisted selection for Fusarium head blight resistance in wheat (*Triticum aestivum* L.). *Molecular Breeding* 36: 84–89.

Ashraf, M. and Harris, P.J.C. (2013). Photosynthesis under stressful environments: an overview. *Photosynthetica* 51: 163–190.

Badger, M.R., Bjorkman, O., and Armond, P.A. (1982). An analysis of photosynthetic response and adaptation to temperature in higher plants – temperature acclimation in the desert evergreen *Nerium oleander* L. *Plant, Cell & Environment* 5: 85–99.

Bita, C.E. and Gerats, T. (2013). Plant tolerance to high temperature in a changing environment: scientific fundamentals and production of heat stress-tolerant crops. *Frontiers in Plant Science* 4: 1–18.

Bitinaite, J., Wah, D.A., Aggarwal, A.K., and Schildkraut, I. (1998). FokI dimerization is required for DNA cleavage. *Biochemistry* 95: 10570–10575.

Boch, J. and Bonas, U. (2010). Xanthomonas AvrBs3 family-type III effectors: discovery and function. *Annual Review of Phytopathology* 48: 419–436.

Bota, J., Medrano, H., and Flexas, J. (2004). Is photosynthesis limited by decreased Rubisco activity and RuBP content under progressive water stress? *New Phytologist* 162: 671–681.

Cabello, J.V., Lodeyro, A.F., and Zurbriggen, M.D. (2014). Novel perspectives for the engineering of abiotic stress tolerance in plants. *Current Opinion in Biotechnology* 26: 62–70.

Carroll, D. (2014). Genome engineering with targetable nucleases. *Annual Review of Biochemistry* 83: 409–439.

Checker, V.G., Chhibbar, A.K., and Khurana, P. (2012). Stress-inducible expression of barley Hva1 gene in transgenic mulberry displays enhanced tolerance against drought, salinity and cold stress. *Transgenic Research* 21: 939–957.

Curtin, S.J., Voytas, D.F., and Stupar, R.M. (2012). Genome engineering of crops with designer nucleases. *The Plant Genome* 5: 42–49.

Duangjit, J., Causse, M., and Sauvage, C. (2016). Efficiency of genomic selection for tomato fruit quality. *Molecular Breeding* 36: 29–35.

Dutta, S., Mohanty, S., and Tripathy, B.C. (2009). Role of temperature stress on chloroplast biogenesis and protein import in pea. *Plant Physiology* 150: 1050–1061.

Fahad, S., Bajwa, A.A., Nazir, U. et al. (2017). Crop production under drought and heat stress: plant responses and management options. *Frontiers in Plant Science* 8: 1147–1151.

Farooq, M., Wahid, A., Kobayashi, N. et al. (2009). Plant drought stress: effects, mechanisms and management. *Agronomy for Sustainable Development* 29: 185–112.

Flexas, J., Galmes, J., and Galle, A. (2010). Improving water use efficiency in grapevines: potential physiological targets for biotechnological improvement. *Australian Journal of Grape and Wine Research* 16: 106–121.

Grant, O.M. (2012). Understanding and exploiting the impact of drought stress on plant physiology. In: *Abiotic Stress Responses in Plants* (ed. P. Ahmad and M.N.V. Prasad), 89–04. New York: Springer.

Gray, S.B. and Brady, S.M. (2016). Plant developmental responses to climate change. *Developmental Biology* 419: 64–77.

Guo, Q., Zhang, J., Gao, Q. et al. (2008). Drought tolerance through overexpression of monoubiquitin in transgenic tobacco. *Journal of Plant Physiology* 165: 1745–1755.

Hasanuzzaman, M., Nahar, K., Alam, M. et al. (2013). Physiological, biochemical, and molecular mechanisms of heat stress tolerance in plants. *International Journal of Molecular Sciences* 14: 9643–9684.

Hayes, B.J., Bowman, P.J., Chamberlain, A.C. et al. (2009). Accuracy of genomic breeding values in multi-breed dairy cattle populations. *Genetics, Selection, Evolution* 41: 51–59.

He, S., Schulthess, A.W., Mirdita, V. et al. (2016). Genomic selection in a commercial winter wheat population. *Theoretical and Applied Genetics* 129: 641–651.

Heffner, E.L., Lorenz, A.J., Jannink, J.L., and Sorrells, M.E. (2010). Plant breeding with genomic selection: gain per unit time and cost. *Crop Science* 50: 1681–1690.

Heffner, E.L., Jannink, J.L., and Sorrells, M.E. (2011). Genomic selection accuracy using multifamily prediction models in a wheat breeding program. *Plant Genome* 4: 65–75.

Hernandez, J., Olmos, E., Corpas, F. et al. (1995). Salt-induced oxidative stress in chloroplasts of pea plants. *Plant Science* 105: 151–167.

Huang, M., Cabrera, A., Hoffstetter, A. et al. (2016). Genomic selection for wheat traits and trait stability. *Theoretical and Applied Genetics* 129: 1697–1710.

IPCC (2013). *Climate Change 2013: The Physical Science Basis. Contribution of Working Group I to the Fifth Assessment Report of the Intergovernmental Panel on Climate Change* (ed. T.F. Stocker, D. Qin, G.-K. Plattner, et al.). Cambridge, UK and New York: Cambridge University Press.

Jaleel, C.A., Manivannan, P., Wahid, A. et al. (2009). Drought stress in plants: a review on morphological characteristics and pigments composition. *International Journal of Agriculture and Biology* 11: 100–105.

Jannink, J.-L., Lorenz, A.J., and Iwata, H. (2010). Genomic selection in plant breeding: from theory to practice. *Briefings in Functional Genomics* 9: 166–177.

Jha, U.C., Bohra, A., and Singh, N.P. (2014). Heat stress in crop plants: its nature, impacts and integrated breeding strategies to improve heat tolerance. *Plant Breeding* 133: 679–701.

Karan, R. and Subudhi, P.K. (2012). Approaches to increasing salt tolerance in crop plants. In: *Abiotic Stress Responses in Plants* (ed. P. Ahmad and M. Prasad), 63–88. New York, NY: Springer.

Läuchli, A. and Grattan, S. (2007). Plant growth and development under salinity stress. In: *Advances in Molecular Breeding Toward Drought and Salt Tolerant Crops* (ed. M.A. Jenks, P.M. Hasegawa and S.M. Jain), 1–32. Dordrecht: Springer.

Levigneron, A., Lopez, F., and Vansuyt, G. (1995). Les plantes face au stress salin. *Cahiers Agricultures* 4: 263–273.

Lisar, S.S.Y. and Agdam, B.H. (2016). Drought stress in plants: causes, consequences, and tolerance. In: *Drought Stress Tolerance in Plants* (ed. M.A. Hossain, S. Wani, D.J. Burritt and L.S.P. Tran), 1–2. Switzerland: Springer.

Los, D.A. and Murata, N. (2004). Membrane fluidity and its roles in the perception of environmental signals. *Biochimica et Biophysica Acta* 1666: 142–157.

Lyzenga, W.J. and Stone, S.L. (2012). Abiotic stress tolerance mediated by protein ubiquitination. *Journal of Experimental Botany* 63: 599–616.

Mansour, M. (2000). Nitrogen containing compounds and adaptation of plants to salinity stress. *Biologia Plantarum* 43: 491–500.

Meuwissen, T.H.E., Hayes, B.J., and Goddard, M.E. (2001). Prediction of total genetic value using genome-wide dense marker maps. *Genetics* 157: 1819–1829.

Mishra, A.K. and Singh, V.P. (2010). A review of drought concepts. *Journal of Hydrology* 391: 202–216.

Mittler, R. and Blumwald, E. (2010). Genetic engineering for modern agriculture: challenges and perspectives. *Annual Review of Plant Biology* 61: 443–462.

Munns, R. and James, R.A. (2003). Screening methods for salinity tolerance: a case study with tetraploid wheat. *Plant and Soil* 253: 201.

Murakeözy, É.P., Nagy, Z., Duhazé, C. et al. (2003). Seasonal changes in the levels of compatible osmolytes in three halophytic species of inland saline vegetation in Hungary. *Journal of Plant Physiology* 160: 395–401.

Muscolo, A., Junker, A., Klukas, C. et al. (2015). Phenotypic and metabolic responses to drought and salinity of four contrasting lentil accessions. *Journal of Experimental Botany* 66: 5467–5480.

Nikus, O., Turk, M.A., and Al-Tawaha, A.M. (2004). Yield response of sorghum (*Sorghum bicolor* L.) to manure supplemented with phosphate fertilizer under semi-arid Mediterranean conditions. *International Journal of Agriculture and Biology* 6: 889–893.

Nongpiur, R.C., Singla-Pareek, S.L., and Pareek, A. (2016). Genomics approaches for improving salinity stress tolerance in crop plants. *Current Genomics* 17: 343–335.

Obata, T. and Fernie, A.R. (2012). The use of metabolomics to dissect plant responses to abiotic stresses. *Cellular and Molecular Life Sciences* 69: 3225–3243.

Odat, N. (2018a). Molecular and biochemical responses of barley (*Hordeum vulgare* L.) to NaCl salinity stress and salicylic acid. *Research on Crops* 19: 101–106.

Odat, N. (2018b). Differential gene expression of durum wheat (*Triticum turgidum* L. var. durum) in relation to genotypic variation under NaCl salinity stress. *Jordan Journal of Biological Sciences* 11: 591–595.

Odat, N. (2020). Intraspecific genetic variation within and between improved cultivars and landraces of durum wheat in germination and root architectural traits under saline conditions. *International Journal of Plant Biology* 11: 7413–7431.

Odat, N., Maen, H., Obeidat, M. et al. (2015). Identifying selection signatures related to domestication process in barley (*Hordeum vulgare* L.) landraces of Jordan using microsatellite markers. *Jordan Journal of Biological Sciences* 8: 307–313.

Onogi, A., Watanabe, M., Mochizuki, T. et al. (2016). Toward integration of genomic selection with crop modelling: the development of an integrated approach to predicting rice heading dates. *Theoretical and Applied Genetics* 129: 805–817.

Orellana, S., Yañez, M., Espinoza, A. et al. (2010). The transcription factor SlAREB1 confers drought, salt stress tolerance and regulates biotic and abiotic stress-related genes in tomato. *Plant, Cell & Environment* 33: 2191–2198.

Parihar, P., Singh, S., Singh, R. et al. (2015). Effect of salinity stress on plants and its tolerance strategies: a review. *Environmental Science and Pollution Research* 22: 4056–4075.

Patakas, A. (2012). Abiotic stress-induced morphological and anatomical changes in plants. In: *Abiotic Stress Responses in plants: Metabolism, Productivity and Sustainability* (ed. P. Ahmad and M.N.V. Prasad), 21–39. Springer Science+Business Media.

Reddy, A.R., Chaitanya, K.V., and Vivekanandan, M. (2004). Droughtinduced responses of photosynthesis and antioxidant metabolism in higher plants. *Journal of Plant Physiology* 161: 1189–1182.

Rivero, R.M., Ruiz, J.M., Garcia, P.C. et al. (2001). Resistance to cold and heat stress: accumulation of phenolic compounds in tomato and watermelon plants. *Plant Science* 160: 315–321.

Röth, S., Paul, P., and Fragkostefanakis, S. (2015). Plant heat stress response and thermotolerance. In: *Genetic Manipulation in Plants for Mitigation of Climate Change* (ed. P. Jaiwal, R. Singh and O. Dhankher), 15–41. New Delhi: Springer.

Siddiqui, M.H., Al-Khaishany, M.Y., Al-Qutami, M.A. et al. (2015). Morphological and physiological characterization of different genotypes of faba bean under heat stress. *Saudi Journal of Biological Sciences* 22: 656–653.

Sorek, R., Lawrence, C.M., and Wiedenheft, B. (2013). CRISPR-mediated adaptive immune systems in bacteria and archaea. *Annual Review of Biochemistry* 82: 237–266.

Sulpanjani, A., Yang, M.S., Tawaha, A.R.M. et al. (2005). Effect of Magnesium application on yield, mineral contents and active components of *Chrysanthemum coronarium* L. under Hydroponics Conditions. *Bioscience Research* 2: 73–79.

Supanjani, A., Tawaha, A.R.M., Yang, M.S. et al. (2005). Role of calcium in yield and medicinal quality of *Chrysanthemum coronarium* L. *Journal of Agronomy* 4: 188–192.

Tardieu, F. (1996). Drought perception by plants. Do cells of droughted plants experience water stress? *Plant Growth Regulation* 20: 93–94.

Tawaha, A.M. and Turk, M.A. (2002). Lentil (*Lens culinaris* Medic.) productivity as influenced by rate and method of phosphate placement in a Mediterranean environment. *Acta Agronomica Hungarica* 50: 197–191.

Tawaha, A.M. and Turk, M.A. (2004). Field pea seeding management for semi-arid Mediterranean conditions. *Journal of Agronomy and Crop Science* 190: 86–92.

Tawaha, A.M., Singh, V.P., and Turk, M.A. (2003). A review on growth, yield components and yield of barley as influenced by genotypes, herbicides and fertilizer application. *Research on Crops* 4: 1–9.

Tester, M. and Langridge, P. (2010). Breeding technologies to increase crop production in a changing environment. *Science* 327: 818–822.

Tremblin, G. and Coudret, A. (1986). Salinité, transpiration et échanges de CO_2 chez Halopeplisamplexicaulis (Vahl.). *Ung Oecologia Plantarum* 21: 417–431.

Turk, M.A. and Tawaha, A.M. (2001). Common vetch (*Vicia sativa* L.) productivity as influenced by rate and method of phosphate fertilization in a Mediterranean environment. *Agricultura Mediterranea* 131: 108–111.

Turk, M.A. and Tawaha, A.M. (2002a). Onion (*Allium cepa* L.) as influenced by rate and method of phosphorus placement. *Crop Research* 23: 105–107.

Turk, M.A. and Tawaha, A.M. (2002b). Impact of seeding rate, seeding date and method of phosphorous application in faba bean (*Viciafaba* L. minor) in the absence of moisture stress. *Biotechnology, Agronomy, Society and Environment* 6: 171–178.

Turk, M.A., Tawaha, A.M., and Shatnawi, M. (2003a). Lentil (*Lens culinaris* Medik) response to plant density, sowing date, phosphorus fertilization and Ethephon application in the absence of moisture stress. *Journal of Agronomy and Crop Science* 189: 1–6.

Turk, M.A., Shatnawi, M.K., and Tawaha, A.M. (2003b). Inhibitory effects of aqueous extracts of black mustard on germination and growth of alfalfa. *Weed Biology and Management* 3: 37–40.

Turk, M.A., Tawaha, A.M., and Lee, K.D. (2004). Seed germination and seedling growth of three lentil cultivars under moisture stress. *Asian Journal of Plant Sciences* 3: 394–397.

Urnov, F.D., Rebar, E.J., Holmes, M.C. et al. (2010). Genome editing with engineered zinc finger nucleases. *Nature Reviews. Genetics* 11: 636–646.

Várallyay, G. (1994). Soil data-base for long-term field experiments and sustainable land use. *Agrokémia és Talajtan* 43: 269–290.

Varshney, R.K., Terauchi, R., and McCouch, S.R. (2014). Harvesting the promising fruits of genomics: applying genome sequencing technologies to crop breeding. *PLoS Biology* 12: 6–9.

Wahid, A. (2007). Physiological implications of metabolite biosynthesis for net assimilation and heat-stress tolerance of sugarcane (*Saccharumofficinarum*) sprouts. *Journal of Plant Research* 120: 219–228.

Wahid, A., Gelani, S., Ashraf, M., and Foolad, R. (2007). Heat tolerance in plants: an overview. *Environmental and Experimental Botany* 61: 199–123.

Waqas, A. (2019). Abiotic stresses and their effects, responses, and adaptations in grapevines (*Vitisvinifera*): overview of modern research. *American Journal of Agricultural Research* 4: 33–39.

Wu, C., Cui, K., Wang, W. et al. (2016). Heat-induced phytohormone changes are associated with disrupted early reproductive development and reduced yield in rice. *Scientific Reports* 6: 34978.

Yadav, S., Irfan, M., Ahmad, A., and Hayat, S. (2011). Causes of salinity and plant manifestations to salt stress: a review. *Journal of Environmental Biology* 32: 667–685.

Zhu, J.K. (2001). Plant salt tolerance. *Trends in Plant Science* 6: 66–71.

6

Innovative Approaches in the Breeding of Climate-Resilient Crops

Sandra Cvejić, Siniša Jocić, Bojan Mitrović, Goran Bekavac, Milan Mirosavljević, Ana Marjanović Jeromela, Miroslav Zorić, Aleksandra Radanović, Ankica Kondić-Špika, and Dragana Miladinović

Institute of Field and Vegetable Crops, Novi Sad, Serbia

CONTENTS

Climate Change and Agriculture: Perspectives, Sustainability and Resilience, First Edition. Edited by Noureddine Benkeblia.

6.1 Importance to Adapt Crops to New and Changing Environments

Climate change is a significant shift in average weather conditions, expressed in extreme weather, higher temperatures, elevated atmospheric CO_2 levels, and water deficiency. It strongly influences agricultural production and cultivation practices of all major crops (Taranto et al. 2018). In addition, climate change provides novel challenges and potential opportunities for cultivated plant species (Chapman et al. 2012) and crop breeding. The challenge posed by climate change has led terms such as climate-smart agriculture (Lipper and Zilberman 2018), climate-smart crops (Kole 2020), or climate-resilient crops (Kole et al. 2015) to become famous (Ceccarelli and Grando 2020). Climate change is undoubtedly increasing both permanent and occasional abiotic stresses (drought, high temperatures, salinity) in a large part of the global agricultural land. As a result, it can drive the abandonment of cultivation and desertification in some regions of the world. In addition, climate change is expected to result in increased biotic stress, as many pests and diseases will potentially migrate to areas where they were not present due to environmental limitations (Bebber et al. 2013).

Adaptation and adoption of climate-smart crop production require implementing practices and technologies based on the knowledge about the type and extent of climatic variables that affect crop production and integrated research on a specific crop (Figure 6.1). One of the critical tools/steps in mitigating the impact of climate change is the creation of physiology-based crop models. So far, the models have been based on limited experimental results. However, there is a growing interest and a tendency to extend this knowledge to several different crop species in broader climatic zones, different soil types, crop management regimens, and different climate change scenarios. These models simplify the natural conditions, but they allow the first assessment of the climate change impact on the production of specific crops. Parent and Tardieu (2012) emphasize the relevance of crop models to predict the genetic variability of yield under water deficit or high temperature.

The development of crops that will be resilient and productive under future climate conditions can only be achieved by exploiting all available breeding strategies (Miladinović et al. 2019) (Figure 6.2). Plant breeding may help develop new cultivars with improved

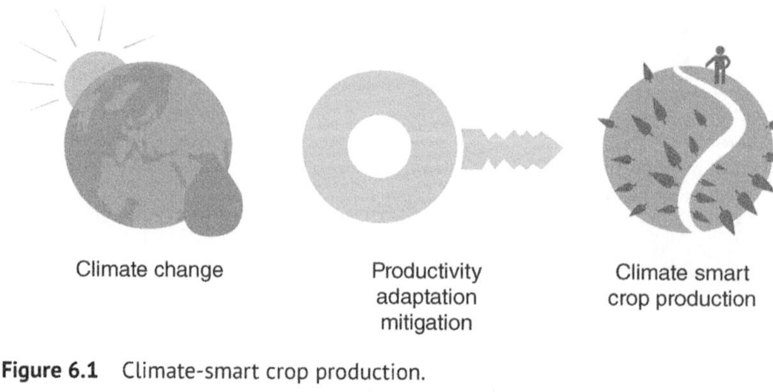

Figure 6.1 Climate-smart crop production.

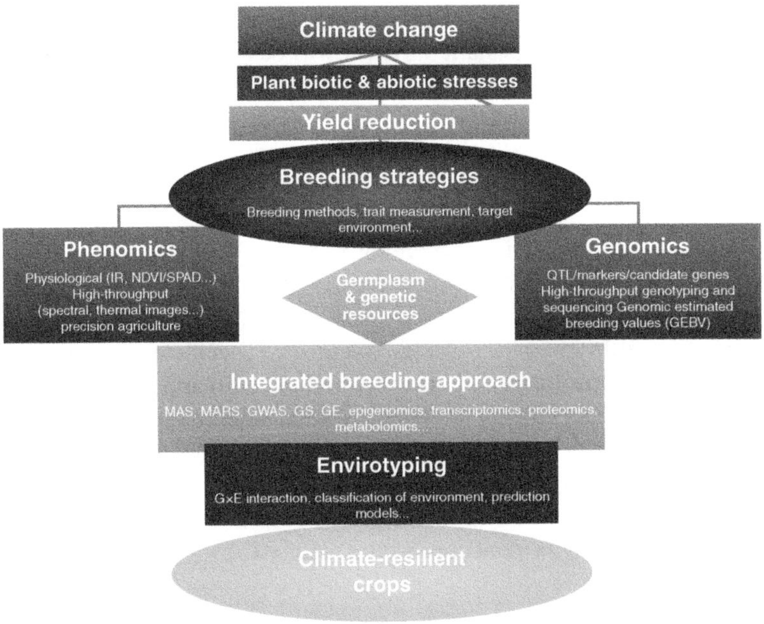

Figure 6.2 Integrated breeding approach for creating climate-resilient crops.

traits better adapted to climate change conditions using conventional and genomic technologies (Habash et al. 2009). These traits include not only drought and temperature stress adaptation but also tolerance to pests and disease, which continue to cause crop losses (Oerke 2006), salinity, and waterlogging (Humphreys et al. 2005). Among the wide-ranging effects of climate change, water deficiency, which causes drought, profoundly impacts agriculture worldwide. Drought is a growth- and yield-limiting factor for crops, influencing flowering, pollination, and grain-filling of most grain crops. Globally, it is estimated that up to 83% of all economic losses caused by drought relate to agriculture, the cost of which amounted to US$ 29 billion in semi-arid and arid areas between 2005 and 2015 (http://www.fao.org/news/story/en/item/1106977/icode/). Hence, breeding for drought resistance is one of the most essential and common objectives of several breeding programs for the entire major food crops in most countries (Ceccarelli et al. 2004, 2007). Opportunities for new cultivars with increased drought tolerance (DT) include changes in phenology or enhanced responses to elevated CO_2. The demands placed on breeding are considerable and depend on the requirements of society and the matching potentials of environments; they include quality and consistency of yield, minimum yield in marginal areas, improvement of neglected species, and efficient use of genetic resources (Stamp and Visser 2012).

This chapter presents an overview of the breeding strategies of major crops to climate conditions in integrating different disciplines/technologies. The chapter will summarize utilizing genetic resources and innovative tools to make crops tolerant of extreme climatic conditions. The focus will be on individual traits to describe the breeding activities with the challenges posed by climate change, innovative solutions in crop breeding, high-throughput biotechnology, molecular techniques, and statistical and biometrical approaches for data analysis. Finally, we will recognize that the advantages that have been introduced into breeding methods of improved climate-resistant varieties will be an option for adapting agriculture to stressful conditions.

6.2 Utilizing Existing and Creating New Genetic Variability

One of the strategic goals in agriculture is overcoming the negative impact of climate change on the productivity of cultivated plants and the creation of tolerant genotypes. Climate change has various effects in different agro-ecological conditions, which requires creating different genotypes within one plant species, adapted to specific environmental conditions and tolerant to stress predominant in the environment, with appropriate adaptability and stability. Therefore, breeders use all available resources and methods to increase genetic variability and thus the chance to create highly productive genotypes. The contemporary challenge is developing sustainable agriculture and increasing the utilization of genetic resources in breeding (Anđelković et al. 2020). For the exploitation of the existing and creation of new genetic variability, several approaches are used:

6.2.1 Target Crossings with Existing Gene Pools

The development of best-performing varieties utilizing the existing plant collections is the most feasible and recommendable method for creating genotypes for specific climatic conditions (Chandramohanan et al. 2016). Creating new genotypes of cultivated plants using classical breeding is still one of the primary research directions. Crosses are made between existing parental lines and introductions of new germplasm carrying potentially valuable traits. The term indirect selection is often used to describe the selection of genotypes in a breeding program based on traits other than those which are specifically of interest to growers of the crop (Chapman et al. 2012). Indirect selection requires that the trait is heritable, can be efficiently measured, and has a genetically correlated relationship with the trait of interest, i.e. yield or quality. This process incrementally improves the average adaptation of the parental pool. The generation of new genetic variability via crossing and introductions increases the opportunity to find better lines than the existing cultivars (Chapman et al. 2012).

Activities in this area also focused on expanding the genetic basis of plant species by cross-breeding genotypes with different genetic bases from existing collections and more intensive genotypes from the so-called traditional gene pool, i.e., domestic and domesticated varieties of wild relatives. The development of new germplasm resources with novel allelic diversity in a helpful background also aids in identifying genes that contribute to critical adaptive traits (Kilian et al. 2021). Furthermore, interspecific hybridization is widely used to transfer beneficial traits from wild relatives to cultivated breeding lines and ultimately to new cultivars, which presents the first step in the pre-breeding process. In many cases, conventional crossing methods have sufficed to produce interspecific hybrids, but some unsuccessful crosses often appear. They are caused by genetic distance and chromosome number and structure (Atlagić et al. 2012). Cross incompatibility can be overcome using embryo culture and somatic hybridization besides conventional hybridization.

Some of the problems that could be overcome by somatic hybridization are sexual incompatibility, nuclear embryogenesis, male or female sterility, and transfer of resistance genes for biotic and abiotic stresses to create novel hybrids with increased yields (Begna 2020).

6.2.2 Induced Mutations

Complement classical breeding methods and contribute to the faster and more efficient creation of new genetic variability. Induced mutations are used to create new mutant populations and combined with biotechnology methods selected to create new genotypes tolerant to different stresses caused by climate change.

Induced mutations emerged in the middle of the last century to artificially develop genetic variability (Cvejić et al. 2014). The use of chemical and physical agents to induce mutations has been successfully adopted worldwide since the 1930s to generate novel alleles. As a result, more than 170 mutant varieties from different plant species have been released today, increasing genetic diversity. However, this approach has been almost abandoned due to high costs and controversial opinions about the consequence of mutagenic agents on human health (Taranto et al. 2018).

An alternative to chemical and physical mutagenesis is represented by techniques based on the use of biological agents. Site-directed mutagenesis represented an alternative forward genetics method to increase genetic diversity (Jankowicz-Cieslak and Till 2015). In the last two decades, induced mutations have been recovered, thanks also to advances in large-scale genome sequencing projects. Targeting Induced Local Lesions in Genomes (TILLING) is a reverse genetic technique based on chemical-induced mutagenesis that allows discovering rare mutations in populations. Traditionally, TILLING protocols were based on enzymatic or physical screening methods and selecting mutagenized lines. Loss-of-function, gain-of-function, and hypomorphic alleles can be identified and possibly associated with corresponding phenotypes (Taranto et al. 2018).

6.2.3 Phenomics

Establishing the connection between genotype and phenotype is currently one of the most significant challenges facing modern plant breeding (Pauli et al. 2016). Significant progress has been made in developing molecular and genetic tools for plant breeding, while quantitative analysis of plant agronomic traits, or phenotyping, has been limited by a lack of efficiency techniques. The extensive collection of phenotypic data for many physiological and developmental traits from many individuals remains onerous (Furbank and Tester 2011). The development of phenotyping platforms is necessary in order to increase the efficiency and accuracy of phenotypic observations. Plant phenomics is a new field of science that connects genomics with plant ecophysiology and agronomy. Phenomics involves gathering appropriate phenotypic data at multiple levels of the organization to progress toward a complete characterization of phenotypic space generated by a particular genome or set of genomes (Dhondt et al. 2013). Thus, plant phenotyping can operate at different levels of resolution and dimensionality, from the molecular to the whole plant, and in different environments, from controlled to field conditions (Lobos et al. 2017). Plant phenomics involves strategies for describing how significant improvements in imaging, sensor technologies, and data analysis are making high-throughput root, shoot, whole-plant, and canopy phenomic studies possible. Phenomic pipelines can help researchers transform large numbers of images and sensor data into knowledge, necessitating novel data handling and modelling methods (Tardieu et al. 2017).

6.3 Development of Climate Resilience in Major Crops

6.3.1 Maize

Global increases in temperature, changes in the quantity and the distribution of rainfall, extreme winds, floods, and the occurrence of new pathogens and pests can all elevate maize production uncertainty, especially in the rain-fed and countries with underdeveloped agriculture. Climate change is projected to cause maize yields reduction of 7.4% as a result of the increase in the mean global temperature of 1 °C (Zhao et al. 2017). Such changes will require adjusting cropping systems and integrating more drought and heat-tolerant maize varieties (Watson 2019). Although plant breeding is a powerful tool, a complex G x E interaction and low heritability of some traits make breeding for better adaptation difficult. Breeding for adaptation implies assessing the available genetic variability and selecting and recombining appropriate resources for projected climate change (Chapman et al. 2012).

6.3.1.1 Major Traits for Adaptation

In breeding for climate-resilient maize, several abiotic and biotic factors have to be considered. Drought is a widespread phenomenon affecting maize yield all around the globe. From an agricultural perspective, drought can be described as a condition in which evapotranspiration overcomes the amount of water taken up by the crop (Swondon et al. 2021). Although detrimental in all stages of growth and development, the reproductive phase is the most susceptible to drought (Grant et al. 1989). Far more, there is a significant difference between female and male flowers. While silking is delayed under drought stress, pollen shed is minimally or unaffected. Several studies confirmed that along with a reduced anthesis-silking interval (ASI), the occurrence of barren plants, a smaller tassel size, a greater harvest index, and delayed leaf senescence are tightly connected to DT (Bolanos and Edmeades 1993a,b; Chapman and Edmeades 1999; Mitrović et al. 2019). Functional stay-green is also a valuable trait for improving crop stress tolerance (Thomas and Ougham 2014) and root architecture and function (Hammer et al. 2009).

DT is one of the most complex quantitative traits to study and improve (Prasanna et al. 2021). Phenotyping for DT is labour and resource-intensive and requires multiple years and locations of field trials (Collins et al. 2008). Under drought stress, direct selection for grain yield is often inefficient due to low heritability (Edmeades et al. 1999). One option to increase breeding efficiency for DT could be an indirect selection for secondary traits (Banzinger et al. 2000). Another way is to select molecular markers associated with grain yield under drought stress (Mikić et al. 2016a). Numerous studies confirmed the presence of quantitative trait loci - QTLs associated with grain yield under different levels of drought. Despite the initial promise of marker-assisted selection (MAS) for quantitative traits, information on QTL has not directly contributed to the development of DT cultivars (Collins et al. 2008). Simulation and empirical results revealed 14-50% larger gains with genome-wide than with QTL-based selection (Bernardo and Yu 2007).

Heat stress is another severe limiting factor to maize production (Cairns et al. 2013) and can be considered by the temperature above a certain level that causes irreversible damage to plant growth and development. Some studies showed that an increase in temperature of 2 °C could cause a more significant reduction in grain yield than a decrease in precipitation of 20% (Lobell and Burke 2010). As opposed to drought, male reproductive tissue is more sensitive to heat than female tissue. High temperatures reduce the quantity and pollen viability and consequently reduce ovules' fertilization. Also, high temperatures at early reproductive stages delay the silk emergence, causing ASI to increase and reduce fertilization (Cicchino et al. 2011). The development of heat-tolerant hybrids is of crucial importance for future maize cropping in temperate regions. In a set of European inbred lines, 607 heat-responsive and 39 heat tolerance genes were discovered, indicating that individual inbred lines developed different genetic mechanisms in response to heat stress (Frey et al. 2016). Although relatively little research has been conducted on heat stress in maize, it is interesting that QTLs for heat tolerance concerning grain yield were identified and dents are more heat tolerant than flints (Frey et al. 2016). Recent studies show that the whole-genome prediction could be the most appropriate approach to improving yield performance under heat or heat combined with drought stress (Prasanna et al. 2021). Due to global temperature increases, heat stress responses assessed at a molecular level can help breed thermos-tolerance (Li and Howell 2021).

A significant maize production area, especially in South and South-Eastern Asia, is annually affected by floods and waterlogging problems (Zaidi and Singh 2010). Waterlogging stress can be described as the stress inhibiting plant growth and development when the soil's water table is above the field capacity (Armstrong and Drew 2002). As a result of the decline in oxygen concentration in the rhizosphere, the plant root suffers from hypoxia and, later on, from anoxia, while carbon dioxide, ethylene, and toxic gases accumulate. An additional effect of water abundance is N, P, and K deficiency and the accumulation of Fe and Mg (Zaidi and Singh 2010). In such conditions, nutrient uptake is significantly reduced (Naveenkumar et al. 2018). Genetic variability for tolerance to waterlogging was observed in temperate (Sachs et al. 1996) and tropical maize (Zaidi and Singh 2010). Some QTLs associated with tolerance to waterlogging have already been identified (Qiu et al. 2007; Zaidi et al. 2015; Mano and Omori 2015). Also, a major QTL named *Subtol6* was mapped within the recombinant inbred lines (RIL) population (Campbell et al. 2015). By pyramiding QTLs controlling several factors responsible for flooding tolerance, it should be possible to develop flooding-tolerant maize inbred lines (Mano and Omori 2015).

Despite the global efforts in the germplasm evaluation, selection, and improvement, an estimation of 200 billion USD is lost annually due to plant diseases alone (Chapman et al. 2012). Most climate change models predict that extreme weather conditions will increase losses due to the presence of diseases. Ear and stalk rots are diseases with an increasing impact in a changing climate (Prasanna et al. 2021). Since the environment directly influences the life cycles of pathogens, climate change could bring some changes in the distribution and the intensity of infection of *Fusarium graminearum, Fusarium verticillioides,* and *Aspergillus flavus* (Savary et al. 2012). Conventional breeding made significant progress in developing genotypes tolerant to diseases. CIMMYT developed some inbred lines with low fumonisin content (Rose et al. 2017), while some resources of resistance to *Aspergillus flavus* were found in temperate (Scott and Zummo 1988) as well as tropical germplasm (Warburton and Williams 2014). Some studies detected QTLs related to resistance to *Fusarium* spp. and *Aspergillus flavus* (Ding et al. 2008; Warburton and Williams 2014). Due to the high number of QTLs responsible for multiple ear rot resistance, genomic selection (GS) rather than MAS are better suited for germplasm improvement (Miedaner and Juroszek 2021).

Climate change models predict that extreme weather conditions will propel diseases and insect pests (Prasanna et al. 2021). Temperature is one of the most important environmental factors influencing insect feeding, behaviour, development, reproduction, and survival. Yield losses of maize, rice, and wheat are projected to increase by 10–25% per one degree of global warming. These losses will be most severe in regions where warming increases insects' population growth and metabolic rate (Deutsch et al. 2018). Moreover, increased insect spreading will increase the incidence and severity of insect-related maize diseases. Consequently, a crucial adaptation measure to climate change will be breeding for tolerance to existing and new pests. Although insect-resistant crops already feed much of the world, more efforts are needed to speed the development of new insect-resistant varieties (Smith 2021).

Maize breeding in the future should be focused on more complex traits controlled by a series of interacting genes and gene blocks such as those involved in defence reactions against pests and diseases (Chapman et al. 2012). New hybrids need to combine multiple diseases and insect resistance, although implementing many different resistance resources in breeding programs will be challenging (Miedaner and Juroszek 2021). Genotyping by sequencing and high-throughput phenotyping could be used to achieve these goals (Smith 2021).

6.3.1.2 Innovative Breeding Strategies for the Increased Heat and Drought Tolerance

As a most complex quantitative trait, breeding for heat and DT has been usually achieved through several approaches such as conventional, molecular, and transgenic breeding (Bänzinger et al. 2000). Within the breeding program, genetic gain estimation is essential in measuring the efficiency of the applied breeding approach (Mitrović et al. 2016). There have been several studies of genetic gain estimation in breeding for heat and DT in the last two decades. Masuka et al. (2017) reported a genetic gain increase of 109.4 $kg\,ha^{-1}\,yr^{-1}$ under optimal conditions, 32.5 $kg\,ha^{-1}\,yr^{-1}$ under managed drought, and 22.7 $kg\,ha^{-1}\,yr^{-1}$ under random drought conditions in Eastern and South Africa. A recent study of genetic gain for eastern Africa performed by CIMMYT estimated a genetic gain of 130.7 $kg\,ha^{-1}\,yr^{-1}$ under optimal conditions and 79.3 $kg\,ha^{-1}\,yr^{-1}$ under managed drought (Prasanna et al. 2021). In order to achieve continuous genetic gain increase, and thus the efficiency of selection, it is necessary to work on improving selection intensity, the accuracy of selection, and genetic variability within the breeding program with the shortening of a breeding cycle at the same time.

The grain yield under heat stress and combined heat and drought stress are complex polygenic traits, and the whole-genome prediction (genomic prediction, GP) proved to be the most promising method for yield improvement (Mitrović et al. 2020; Prasanna et al. 2021). The GP method can increase selection intensity and shorten the cycle time, leading to genetic gain increase. GS has recently become a most promising tool in maize breeding, mainly due to several important facts: the development of cheap and efficient methods for genotyping individuals with a large number of markers (mainly Single nucleotide polymorphisms -SNPs), the development of efficient statistical methods for processing very complex and multidimensional data, as well as the development of strong software packages and computing capacities that can cope with such a volume of data (Kadam and Lorenz 2018).

A large number of studies, based mainly on simulations or historical phenotypic data (Mitrović et al. 2016), examined the effects of various factors on the efficiency of GS, primarily the size and type of training population (Zhao et al. 2012), the number and type of molecular markers (Zhang et al. 2015), the choice of the optimal statistical model (Technow et al. 2012), as well as the method of integration into existing breeding programs. In one of the first studies comparing the efficiency of GS to the pedigree method in eight bi-parental populations under drought stress (Beyene et al. 2015), after three cycles of GS, a 7.3% higher efficiency was found comparing to one cycle of phenotypic selection in the same time frame. The GS also affects the shortening of the selection cycle, as presented in Zhang et al. (2017). The genetic gain from a selection of 100 kg ha^{-1} yr^{-1} was established using the rapid cycling GS method. Combining GS methods and double haploids (DH) enables the strategy of testing-half-and-predicting-the-remaining based on the marker data, which can also enhance the efficiency of breeding programs (Prasanna et al. 2021).

6.3.1.2.1 Double Haploids (DH)

In recent years, DHs technology has become an effective alternative to the traditional method of inbred line development (Prigge and Melchinger 2012). Major advantages of this technology over the pedigree method include the development of completely homozygous lines in two crop seasons (shortening of the breeding cycle), reduced costs in maintenance breeding, higher efficiency in molecular breeding methods, complete fulfilment of Distinctness, Uniformity, and Stability (DUS) criteria for cultivar registration and protection (Sserumaga et al. 2016). The DH production involves in vivo induction of maternal haploids with the specific male haploid inducer genotype, identification of haploids from diploids at the seed or seedling stage, and chromosome doubling of haploid seedlings using colchicine and selfing of the fertile plants (Chaikam et al. 2019).

The use of DHs technology can improve the selection gain, which is crucial in breeding for drought and heat tolerance. Higher genetic gain in breeding programs that use the DH technology results from reduced time in the inbreeding and more accurate phenotyping of DH lines compared to lines in early selfing generations over multiple locations and years (Yan et al. 2017; Sserumaga et al. 2016). Integration of DH with other technologies such as molecular breeding, bioinformatics, and high-throughput phenotyping offers further opportunities to develop better commercial products. In the past decade, the most promising route to accelerate generation cycles and increase the precision of selection is combination of DH technology and GS method (Liu et al. 2017).

Another great potential of DH technology is allowing for more effective access to the genetic diversity from the landraces and the open-pollinated varieties. The DH lines from landraces and the open-pollinated varieties can be successfully evaluated in replicated yield trials. After that, the lines with desirable traits can be easily incorporated into the elite germplasm (Strigens et al. 2013). One of the first studies which examined the potential use of DH lines for tolerance to drought and heat stress was performed by Beyene et al. (2013). The obtained results showed that the use of the DH lines developed from drought-tolerant maize source populations performed much better in testcross than the commercial hybrids in stressed and non-stressed environments.

6.3.1.2.2 High Throughput Phenotyping

Quality phenotyping is essential for reducing the gap between genotype and phenotype, especially for quantitative traits such as drought and heat tolerance (Tuberosa 2012). Recent advances in sensor technology and data processing made possible phenotyping of numerous secondary traits that can add value to the breeding for targeted environments and contribute to a better crop performance (Reynolds et al. 2020). High-throughput phenotyping methods cover the ground base and aerial sensing methods, measuring crop characteristics throughout the season. Thus, in breeding for drought and heat stress tolerance, key traits like early vigour or canopy senescence can be easily quantified and predicted by the appropriate statistical model. High-throughput phenotyping also plays a key role when evaluating a training population in GS models. Such datasets provide the basis for developing a statistical model that will predict the phenotypic performance of the untested individuals in a breeding population (Cobb et al. 2013). That way, integrating GS and high-throughput phenotyping provides an opportunity to increase selection intensity, improve selection accuracy, and improve the decision support system (Araus et al. 2018).

6.3.2 Small Grains

Small grain cereals are important constituents of human and non-ruminant animal nutrition, providing calories, proteins, and different nutrients (Shewry and Hey 2015; Langridge 2018). Among them, wheat is the most widely grown cereal crop in the world, while barley ranks fourth in the global cereal production, after maize, rice, and wheat (FAOSTAT 2021). Both wheat and barley are temperate crops, and weather conditions in Europe, especially Western Europe, are favourable for their production. In Europe, wheat and barley are the two most important cereals, contributing 35% and 60% to the global wheat and barley production, respectively (FAOSTAT 2021).

From the early 1990s onwards, the rate of yield increase of European major crops has flattened (Schauberger et al. 2018). The causes of grain yield stagnation may be related to agronomic, genetic, and climatic factors (Le Gouis et al. 2020). Generally, different studies with historical sets of cultivars showed that the genetic progress of major crops was constant, indicating that grain yield did not reach the plateau (Mitrović et al. 2016; Mirosavljević et al. 2020a). Also, the influence of agronomic causes, such as changes in soil quality, crop rotations, and application of fertilizers and pesticides, is unlikely the main cause for yield stagnation (Brisson et al. 2010). Therefore, grain yield stagnations are mostly driven by changes in environmental conditions during growing seasons. Ongoing climate changes, characterized by the increase in frequency and intensity of climate variability, as well as the trend of global warming, endanger worldwide cereal production (Zampieri et al. 2017). The recent projections showed that the risk of adverse weather events in the main European wheat-growing area would increase more than twofold by the end of the century (Trnka et al. 2015). The effect of climate change will vary across different European regions. In the northern parts, there is a possible positive effect on crop production, while in the southern areas climate change will result in a decrease in grain yield and an increase in yield variability (Bindi and Olesen 2011).

Small grain cereals are highly sensitive to the influence of temperature and precipitation extremes, and different environmental conditions unfavourable for European wheat, such as drought and heat stress, water lodging, or frost, will substantially increase with climate change (Trnka et al. 2014). Heat stress during the anthesis period significantly decreases grain number per spike, while during the grain-filling period notably reduces grain weight (Mirosavljević et al. 2021). Moreover, the appearance of late frosts during the stem elongation or heading period significantly reduces grain setting, decreasing grain number per unit area and grain yield (Martino and Abbate 2019). Drought periods during the stem elongation, anthesis, and grain-filling negatively affect crop development, reducing grain number per unit area and grain weight (Farooq et al. 2014; Hlaváčová et al. 2018). On the other hand, storms with heavy rainfalls during the grain-filling period, especially near the harvest time, result in crop lodging and a decrease in both grain yield and grain quality (Niu et al. 2016).

6.3.2.1 Major Traits for Adaptation

Crop production needs to be doubled by 2050 (Pastor et al. 2019) and climate changes represent the major challenge in achieving that goal. However, uncertain prediction of climate variability represents a notable problem for cereal breeders regarding the definition of appropriate traits of an ideotype for climate change. The most important traits include adjustment of crop phenology, improvement of crop water status, photosynthetic traits, root architecture, and the major grain yield traits.

An adjustment of small grain's phenological pattern to avoid stressful conditions during critical developmental stages has been recognized as one of the most valuable options for grain yield improvement (Mirosavljević et al. 2019). Early canopy closure, a result of higher leaf appearance rate and leaf number, is related to improved radiation interception and decreased evapotranspiration, representing a valuable mechanism for drought-prone environments (Foulkes et al. 2009). Furthermore, the introduction of early maturity cultivars could decrease the risk of drought and heat stress during the grain-filling period. However, this drought avoidance strategy during the grain-filling is insufficient since unfavourable conditions during pre-anthesis period are becoming more frequent. In addition, early cultivars intercept less global radiation, resulting in a decrease of grain yield potential. Therefore, the development and cultivation of heat- and drought-tolerant cultivars could be a more promising strategy.

Canopy temperature (CT) is one of the most important physiological traits, used to identify heat and drought-tolerant genotypes (Mason and Singh 2014). CT provides information about stomata conductance and crop water status (Rebetzke et al. 2013). Under drought conditions, cooler canopy is associated with better hydration status; while under irrigated conditions, it indicates greater gas exchange. Genotypic differences in CT are a result of variation in stomata aperture, transpiration, root architecture, and sink balance (Reynolds et al. 2012). CT measurements integrate fast, low-cost, and non-destructive methods, using hand-held or airborne measurements, showing a high relationship with grain yield.

In cereals, stay green traits are closely related to grain yield potential under both favourable and unfavourable conditions (Christopher et al. 2016). The stay green, maintenance of leaf photosynthetic capacity and chlorophyll content, is an important adaptive trait for heat and drought-prone environments (Gregersen et al. 2013). Further improvement of stay green duration, as well as crop greenness at different grain-filling stages, should be related with increased N uptake and grain remobilization in order to increase both grain yield and grain nitrogen concentration (Mirosavljević et al. 2020b). Recently, the introduction of unmanned aerial vehicles in precision farming and regular breeding programs became common activity and it could be used for easy crop greenness estimation due to a high resolution, easy and fast operation.

Root architecture alternation related to the adaptation to higher temperature has been recognized as an important goal in order to confront climate change effects. Identifying root structure that would lead to higher soil moisture capture under water stress should be closely related to target environments. In the environments that retain moisture in upper soil layers, shallow root systems have advantages for rainfall extraction that do not infiltrate too deep (Ehdaie et al. 2012). On the other hand, root traits, such as deep roots with high biomass, longer seminal roots, and higher number of small diameter roots, are related to the increased water extraction capacity from deeper soil layers maintaining high productivity (Becker et al. 2016). To mitigate the negative effects of climate changes on wheat production, combination of root system enhancement and adjustment of production technology is required to improve wheat yield and profits (Woo et al. 2020).

Grain yield genetic gain in wheat and barley during the past century was highly related to the improvement of harvest index – the ratio of grain yield to total plant yield (Mirosavljević et al. 2020c). Harvest index in modern cultivars is close to upper theoretical values (Fischer and Edmeades 2010), and further biomass increase is often followed by harvest index decrease due to sink limitation (Aisawi et al. 2015). Therefore, one of the main strategies for grain yield increase under conditions of climate change should be related to spike fertility and grain number improvement. Generally, grain number per unit area is the main grain yield determinant in main cereal crops (Mirosavljević et al. 2018), and under future climate changes drought tolerant 'wheat ideotype' should be linked to higher number of fertile florets per spike (Senapati et al. 2019) (Table 6.1).

6.3.2.2 Innovative Breeding Approaches

To meet the demands of future global food population, genetic gain in grain yield should be accelerated from the current rate (less than 1% per year) up to 2.4% (Ray et al. 2013). On the other hand, climate change threatens to additionally decrease yield gains. Also, the development of climate-smart small grains cultivars represents a difficult task due to complex polygenic control of target traits, low heritability, presence of significant genotype by environment interaction, and expensive and time consuming genotyping of large set of genotypes (Hu and Xiong 2014).

One of the new approaches in the development of new cereal cultivars is combining commercial breeding activities with crop physiology studies. The involvement of crop

Table 6.1 Traits associated with yield adaptation under heat and drought stress in cereals.

Traits		References
Phenology/ morphology	Adjusting heading date	Lopes et al. (2018)
	Early ground cover	Richards et al. (2010)
	Leaf rolling	Reynolds et al. (2007)
	Pigments	Salika and Riffat (2021)
	Leaf wax layer	Xue et al. (2017)
Dry matter partitioning	Harvest index	Sabella et al. (2020)
	Dry matter accumulation	Cosentino et al. (2019)
	Spike fertility	Reynolds et al. (2009); Semenov and Stratonovitch (2013)
Root architecture	Shallow root systems – moisture in upper soil layers	Ehdaie et al. (2012)
	Deep root systems – moisture in deep soil layers	Becker et al. (2016)
Radiation/water use efficiency	Stay green	Mirosavljević et al. (2020a,b,c)
		Latif et al. (2020)
	Radiation use efficiency	Taube et al. (2020)
	Water use efficiency	Hatfield and Dold (2019), El Habti et al. (2020)
	Canopy temperature	Rebetzke et al. (2013)
	Grain filing	Ma et al. (2017)

physiology in breeding activities already plays an important role in CIMMYT programs, showing significant genetic gain in the International Wheat Improvement Network testing environments. According to Reynolds and Langridge (2016), the physiological approach could be involved in different breeding activities, such as designing the plant ideotype, selection of parents and progeny, genetic resource characterization, and genetic dissection of valuable physiological traits.

Recent progress in high-throughput phenotyping methods based on the modern ground-mounted and aerial vehicles alleviated the field phenotyping bottleneck (Crain et al. 2016; Haghighattalab et al. 2016). The involvement of the phenotyping system provides different information including CT, water and nutrient status, photosynthesis, biomass accumulation, and main agronomic traits. The integration of high-throughput phenotyping data for GS provides a great opportunity to improve prediction and selection accuracy in wheat breeding (Crain et al. 2018). The prediction accuracies of GS models in multi-environments could be increased by including correlated traits to the trait of interest (Guo et al. 2020).

6.3.3 Oil Crops

Oil crops are grown worldwide and in many economies as a vital part of the agricultural sector. For European agro-ecological conditions, traditional oil crops are mostly annual and biannual crops such as sunflower, rapeseed, soybean, castor, poppy, or pumpkin seed, but novel or speciality oil crops such as linseed, camelina, groundnut, sesame, safflower, and others could be of particular interest regionally. Given their value for diverse food and non-food applications, the total production of oil crops has increased over the past 30 years due to a steadily growing demand for food, feed, and non-food uses (e.g. biofuel) of vegetable oils and their by-products. Apart from improved agronomic procedures and better plant protection, plant breeding has made significant contributions to increasing harvest yield per unit area for all major oil crops (Marjanović-Jeromela et al. 2019a,b). Besides yield, plant breeding has contributed to better oil quality as illustrated by low erucic acid rapeseed, high oleic acid sunflower, or low linolenic acid soybean as most prominent examples. Moreover, biotechnologies have been implemented during the last decade to develop new oil qualities and fatty acid compositions unknown before, such as long-chain polyunsaturated fatty acids for various speciality applications of vegetable oils (Cahoon et al. 2009; Ruiz-Lopez et al. 2014).

Climate change, which is characterized by higher temperatures, elevated atmospheric CO_2 concentrations, extreme climatic hazards, and low water availability, is posing serous treat to oil crops production worldwide. Even though most oil crops are currently grown at temperatures that meet their needs, temperatures may rise in the foreseeable future due to global warming and climate change. In that case, the critical thresholds for growth, flowering, seed set, and oil production of oil crops could be reached and even crossed. An increase in temperature and a decrease in precipitation are significant factors that regulate yield and seed oil concentration (Jaime et al. 2018; Namazkar et al. 2016). These changes also affect crop productivity and seed or oil quality and how biotic stresses impact such crops (Ortiz et al. 2014). The yield of oil crops worldwide is affected also by drought caused by climate change, which is the subject of many studies regarding abiotic stress in crops generally (Snowdon et al. 2021).

Accompanying effects of heat impose additional stress on oil crop productivity, which can also be followed by high light stress (Zandalinas et al. 2020). Differences among oil crop cultivars in their reaction to temperature extremes seem larger than other abiotic stresses (Rodríguez et al. 2014), thus highlighting the importance of genetic manipulation and plant breeding as essential tools for adapting these crops to future climate change, since the frequency and intensity of heatwaves are rising and have a more deleterious effect on crop yield than floods, frost, and other weather extremes (Lesk et al. 2016). High temperatures harm seed forming and development, thus decreasing final yield from 15% to 77% depending on the plant developing phase at the moment of occurrence (Gan et al. 2004). Increasing CO_2 concentration may affect oil crops production positively, but this may be annulled by higher temperatures, lower water use efficiency, and accelerated development, causing lower seed and oil yields.

The ability of an oil crop plant to adjust to seasonal variations during development is an indicator of its phenological plasticity (Bloomfield et al. 2014), which is a desirable trait in the context of breeding crops resilient to climate change. According to Nicotra et al. (2010), functional traits that can be used in investigating phenotypic plasticity in the context of climate resilience include:

- root traits (specific length as relates to precipitation pattern shift under climate change)
- leaf traits (specific leaf area as an indicator of relative growth rate, photosynthesis capacity, leaf duration, and leaf nitrogen content; leaf size, shape, thickness, and pigmentation)
- flowering (time, flower size, and phenology as indicators of species' ability to respond to climate change)
- seed (size and number per fruit as an indicator of fitness).

Improving these traits through breeding and modification of crop management practices could enable oil crops to cope with the negative impacts of climate change successfully. Thus, a better understanding of plant adaptation to drought and other treats is of great interest thought prioritizing climate-smart traits. For example, in sunflower, growing early mature hybrids to avoid drought in flowering time, a shift in planting dates, and displace of growing areas of sunflower could be developed, assessed, and combined to partly cope with adverse impacts of climate change (Debaeke et al. 2017; Jocković et al. 2019; Cvejić et al. 2019). Since many pathogens attack sunflower, increasing resistance is also an essential task of sunflower breeders. That is why it is necessary to achieve long-term tolerance or resistance to a particular pathogen (Jocić et al. 2015). In oilseed rape, specific crop physiological traits can become increasingly important under climate change to enhance productivity and reduce yield variability. Such as the extension of the duration of the late reproductive phase, increasing the number of seeds per pod, which is the result of higher photoperiodic sensitivity in the vegetative phase, an increase of the harvest index, for example by increasing the number of seeds per pod and pod length, and greater light use efficiency through changes in plant architecture (Marjanović-Jeromela et al. 2016, 2019a,b; Pullens et al. 2021). Among minor oilseeds crops, camelina can become attractive as a new crop in a crop rotation that will contribute to agro-biodiversity considering modest requirements for agro-ecological conditions and low input features of its production (Royo-Esnal and Valencia-Gredilla 2018; Zanetti et al. 2021; Marjanović Jeromela et al. 2021).

6.3.3.1 Innovative Approaches to Improvement of Climate-Resilience

In order to better understand the genetic basis of a complex traits associated with climate resilience, a detailed and precise assessment of the phenotype is necessary. However, although much effort has been put into advancing genotyping technologies, phenotyping has been neglected and progressed more slowly, leading to limited capability to precisely and efficiently characterize of complex quantitative traits. Recent advances in phenomics, thorough development of tools for high-throughput phenotyping, enabled its use to explore multivariate phenotypic information (Jocković et al. 2021).

Plant tolerance, oil crops included, to stressful conditions induced by drought largely depends on the condition of the root system. Hence, phenotyping platforms for non-invasive root analysis can be very helpful in breeding for modified root characteristics. In this context, the first attempts of sunflower root phenotyping were made, using an automated phenotyping platform, GROWSCREEN-Rhizo (Dimitrijević et al. 2018).

High-throughput root phenotyping could be further combined with the use of DH mapping populations, as it was done in rapeseed for the identification of root architectural traits which correlate with shoot biomass and QTLs linked to root traits (Shi et al. 2013). This approach enables the simultaneous identification of traits and QTLs of interest that could be further used in breeding. The same stands for the combination of root phenotyping with genome-wide association studies (GWAS) that was used to detect marker/trait association in rapeseed and for the development of SNP markers associated with different root architectural traits that can be used MAS for root traits in rapeseed (Arifuzzaman et al. 2019).

Another integrated approach to climate resilience improvement could be a combined analysis of gas exchange, physiology, and morphology. This approach was used in both sunflower (C3 crop) and maize (C4 crop) in order to assess the impact of heat and drought stress on those two crops with conclusion that selection for heat and DT could lead to improvement of the yields of C3 crops grown in semi-arid regions (Killi et al. 2017).

6.4 Genomics Strategies for the Development of Climate Resilient Crops

Biotechnology scientists believe that new tools and methods can work in conjunction with better breeding to give maximum benefit to human being. They expect that, by using conventional and biotech methods, crops can be improved to yield more in optimum as well as harsh climatic conditions can be made healthier, pest resistant, and can be developed in ways that create more energy for use in ethanol production. For example, there is a prediction that by using modern tools corn and soybean yields can be increased by 40% over the next decade. Having this in mind, it is felt, that biotechnology in agriculture would become the key to feeding the growing population; take care of agricultural production in the adverse climatic conditions for food, feed, and production of ethanol.

During the last decade, significant progress in developing the resources, tools, and approaches to allow more rapid development of improved crops has been achieved. These novel approaches to germplasm characterization and population improvement include genomics, transcriptomics, and metabolomics as well as non-destructive, dynamic high-throughput phenotyping (phenomics) (Kole et al. 2015). The effective application of all these modern tools and strategies in the development of climate-resilient crops adapted to a changing environment can also be assisted by an evolutionary perspective. Variations of the genome found in wild crop relatives could be used for the creation of environmentally adapted and climate-resilient crops (Henry 2014).

6.4.1 QTL Mapping and MAS

Present knowledge related to molecular marker technologies offers a range of possible novel approaches to improve the efficiency of crop breeding programs worldwide. Molecular markers closely linked to genes of interest could be used in MAS and for pyramiding of two or more desirable genes within a genotype. To ensure their effective application in breeding, molecular markers used for MAS should have a high level of reproducibility, as well as the capacity to enable screening of large 'mapping' populations. The close interaction between breeders and biotechnologists may also improve and accelerate the effective implementation of MAS in breeding programs for the development of climate-resilient crops.

The procedure that precedes the MAS is mapping, which involves detecting loci associated with targeted traits, and determining their positions on the genetic map, and effects. The creation of genetic maps requires a large scale genotyping and phenotyping of many targeted populations, as well as use of appropriate statistical models that enable the detection of significant associations between phenotypic and genotypic variability for targeted traits (Brbaklić et al. 2015b; Mikić et al. 2016a; Trkulja et al. 2019).

Nowadays high-throughput genotyping platforms are available for the detection of SNP (Loman et al. 2012). Some of the most commonly used genotyping approaches based on sequencing are genotyping-by-sequencing (GBS), restriction-site-associated DNA sequencing, double digest restriction associated DNA, skim-based genotyping by sequencing, repeat amplification sequencing, exome sequencing, and whole-genome re-sequencing (WGRS). Fixed SNP genotype sequences may be more desirable than NGS-based ones due to higher throughput at a lower cost per sample with minimal data analysis required. Recently, Rasheed et al. (2017) compiled a review of chips and genotyping tools that could be used for crop improvement.

Robust, precise, and high-throughput phenotyping systems are also available for non-destructively measuring of quantitative phenotypic variations across cells, organs, and tissues, developmental stages, years, environments, species, and research programs (Fahlgren et al. 2015; Coppens et al. 2017). These platforms for high-throughput phenotyping use recent advances in sensors for imaging plants, ranging from remote sensing including spectroradiometry, light detection and ranging (LiDAR), visible to far-infrared, hyperspectral, thermal, fluorescence, and 3D laser scanning to trichromatic imaging in conjunction with advanced autonomous vehicles (Lobet 2017; Tardieu et al. 2017). Autonomous platforms such as unmanned aerial vehicles and ground robots that are equipped with multiple sensors can take near real-time images of the experimental plots both several times per day and over the entire season (Perez-Sanz et al. 2017). The use of such phenotyping platforms in breeding programs for resilience screening could facilitate the identification of genes and genotypes that could cope with future extreme climatic conditions (Varshney et al. 2018; Paul et al. 2019; Mirosavljević et al. 2021).

One of many definitions of climate-resilient crops and crop varieties indicates that they are adopted to cope with abiotic stresses such as drought, heat, flooding, salinity, and shorter growing season (early-maturing crops), as well as pests associated with changes in weather or climate patterns (disease and pest resistance) (Acevedo et al. 2020). Most of these traits are determined by a large number of genes (QTLs) and are significantly influenced by environmental factors. In QTL mapping, early generation crosses (F1 or F2) are used to parse quantitative variability, based on the distribution of individuals relative to the parent generation. QTL mapping has been shown to be extremely useful in identifying genomic regions that determine certain quantitative traits by developing and use of various mapping populations such as RILs, near-isogenic lines (NILs), DHs, etc. (Trkulja et al. 2012; Salgotra and Stewart 2020).

However, this approach has a number of drawbacks. First, allelic variability at each crossing is limited because typically two parents were used to create a mapping population. Second, since the early generations of crosses are analyzed, the number of recombinations per chromosome is generally small, limiting the resolution of genetic maps. In addition, for many organisms, the creation of mapping populations is either not possible (e.g. in the human population) or takes a very long time (e.g. in most tree species) (Hall et al. 2010).

To overcome at least some of these barriers, an alternate approach, association mapping, has been used for determining the genetic basis of complex traits in natural or breeding germplasm. It is based on the use of evident variability in germplasm of different origin, which is significantly higher compared to mapping populations. In addition, existing recombination occurred naturally throughout evolutionary history, so that blocks of bound genes are significantly smaller in the associative, compared to the QTL mapping population, which greatly affects the mapping resolution (Nordborg and Tavare 2002). While linking analysis is based on the detection of non-random relationships between genotypes and phenotypes in populations of offspring produced by crossing two genetically divergent varieties, associative mapping focuses on establishing relationships within populations of unrelated individuals. In other words, the time period to the first common ancestor of any two individuals from a population of unrelated individuals will be longer than in populations of relatives. This is what makes linkage disequilibrium (LD – gamete imbalance) mapping more suitable for 'finer' mapping: the higher the number of generations, the more likely it is that recombination of genetic material within a given species will occur, while within mapping populations only a few generations of material exchange occur within homologous chromosomes. An increase in the recombination rate leads to regrouping of chromosomal segments, and as a consequence, there is a reduction of LD within short distances, which increases the mapping resolution. For comparison, in mapping populations, cross-over occurs in several generations, resulting in the exchange of genetic material at the level of several megabases, while the material in associative studies involves recombination of segments measured in kilobases (Ersoz et al. 2007).

Associative analysis is built on the concept of gamete imbalance, i.e. non-random related inheritance of alleles of two loci. It identifies polymorphisms near or within a gene of interest that controls the phenotypic differences between genotypes (Brbaklić et al. 2013, 2015a; Mikić et al. 2018; Soto-Cerda and Cloutier 2012) and facilitates the search for functional variations in genes of interest (Breseghello and Sorrels 2006). Linkage disequilibrium with its ability to be maintained over many generations between loci that are genetically linked enables MAS (Rostoks et al. 2006).

Advances in genomics have led to the development of next-generation sequencing (NGS) tools, which have significantly shortened the time needed for trait mapping using old approaches, such as bulked segregant analysis, for instance, that could be replaced with whole-genome sequencing-based high-resolution trait mapping (Schlotterer et al. 2014). The availability of draft or whole-genome sequences for a number of crop species and reduced sequencing costs enabled resequencing of hundreds to thousands of individuals and deployment of WGRS or low coverage sequencing of the entire population or bulks of the extremes. GBS and WGRS of entire mapping populations or extreme pools provide large-scale genome-wide SNPs for conducting high-resolution trait mapping that has already been used in many crop species (Pandey et al. 2016). On the other hand, bulk or pool-based sequencing has become the method of choice for rapid trait mapping in recent years (Pandey et al. 2017; Singh et al. 2017).

In the last few decades, the location of a large number of genes within specific chromosomes, chromosomal arms, or regions has been determined in all major crop species, enabling the potential use of molecular markers in crop genotyping, mapping, and breeding (Nadeem et al. 2018; Salgotra and Stewart, 2020). Using different approaches and methodologies QTLs have been identified in important crops for different traits associated with climate resilience such as:

- DT in maize (Mikić et al. 2016b, 2017; Brown et al. 2011), rice (Huang et al. 2010), wheat (Kollers et al. 2013), soybean (Hwang et al. 2014), sorghum (Morris et al. 2013), foxtail millet (Jia et al. 2013), pearl millet (Bidinger et al. 2007), among other crops.
- Heat tolerance in rice (Zafar et al. 2018), wheat (Sharma et al. 2021), maize (Frey et al. 2016), tomato (Wen et al. 2019), cowpea (Lucas et al. 2013), sorghum (Xu et al. 2000), etc.
- Salt tolerance in durum wheat (Munns et al. 2000), tomato (Fooland et al. 2001), rice (Sabouri et al. 2009), barley (Siahsar and Aminfar 2010), etc.
- Flooding tolerance in different crops (for review see Mustroph 2018).

After obtaining the genetic maps with identified candidate genes and associated markers, a number of molecular breeding approaches could be used to introgress genomic regions into elite lines (Huang et al. 2020). MAS is useful for the introgression of a few loci (<10 loci) for improving elite varieties. It has been widely used in a number of crops to incorporate desired traits into elite cultivars through marker-assisted backcrossing (MABC). Nowadays there is an emphasis on early generation selection by using the forward breeding (FB) approach (Williams et al. 1990). In this approach, a set of diagnostic markers for must-have traits is used to screen early generations and a subset of lines is advanced to the next generation. This approach is useful for enhancing the selection intensity to accelerate the genetic gains. Marker-assisted recurrent selection may be appropriate for more complex traits controlled by up to 40 loci, such as drought, heat, and salt tolerance. This approach can be used to develop superior climate-resilient lines with an optimum combination of superior alleles through repeated inter-crossing (Conner 2004). The main aim of a recurrent selection program is to increase the frequency of favourable alleles for traits of interest, or to gradually accumulate, by recombination cycles, the desirable and available alleles in different parents while conserving the genetic variability (Mahalingam et al. 2018). For instance, potato heat stress-tolerant genotypes were successfully developed using this approach (Benites and Pinto 2011).

6.4.2 Genome-Wide Association Studies and Genomic Selection

High-throughput genotyping technologies and large-scale genomic resources enabled identification of a large number of SNP candidates for high-resolution linkage map production (Huang et al. 2017). Furthermore, GWAS are now an efficient replacement for QTL mapping for the study of quantitative traits, important for crop adaptation to adverse climatic conditions. GWAS could provide efficient screening of representative populations and enables high-density genotyping to identify genomic regions linked to the phenotypic traits. It has been successfully applied for the identification of genomic regions associated with traits related in climate change adaptation in cereals and other crop species (Kole et al. 2015; Nadeem et al. 2018). In these crops, GWAS has also provided information on marker-trait associations, of importance for its efficient use in plant breeding programs (Huang et al. 2017).

GS is one of the latest methods in plant breeding. It involves combining data obtained by genome analysis using a large number of molecular markers and phenotypic data of significant agronomic traits, on a selected set of genotypes. The combined data are used to calculate the parameters of the prognostic model, which are used to select new genotypes only on the basis of genomic data. GS makes the selection process cheaper and faster, because the selection of new genotypes is based only on genomic data, and not on phenotypic data. Unlike the classical approach of MAS, where individual molecular markers are used in which a significant association with agronomic traits has been established, GP involves the simultaneous use of all molecular markers across the genome during selection (Mahalingam et al. 2018). This approach is especially desirable when selecting quantitative traits, such as tolerance to abiotic stresses, which allows the characterization of the effects of all, even genes with small effects, on phenotype expression, which will increase the precision and efficiency of breeding crops. Simulation studies indicated that GS could increase the correlations between predicted and real breeding value over several generations, resulting in cost reduction and increased rates of genetic gain (Habier et al. 2009; Hefner et al. 2009).

GS is currently used for DT improvement in maize in CIMMYT (Crossa et al. 2014). Initial efforts towards the introduction of GS are made for other crops as well, including sugarcane, chickpea, pigeon pea, groundnut, white spruce, cassava, apple, and wheat, among others (Khan et al. 2016). GS has so far been used in sunflower for the prediction of hybrid performance (Reif et al. 2013), hybrid oil content (Mangin et al. 2017), and *Sclerotinia* mid-stalk rot (Livaja et al. 2016). The results obtained in these studies showed that GS could lead to a major improvement in breeding efficiency compared to the classical general combining ability (GCA) modelling when either one or both parents are not well-characterized. However, time is needed for validation of the results of GS breeding programs, accuracy of prediction of performance, and genetic gains obtained.

6.4.3 Epigenomics

Modern sequencing technologies have opened the way for the study of the epigenetic events at a whole-genome level, with genome-wide mapping and identification of epigenetic marks being most frequently used in crops. Epigenome characterization, along with the new (epi)genome editing (GE) tools, could lead to the creation of new epiallelic variants by inducing the changes in chromatin or DNA methylation and crop improvement through epigenome engineering (Springer and Schmitz 2017).

Natural or induced epigenetic modifications can lead to the appearance of new epi-allele variants in crops that have an effect on stress resilience. If heritable, these epigenetic variants could be used as breeding targets for crop improvement (Mladenov et al. 2021). For instance, in maize, Mao et al. (2015) found a MITE (miniature inverted-repeat transposable element) insertion in the promoter of a NAC transcription factor encoding gene to be associated with variation in drought-tolerant maize phenotypes. The authors identified *ZmNAC111* gene as a potential candidate for GE, and its MITE–allele as a potential selection target for the genetic improvement of DT in maize.

The emergence of CRISPR/Cas9 and dCAS technology has provided new routes for the creation and exploitation of epigenetic variability in crops. These new tools enable precise manipulation and/or modification of epigenetic traits and modification of plant phenotype (Hilton et al. 2015; Moradpour and Akmar Abdulah 2020). The engineered DNA-binding domains such as zinc fingers (ZFs), transcription activator-like effectors (TALEs), and the endonuclease-deficient Cas9 (dCas9) protein can be used to introduce specific chromatin marks at particular loci and enable elucidation of how the activation or repression of particular chromatin regions affect the crop phenotype during exposure to stress (Bilichak and Kovalchuk 2016). The knowledge gained, along with the use of epigenetic tools such as QTLsepi or SNPsepi, could be further used for trait improvement purposes in crops and the development of new, efficient, and GMO-free crop improvement method (Bilichak and Kovalchuk 2016).

For optimal use of those new epigenetic and epigenomic tools in crop improvement, further studies are needed on the association of stress-induced gene expression changes with modifications in histone and DNA methylation, the mode of inheritance of these modifications, as well as their value regarding crop adaptation to stress (Chinnusamy and Zhu 2009; Varotto et al. 2020).

6.4.4 Integration of -Omics Technologies

The latest developments made in the field of -omics have opened new paths to understand and study the complex mechanisms of abiotic stress tolerance in plants (Ozturk Gokce et al. 2020). In most of the complex traits, such as DT, using only one -omics approach is not sufficient to develop novel resilient crop varieties. Hence, efficient integration of different tools, techniques, and approaches looks like a promising strategy (Chaudhary et al. 2019). Dissecting the relationship between differentially expressed transcriptome, proteome, and metabolome could potentially improve the understanding of the molecular mechanisms underpinning abiotic stress tolerance in plants. The successful integration of the transcriptomics (gene), proteomics (proteins), metabolomics (metabolite), ionomics (analysis of elemental compositions), epigenomics (inheritance), interactomics (protein–protein or protein–DNA interactions) will facilitate the breeder to select the potential candidates and best traits to generate and improve the crop productivity under abiotic stress (Shah et al. 2018).

The integration of information from various -omics analyses is a prerequisite for a comprehensive understanding of the molecular mechanisms of many important quantitative traits (Jocković et al. 2021). Combining information from different -omics technologies, also known as systems biology, offers novel possibilities for deciphering the complex pathways related to crop response to the environment and provides important answers that can be used for more efficient crop breeding for climate resilience (Wu et al. 2017).

Moschen et al. (2016, 2017) used systems biology approach and combined transcriptomic and metabolomic study in order to characterize the process of leaf senescence in sunflower as well as to identify pathways related to hub transcription factors involved in sunflower drought stress response. An integrated approach to study sunflower response to water-deficit conditions was described by Sarazin et al. (2017) who used genomics (RNA-seq and quantitative RT-PCR), physiological (growth, water statute, stomatal conduction evaluations, and transpiration rate), and biochemical analyses (LC–MS). The authors concluded that sunflower water stress tolerance is correlated with a transcriptome fine-tuning that triggers activation of ABA-dependent genes and not to ABA overproduction. In wheat, although progress has been made in genomics and transcriptomics, the other major -omics techniques like proteomics, metabolomics, and phenomics are still lagging behind (Shah et al. 2018). Furthermore, the existing databases of wheat microarray information that are publicly available, open the way for gene co-expression network analysis, through utilization of standardized platforms, such as the Affymetrix Wheat GeneChip.

6.4.5 Genome Editing

The development of new breeding techniques, such as GE, could provide new perspectives for more efficient crop breeding. Generally, traits related to climate resilience, such as tolerance to various abiotic stresses, are complex phenotypic traits controlled by polygenes, and it is usually necessary to study more than a single gene or single class of genes to understand molecular mechanisms underlying respective tolerance. GE could be valuable tool for the evaluation and validation predictive value of a given candidate gene by transferring its best alleles into different genetic backgrounds (Nogue et al. 2016). Since multiple genes can be individually engineered at the same time, GE also could be used to modify linked genes or QTLs (Flavell 2010). Hence, GE technologies provide more precise and more efficient tools to increase crop abiotic stress resilience by means of trait engineering (Miladinović et al. 2021).

As in most of tissue culture techniques, regeneration efficiency could be a bottleneck for the effective application of GE techniques in crop breeding (Miladinović et al. 2019). Many crops are either recalcitrant or have difficult and long regeneration protocols (Taški-Ajduković et al. 2010; Gao et al. 2014). In addition to the already-mentioned problems, in polyploid crops, the GE protocols should enable simultaneous targeting of multiple alleles, which could be an obstacle for GE of wheat and other crops with complex genomes. Thus, it is crucial to determine the efficiency of the sgRNAs selected for the CRISPR/Cas9 system in advance, as the sequence of the target site has a strong influence on the efficiency of the sgRNA (Wang et al. 2014).

In sunflower, up to date, there is only one report on the use of GE. Ynet and Yilancioglu (2018) described CRISPR/Cas9 model for sunflower resistance for biotic and abiotic stresses. The authors determined the suitable genetic changes to perform and design of relevant plasmids, plasmid transfer method into the sunflower genome, as well as the control for whether the plasmid is transferred and for whether GE is effective. Similar to sunflower, wheat is also lagging behind in the utilization of CRISPR-based genome modifications. The hexaploid, large genome size, and the recalcitrant nature in terms of tissue culture are the major obstacles to CRISPR utilization in wheat (Kumar et al. 2019). In maize, grain yield under drought has been increased by negative regulatory gene knockout ARGOS8 locus (Shi et al. 2017).

6.5 Biometrical Tools for Envirotyping

6.5.1 Phenotypic Data Analysis and GE Interaction

In early- and late-stage trials, plant breeders tested the genotypes under different environmental conditions which can be defined as the target population of environments (TPE) (Comstock and Moll 1963). The multi-environment trials (MET) are the important breeder tool for ranking and comparisons of genotype performances and to identify those that are superior in terms of stability and adaptability (Patterson 1997). Focal point of the MET is the occurrence of genotype by environment (GE) interaction or differential response to a set of genotypes across the TPE. The environments have a space and time dimensions. For example, the genotype by location interactions is caused by the variability among the locations, while the genotype by year interaction is influenced by year-to-year variation in uncontrolled weather conditions (Malosetti et al. 2016). The GE interactions are important for plant breeders mainly if they involve genotype rank changes from one to another environment (Cooper and De Lacy 1994; Huhn 1996). In addition, the METs data analysis enables the examination of environment relationships in order to identify the highly correlated and if possible repeatable clusters of environments over years (Yan 2014). If the METs data are integrated with weather data the environmental causes of the GE interaction can be identified (Denis 1988; Van Eeuwijk et al. 1996; Voltas et al. 2002). The collections of external environmental variables that affect the performance of the genotypes under the METs have been referred to as the envirotyping concept (Xu 2016). The envirotyping collectively integrates the genotyping, and the phenotyping allowing better understanding of the GE interactions. Such characterization and prediction can be coupled with phenomic, crop-growth modelling, and genome-wide prediction (Cooper et al. 2014). For all these reasons, the analysis and studying of the datasets obtained from plant breeding METs are of utmost importance (Smith et al. 2001, 2005; Piepho et al. 2012).

There are a number of statistical and biometrical models for the analysis of datasets from the METs. An important question prior to the METs analysis is how to treat the genotype and environment effects in the model? All levels of the population parameters are included for the fixed effects while the random factor only takes a random sample of population levels (McCullagh and Nelder 2008). Usually, the genotypes are considered as a random sample from the breeding population, and the environments are often fixed since repeated across years and locations in a linear model (Smith et al. 2001; Piepho et al. 2008).

A special form of a linear model that contained both linear and random effects is called the linear mixed model (LMM). The LMM is applicable to large scale situations in plant breeding (Piepho et al. 2003, 2008, 2012; Smith et al. 2001; Smith and Cullis 2018). From a wider statistical perspective, the LMM is a special case of more general statistical framework called the generalized linear mixed models (GLMMs). Both, the LMMs and the GLMMs can accommodate the fixed and random effect except that latter can be efficiently used for the analysis of discrete (or non-normal) data which are common in plant breeding experiments (Stroup 2015). The LMMs represent a natural framework for analysis of the METs due to the fact that these data are frequently noisy with heterogeneous genotypic variances across environments. The genetic correlations among pairs of environments can be heterogeneous (Bustos-Korts et al. 2016). In the LMMs, the variance–covariance structures (VCOV) can be incorporated into the random effects of the GE interaction and the residual variation to account for the heterogeneity. In that sense, the most realistic VCOV of the GE effects is the factor analytic (FA) (Smith et al. 2001) for parsimonious description of the interaction effects. The FA structure can be used as a basis for conducting the stability analysis through the Factor Analytic Selection Tools (FAST) approach (Smith and Cullis 2018). Variance heterogeneity can induce the bias in the prediction of breeding values and estimation of genetic variances, especially if the genotypes are not replicated across the environments (Hill 1984). The LMM with an appropriate VCOV structure for accounting for variance heterogeneity problem can improve genotype predictions by giving a more weight to the environments with a lower error variance (Smith et al. 2001; Piepho et al. 2008).

The presence of field spatial variation is the consequence of the dependence or correlation among neighbouring plots and/or extraneous field variation that affects whole columns or row directions (Gilmour et al. 1997). The LMM involves a direct modelling of spatial variation through the auto-regressive correlation structure of order one (AR1) along with row and column directions (Gilmour et al. 1997). It is based on the postulate that two neighbouring plots are more similar (or more correlated) than the plots further apart (Cullis and Gleeson 1991). In some instances, the neighbouring plots can be negatively correlated due to the phenomenon of inter-plot interference (Martin et al. 2004). The inter-plot interference is a consequence of various genetic and environmental factors (Magnussen 1989).

The LMMs can efficiently estimate fixed effects (i.e. Best Linear Unbiased Estimates, BLUEs) and predictions for random effects (i.e. Best Linear Unbiased Predictions, BLUPs) (Piepho et al. 2008; Hu and Spilke 2011). The BLUPs give accurate predictions of genotype performance with respect to regression towards the mean-shrinkage property (Smith et al. 2001; Piepho et al. 2008). The BLUPs and BLUEs can rank genotypes differently because of shrinkage property that makes them closer to the overall mean than their respective BLUEs. The amount of shrinkage depends on the trait heritability and the number of information available for a specific genotype (Piepho et al. 2008). When pedigree and genomic data are available, the BLUPs can account for the genotype relatedness based on pedigree (A-BLUPs) and genomic (G-BLUP) data matrices.

In addition to LMMs, there are additional approaches to get further insight into the GE interaction (Malosetti et al. 2013). They can be used for grouping of genotypes and environments based on interaction patterns or to integrate the genotypic and/or environmental informations to explain the causes of the interaction. The additive main effects and multiplicative interaction (AMMI) model is used for two-way genotype-by-environment data (Gauch 2006). The AMMI model approximates the GE interaction pattern into a lower number of dimensions while preserving the majority of the interaction pattern. Visual inspection of the results through the biplots is widely used for assessing GE interactions. While the AMMI model is based only on the residuals from the additivity, the genotype main effects and genotype by environment interaction effects (GGE) model (also known as the Sites Regression model after [Crossa and Cornelius 1997] also visualizes the interaction patterns in the METs data [Yan et al. 2000] by its integration with genotype main effect). The biplot display based on the GGE model show which genotype performs best and where (which-won-where pattern), and efficiently assesses the representativeness and discriminating ability of testing environments. The GGE biplot can also assist in visualizing the representativeness and repeatability of testing environments (Yan et al. 2011), which should discriminate genetic differences among genotypes included in the trial and represents the target environments where selected genotypes can repeat their performance.

The factorial regression (FR) (Van Eeuwijk et al. 1996) and the partial least squares (PLS) regressions (Vargas et al. 1999) incorporate external environmental and genetic variables directly into the model for interpreting GE interaction (Vargas et al. 1999). The FR is a linear model that allows the inclusion of external variables such as weather or soil data. Van Eeuwijk et al. (1996) described the catalogue of modelling possibilities for the included external variables into the FR model to interpret the GE interaction space, while Malosetti et al. (2004) fitted the FR model in the LMMs framework. One obstacle in the FR application is the high correlation (or multi-collinearity) among the external variables. Under the presence of the multi-collinearity problem, the estimation of the parameters becomes unreliable and unstable which can lead to erroneous interpretations. The PLS regression (Aastveit and Martens 1986; Vargas et al. 1998) is a bilinear model that is robust to the multi-collinearity problem. The PLS model is a dimension reduction approach that can include a large set of the genetic and environmental variables to extract the latent dimensions that account the most of the variability in the X matrix (matrix with environmental or genetic variables) and Y matrix (matrix with the interaction GE effects).

6.5.2 Genomic Prediction Modelling

With the increased requirements an improved accuracy of selection and better selection response, the procedures for GP become very important for many theoretical and applied research studies (Piepho 2009; Heslot et al. 2012; Li and Sillanpää 2012; Crossa et al. 2017). The developed GP models differ in computational efficiency, degree of complexity, and level of the prediction accuracy (Ogutu et al. 2012). The choice of GP approach is influenced by the genetic architecture of the traits of interest, the accuracy of estimation of variance components and heritability, degree of LD, and sample size (Nakaya and Isobe 2012).

A straightforward method was first suggested by Meuwissen et al. (2001). This method is based on multiple regressions where the phenotypic response of the interest is the dependent variable and SNP markers are explanatory variables (predictors). However, as the number of markers increases rapidly through time and number of variables (p) exceeds the number of observations (n) (so-called $n < p$ problem or curse of the dimensionality problem; Bishop (2006) the accurate estimation of the individual effects of the SNP markers becomes impossible by ordinary least squares estimation method (Fahrmeir et al. 2013).

There are many GP approaches adopted from statistical theory for the prediction of phenotypic traits in plant breeding. One of the first and most widely used is the GBLUP model based on a mixed model. The genomic estimated breeding value is predicted by incorporating the additive genomic relationship matrix (GRM) into the mixed model equation. The GRM estimates the proportion of genome identity-by-descent among individuals in a population using whole-genome identity-in-state marker information (Van Raden 2008; Endelman and Jannink 2012). In the approach, the GRM is computed assuming that all markers are associated with an equal amount of genetic variance. The model based on additive GRM resulted in higher predictive accuracy than the models based on pedigree data (Heslot et al. 2014; Crossa et al. 2017; Juliana et al. 2019). Another possibility is the inclusion of a dominance genomic relationship to predict the non-additive genetic variance component (Munoz et al. 2014), and prediction of hybrid performance (Kadam et al. 2016; Ramstein et al. 2020).

In contrast to the GBLUP approach when all markers have the same or equal contribution to genetic variance, the regularization techniques can penalize the effect of some markers in the model and exclude them (De los Campos et al. 2013). For example, the ridge regression (Hoerl and Kennard 1970) equally penalizes all markers assuming that the trait is in the function of large number of markers with small variance (Gianola et al. 2013). Ridge regression BLUP is similar to the GBLUP since their model assumptions (Habier et al. 2007; Endelman 2011). Both models are equally good when predicting the complex traits, but might not be efficient for more oligogenic inheritance architecture. In such cases, the LASSO (least absolute shrinkage and selection operator) regression (Tibshirani 1996) can be used with the relaxed assumption that all markers have the same effect on the phenotypic trait. The LASSO regression is penalized variable selection model that can be forcing some markers out of the model thus leading more interpretative model than those obtained by ridge regression (Tibshirani 1996).

Another group of models known as 'Bayesian alphabet' (Meuwissen et al. 2001) incorporates the genetic architecture via priors describing the distribution of marker effects. BayesA model allows markers to vary in contribution to genetic variance. When compared with RR-BLUP, BayesA model applies the marker-specific shrinkage. BayesB model is similar to BayesA but incorporates hyper-parameter which is specified during each iteration of the Gibbs samples what proportion of markers is allowed to have non-zero effect. In BayesianCpi model (Habier et al. 2011) treated the hyper-parameter as unknown value that should be estimated from marker data and the markers have a common marker variance (Habier et al. 2011).

In recent years, the GP models with different parametric and non-parametric kernels for estimating the genetic distance or similarity between the individuals in the training population were developed (Gianola and Van Kaam 2008; De los Campos et al. 2009; Gota and Gianola 2014; Sousa et al. 2017; Cuevas et al. 2017). Gaussian kernel (GK) methods have been shown to increase the prediction accuracy depending on the trait and the model (Cuevas et al. 2017, 2018). The GK methods can potentially capture complex marker effects in addition to major marker effects and interaction effects (Granato et al. 2018).

The prediction accuracy of the GP models can be improved by the inclusion of the GE interaction component and by allowing the breeders to borrow the information from the genetically related material and from the correlated environments (Burgueño et al. 2012). Jarquín et al. (2014) developed the reaction norm model for GP, where the genetic and environmental values are replaced by the regression of the markers and on the interaction between the markers and the environmental variables. Heslot et al. (2015) proposed the GBLUP model which partitions the GE model into components due to genetic main effects and environment-specific genetic effects. The variance component related to the genetic main effects measures the stability of genetic performance across the TSE, whereas the environment-specific components are deviations from stable effects (Cuevas et al. 2018). One serious obstacle to this approach is that it requires the positive correlation among the environments within the TSE (Bandeira e Sousa et al. 2017; Cuevas et al. 2018). It implies that the training set of environments should reflect the TPEs in order to obtain the reliable environment-specific predictions (Lorenz et al. 2011; Van Eeuwijk et al. 2016). When the association among some of the environments is negative indicating the presence of the important GE crossover interactions more flexible computationally demanding models should be used to approximate the environmental covariance (Crossa et al. 2017; Cuevas et al. 2018). The GGE-based model called 3GS (Jighly et al. 2021) is proposed to overcome this problem and maximize the benefits of the GE interaction to achieve better prediction accuracy.

In recent years, thankful to the increased complexity and volume of the data generated in plant breeding programs open the possibility for application of the advanced algorithms outside the field of statistics and biometrics. The subfield of artificial intelligence such as machine learning and deep learning becomes a standard algorithmic approach for GP tasks. The most common prediction tool in machine learning is the neural networks (NN) and they perform on a similar way as the biological NNs. The NN can efficiently capture the interaction among the input layers, i.e. the markers which resulted in higher prediction accuracy (González-Camacho et al. 2016). In some traits like the plant diseases, the classification problem becomes valuable (Ornella et al. 2014). The probabilistic NNs are used in the classification of the genotypes into groups based on the estimated probabilities.

When selecting and implementing the GP models users should be having in mind that many of the models including the mixed and Bayesian models share the same or similar assumptions and that they are dependent on the genetic architecture of the quantitative trait. Frequently, the complex machine learning GP algorithms resulted in higher prediction accuracy when compared with classical ones, but they have the problem with lack of the biological inference and interpretability (Bayer et al. 2021).

6.6 Final Remarks

Crop production needs to be doubled by 2050 (Pastor et al. 2019) and climate changes represent the major challenge in achieving that goal. However, uncertain prediction of climate variability represents a significant problem for plant breeders regarding the definition of appropriate traits of an ideotype for climate change. Furthermore, climate changes have potentially different impacts on crop performance due to different time points, levels, duration, and stress combinations, and there is no single trait or solution for climate resilience. However, the further alterations in the climate will be mostly related to the warmer temperature, less precipitation, and higher incidence of weather extremes (Dempewolf et al. 2014). Therefore, the most important traits related to heat and drought adaptation will include adjusting crop phenology, improving crop water status, photosynthetic traits, root architecture, and the major grain yield traits. Moreover, breeding strategies should mainly aim at proper exploration of germplasm and harnessing of novel alleles from the wild gene pool (Jha et al. 2014).

An integrated breeding approach, a combination of classical breeding with omics technologies and biometrical tools, can provide new environmental challenges. Facing this challenge involves considering the intimate interaction between environmental conditions and plant structure, functions, and metabolism, which require environmental characterization to be part of all steps of plant phenotyping, from data collection to meta-analyses (Tardieu et al. 2017). However, linking genotype to phenotype remains a considerable challenge and is hampering the optimal application of high-throughput genomics and phenomics to advanced breeding. Critical to success is the need to assimilate large amounts of data into biologically meaningful interpretations (Harfouche et al. 2019). As high-throughput and sequence-based genotyping become routine and high-throughput and precision phenotyping becomes achievable, a complete set of next-generation high-throughput precision envirotyping technologies will also show up (Xu 2016).

6.7 Summary

Climate change is strongly influencing agricultural production and cultivation practices of all major crops. Breeding crop varieties for environmental stresses is a slow and challenging process, as the effects of these stresses on crops are variable and complex. In this chapter, we describe different breeding strategies for tackling the challenges imposed by climate change, including efficient exploration of genetic resources, innovative solutions in crop breeding, high-throughput phenotyping and molecular techniques, and the biometrical data management tools.

Acknowledgment

This work is supported by the Ministry of Education, Science and Technological Development of the Republic of Serbia, grant number 451-03-68/2022-14/ 200032, by the Science Fund of the Republic of Serbia through IDEAS project "Creating climate smart sunflower for future challenges" (SMARTSUN), grant number 7732457, by the European Commission through Twinning Western Balkans project CROPINNO, grant number 101059784, by Center of Excellence for Innovations in Breeding of Climate-Resilient Crops - Climate Crops, Institute of Field and Vegetable Crops, Novi Sad, Serbia, and based upon work from COST Actions PlantEd (CA18111) and EPI-CATCH (CA19125), supported by COST (European Cooperation in Science and Technology). www.cost.eu.

References

Aastveit, H. and Martens, H. (1986). ANOVA interactions interpreted by partial least squares regression. *Biometrics* 42: 829–844.

Acevedo, M., Pixley, K., Zinyengere, N. et al. (2020). A scoping review of adoption of climate-resilient crops by small-scale producers in low- and middle-income countries. *Nature Plants* 6: 1231–1241.

Aisawi, K.A.B., Reynolds, M.P., Singh, R.P., and Foulkes, M.J. (2015). The physiological basis of the genetic progress in yield potential of CIMMYT spring wheat cultivars from 1966 to 2009. *Crop Science* 55 (4): 1749–1764.

Anđelković, V., Cvejić, S., Jocić, S. et al. (2020). Use of plant genetic resources in crop improvement–example of Serbia. *Genetic Resources and Crop Evolution* 67: 1–14.

Araus, J.L., Kefauver, S.C., Zaman-Allah, M. et al. (2018). Translating high-throughput phenotyping into genetic gain. *Trends in Plant Science* 23 (5): 451–466.

Arifuzzaman, M., Oladzadabbasabadi, A., McClean, P., and Rahman, M. (2019). Shovelomics for phenotyping root architectural traits of rapeseed/canola (*Brassica napus* L.) and genome-wide association mapping. *Molecular Genetics and Genomics* 294 (4): 985–1000.

Armstrong W, & Drew, MC 2002, Root growth and metabolism under oxygen deficiency, Eshel A & Kafkafi U (eds.), *Plant Roots: The Hidden Half* (3), Marcel Dekker, New York, USA, pp. 729–61.

Atlagić, J., Terzić, S., and Marjanović-Jeromela, A. (2012). Staining and fluorescent microscopy methods for pollen viability determination in sunflower and other plant species. *Industrial Crops and Products* 35 (1): 88–91.

Bandeira e Sousa, M., Cuevas, J., Couto, E.G. et al. (2017). Genomic-enabled prediction in maize using kernel models with genotype × environment interaction. *G3: Genes, Genomes, Genetics* 7: 1995–2014.

Bänzinger, M., Edmeades, G.O., Beck, D., and Bellon, M. (2000). *Breeding for Drought and Nitrogen Stress Tolerance in Maize: From Theory to Practice*. Cimmyt: Mexico.

Bayer, P.E., Petereit, J., Danilevicz, M.F. et al. (2021). The application of pangenomics and machine learning in genomic selection in plants. *Plant Genome* 14 (3): e20112.

Bebber, DP, Ramotowski, MA, & Gurr, SJ 2013, 'Crop pests and pathogens move polewards in a warming world', *Nature Climate Change*, vol. 3(11), pp. 985–8.

Becker, S.R., Byrne, P.F., Reid, S.D. et al. (2016). Root traits contributing to drought tolerance of synthetic hexaploid wheat in a greenhouse study. *Euphytica* 207 (1): 213–224.

Begna, T 2020, 'Review on somatic hybridization and its role in crop improvement', *Journal of Biology, Agriculture and Healthcare*, vol. 10(11), pp. 29–40.

Benites, F.R.G. and Pinto, C.A.B.P. (2011). Genetic gains for heat tolerance in potato in three cycles of recurrent selection. *Crop Breeding and Applied Biotechnology* 11: 133–140.

Bernardo, R. and Yu, J. (2007). Prospects for genome wide selection for quantitative traits in maize. *Crop Science* 47 (3): 1082–1090.

Beyene, Y., Mugo, S., Semagn, K. et al. (2013). Genetic distance among doubled haploid maize lines and their testcross performance under drought stress and non-stress conditions. *Euphytica* 192 (3): 379–392.

Beyene, Y., Semagn, K., Mugo, S. et al. (2015). Genetic gains in grain yield through genomic selection in eight bi-parental maize populations under drought stress. *Crop Science* 55 (1): 154.

Bidinger, F.R., Nepolean, T., Hash, C.T. et al. (2007). Quantitative trait loci for grain yield in pearl millet under variable post-flowering moisture conditions. *Crop Science* 47: 969–980.

Bilichak, A. and Kovalchuk, I. (2016). Transgenerational response to stress in plants and its application for breeding. *Journal of Experimental Botany* 67: 2081–2092.

Bindi, M. and Olesen, J.E. (2011). The responses of agriculture in Europe to climate change. *Regional Environmental Change* 11 (1): 151–158.

Bishop, C.M. (2006). *Pattern Recognition and Machine Learning*. New York, NY: Springer.

Bloomfield, J., Rose, T., and King, G. (2014). Sustainable harvest: managing plasticity for resilient crops. *Plant Biotechnology Journal* 12: 517–533.

Bolanos, J. and Edmeades, G.O. (1993a). Eight cycles of selection for drought tolerance in lowland tropical maize. I. Responses in grain yield, biomass and radiation utilization. *Field Crop Research* 31: 233–252.

Bolanos, J. and Edmeades, G.O. (1993b). Eight cycles of selection for drought tolerance in lowland tropical maize. II. Responses in reproductive behavior. *Field Crops Research* 31: 253–268.

Brbaklić, L., Trkulja, D., Kondić-Špika, A. et al. (2013). Detection of QTLs for important agronomical traits in hexaploid wheat using association analysis. *Czech Journal of Genetics and Plant Breeding* 49: 1–8.

Brbaklić, L., Trkulja, D., Kondić-Špika, A. et al. (2015a). Genetic associations in the detection of QTLs for wheat spike-related traits. *Pesquisa Agropecuária Brasileira* 50: 149–159.

Brbaklić, L., Trkulja, D., Kondić-Špika, A. et al. (2015b). Determination of population structure of wheat core collection for association mapping. *Cereal Research Communications* 43: 22–28.

Breseghello, F. and Sorrels, M.E. (2006). Association analysis as a strategy for improvement of qualitative traits in plants. *Crop Science* 46: 1323–1330.

Brisson, N., Gate, P., Gouache, D. et al. (2010). Why are wheat yields stagnating in Europe? A comprehensive data analysis for France. *Field Crops Research* 119 (1): 201–212.

Brown, P.J., Upadyayula, N., Mahone, G.S. et al. (2011). Distinct genetic architectures for male and female inflorescence traits of maize. *PLoS Genetics* 7: e1002383.

Burgueño, J., De los Campos, G., Weigel, K., and Crossa, J. (2012). Genomic prediction of breeding values when modeling genotype × environment interaction using pedigree and dense molecular markers. *Crop Science* 52: 707–719.

Bustos-Korts, D., Malosetti, M., Chapman, S., and Van Eeuwijk, F. (2016). Modelling of genotype by environment interaction and prediction of complex traits across multiple environments as a synthesis of crop growth modelling, genetics and statistics. In: *Crop Systems Biology* (ed. X. Yin and P. Struik). Springer.

Cahoon, E.B., Clemente, T.E., Damude, H.G., and Kinney, A.J. (2009). Modifying vegetable oils for food and non-food purposes. In: *Oil Crops*, 31–56. New York, NY: Springer.

Cairns, J.E., Hellin, J., Sonder, K. et al. (2013). Adapting maize production to climate change in sub-Saharan Africa. *Food Security* 5 (3): 345–360.

Campbell, M.T., Proctor, C.A., Dou, Y. et al. (2015). Genetic and molecular characterization of submergence response identifies *Subtol6* as a major submergence tolerance locus in maize. *PloS ONE* 10 (3): e0120385.

Ceccarelli, S. and Grando, S. (2020). Evolutionary plant breeding as a response to the complexity of climate change. *Iscience* 23 (12): 101815.

Ceccarelli, S, Grando, S, Baum, M & Udupa, SM 2004, 'Breeding for drought resistance in a changing climate', *Challenges and Strategies of Dryland Agriculture*, vol. 32, pp. 167–90.

Ceccarelli, S, Grando, S, & Baum, M 2007, 'Participatory plant breeding in water-limited environments', *Experimental Agriculture*, vol. 43(4), pp. 411–35.

Chaikam, V., Molenaar, W., Melchinger, A.E., and Boddupalli, P.M. (2019). Doubled haploid technology for line development in maize: technical advances and prospects. *Theoretical and Applied Genetics* 132 (12): 3227–3243.

Chandramohanan, KT, Radhakrishnan, VV, & Mohanan, KV 2016, 'Performance evaluation of some native rice cultivars of the Kaippad farming system of Kerala state of India', *University Journal of Agricultural Research*, vol. 4, pp. 37–41.

Chapman, S.C. and Edmeades, G.O. (1999). Selection improves drought tolerance in tropical maize populations: II. Direct and correlated responses among secondary traits. *Crop Science* 39: 1315–1324.

Chapman, S.C., Chakraborty, S., Dreccer, M.F., and Howden, S.M. (2012). Plant adaptation to climate change—opportunities and priorities in breeding. *Crop and Pasture Science* 63 (3): 251–268.

Chaudhary, J., Khatri, P., Singla, P. et al. (2019). Advances in omics approaches for abiotic stress tolerance in tomato. *Biology* 8: 90.

Chinnusamy, V. and Zhu, J.K. (2009). Epigenetic regulation of stress responses in plants. *Current Opinions in Plant Biology* 12: 133–139.

Christopher, J.T., Christopher, M.J., Borrell, A.K. et al. (2016). Stay-green traits to improve wheat adaptation in well-watered and water-limited environments. *Journal of Experimental Botany* 67 (17): 5159–5172.

Cicchino, M., Edreria, J.I., Uribelarrea, M., and Otegui, M.E. (2011). Heat stress in field-grown maize: response of physiological determinants of grain yield. *Crop Science* 50: 1438–1448.

Cobb, J.N., DeClerck, G., Greenberg, A. et al. (2013). Next-generation phenotyping: requirements and strategies for enhancing our understanding of genotype–phenotype relationships and its relevance to crop improvement. *Theoretical and Applied Genetics* 126 (4): 867–887.

Collins, N.C., Tardieu, F., and Tuberosa, R. (2008). Quantitative trait loci and crop performance under abiotic stress: where do we stand? *Plant Physiology* 147 (2): 469–486.

Comstock, R.E. and Moll, R.H. (1963). Genotype-environment interactions. In: *Statistical Genetics and Plant Breeding* (ed. W.D. Hanson and H.F. Robinson), 164–196. Washington, DC: NASNRC.

Conner, T. (2004). Precision breeding: a new genetic technique providing international opportunities for crop improvement. *Seed Quest* https://www.seedquest.com (accessed 20 May 2022).

Cooper, M. and De Lacy, I.H. (1994). Relationships among analytical methods used to study genotypic variation and genotype by environment interaction in plant breeding multi-environment experiments. *Theoretical and Applied Genetics* 88: 561–572.

Cooper, M.D., Messina, C.D., Podlich, D. et al. (2014). Predicting the future of plant breeding: complementing empirical evaluation with genetic prediction. *Crop and Pasture Science* 65: 311–336.

Coppens, F., Wuyts, N., Inzé, D., and Dhondt, S. (2017). Unlocking the potential of plant phenotyping data through integration and data-driven approaches. *Current Opinion in Plant Biology* 4: 58–63.

Cosentino, S.L., Sanzone, E., Testa, G. et al. (2019). Does post-anthesis heat stress affect plant phenology, physiology, grain yield and protein content of durum wheat in a semi-arid Mediterranean environment? *Journal of Agronomy and Crop Science* 205 (3): 309–323.

Crain, J.L., Wei, Y., Barker, J. et al. (2016). Development and deployment of a portable field phenotyping platform. *Crop Science* 56 (3): 965–975.

Crain, J., Mondal, S., Rutkoski, J. et al. (2018). Combining high-throughput phenotyping and genomic information to increase prediction and selection accuracy in wheat breeding. *The Plant Genome* 11 (1): 170043.

Crossa, J. and Cornelius, P.L. (1997). Sites regression and shifted multiplicative model clustering of cultivar trial sites under heterogeneity of error variances. *Crop Science* 37: 405–415.

Crossa, J., Pérez, P., Hickey, J. et al. (2014). Genomic prediction in CIMMYT maize and wheat breeding programs. *Heredity* 112: 48–60.

Crossa, J., Pérez-Rodríguez, P., Cuevas, J. et al. (2017). Genomic selection in plant breeding: methods, models, and perspectives. *Trends in Plant Science* 22: 961–975.

Cuevas, J., Crossa, J., Montesinos-López, O.A. et al. (2017). Bayesian genomic prediction with genotype × environment interaction kernel models. *G3: Genes, Genomes, Genetics* 7: 41–53.

Cuevas, J., Granato, I., Fritsche-Neto, R. et al. (2018). Genomic-enabled prediction kernel models with random intercepts for multi-environment trials. *G3: Genes, Genomes, Genetics* 8: 1347–1365.

Cullis, B.R. and Gleeson, A.C. (1991). Spatial analysis of field experiments-an extension to two dimensions. *Biometrics* 47: 1449–1460.

Cvejić, S, Miladinović, D, & Jocić, S 2014, 'Mutation breeding for changed oil quality in sunflower', in: Tomlekova NB, Kozgar MI, Wani MR (eds.) *Mutagenesis: Exploring Genetic Diversity of Crops*. Wageningen Academic Publishers, Wageningen, pp. 77–96.

Cvejić, S., Jocić, S., Mladenov, V. et al. (2019). Selection of sunflower hybrids based on stability across environments. *Genetika-Belgrade* 51 (1): 81–92.

De Los Campos, G, Gianola, D, & Rosa, GJ 2009, 'Reproducing kernel Hilbert spaces regression: a general framework for genetic evaluation', *Journal of Animal Science*, vol. 87, pp. 1883–87.

De los Campos, G., Hickey, J.M., Pong-Wong, R. et al. (2013). Whole-genome regression and prediction methods applied to plant and animal breeding. *Genetics* 193: 327–345.

Debaeke, P., Casadebaig, P., Flenet, F., and Langlade, N. (2017). Sunflower crop and climate change: vulnerability, adaptation, and mitigation potential from case-studies in Europe. *OCL Oilseeds and Fats Crops and Lipids* 24 (1): 15.

Dempewolf, H., Eastwood, R.J., Guarino, L. et al. (2014). Adapting agriculture to climate change: a global initiative to collect, conserve, and use crop wild relatives. *Agroecology and Sustainable Food Systems* 38 (4): 369–377.

Denis, J.B. (1988). Two-way analysis using covariables. *Statistics* 19: 123–132.

Deutsch, C.A., Tewksbury, J.J., Tigchelaar, M. et al. (2018). Increase in crop losses to insect pests in a warming climate. *Science* 361 (6405): 916–919.

Dhondt, S, Wuyts, N, & Inzé, D 2013, 'Cell to whole-plant phenotyping: the best is yet to come', *Trends in Plant Science*, vol. 18(8), pp. 428–39.

Dimitrijević, A, Galinski, A, Miladinović, D, Cvejić, S, Jocić, S, Terzić, S, Nagel K, & Fiorani, F 2018, Root phenotyping of NS sunflower'. in: Anačkov G & Zorić L, (eds.) *Agronomy and Forestry*. Seventh Balkan Botanical Congress. Novi Sad, Serbia, 10–14 September 2018 Abstract Book in Botanica Serbica 42(supplement 1). Belgrade: University of Belgrade, Institute of Botany, pp. 164–5.

Ding, J.Q., Wang, X.M., Chander, S. et al. (2008). QTL mapping of resistance to *Fusarium* ear rot using a RIL population in maize. *Molecular Breeding* 22 (3): 395–403.

Edmeades, G.O., Bolaños, J., Chapman, S.C. et al. (1999). Selection improves drought tolerance in tropical maize populations: I. Gains in biomass, grain yield, and harvest index. *Crop Science* 39 (5): 1306–1315.

van Eeuwijk, F.A., Denis, J.B., and Kang, M.S. (1996). Incorporating additional information on genotypes and environments in models for two-way genotype by environment tables. In: *Genotype-by-Environment Interaction* (ed. M.S. Kang and H.G. Gauch), 15–50. Boca Raton, FL: CRC Press Inc.

van Eeuwijk, F.A., Bustos-Korts, D.V., and Malosetti, M. (2016). What should students in plant breeding know about the statistical aspects of genotype × environment interactions? *Crop Science* 56: 2119–2140.

Ehdaie, B., Layne, A.P., and Waines, J.G. (2012). Root system plasticity to drought influences grain yield in bread wheat. *Euphytica* 186: 219–232.

El Habti, A., Fleury, D., Jewell, N. et al. (2020). Tolerance of combined drought and heat stress is associated with transpiration maintenance and water soluble carbohydrates in wheat grains. *Frontiers in Plant Science* 11: 1555.

Endelman, J.B. (2011). Ridge regression and other kernels for genomic selection with R Package rrBLUP. *The Plant Genome* 4: 250–255.

Endelman, J.B. and Jannink, J.L. (2012). Shrinkage estimation of the realized relationship matrix. *G3 Genes, Genomes, Genetics* 2: 1405–1413.

Ersoz, E.S., Yu, J., and Buckler, E.S. (2007). Applications of linkage disequilibrium and association mapping in crop plants. In: *Genomics-Assisted Crop Improvement* (ed. R.K. Varshney and R. Tuberosa). Dordrecht: Springer.

Fahlgren, N., Gehan, M.A., and Baxter, I. (2015). Lights, camera, action: highthroughput plant phenotyping is ready for a close-up. *Current Opinion in Plant Biology* 24: 93–99.

Fahrmeir, L., Kneib, T., Lang, S., and Marx, B. (2013). *Regression: Models, Methods and Applications*. Germany: Springer.

FAOSTAT (2021). Food and Agriculture Organization of the United Nations. http://www.fao.org/faostat/en/#data/QC (accessed 1 February 2021).

Farooq, M., Hussain, M., and Siddique, K.H. (2014). Drought stress in wheat during flowering and grain-filling periods. *Critical Reviews in Plant Sciences* 33 (4): 331–349.

Fischer, R.A. and Edmeades, G.O. (2010). Breeding and cereal yield progress. *Crop Science* 50: 85–98.

Flavell, R. (2010). From genomics to crop breeding. *Nature Biotechnology* 28: 144–145.

Fooland, M.R., Zhang, L.P., and Lin, G.Y. (2001). Identification and validation of QTLs for salt tolerance during vegetative growth in tomato by selective genotyping. *Genome* 44: 444–454.

Foulkes, M.J., Reynolds, M.P., and Sylvester-Bradley, R. (2009). Genetic improvement of grain crops: yield potential. In: *Crop Physiology: Applications for Genetic Improvement and Agronomy* (ed. V. Sadras and D. Calderini), 285–319. Amsterdam: Elsevier.

Frey, F.P., Presterl, T., Lecoq, P. et al. (2016). First steps to understand heat tolerance of temperate maize at adult stage: identification of QTL across multiple environments with connected segregating populations. *Theoretical and Applied Genetics* 129: 945–961.

Furbank, RT, & Tester, M 2011, 'Phenomics–technologies to relieve the phenotyping bottleneck', *Trends in Plant Science*, vol. 16(12), pp. 635–44.

Gan, Y., Angadi, S.V., Cutforth, H. et al. (2004). Canola and mustard response to short periods of temperature and water stress at different developmental stages. *Canadian Journal of Plant Science* 84 (3): 697–704.

Gao, G., Li, J., Li, H. et al. (2014). Comparison of the heat stress induced variations in DNA methylation between heat-tolerant and heat-sensitive rapeseed seedlings. *Breeding Science* 64: 125–133.

Gauch, H.G. (2006). Statistical analysis of yield trials by AMMI and GGE. *Crop Science* 46 (4): 1488–1500.

Gianola, D. (2013). Priors in whole-genome regression: the Bayesian alphabet returns. *Genetics* 194: 573–596.

Gianola, D. and Van Kaam, J.B.C.H.M. (2008). Reproducing kernel Hilbert spaces regression methods for genomic assisted prediction of quantitative traits. *Genetics* 178: 2289–2303.

Gilmour, A.R., Cullis, B.R., and Verbyla, A.P. (1997). Accounting for natural and extraneous variation in the analysis of field experiments. *Journal of Agricultural Biological and Environmental Statistics* 2: 269–273.

González-Camacho, J.M., Crossa, J., Pérez-Rodríguez, P. et al. (2016). Genome-enabled prediction using probabilistic neural network classifiers. *BMC Genomics* 17 (1): 1–16.

Gota, M. and Gianola, D. (2014). Kernel-based whole-genome prediction of complex traits: a review. *Frontiers in Genetics* 5: 1–13.

Granato, I., Cuevas, J., Luna-Vázquez, F. et al. (2018). BGGE: A new package for genomic-enabled prediction incorporating genotype × environment interaction models. *G3: Genes, Genomes, Genetics* 8: 3039–3047.

Grant, RF, Jackson, BS, Kiniry, JR, & Arkin, GF 1989, 'Water deficit timing effects on yield components in maize', *Agronomy Journal*, vol. 81, pp. 61–5.

Gregersen, P.L., Culetic, A., Boschian, L., and Krupinska, K. (2013). Plant senescence and crop productivity. *Plant Molecular Biology* 82: 603–622.

Guo, J., Pradhan, S., Shahi, D. et al. (2020). Increased prediction accuracy using combined genomic information and physiological traits in a soft wheat panel evaluated in multi-environments. *Scientific Reports* 10 (1): 1–12.

Habash, DZ, Kehel, Z, & Nachit, M 2009, 'Genomic approaches for designing durum wheat ready for climate change with a focus on drought, *Journal of Experimental Botany*, vol. 60(10), pp. 2805–15.

Habier, D., Fernando, R.L., and Dekkers, J.C.M. (2007). The impact of genetic relationship information on genome-assisted breeding values. *Genetics* 177: 2389–2397.

Habier, D., Fernando, R.L., and Dekkers, J.C.M. (2009). Genomic selection using low-density marker panels. *Genetics* 182: 343–353.

Habier, D., Fernando, R.L., Kizilkaya, K., and Garrick, D.J. (2011). Extension of the Bayesian alphabet for genomic selection. *BMC Bioinformatics* 12: 186.

Haghighattalab, A., Pérez, L.G., Mondal, S. et al. (2016). Application of unmanned aerial systems for high throughput phenotyping of large wheat breeding nurseries. *Plant Methods* 12 (1): 1–15.

Hall, D., Tegström, C., and Ingvarsson, P.K. (2010). Using association mapping to dissect the genetic basis of complex traits in plants. *Briefings in Functional Genomics* 9: 57–165.

Hammer, G.L., Dong, Z., McLean, G. et al. (2009). Can changes in canopy and/or root system architecture explain historical maize yield trends in the US corn belt? *Crop Science* 49 (1): 299–312.

Harfouche, A.L., Jacobson, D.A., Kainer, D. et al. (2019). Accelerating climate resilient plant breeding by applying next-generation artificial intelligence. *Trends in Biotechnology* 37 (11): 1217–1235.

Hatfield, J.L. and Dold, C. (2019). Water-use efficiency: advances and challenges in a changing climate. *Frontiers in Plant Science* 10: 103.

Hefner, E.L., Sorrells, M.E., and Jannink, J.L. (2009). Genomic selection for crop improvement. *Crop Science* 49: 1–12.

Henry, R.J. (2014). Genomics strategies for germplasm characterization and the development of climate resilient crops. *Frontiers in Plant Science* 5: 68.

Heslot, N., Yang, H.P., Sorrells, M.E., and Jannink, J.-L. (2012). Genomic selection in plant breeding: a comparison of models. *Crop Science* 52: 146–160.

Heslot, N., Akdemir, D., Sorrells, M.E., and Jannink, J.-L. (2014). Integrating environmental covariates and crop modeling into the genomic selection framework to predict genotype by environment interactions. *Theoretical and Applied Genetics* 127: 463–480.

Heslot, N., Jannink, J.-L., and Sorrells, M.E. (2015). Perspectives for genomic selection applications and research in plants. *Crop Science* 55: 1–12.

Hill, W.G. (1984). On selection among groups with heterogeneous variance. *Animal Science* 39 (3): 473–477.

Hilton, I.B., D'ippolito, A.M., Vockley, C.M. et al. (2015). Epigenome editing by a CRISPR-Cas9-based acetyltransferase activates genes from promoters and enhancers. *Nature Biotechnology* 33: 510–517.

Hlaváčová, M., Klem, K., Rapantová, B. et al. (2018). Interactive effects of high temperature and drought stress during stem elongation, anthesis and early grain filling on the yield formation and photosynthesis of winter wheat. *Field Crops Research* 221: 182–195.

Hoerl, A.E. and Kennard, R.W. (1970). Ridge regression: biased estimation for nonorthogonal problems. *Technometrics* 12: 55–57.

Hu, X. and Spilke, J. (2011). Variance–covariance structure and its influence on variety assessment in regional crop trials. *Field Crops Research* 120: 1–8.

Hu, H. and Xiong, L. (2014). Genetic engineering and breeding of drought-resistant crops. *Annual Review of Plant Biology* 65: 715–741.

Huang, X, Wei, X, Sang, T, Zhao, Q, Feng, Q, Zhao, Y, Li, C, Zhu, C, Lu, T, Zhang, Z, Li, M, Fan, D, Guo, Y, Wang, A, Wang, L, Deng, L, Lu, Y, Weng, Q, Liu, K, Huang, T, Zhou, T, Jing, Y, Li, W, Lin, Z, Buckler, E.S, Qian, Q, Zhang, Q, Li, J, & Han, B 2010, Genome-wide association studies of 14 agronomic traits in rice landraces, *Nature Genetics*, vol. 42, pp. 961–7.

Huang, C., Nie, X., Shen, C. et al. (2017). Population structure and genetic basis of the agronomic traits of upland cotton in China revealed by a genome-wide association study using high-density SNPs. *Plant Biotechnology Journal* 15: 1374–1386.

Huang, C., Chu, P., Wu, Y. et al. (2020). Identification of functional SSR markers in fresh water ornamental Shrimps Neocaridinadenticulata using transcriptome sequencing. *Marine Biotechnology* 22: 772–785.

Huhn, M. (1996). Nonparametric analysis of genotype x environment interactions by ranks. In: *Genotype–by–Environment Interaction* (ed. M.S. Kang and H.G. Gauch Jr.), 235–271. Boca Raton, FL: CRC Press.

Humphreys E, Meisner C, Gupta R, Timsina J, Beecher HG, Tang Yong Lu, Yadvinder-Singh Y.-S., Gill MA, Masih I, Zheng Jia Guo & Thompson JA (2005) Water saving in rice-wheat systems, *Plant Production Science*, 8:3, 242–258, DOI: https://doi.org/10.1626/pps.8.242

Hwang, E.-Y., Song, Q., Jia, G. et al. (2014). Genome-wide association study of seed protein and oil content in soybean. *BMC Genomics* 15: 1.

Jaime, R., Alcántara, J.M., Manzaneda, A.J., and Rey, P.J. (2018). Climate change decreases suitable areas for rapeseed cultivation in Europe but provides new opportunities for white mustard as an alternative oilseed for biofuel production. *PLoS ONE* 13 (11): e0207124.

Jankowicz-Cieslak, J, & Till, BJ 2015, 'Forward and reverse genetics in crop breeding', in: J.M. Al-Khayri et al. (eds.), *Advances in Plant Breeding Strategies: Breeding, Biotechnology and Molecular Tools*, Springer International Publishing Switzerland, pp. 215–40.

Jarquín, D, Crossa, J, Lacaze, X, Cheyron, P, Daucourt, J, Lorgeou, J, Piraux, F, Guerreiro, L, Pérez, P, Calus, M, Burgueño, J, & De los Campos, G 2014, 'A reaction norm model for genomic selection using high-dimensional genomic and environmental data', *Theoretical and Applied Genetics*, vol. 127, pp. 595–607.

Jha, UC, Bohra, A, & Singh, NP 2014, 'Heat stress in crop plants: its nature, impacts and integrated breeding strategies to improve heat tolerance', *Plant Breeding* 133(6), pp. 679–701.

Jia, G., Huang, X., Zhi, H. et al. (2013). A haplotype map of genomic variations and genome-wide association studies of agronomic traits in foxtail millet (*Setaria italica*). *Nature Genetics* 45: 957–961.

Jighly, A, Hayden, M, & Daetwyler, H 2021, 'Integrating genomic selection with a genotype plus genotype × environment (GGE) model improves prediction accuracy and computational efficiency', *Plant Cell Environment*, pp. 1–12.

Jocić, S, Miladinović, D, & Kaya, Y 2015, Breeding and genetics of sunflower', in Martínez-Force, E, Dunford, NT & Salas, JJ (eds.), *Sunflower: Chemistry, Production, Processing, and Utilization*, Elsevier/AOCS Press, Urbana, IL, USA, pp. 1–25.

Jocković, M., Cvejić, S., Jocić, S. et al. (2019). Evaluation of sunflower hybrids in multi-environment trial (MET). *Turkish Journal of Field Crops* 24 (2): 202–210.

Jocković, M, Jocić, S, Cvejić, S, Marjanović-Jeromela, A, Jocković, J, Radanović, A, & Miladinović, D 2021, 'Genetic improvement in sunflower breeding-integrated omics approach', *Plants*, vol. 10, pp. 1150.

Juliana, P, Montesinos-López, OA, Crossa, J, Mondal, S, González Pérez, L, Poland, J, Huerta-Espino, J, Crespo-Herrera, L, Govindan, V, Dreisigacker, S, Shrestha, S, Pérez-Rodríguez, P, Espinosa, FP, & Singh, RP 2019, 'Integrating genomic-enabled prediction and high-throughput phenotyping in breeding for climate-resilient bread wheat', *Theoretical and Applied Genetics*, vol. 132, pp. 177–94.

Kadam, D.C. and Lorenz, A.J. (2018). Toward redesigning hybrid maize breeding through genomics-assisted breeding. In: *The Maize Genome, Compendium of Plant Genomes* (ed. J. Bennetzen et al.), 367–388. Cham: Springer.

Kadam, DC, Potts, SM, Bohn, OM, Lipka, AE, & Lorenz, AJ 2016, 'Genomic prediction of single crosses in the early stages of a maize hybrid breeding pipeline', *G3: Genes, Genomes, Genetics*, vol. 6, pp. 3443–53.

Khan, A., Sovero, V., and Gemenet, D. (2016). Genome-assisted breeding for drought resistance. *Current Genomics* 17: 330–342.

Kilian, B, Dempewolf, H, Guarino, L, Werner, P, Coyne, C, & Warburton, ML 2021, 'Crop science special issue: adapting agriculture to climate change: a walk on the wild side', *Crop Science*, vol. 61, pp. 32–6.

Killi, D, Bussotti, F, Raschi, A & Haworth, M 2017, 'Adaptation to high temperature mitigates the impact of water deficit during combined heat and drought stress in C3 sunflower and C4 maize varieties with contrasting drought tolerance', *Physiol Plantarum*, vol. 159, pp. 130–47.

Kole, C. (2020). *Genomic Designing of Climate-Smart Fruit Crops*. Springer Nature.

Kole, C., Muthamilarasan, M., Henry, R. et al. (2015). Application of genomics-assisted breeding for generation of climate resilient crops: progress and prospects. *Frontiers in Plant Science* 6: 563.

Kollers, S., Rodemann, B., Ling, J. et al. (2013). Whole genome association mapping of *Fusarium* head blight resistance in European winter wheat (*Triticumaestivum* L.). *PLoS One* 8: e57500.

Kumar, R., Kaur, A., Pandey, A. et al. (2019). CRISPR-based genome editing in wheat: a comprehensive review and future prospects. *Molecular Biology Reports* 46 (3): 3557–3569.

Langridge, P. (2018). Economic and academic importance of barley. In: *The Barley Genome*, vol. 2018 (ed. N. Stein, G.J. Muehlbauer and J. Gary), 1–10. Cham: Springer.

Latif, S., Wang, L., Khan, J. et al. (2020). Deciphering the role of stay-green trait to mitigate terminal heat stress in bread wheat. *Agronomy* 10 (7): 1001.

Le Gouis, J., Oury, F.X., and Charmet, G. (2020). How changes in climate and agricultural practices influenced wheat production in Western Europe. *Journal of Cereal Science* 93: 102960.

Lesk, C., Rowhani, P., and Ramankutty, N. (2016). Influence of extreme weather disasters on global crop production. *Nature* 529 (7584): 84–87.

Li, Z. and Howell, S.H. (2021). Heat stress responses and thermotolerance in maize. *International Journal of Molecular Sciences* (2): 22, 948.

Li, Z. and Sillanpää, M.J. (2012). Overview of LASSO-related penalized regression methods for quantitative trait mapping and genomic selection. *Theoretical and Applied Genetics* 125: 419–435.

Lipper, L. and Zilberman, D. (2018). A short history of the evolution of the climate smart agriculture approach and its links to climate change and sustainable agriculture debates. In: *Climate Smart Agriculture, Building Resilience to Climate Change* (ed. L. Lipper, N. McCarthy, D. Zilberman, et al.), 13–30. Cham: Springer.

Liu, C., Li, X., Meng, D. et al. (2017). A 4-bp insertion at ZmPLA1 encoding a putative phospholipase A generates haploid induction in maize. *Molecular Plant* 10 (3): 520–522.

Livaja, M., Unterseer, S., Erath, W. et al. (2016). Diversity analysis and genomic prediction of Sclerotinia resistance in sunflower using a new 25 K SNP genotyping array. *Theoretical and Apply Genetics* 129 (2): 317–329.

Lobell, D.B. and Burke, M.B. (2010). On the use of statistical models to predict crop yield responses to climate change. *Agricultural and Forest Meteorology* 150 (11): 1443–1452.

Lobet, G. (2017). Image analysis in plant sciences: publish then perish. *Trends in Plant Science* 22: 559–566.

Lobos, GA, Camargo, AV, Del Pozo, A, Araus, JL, Ortiz, R, & Doonan, JH 2017, 'Plant phenotyping and phenomics for plant breeding', *Frontiers in Plant Science*, vol. 8, pp. 2181.

Loman, N.J., Misra, R.V., Dallman, T.J. et al. (2012). Performance comparison of benchtop high-throughput sequencing platforms. *Nature Biotechnology* 30: 434–439.

Lopes, M.S., Royo, C., Alvaro, F. et al. (2018). Optimizing winter wheat resilience to climate change in rain fed crop systems of Turkey and Iran. *Frontiers in Plant Science* 9: -563.

Lorenz, A.J., Chao, S., Asoro, F.G. et al. (2011). Genomic selection in plant breeding: knowledge and prospects. *Advances in Agronomy* 110: 77–123.

Lucas, M.R., Ehlers, J.D., Huynh, B.L. et al. (2013). Markers for breeding heat-tolerant cowpea. *Molecular Breeding* 31: 529–536.

Ma, J., Li, R., Wang, H. et al. (2017). Transcriptomics analyses reveal wheat responses to drought stress during reproductive stages under field conditions. *Frontiers in Plant Science* 8: 592.

Magnussen, S. (1989). Inter-plant interactions and their influence on within- and among-plot variances. *Scandinavian Journal of Forest Research* 4: 369–377.

Mahalingam, G., Pattanashetti, S.K., Patne, N., and Kanatti, A.A. (2018). Breeding cultivars for heat stress tolerance in staple food crops. In: *Next Generation Plant Breeding*. IntechOpen Limited. DOI: https://doi.org/10.5772/intechopen.76480.

Malosetti, M., Voltas, J., Romagosa, I. et al. (2004). Mixed models including environmental covariables for studying QTL by environment interaction. *Euphytica* 137: 139–145.

Malosetti, M., Ribaut, J.M., and van Eeuwijk, F.A. (2013). The statistical analysis of multi-environment data: modeling genotype-by-environment interaction and its genetic basis. *Frontiers in Physiology* 4: 1–17.

Malosetti, M., Bustos-Korts, D., Boer, M.P., and van Eeuwijk, F.A. (2016). Predicting responses in multiple environments: Issues in relation to genotype × environment interactions. *Crop Science* 56: 2210–2222.

Mangin, B., Bonnafous, F., Blanchet, N. et al. (2017). Genomic prediction of sunflower hybrids oil content. *Frontiers in Plant Science* 8: 1633.

Mano, Y. and Omori, F. (2015). Flooding tolerance in maize (*Zea mays* subsp. mays) F 1 hybrids containing a QTL introgressed from teosinte (*Zea nicaraguensis*). *Euphytica* 205 (1): 255–267.

Mao, H., Wang, H., Liu, S. et al. (2015). Transposable element in a NAC gene is associated with drought tolerance in maize seedlings. *Nature Communications* 6: 8326. https://doi.org/10.1038/ncomms9326.

Marjanović Jeromela, A., Cvejić, S., Mladenov, V. et al. (2021). Technological quality traits phenotyping of Camelina across multienvironment trials. *Acta Agriculturae Scandinavica, Section B—Soil & Plant Science* 1–7.

Marjanović-Jeromela, A., Dimitrijević, A., Terzić, S. et al. (2016). Applying Mendelian rules in rapeseed (*Brassica napus*) breeding. *Genetika* 48: 1077–1086.

Marjanović-Jeromela, A., Cvejić, S., Jocić, S. et al. (2019a). Achievements and future trends in oil crop breeding/Risultati e prospettive nel miglioramento genetico delle colture de olio. In: *Oils and fats/Oli e grassi* (ed. G. Mosca), 197–205. Milano, Italia: Edagricole di New Business Media srl. ISBN: 978-88-506-5564-9 (In Italian).

Marjanović-Jeromela, A., Terzić, S., Jankulovska, M. et al. (2019b). Dissection of year related climatic variables and their effect on winter rapeseed (*Brassica napus* L.) development and yield. *Agronomy* 9 (9): 517.

Martin, R.J., Eccleston, J.A., and Chan, B.S.P. (2004). Efficient factorial experiments when the data are spatially correlated. *Journal of Statistical Planning and Inference* 126 (1): 377–395.

Martino, D.L. and Abbate, P.E. (2019). Frost damage on grain number in wheat at different spike developmental stages and its modelling. *European Journal of Agronomy* 103: 13–23.

Mason, R.E. and Singh, R.P. (2014). Considerations when deploying canopy temperature to select high yielding wheat breeding lines under drought and heat stress. *Agronomy* 4 (2): 191–201.

Masuka, B., Atlin, G.N., Olsen, M. et al. (2017). Gains in maize genetic improvement in Eastern and Southern Africa: I. CIMMYT hybrid breeding pipeline. *Crop Science* 57: 168–179.

McCullagh, P. and Nelder, J.A. (2008). *Generalized, Linear, and Mixed Models*, 2e. Wiley.

Meuwissen, T.H.E., Hayes, B.J., and Goddard, M.E. (2001). Prediction of total genetic value using genome-wide dense marker maps. *Genetics* 157: 1819–1829.

Miedaner, T. and Juroszek, P. (2021). Global warming and increasing maize cultivation demand comprehensive efforts in disease and insect resistance breeding in north-western Europe. *Plant Pathology* 70 (5): 1032–1046.

Mikić, S., Kondić-Špika, A., Brbaklić, L. et al. (2016a). Multiple marker-traits associations for maize agronomic traits. *Chilean Journal of Agricultural Research* 76: 300–306.

Mikić, S., Zorić, M., Stanisavljević, D. et al. (2016b). Agronomic and molecular evaluation of maize inbred lines for drought tolerance. *Spanish Journal of Agricultural Research* 14: e0711.

Mikić, S., Kondić-Špika, A., Brbaklić, L. et al. (2017). Molecular and phenotypic characterisation of diverse temperate maize inbred lines in Southeast Europe. *Zemdirbyste-Agriculture* 104: 31–40.

Mikić, S., Kondić-Špika, A., Brbaklić, L. et al. (2018). Association analysis of agronomic traits with microsatellites in maize inbred lines. *Genetika* 50 (2): 379–394.

Miladinović, D, Hladni, N, Radanović, A, Jocić, S, & Cvejić, S 2019, 'Sunflower and climate change: possibilities of adaptation through breeding and genomic selection, in: Kole C (ed.), Genomic Designing of Climate-smart Oilseed Crops, Springer, Cham, pp. 173–238.

Miladinović, D., Antunes, D., Yildirim, K. et al. (2021). Targeted plant improvement through genome editing: from laboratory to field. *Plant Cell Reports* 40: 935–951.

Mirosavljević, M., Momčilović, V., Denčić, S. et al. (2018). Grain number and grain weight as determinants of triticale, wheat, two-rowed and six-rowed barley yield in the Pannonian environment. *Spanish Journal of Agricultural Research* 16 (3): 17.

Mirosavljević, M., Momčilović, V., Mikić, S. et al. (2019). Changes in leaf appearance and developmental phases associated with breeding progress in six-rowed barley in the Pannonian Plain. *Crop Breeding and Applied Biotechnology* 19 (3): 300–308.

Mirosavljević, M., Momčilović, V., Živančev, D. et al. (2020a). Genetic improvement of grain yield and bread-making quality of winter wheat over the past 90 years under the Pannonian Plain conditions. *Euphytica* 216 (12): 1–16.

Mirosavljević, M., Momčilović, V., Mikić, S. et al. (2020b). Changes in stay-green and nitrogen use efficiency traits in historical set of winter barley cultivars. *Field Crops Research* 249: 107740.

Mirosavljević, M., Momčilović, V., Mikić, S. et al. (2020c). Breeding progress in grain filling and grain yield components of six-rowed winter barley. *Zemdirbyste-Agriculture* 107 (3): 271–278.

Mirosavljević, M., Mikić, S., Kondić-Špika, A. et al. (2021). The effect of heat stress on some main spike traits in 12 wheat cultivars at anthesis and mid-grain filling stage. *Plant, Soil and Environment* 67 (2): 71–76.

Mitrović, B., Stojaković, M., Zorić, M. et al. (2016). Genetic gains in grain yield, morphological traits and yield stability of middle-late maize hybrids released in Serbia between 1978 and 2011. *Euphytica* 211 (3): 321–330.

Mitrović, B., Zorić, M., Stanisavljević, D. et al. (2019). Estimation of grain filling parameters of maize hybrids from different breeding periods. Book of Abstracts, XXIVth EUCARPIA Maize and Sorghum Conference - Integrating Quantitative and Molecular Genetics to Develop New Breeding Methods in Maize and Sorghum, Freising, Germany, 7–9 October 2019.

Mitrović, B., Drašković, B., Stanisavljević, D. et al. (2020). Environmental modeling of interaction variance for grain yield of medium early maturity maize hybrids. *Genetika* 52 (1): 367–378.

Mladenov, V., Fotopoulos, V., Kaiserli, E. et al. (2021). Deciphering the epigenetic alphabet involved in transgenerational stress memory in crops. *International Journal of Molecular Sciences* 22 (13): 7118.

Moradpour, M. and Akmar Abdulah, S.N. (2020). CRISPR/dCas9 platforms in plants: strategies and applications beyond genome editing can induced epigenetic modifications. *Plant Biotechnology Journal* 18: 32–44.

Morris, G.P., Ramu, P., Deshpande, S.P. et al. (2013). Population genomic and genome-wide association studies of agroclimatic traits in sorghum. *Proceedings of the National Academy of Sciences* 110: 453–458.

Moschen, S., Bengoa Luoni, S., Di Rienzo, J.A. et al. (2016). Integrating transcriptomic and metabolomic analysis to understand natural leaf senescence in sunflower. *Plant Biotechnology Journal* 14: 719–734.

Moschen, S., Luoni, S.B., Di Rienzo, J. et al. (2017). Integration of transcriptomic and metabolic data reveals hub transcription factors involved in drought stress response in sunflower (*Helianthus annuus* L.). *Plant Molecular Biology* 94: 549–564.

Munns, R., Hare, R.A., James, R.A., and Rebetzke, G. (2000). Genetic variation for improving the salt tolerance of durum wheat. *Australian Journal of Agricultural Research* 51: 69–74.

Munoz, P.R., Resende, M.F.R., Gezan, S.A. et al. (2014). Unraveling additive from nonadditive effects using genomic relationship matrices. *Genetics* 198: 1759–1768.

Mustroph, A. (2018). Improving flooding tolerance of crop plants. *Agronomy* 8: 160.

Nadeem, M.A., Nawaz, M.A., Shahid, M.Q. et al. (2018). DNA molecular markers in plant breeding: current status and recent advancements in genomic selection and genome editing. *Biotechnology & Biotechnological Equipment* 32: 261–285.

Nakaya, A. and Isobe, S.N. (2012). Will genomic selection be a practical method for plant breeding? *Annals of Botany* 110: 1303–1316.

Namazkar, S., Stockmarr, A., Frenck, G. et al. (2016). Concurrent elevation of CO_2, O_3 and temperature severely affects oil quality and quantity in rapeseed. *Journal of Experimental Botany* 67: 4117–4125.

Naveenkumar KL, Sen D, & Khanna VK 2018, Effect of maize production in a changing climate: its impacts, adaptation, and mitigation strategies through breeding, *Open Access Journal of Oncology and Medicine* vol. 2 (4). 000142. DOI: https://doi.org/10.32474/OAJOM.2018.02.000142

Nicotra, A., Atkin, O., Bonser, S. et al. (2010). Plant phenotypic plasticity in a changing climate. *Trends in Plant Science* 15: 684–692.

Niu, L., Feng, S., Ding, W., and Li, G. (2016). Influence of speed and rainfall on large-scale wheat lodging from 2007 to 2014 in China. *PloS ONE* 11 (7): 0157677.

Nogue, F., Mara, K., Collonnier, C., and Casacuberta, J.M. (2016). Genome engineering and plant breeding: impact on trait discovery and development. *Plant Cell Reports* 35: 1475–1486.

Nordborg, M. and Tavaré, S. (2002). Linkage disequilibrium: what history has to tell us. *TRENDS in Genetics* 18: 83–90.

Oerke, EC 2006, 'Crop losses to pests', *The Journal of Agricultural Science*, vol. 144(1), pp. 31–43.

Ogutu, JO, Schulz-Streeck, T, & Piepho, HP 2012, 'Genomic selection using regularized linear regression models: ridge regression, lasso, elastic net and their extensions'. *BMC Proceedings* vol. 6: pp. S10.

Ornella, L, Pérez, P., Tapia, E., González-Camacho, J, Burgueño, J, Zhang, X, Singh, S, Vicente, FS, Bonnett, D, Dreisigacker, S, Singh, R, Long, N, & Crossa, J 2014, 'Genomic-enabled prediction with classification algorithms'. *Heredity*, vol. 112 (6), pp. 616–26.

Ortiz, R., Jarvis, A., Fox, P. et al. (2014). Plant genetic engineering climate change and food security. *CCAFS Working Paper no 72, CGIAR Research Program on Climate Change Agriculture and Food Security (CCAFS)*, Copenhagen Denmark.

Ozturk Gokce, Z.N., Akbas, S., Ayten, S. et al. (2020). Abiotic stress tolerance in field crops: integration of omics approaches. In: *Agronomic Crops* (ed. M. Hasanuzzaman), 503–526. Singapore: Springer.

Pandey, M.K., Roorkiwal, M., Singh, V.K. et al. (2016). Emerging genomic tools for legume breeding: current status and future prospects. *Frontiers in Plant Science* 7: 455.

Pandey, M.K., Khan, A.W., Singh, V.K. et al. (2017). QTL-seq approach identified genomic regions and diagnostic markers for rust and late leaf spot resistance in groundnut (*Arachishypogaea* L.). *Plant Biotechnology Journal* 15: 927–941.

Parent, B, & Tardieu, F 2012, 'Temperature responses of developmental processes have not been affected by breeding in different ecological areas for 17 crop species, *New Phytologist*, vol. 194 (3), pp. 760–74.

Pastor, A.V., Palazzo, A., Havlik, P. et al. (2019). The global nexus of food–trade–water sustaining environmental flows by 2050. *Nature Sustainability* 2 (6): 499–507.

Patterson, H.D. (1997). Analysis of series of variety trials. In: *Statistical Methods for Plant Variety Evaluation* (ed. R.A. Kempton and P.N. Fox), 139–161. London: Chapman & Hall.

Paul, K., Pauk, J., Kondic-Spika, A. et al. (2019). Co-occurrence of mild salinity and drought synergistically enhances biomass and grain retardation in wheat. *Frontiers in Plant Science* 10: 501.

Pauli, D., Chapman, S.C., Bart, R. et al. (2016). The quest for understanding phenotypic variation via integrated approaches in the field environment. *Plant Physiology* 172 (2): 622–634.

Perez-Sanz, F., Navarro, P.J., and Egea-Cortines, M. (2017). Plant phenomics: an overview of image acquisition technologies and image data analysis algorithms. *GigaScience* 6: 1–18.

Piepho, H.P. (2009). Ridge regression and extensions for genomewide selection in maize. *Crop Science* 49: 1165–1176.

Piepho, H.P., Büchse, A., and Emrich, K. (2003). A Hitchhiker's guide to mixed models for randomized experiments. *Journal of Agronomy and Crop Science* 189 (5): 310–322.

Piepho, J.P., Möhring, J., Melchinger, A.E., and Büchse, A. (2008). BLUP for phenotypic selection in plant breeding and variety testing. *Euphytica* 161: 209–228.

Piepho, H.P., Möhring, J., Schulz-Streeck, T., and Ogutu, J.O. (2012). A stage-wise approach for the analysis of multi-environment trials. *Biometrical Journal* 54 (6): 844–860.

Prasanna, B.M., Cairns, J.E., Zaidi, P.H. et al. (2021). Beat the stress: breeding for climate resilience in maize for the tropical rainfed environments. *Theoretical and Applied Genetics* https://doi.org/10.1007/s00122-021-03773-7.

Prigge, V. and Melchinger, A.E. (2012). Production of haploids and doubled haploids in maize. In: *Plant Cell Culture Protocols, Methods in Molecular Biology (Methods and Protocols)*, vol. 877 (ed. V. Loyola-Vargas and N. Ochoa-Alejo), 161–172. Totowa, NJ: Humana Press.

Pullens, J.W.M., Kersebaum, K.C., Böttcher, U. et al. (2021). Model sensitivity of simulated yield of winter oilseed rape to climate change scenarios in Europe. *European Journal of Agronomy* 129: 126341.

Qiu, F., Zheng, Y., Zhang, Z., and Xu, S. (2007). Mapping of QTL associated with waterlogging tolerance during the seedling stage in maize. *Annals of Botany* 99 (6): 1067–1081.

Ramstein, G.P., Larsson, S.J., Cook, J.P. et al. (2020). Dominance effects and functional enrichments improve prediction of agronomic traits in hybrid maize. *Genetics* 215: 215–230.

Rasheed, A., Hao, Y., Xia, X. et al. (2017). Crop breeding chips and genotyping platforms: progress, challenges, and perspectives. *Molecular Plant* 10: 1047–1064.

Ray, D.K., Mueller, N.D., West, P.C., and Foley, J.A. (2013). Yield trends are insufficient to double global crop production by 2050. *PLoS ONE* 8 (6): 66428.

Rebetzke, G.J., Rattey, A.R., Farquhar, G.D. et al. (2013). Genomic regions for canopy temperature and their genetic association with stomatal conductance and grain yield in wheat. *Functional Plant Biology* 40 (1): 14–33.

Reif, J.C., Zhao, Y., Würschum, T. et al. (2013). Genomic prediction of sunflower hybrid performance. *Plant Breeding* 132: 107–114.

Reynolds, M.P. and Langridge, P. (2016). Physiological breeding. *Current Opinion in Plant Biology* 31: 162–171.

Reynolds, M.P., Pierre, C.S., Saad, A.S. et al. (2007). Evaluating potential genetic gains in wheat associated with stress-adaptive trait expression in elite genetic resources under drought and heat stress. *Crop Science* 47, S-172.

Reynolds, M., Manes, Y., Izanloo, A., and Langridge, P. (2009). Phenotyping approaches for physiological breeding and gene discovery in wheat. *Annals of Applied Biology* 155 (3): 309–320.

Reynolds, M.P., Pask, A.J.D., and Mullan, D.M. (2012). *Physiological Breeding I: Interdisciplinary Approaches to Improve Crop Adaptation*. Mexico City, Mexico: CIMMYT.

Reynolds, M., Chapman, S., Crespo-Herrera, L. et al. (2020). Breeder friendly phenotyping. *Plant Science* 295: 110396.

Richards, R.A., Rebetzke, G.J., Watt, M. et al. (2010). Breeding for improved water productivity in temperate cereals: phenotyping, quantitative trait loci, markers and the selection environment. *Functional Plant Biology* 37 (2): 85–97.

Rodríguez, V., Soengas, P., Cartea, E. et al. (2014). Suitability of a European nuclear collection of *Brassica oleracea* L. landraces to grow at high temperatures. *Journal of Agronomy and Crop Science* 14: 183–190.

Rose, L.J., Okoth, S., Beukes, I. et al. (2017). Determining resistance to Fusarium verticillioides and fumonisin accumulation in African maize inbred lines resistant to Aspergillus flavus and aflatoxins. *Euphytica* 213 (4): 93.

Rostoks, N.T., Ramsay, L., Mackenzie, K. et al. (2006). Recent history of artificial outcrossing facilitates whole-genome association mapping in elite inbred crop varieties. *Proceedings of the National Academy of Sciences* 103: 18656–18661.

Royo-Esnal, A. and Valencia-Gredilla, F. (2018). Camelina as a rotation crop for weed control in organic farming in a semiarid Mediterranean climate. *Agriculture* 8: 156–110.

Ruiz-Lopez, N., Haslam, R.P., Napier, J.A., and Sayanova, O. (2014). Successful high-level accumulation of fish oil omega-3 long-chain polyunsaturated fatty acids in a transgenic oilseed crop. *The Plant Journal* 77 (2): 198–208.

Sabella, E., Aprile, A., Negro, C. et al. (2020). Impact of climate change on durum wheat yield. *Agronomy* 10 (6): 793.

Sabouri, H., Rezai, A.M., Moumeni, A. et al. (2009). QTLs mapping of physiological traits related to salt tolerance in young rice seedlings. *Biologia Plantarum* 53: 657–662.

Sachs, M.M., Subbaiah, C.C., and Saab, I.N. (1996). Anaerobic gene expression and flooding tolerance in maize. *Journal of Experimental Botany* 47 (1): 1–15.

Salgotra, R.K. and Stewart, C.N. Jr. (2020). Functional markers for precision plant breeding. *International Journal of Molecular Science* 21: 4792.

Salika, R. and Riffat, J. (2021). Abiotic stress responses in maize: a review. *Acta Physiologiae Plantarum* 43 (9): 1–22.

Sarazin, V., Duclercq, J., Guillot, X. et al. (2017). Water-stressed sunflower transcriptome analysis revealed important molecular markers involved in drought stress response and tolerance. *Environmental Experimental Botany* 142: 45–53.

Savary, S., Ficke, A., Aubertot, J.N., and Hollier, C. (2012). Crop losses due to diseases and their implications for global food production losses and food security. *Food Security* 4: 519–537.

Schauberger, B., Ben-Ari, T., Makowski, D. et al. (2018). Yield trends, variability and stagnation analysis of major crops in France over more than a century. *Scientific Reports* 8 (1): 1–12.

Schlotterer, C., Tobler, R., Kofler, R., and Nolte, V. (2014). Sequencing pools of individuals — mining genome-wide polymorphism data without big funding. *Nature Reviews Genetics* 15: 749–763.

Scott, G.E. and Zummo, N. (1988). Sources of resistance in maize to kernel infection by *Aspergillus flavus* in the field. *Crop Science* 28 (3): 504–507.

Semenov, M.A. and Stratonovitch, P. (2013). Designing high-yielding wheat ideotypes for a changing climate. *Food and Energy Security* 2 (3): 185–196.

Senapati, N., Stratonovitch, P., Paul, M.J., and Semenov, M.A. (2019). Drought tolerance during reproductive development is important for increasing wheat yield potential under climate change in Europe. *Journal of Experimental Botany* 70 (9): 2549–2560.

Shah, T., Xu, J., Zou, X. et al. (2018). Omics approaches for engineering wheat production under abiotic stresses. *International Journal of Molecular Sciences* 19 (8): 2390.

Sharma, P., Mehta, G., Shefali, A.K. et al. (2021). Development and validation of heat-responsive candidate gene and miRNA gene based SSR markers to analysis genetic diversity in wheat for heat tolerance breeding. *Molecular Biology Reports* 48: 381–393.

Shewry, P.R. and Hey, S.J. (2015). The contribution of wheat to human diet and health. *Food and Energy Security* 4: 178–202.

Shi, L., Shi, T., Broadley, M.R. et al. (2013). High-throughput root phenotyping screens identify genetic loci associated with root architectural traits in *Brassica napus* under contrasting phosphate availabilities. *Annals of Botany* 112 (2): 381–389. https://doi.org/10.1093/aob/mcs245. Epub 2012 Nov 21. PMID: 23172414; PMCID: PMC3698377.

Shi, J., Gao, H., Wang, H. et al. (2017). ARGOS 8 variants generated by CRISPR-Cas9 improve maize grain yield under field drought stress conditions. *Plant Biotechnology Journal* 15: 207.

Siahsar, B.A. and Aminfar, Z. (2010). Mapping QTLs of physiological traits associated with salt tolerance in 'Steptoe' × 'Morex' doubled haploid lines of barley at seedling stage. *Food, Agriculture and Environment* 8: 751–759.

Singh, V.K., Khan, A.W., Saxena, R.K. et al. (2017). Indel-seq: a fast-forward genetics approach for identification of trait-associated putative candidate genomic regions and itsapplication in pigeonpea (*Cajanus cajan*). *Plant Biotechnology Journal* 15: 906–914.

Smith, C.M. (2021). Conventional breeding of insect-resistant crop plants: still the best way to feed the world population. *Current Opinion in Insect Science* 45: 7–13.

Smith, A.B. and Cullis, B.R. (2018). Plant breeding selection tools built on factor analytic mixed models for multi-environment trial data. *Euphytica* 214: 143.

Smith, A.B., Cullis, B.R., and Thompson, R. (2001). Analyzing variety by environment data using multiplicative mixed models and adjustments for spatial field trend. *Biometrics* 57: 1138–1147.

Smith, A.B., Cullis, B.R., and Thompson, R. (2005). The analysis of crop cultivar breeding and evaluation trials: an overview of current mixed model approaches. *Journal of Agricultural Science* 143: 449–462.

Snowdon, R.J., Wittkop, B., Chen, T.W., and Stahl, A. (2021). Crop adaptation to climate change as a consequence of long-term breeding. *Theoretical and Applied Genetics* 134 (6): 1613–1623.

Soto-Cerda, B.J. and Cloutier, S. (2012). Association mapping in plant genomes. In: *Genetic Diversity in Plants* (ed. M. Caliskan), 29–54. Rijeka: InTech.

Sousa, M.B.E., Cuevas, J., Couto, E.G.O. et al. (2017). Genomic-enabled prediction in maize using kernel models with genotype × environment interaction. *G3: Genes, Genomes, Genetics* 7: 1995–2014.

Springer, N.M. and Schmitz, R.J. (2017). Exploiting induced and natural epigenetic variation for crop improvement. *Nature Reviews in Genetics* 18: 563–575.

Sserumaga, J.P., Oikeh, S.O., Mugo, S. et al. (2016). Genotype by environment interactions and agronomic performance of doubled haploids testcross maize (*Zea mays* L.) hybrids. *Euphytica* 207 (2): 353–365.

Stamp, P. and Visser, R. (2012). The twenty-first century, the century of plant breeding. *Euphytica* 186 (3): 585–591.

Strigens, A., Schipprack, W., Reif, J.C., and Melchinger, A.E. (2013). Unlocking the genetic diversity of maize landraces with doubled haploids opens new avenues for breeding. *PloS ONE* 8 (2): e57234.

Stroup, W.W. (2015). Rethinking the analysis of non-normal data in plant and soil science. *Agronomy Journal* 107 (2): 811–827.

Taranto, F., Nicolia, A., Pavan, S. et al. (2018). Biotechnological and digital revolution for climate-smart plant breeding. *Agronomy* 8 (12): 277.

Tardieu, F., Cabrera-Bosquet, L., Pridmore, T., and Bennett, M. (2017). Plant phenomics, from sensors to knowledge. *Current Biology* 27 (15): R770–R783.

Taški-Ajduković, K., Nagl, N., and Miladinović, D. (2010). Towards reducing genotype specificity in regeneration protocols after somatic hybridization between cultivated sunflower and wild Helianthus species. *Acta Biologica Hungarica* 61: 214–223.

Taube, F., Vogeler, I., Kluß, C., Herrmann, A., Hasler, M., Rath, J., Loges, R. and Malisch, C.S., 2020. Yield progress in forage maize in NW Europe—breeding progress or climate change effects?. *Frontiers in Plant Science*, p.1214, doi: https://doi.org/10.3389/fpls.2020.01214.

Technow, F., Riedelsheimer, C., Schrag, T.A., and Melchinger, A.E. (2012). Genomic prediction of hybrid performance in maize with models incorporating dominance and population specific marker effects. *Theoretical and Applied Genetics* 125: 1181–1194.

Thomas, H. and Ougham, H. (2014). The stay-green trait. *Journal of Experimental Botany* 65 (14): 3889–3900.

Tibshirani, R. (1996). Regression shrinkage and selection via the Lasso. *Journal of the Royal Statistical Society* 58: 267–288.

Trkulja, D., Kondić-Špika, A., Brbaklić, L. et al. (2012). Marker-trait associations for spike-related characters in a doubled haploid population of wheat. *Romanian Agricultural Research* 29: 9–16.

Trkulja, D., Kondić-Špika, A., Brbaklić, L. et al. (2019). Genetic structure and allelic richness of the wheat core collection for association mapping of yield. *Zemdirbyste-Agriculture* 106: 257–264.

Trnka, M., Rötter, R.P., Ruiz-Ramos, M. et al. (2014). Adverse weather conditions for European wheat production will become more frequent with climate change. *Nature Climate Change* 4 (7): 637–643.

Trnka, M., Hlavinka, P., and Semenov, M.A. (2015). Adaptation options for wheat in Europe will be limited by increased adverse weather events under climate change. *Journal of the Royal Society Interface* 12 (112): 20150721.

Tuberosa, R. (2012). Phenotyping for drought tolerance of crops in the genomics era. *Frontiers in Physiology* 3: 347.

VanRaden, P.M. (2008). Efficient methods to compute genomic predictions. *Journal of Dairy Science* 91: 4414–4423.

Vargas, M., Crossa, J., Sayre, K. et al. (1998). Interpreting genotype × environment interaction in wheat using partial least squares regression. *Crop Science* 38: 679–689.

Vargas, M., Crossa, J., van Eeuwijk, F.A. et al. (1999). Using partial least square regression factorial regression, and AMMI models for interpreting genotype x environment interaction. *Crop Science* 39: 955–967.

Varotto, S, Tani, E, Abraham, E, Krugman, T, Kapazoglou, A, Melzer, R, Radanović, A, & Miladinović, D 2020, Epigenetics: possible applications in climate-smart crop breeding. *Journal of Experimental Botany*, vol. 71, pp. 5223–36. doi: https://doi.org/10.1093/jxb/eraa188.

Varshney, R.K., Sing, V.K., Kumar, A. et al. (2018). Can genomics deliver climate-change ready crops? *Current Opinion in Plant Biology* 45: 205–211.

Voltas, J., van Eeuwijk, F., Igartua, E. et al. (2002). Genotype by environment interaction and adaptation in barley breeding: basic concepts and methods of analysis. In: *Barley Science: Recent Advances from Molecular Biology to Agronomy of Yield and Quality* (ed. G. Slafer), 205–241. Binghampton, NY: Food Products Press.

Wang, T., Lin, J., Chen, Z. et al. (2014). Green synthesized iron nanoparticles by green tea and eucalyptus leaves extracts used for removal of nitrate in aqueous solution. *Journal of Cleaner Production 83*: 413–419.

Warburton, M.L., Williams, W.P. 2014. Aflatoxin resistance in maize: what have we learned lately?. *Advances in Botany* 2014:352831, doi:https://doi.org/10.1155/2014/352831.

Watson, D 2019, 'Adaption to climate change: climate adaptive breeding of maize, wheat and rice', in: A. Sarkar, SR. Sensarma, & GW. vanLoon, (eds.), *Sustainable Solutions for Food Security Combating Climate Change by Adaptation*. Springer Nature Switzerland, pp. 67–89.

Wen, J., Jiang, F., Weng, Y. et al. (2019). Identification of heat-tolerance QTLs and high-temperature stress-responsive genes through conventional QTL mapping, QTL-seq and RNA-seq in tomato. *BMC Plant Biology* 19: 398.

Williams, J., Kubelik, A., Livak, K. et al. (1990). DNA polymorphisms amplified by arbitrary primers are useful as genetic markers. *Nucleic Acids Research* 18: 6531–6535.

Woo, D.K., Riley, W.J., and Wu, Y. (2020). More fertilizer and impoverished roots required for improving wheat yields and profits under climate change. *Field Crops Research* 249: 107756.

Wu, S, Ning, F, Zhang, Q, Wu, X & Wang, W 2017, Enhancing omics research of crop responses to drought under field conditions. *Frontiers in Plant Science* vol. 8, p. 174. doi: https://doi.org/10.3389/fpls.2017.00174

Xu, Y. (2016). Envirotyping for deciphering environmental impacts on crop plants. *Theoretical and Applied Genetics* 129 (4): 653–673.

Xu, W., Subudhi, P.K., Crasta, O.R. et al. (2000). Molecular mapping of QTLs conferring stay-green in grain sorghum (*Sorghum bicolor* L. Moench). *Genome* 43: 461–469.

Xue, D., Zhang, X., Lu, X. et al. (2017). Molecular and evolutionary mechanisms of cuticular wax for plant drought tolerance. *Frontiers in Plant Science* 8: 621.

Yan, W. (2014). *Crop Variety Trials: Data Management and Analysis*. Wiley.

Yan, W., Hunt, L.A., Sheng, Q., and Szlavnics, Z. (2000). Cultivar evaluation and mega-environment investigation based on the GGE biplot. *Crop Science* 40 (3): 597–605.

Yan, W., Pageau, D., Frégeau-Reid, J., and Durand, J. (2011). Assessing the representativeness and repeatability of test locations for genotype evaluation. *Crop Science* 51 (4): 1603–1610.

Yan, G., Liu, H., Wang, H. et al. (2017). Accelerated generation of selfed pure line plants for gene identification and crop breeding. *Frontiers in Plant Science* 8: 1786.

Ynet, N. and Yilancioglu, K. (2018). A CRISPR/Cas9 model of sunflower (*Helianthus annuus* L.) resistance for biotic and abiotic stresses. *New Biotechnologist* 44: S84.

Zafar, S.A., Hameed, A., Nawaz, M.A. et al. (2018). Mechanisms and molecular approaches for heat tolerance in rice (*Oryza sativa* L.) under climate change scenario. *Journal of Integrative Agriculture* 17: 726–738.

Zaidi, P.H. and Singh, N.N. (2010). Effect of waterlogging on growth, biochemical compositions and reproduction in maize. *Journal of Plant Biology* 28: 61–69.

Zaidi, P.H., Rashid, Z., Vinayan, M.T. et al. (2015). QTL mapping of agronomic waterlogging tolerance using recombinant inbred lines derived from tropical maize (*Zea mays* L.) germplasm. *PLoS ONE* 10 (4): e0124350.

Zampieri, M., Ceglar, A., Dentener, F., and Toreti, A. (2017). Wheat yield loss attributable to heat waves, drought and water excess at the global, national and subnational scales. *Environmental Research Letters* 12: 064008.

Zandalinas, S.I., Fichman, Y., Devireddy, A.R. et al. (2020). Systemic signaling during abiotic stress combination in plants. *Proceedings of the National Academy of Sciences of the United States of America* 117: 13810–13820. https://doi.org/10.1073/pnas.2005077117.

Zanetti, F., Alberghini, B., Marjanović-Jeromela, A. et al. (2021). Camelina, an ancient oilseed crop actively contributing to the rural renaissance in Europe. A review. *Agronomy for Sustainable Development* 41: 1–18.

Zhang, X., Pérez-Rodríguez, P., Semagn, K. et al. (2015). Genomic prediction in biparental tropical maize populations in water-stressed and well-watered environments using low-density and GBS SNPs. *Heredity* 114: 291–299.

Zhang, X., Pérez-Rodríguez, P., Burgueño, J. et al. (2017). Rapid Cycling genomic selection in a multiparental tropical maize population. *G3 (Bethesda, MD)* 7 (7): 2315–2326.

Zhao, Y., Gowda, M., Liu, W. et al. (2012). Accuracy of genomic selection in European maize elite breeding populations. *Theoretical and Applied Genetics* 124 (4): 769–776.

Zhao, C., Liu, B., Piao, S. et al. (2017). 'Temperature increase reduces global yields'. *Proceedings of the National Academy of Sciences* 114 (35): 9326–9331.

7

Challenges of Soil Fertility Under Changing Climate and Its Opposing Components

Imran[1,2], Amanullah[1], Ibrahim Ortas[3], Tariq Mahmood[4], Muhammad Arif[1], Abdel Rahman Mohammad Said Al-Tawaha[5], and Mohammad Ilyas[6]

[1] *Department of Agronomy, The University of Agriculture, Peshawar, Pakistan*
[2] *Ministry of Agriculture, Agriculture Extension Wing, Merged Area, Khyber Pakhtunkhwa, Pakistan*
[3] *Department of Soil Science and Plant Nutrition, Faculty of Agriculture, University of Çukurova, Adana, Turkey*
[4] *Department of Bioinformatics/Agriculture, Hazara University, Mansehra, Pakistan*
[5] *Department of Biological Sciences, Al-Hussein Bin Talal University, Ma'an, Jordan*
[6] *Ministry of Agriculture, Research Wing, Agriculture Research Institute, Mingora Swat, Pakistan*

CONTENTS

Climate Change and Agriculture: Perspectives, Sustainability and Resilience,
First Edition. Edited by Noureddine Benkeblia.

7.1 Introduction

Soil microbial population plays a vital role in the retention, breakdown, and incorporation of plant remains nutrient cycling, and their influence on soil structure and porosity (Imran et al. 2020a). Climate change has both direct and indirect effects on soil health and on the activities of soil microbes that feedback greenhouse gases to the atmosphere and contribute to global warming (Imran et al. 2020b). Direct effects include the influence on soil microbes and greenhouse gas production of temperature, changing precipitation, and extreme climatic events, whereas indirect effects result from climate-driven changes in plant productivity and diversity which alter soil physicochemical conditions, the supply of carbon to the soil, and the structure and activity of microbial communities involved in decomposition processes and carbon release from soil (Amanullah et al. 2020a). Indirect climate-microbe feedback is the indirect effects on soil microbial communities and their activity and hence the potential for microbial feedback to climate change through its influence on plant growth and vegetation composition (Khan et al. 2016). Climate change may have stronger or weaker, permanent or periodical, favourable or unfavourable, harmful (sometimes catastrophic), primary (direct), or secondary (indirect) impact on soil processes (Amanullah et al. 2015). Among these processes, soil moisture regime and beneficial microbes play a distinguished role. It determines the water supply of plants and influences the air and heat regimes, biological activity, and plant nutrient status of soil (Imran et al. 2020c). In most cases, it determines the agro-ecological potential, the biomass production of various natural and agroecosystems, and the hazard of soil and water pollution (Amanullah et al. 2016).

Excessive precipitation in changing climate causes a significant source of soil nutrient loss like nitrate leaching (Amanullah et al. 2020b). Agricultural areas with poorly drained soils or that experience frequent or intense rainfall events can have waterlogged soils that become hypoxic (Imran et al. 2020d,e). The change in soil redox status under low oxygen can lead to elemental toxicities of Mn, Fe, Al, and B that reduce crop yields and the production of phytotoxic organic solutes that impair root growth and function (Imran et al. 2019c).

Hypoxia can also result in nutrient deficiency since the active transport of ions into root cells is driven by ATP synthesized through the oxygen-dependent mitochondrial electron transport chain. Significant nitrogen losses can also occur under hypoxic conditions through denitrification as nitrate is used as an alternative electron acceptor by microorganisms in the absence of oxygen (Havlin et al. 2005). Soil warming can increase nutrient uptake from 100 to 300% by enlarging the root surface area and increasing rates of nutrient diffusion and water influx (Hoomin et al. 2019). Since warmer temperatures increase rates of transpiration, plants tend to acquire water-soluble nutrients (nitrate, sulfate, Ca, Mg primarily move towards roots through transpiration-driven mass flow) more readily as temperature increases (Hussain et al. 2020). Temperature increases in the rhizosphere can also stimulate nutrient acquisition by increasing nutrient uptake via faster ion diffusion rates and increased root metabolism. However, any positive effects of warmer temperatures on nutrient capture are dependent on adequate soil moisture (Imran et al. 2015a). If under dry conditions, higher temperatures result in extreme vapour pressure deficits that trigger stomatal closure (reducing the water diffusion pathway in leaves), then nutrient acquisition driven by mass flow will decrease (Amanullah et al. 2020c).

Various greenhouse gases (GHGs) emissions from crop and livestock agriculture have risen from 4.7 billion tonnes of CO_2 equivalent in 2001 to more than 5.3 billion today, an increase of more than 14%. Organic agriculture including biofertilization can help to tackle climate change by reducing greenhouse gas emissions (Imran et al. 2016a). There is a direct correlation between nitrous oxide emissions and the amount of nitrogen fertilizer applied to agricultural land. Nitrous oxide emissions from managed soils account for almost 40% of agricultural emissions in the European Union (EU) (Hussain et al. 2013). This is particularly important because the impact of 1 kg of nitrous oxide on warming the atmosphere is about 300 times greater than the impact of 1 kg of carbon dioxide. Organic farming does not allow the use of synthetic nitrogen fertilizers, focusing instead on establishing closed nutrient cycles, minimizing losses via runoff, volatilization, and emissions, nitrogen levels on organic farms tend to be lower per hectare than on conventional farms which can contribute to a sustainable climate-friendly production system that delivers enough food (Ibrahim and Khan 2017; Imran et al. 2016d; Imran and Amanullah 2018). Conventional agriculture uses vast quantities of synthetic fertilizers and pesticides. It takes significant amounts of energy to manufacture these chemicals. Organic agriculture and biofertilizer application minimize energy consumption by 30–70% per unit of land by eliminating the energy required to manufacture synthetic fertilizers, and by using internal farm inputs. Organic agriculture can also help combat global warming by storing carbon in the soil (Imran et al. 2020a). Many management practices used by organic agriculture (e.g. beneficial microbes application, minimum tillage, returning crop residues to the soil, the use of cover crops and rotations, and the greater integration of nitrogen-fixing legumes), increase the return of carbon to the soil. This raises productivity and favours carbon storage.

Crop nutrient management is an important aspect provided mainly by chemical fertilizers (Imran et al. 2021a). However, it is widely accepted that the application of a balanced fertilizer with the effective use of organic sources and beneficial microbes is key to achieving higher crop production (Imran et al. 2019b). Repeated applications of chemical fertilizers and heavy metals lead to soil degradation and causing environmental damage. The application of organic and biofertilizers can reduce these problems as useful chemical fertilizers to improve soil fertility (Imran et al. 2021b). Thus, it was considered appropriate to merge the three different sources of nutrients, organic, bio, and chemicals to get more efficient and economical results in the long term (Diep et al. 2016a,b). The literature on the use of organic substances along with a bio and chemical fertilizers on the yield of agronomic crops is very limited (Aise et al. 2011; Akpalu et al. 2014). The challenges for agriculture production in the harsh and hostile environment with climate change in the twenty-first century for food and fibre is only could be overcome with the application of biofertilizer (Imran 2017a,b; Diep et al. 2016a,b). Beneficial microbes are favoured by the presence of high levels of plant roots, which they colonize readily. Some strains are highly rhizosphere competent, i.e. able to colonize and grow on roots as they develop (Imran et al. 2020c). The most strongly rhizosphere competent strains can be added to soil or seeds by any method. Once they come into contact with roots, they colonize the root surface or cortex, depending on the strain (Elkoca et al. 2008). Thus, if added as a seed treatment, the best strains will colonize root surfaces even when roots are a meter or more below the soil surface and they can persist at useful numbers up to 18 months after application (FAO (Food and Agriculture Organization of the United Nations) 2014). However, most strains lack this ability. The use of mineral fertilizers is faster and safer to increase grain production; however, cost and other restrictions discourage farmers to use mineral fertilizers (Dubey et al. 1997). So farmers need to take full advantage of the potential of alternative sources of plant nutrients (Akter et al. 2013). So, we should pay more attention to the efficient use of chemical fertilizers and use these low-cost sources of nutrients such as organic matter and biological fertilizer to reduce production costs along with soil productivity and health support (Drevon and Hartwig 1997).

Also *Trichoderma* Species the attack, parasitize and otherwise gain nutrition from other fungi. Since *Trichoderma* spp. grow and proliferate best when there are abundant healthy roots, they have evolved numerous mechanisms for both attacks of other fungi and for enhancing plant and root growth (Shah and Khalil 2016). Several new general methods for both biocontrol and for causing the enhancement of plant growth have recently been demonstrated, and it is now clear that there must be hundreds of separate genes and gene products involved in these processes (Imran et al. 2015c,d,e,f, 2017).

7.2 Soil Productivity and Soil Health with Climate Change

The capacity of a specific kind of soil to function – within natural or managed ecosystem boundaries, to sustain plant and animal productivity – maintains or enhances water quality and support human health and habitation is known as productive soil and healthy soil (Muhammad et al. 2017a). Soil quality could, in part, be viewed as a static (qualitative) measure of the capability of the soil, whereas 'Soil health' infers a dynamic state, where human impact causes a shift in quality (Anchal et al. 1997; Ankomah et al. 1996; Argaw 2012). Biofertilizers and Organic matter are vital because it supports many soil processes that are associated with fertility and physical stability of soil across the various ecosystem services under changing climate (Muhammad et al. 2012, 2017b; Naveed et al. 2015). In particular, organic matter provides an energy source for microbes structurally stabilizes soil particles, stores, and supplied plant essential nutrients such as nitrogen, phosphorus, and sulfur, and provides cation/anion exchange for retention of ions and nutrients under harsh environment (Assiouty and Abo-Sedera 2005; Ayoub et al. 1994). The carbon within the terrestrial biosphere can also behave as either a source or sink for atmospheric CO_2 depending on land management, thus potentially mitigating or accelerating the greenhouse effect. Cycling of soil organic carbon is also strongly influenced by moisture and temperature, these two factors which are predicted to change under global warming (Imran et al. 2016b).

7.3 No Life on Earth Without Soil Biota

Microbes quite literally support all of life on Earth and the world population depends on soil biota and beneficial microbes (Imran et al. 2020d,e). They are at the very bottom of the food chain, no matter where you look and regulate several processes critical to all of us. They are also the vast majority of life forms on the planet, outnumbering every other living thing by staggering amounts. But in the vast majority of cases, we have yet to understand how they will be impacted by climate change and how in turn they will affect how climate change unfolds (Imran et al. 2016b). Soil fertility is vital to help feed growing global populations in the face of a changing climate and pressures on land. Soil is a non-renewable and finite natural asset (Azarmi et al. 2009; Bardan 2003; Begum et al. 2015). Yet the world's expanding populations are destroying land capital at alarming rates. This reduces opportunities for food production, spaces for recreation, and the wilderness needed by nature. We all benefit from conserving soil, while agriculture depends on it (Imran et al. 2016b; Khan 2015a; Khan et al. 2015a,b). In turn, fertility relies on the populations of microbes of which our understanding is very limited. Soils are systems of pores and aggregated mineral and organic particles of differing sizes (Begum et al. 2015; Bhattacharyya et al. 2010).

Within the pores microbes are continuously eating, respiring, reproducing, competing, cooperating, and responding to their environment (Imran et al. 2021c). They concentrate mainly around roots, which exude sugars providing them with energy (Chiezey and Odunze 2012). Soil temperature, moisture content, acidity/alkalinity, and nutrient availability regulate the microbial activity, which changes with the seasons and the way land is used. Adding nitrogen decreases populations of fungi compared with bacteria, while liming tends to favour fungi (Imran et al. 2019a). Fungi are encouraged by good practices such as reducing the intensity of cultivation, increasing rotations, and using cover cropping. Intensive cropping excludes all but one plant from an area of land and reduces its microbial diversity (Imran and Amanullah 2021; Imran 2021; Imran et al. 2021c). Subsequent husbandry should aim at restoring fertility because the numbers of beneficial and benign microbes far outweigh and suppress those causing diseases. Additionally, beneficial microbes enhance crop water-use efficiency and nutrient uptake, particularly phosphorus (Crews 1993; Devi et al. 2012, 2013). Maintaining soil fertility reduces the need for synthetic fertilizers and pesticides and saves growers money (Iqbal et al. 2016). Degrading dead plants and animals into useful organic matter are a key microbial function. Decomposition underpins sustainable land-use in all managed, agricultural and horticultural, semi-natural and natural soils, part of the natural carbon, and nitrogen cycles that are driven by free solar energy captured by plant photosynthesis.

7.4 Exogenous Application of Beneficial Microbes

Commercial use of beneficial bacteria and fungi is gaining ground as part of integrated pest management strategies. Belchim contains *Coniothyrium minitans*, which helps control soil-borne pathogens such as *Sclerotinia sclerotiorum* and *Sclerotinia minor* (Hasanuzzaman 2019). According to specialist Fargro technical and energy director Paul Sopp: "Growers are experimenting with biopesticides such as Serenade *(Bacillus subtilis* QST713) because this widens their options for pest and disease control." This is especially the case with diseases such as downy mildew on brassicas and onions (Hussain et al. 2013). Several formulations of beneficial microbes antagonistic to pathogens are available including *Agrobacterium radiobacter* active against *Agrobacterium tumefaciens (crown gall, NoGall), Pseudomonas fluorescens* active against *Sclerotinia, Rhizoctonia,* and *Pythium* (*Biomonas*), *Streptomyces griseoviridis* active against *Fusarium, Phytophthora, Pythium,* and *Rhizoctonia* (*Mycostop*), *Trichoderma atroviride* active against *Sclerotium cepivorum* (*Plantmate*) and *Trichoderma harzianum* active against *Pythium, Rhizoctonia* and *Fusarium* (*Rootshield, PlantShield*).

7.5 Plants and Microbe are Thick as Thieves

Microbes and plant roots communicate by exchanging chemical signals. When necessary, they activate their defences and those of host plants – for example, in mycorrhizal combinations (Diacono and Montemurro 2010; Drevon and Hartwig 1997). Exploiting these natural communications offers prospects for developing new generations of environmentally benign fertilizers, crop stimulants, and protectants (Imran et al. 2016d). For example, isopyrazam (IZM), Syngenta's new cereal fungicide, is a move in this direction. Productive soil cares for the conservation of beneficial microbes maintaining fertility is of prime importance. We must feed a world population of 10 billion by 2080 from diminishing land assets and in increasingly adverse climates (Fageria et al. 1995). Microbes and soil fertility are some of our best friends in grappling with these problems (Imran et al. 2015a, 2016a, 2019c). Microbes are also known as microorganisms. They are very small and tiny living organisms that you cannot see with your naked eyes (Iqbal et al. 2016; Khan et al. 2016). They are everywhere in this world; they are in the air, water, soil, and even in the animal and human bodies. Their numbers in this world are greatly higher than any other organism and they play important roles in our daily lives (Imran et al. 2021d). Some of the well-known microbes groups are bacteria, viruses, fungi, and algae. Some of the microbes are beneficial to us and some of them are our enemies. In terms of agriculture, microbe plays a very important role in plant growth development (Imran and Amanullah 2019; Imran 2011, 2017a,b). Plants even help in taking care of them, for your information, 50% of carbon created from the photosynthesis process is released through the plants' roots for the microbes to be consumed as food.

7.6 Emerging Trends of Biofertilization

The current generation of commercially available microbes is actively antagonistic to specific disease-causing pathogens. An alternative and more effective approach is using combinations with differing modes of action (Imran and Amanullah 2019;

Imran 2011, 2017a,b). For instance, this might include combining non-pathogenic forms such as strains of *Fusarium oxysporum* that compete with pathogens for energy sources along with bacterial antagonists that block access to iron. Alternatively, there is scope for integrating microbial treatments with beneficial husbandry. This approach could be valuable where biofumigant crops are used (Imran et al. 2021d). Natural suppression of soil-borne pests and pathogens follows the incorporation of green manure or harvest residues from brassicas with high-glucosinolate contents such as Caliente mustard into the soil. Adding compost containing growth-promoting rhizobacteria strengthens the control of diseases such as *Fusarium* wilt and *Rhizoctonia* damping-off and builds soil fertility further.

7.7 Microorganisms Convert Soil Carbon into Stable Forms

The stable forms of soil carbon such as humus and glomalin are manufactured by microorganisms (Imran et al. 2020a). They convert the carbon compounds that are readily oxidized into CO_2 into stable polymers that can last thousands of years in the soil. The process of making composts uses microbes to build humus and other stable carbons (Imran et al. 2021a). The microorganisms that create compost continue working in the soil after compost applications, converting the carbon gifted by plant roots into stable forms. Regular applications of compost or compost teas will inoculate the soil with beneficial organisms that build humus and other long-lasting carbon polymers. Synthetic nitrogen fertilizers are one of the major causes of the decline of soil carbon (Imran et al. 2014, 2015b, 2016c). This is because it stimulates a range of bacteria that feed on nitrogen and carbon to form amino acids for their growth and reproduction. These bacteria have a carbon to nitrogen ratio of around 30-1. In other words, every ton of nitrogen applied results in the bacteria consuming 30 tonnes of carbon (Imran 2021). The quick addition of these nitrogen fertilizers causes the nitrogen-feeding bacteria to rapidly multiply, consuming the soil carbon to build their cells (Amanullah et al. 2021; Ilyas et al. 2020).

7.8 Soil Microbes and Carbon, Nitrogen Cycles

Carbon from dead plants and animals is recycled naturally back into the atmosphere as carbon dioxide by microbes using chemical energy. Nitrogen is taken by free-living soil bacteria from organic matter and added to the soil (Imran et al. 2014, 2015b, 2016c). Associations of bacteria living with legume roots add further nitrogen supplies into soils. Fixation of atmospheric nitrogen by legume-rhizobia deposits 33–46 million tonnes of nitrogen annually into the world's soils. At current fertilizer prices, this is worth the US$50–70 billion yearly. Using processes understood only in outline, microbes make soil inorganic phosphorus soluble and available to plant roots (Islam et al. 2015). Additionally, the biochemical diversity and adaptability of microbes can be used in repairing polluted land. These processes are further stimulated by adding substances such as active charcoal or biochar.

7.9 Biofertilizer Act as a Suppressing Agent for Pests and Pathogens

Fertile soils rich in microbes suppress plant pathogens. Actively beneficial bacteria include *Bacillus* spp., *Enterobacter* spp., *Flavobacterium balustinum*, and *Pseudomonas* spp. and fungi such as *Penicillium* spp., *Gliocladium virens*, and several *Trichoderma* spp. Microbes such as *Trichoderma* spp. offer several beneficial effects (Iqbal et al. 2016; Islam et al. 2015). They compete with pathogens for nutrients and soil space, parasitize pathogens, and induce plant disease resistance (Ilyas et al. 2020). The soil-borne diseases that are reduced and crops that benefit from this natural biocontrol include *Pythium* in turfgrass, snow moulds in turfgrass, and *Fusarium*, *Pythium*, and *Rhizoctonia* in tomato and cucumber. Suppression of *Rhizoctonia solani* attacking cucumber resulted from inoculating seedling compost with *Trichoderma asperellum*, while a reduction of bacterial leaf spots of radish, lettuce, and tomato followed inoculation of compost with *Trichoderma hamatum*. Some free-living soil fungi diminish soil nematode populations very effectively. Their fungal hyphae form nooses that capture the nematodes and then they exude enzymes that dissolve them (Karim 2019).

7.10 Beneficial Microbes Application Enhance Nitrogen Capturing and Fixation

Some microbes are helping to prepare the nitrogen element within the soil for plants to use. These types of microbes contribute and play important roles in ensuring the healthy growth of plants. An example of a very well-known and common microbe that is doing this type of job is the nitrogen fixation bacteria (Karim 2019; Khan 2015a,b). This type of bacteria helps to fix the nitrogen in the atmosphere into the form that can be absorbed by the plants. This is due to plants cannot absorb the nitrogen in its original form and can only absorb the nitrogen in the form of ammonia or nitrate after the fixation process. A few examples of bacteria species that are doing this type of job are *Azotobacter* and *Clostridium*. Nitrogen fixation bacteria play a very important role in the nitrogen circle in the atmosphere as 90% of nitrogen fixation is being contributed by these microbes.

7.11 Beneficial Microbe's Application Improve Soil Structure

Some microbes help to improve the soil structure. As we all know soil structure is very important to have healthy and vigorous plant growth. The good news is that microbes can also help you in achieving this. An example of the soil microbe that can help in improving the soil structure is the algae. Algae is a microbe that can be categorized as micro-flora. Like any other floras, algae can produce their food through the photosynthesis process (Karim 2019; Khalil et al. 2016; Khan 2015a,b). In terms of helping to improve the soil structure, algae will usually produce some type of slimy substance that can fix the soil particle together that is water stable and soluble. Another type of microbe that can help in improving soil structure same as algae is cyanobacteria. The soil particles that were fixed together with the help of the above said microbes help in increasing the soil moisture content and also reducing the effect of erosion from the running water.

7.12 Beneficial Microbe's Application Digest Nutrients in the Soil

Another function of soil microbes is to digest the nutrients in the organic matter within the soil before releasing them back in the form that can be taken by the plants. It is very well known that microbe plays a major role in the decaying process (Karim 2019; Khan 2015b,c). The nutrients that are available in the organic matter will not be available to the plants until that organic matters are fully decayed and released back all the nutrients that they contain into the soil. The decaying process is highly dependent on the availability of the microbes in the soils. The more beneficial microbes you have in your soil, the decaying process will be much faster.

7.13 Beneficial Microbes' Application Preventing Diseases and Pests Attack

This might be a shocking fact for some of you as we were usually informed that microbes are usually the source of many diseases not only to the human but also to the plants. However many studies have shown that some microbes can help in preventing some types of diseases and pests from ruining your plants. In the wild, some bacteria feed on other bacteria. With this fact, we can leverage it by introducing beneficial bacteria that feed on other bacteria into the soils that are highly populated by harmful bacteria so that they can feed on them. Studies also revealed that some microbes could excrete some sort of antibiotic substance that helps plants to fight pathogens that can cause certain diseases. Mycorrhizal fungi are one of the microbes that are functioning like this. Some studies have been done that showing plants and microbes are working closely with each other for mutual benefits.

From the studies, they found that plants can purposely excrete substances that are attracting beneficial microbes to be near them and protect them from the attack of other microbes or pests. It is some sort of mutual understanding where plants will excrete the substance to be as the food for the beneficial microbes and the microbes will return the favour by protecting the plants from their enemies (Figures 7.1 and 7.2).

7.14 Beneficial Microbes Application Create Organic Matter for Soil

As mentioned earlier, microbe plays important role in the decaying process which means they help in preparing the organic matter for the soils. As we all know, organic matters help to increase soil fertility by increasing the water abortion capacity of the soil and improve the air circulation within the soil. Soils that contain more beneficial microbes will be more fertile than the soils that contain less beneficial microbes (Imran et al., 2020d,e, 2021e).

Figure 7.1 Tangible view of mungbean (green gram) experimental plots at ARI with (+) and without (−) biofertilization.

Figure 7.2 Tangible view of soybean experimental plots at ARI with (+) biofertilization.

7.15 Biofertilizers, Nutrients Availability, and Crop Responses

Microbes are also important in preparing the availability of other soil nutrients in the soil such as phosphorus and potassium in their daily activities. Some of them can also help in the mobilization of nutrients in the soil. Other microbes can also excrete some type of hormone that can stimulate the plant's growth (Figures 7.3 and 7.4).

Figure 7.3 Tangible view of maize experimental plots at ARI without (−) and with (+) bio fertilization.

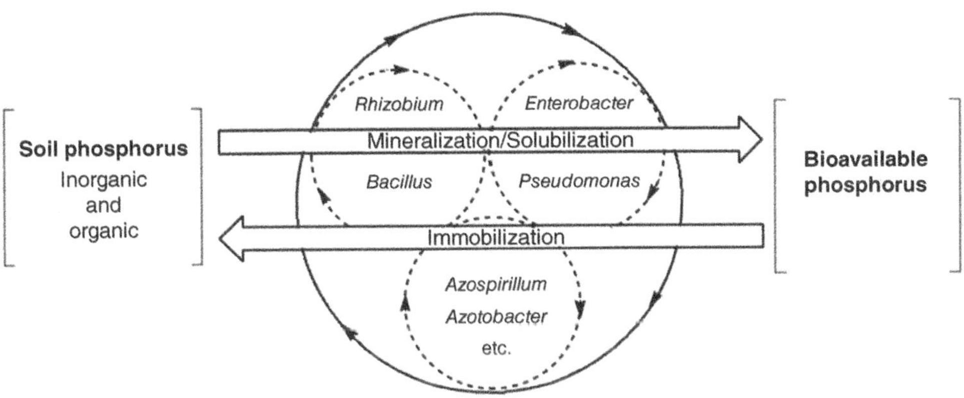

Figure 7.4 Schematic diagram of soil phosphorus mobilization and immobilization by bacteria.

7.15.1 Biofertilizers and Cereal Crops

Bio-Power vaccination was assessed for the growth and yield of wheat and maize by Bekere et al. (2012) in field conditions. Bio-Power by half and the full rate of N fertilizer increased grain yields of wheat and maize by over half and the full rate of N fertilizer. Biofertilization may enhance nutrients availability in soil and proliferate uptake and biofortification of various nutrients in crop various parts. Bellore and Mall (1975) reported that inoculation of *Rhizobia* alone or in combination with chemical fertilizers produced a much larger number of nodules, number of the pod, pod weight, and yields as compared with those who were not vaccinated. Combining treatments with beneficial microbes gave the highest yield.

Berg and Lynd (1985) studied the effectiveness of biofertilizers in improving the availability of phosphorous. The results indicated that P concentration in maize leaf, stalk, grain, and total P uptake was much higher than no inoculated plants. Comparison of control with treatments asserted plots, wheat production has increased by 21 and 35% respectively, due to increased availability of phosphorus. Maize responded positively to inoculation in terms of increasing residual biomass of plants, number of cobs, and grain yield (Bhattacharyya et al. 2010).

Gao et al. (2007) determined the effect of arbuscular mycorrhizal fungi (AMF) inoculation on growth performance and Zn uptake by rice genotypes. A pot experiment was conducted with six aerobic rice genotypes inoculated with *Glomus mosseae* or *G. etunicatum* or without AMF on a low Zn soil. Plant growth, Zn uptake, and mycorrhizal responsiveness were determined. AMF inoculated plants produced more biomass and took up more Zn than nonmycorrhizal controls. Mycorrhizal inoculation, however, significantly increased Zn uptake only in genotypes that had a low Zn uptake in the nonmycorrhizal condition (Cassman et al. 1981a,b; Brady 2002). They concluded that genotypes that are less efficient in Zn uptake are more responsive to AMF inoculation.

7.15.2 Biofertilizers and Pulses

Mungbean (green gram) inoculation with VAM gave the highest green forage, dry matter, and crude protein. It has been reported that the highest number of plants per square meter, plant height (cm), number of leaves per plant, pods length (cm), numbers of seeds per pod, number of pods per plants, thousand seed weight, seed yield, and harvest index was noted with the inoculation of biofertilizer (phosphate solubilizing bacteria (PSB)) (Imran et al. 2016d). Alam et al. (2007) studied beneficial microbe inoculation with three different cultivars of mung bean and reported that growth and yield components were highest in cv. NM-92 with inoculation of *Trichoderma*. Recently, the author (Imran et al. 2020a) conducted two years (Imran et al. 2016d; Imran 2017a,b) consecutive experiments on soybean, maize, and subsequent wheat to evaluate soybean, maize, and subsequent wheat response to biofertilizers (PSB, *Trichoderma*) and their interactions with organic sources and phosphorous levels. The author not yet published the work and will be submitted to the higher education commission of Pakistan. He reported that soybean's phenological and growth characteristics were holistically improved with the inoculation of *Trichoderma* (Imran 2021). Dry matter portioning was much higher with PSB inoculation. He concluded that *Trichoderma* is favoured by the presence of high levels of plant roots, which they colonize readily. Some strains are highly rhizosphere competent, i.e., able to colonize and grow on roots as they develop. The most strongly rhizosphere competent strains can be added to soil or seeds by any method (Ansari and Sukhraj 2010). Once they come into contact with roots, they colonize the root surface or cortex, depending on the strain. Thus, if these microbes added as a seed treatment, the best strains will colonize root surfaces even when roots a meter or more below the soil surface and they can persist at useful numbers up to 18 months after application. However, most strains lack this ability. By fixing atmospheric nitrogen and solubilizing phosphates, pulses contribute to reducing the need for synthetic fertilizers and, in doing so, greatly contribute to reducing the risk of soil and water pollution, supporting soil biodiversity, and combating and building resilience to climate change (Aziz et al. 2016). Biological nitrogen fixation is particularly important for global agricultural productivity and might be considered one of the most important biological processes on the planet. It provides circa 100 million metric tonnes of N which leads to an annual saving of around US$10 billion in N fertilizer (Arancon et al. 2005). Lentils alone could fix nitrogen in the range of 35–100 kg ha^{-1}. Furthermore, the reduced need for (or use of) synthetic fertilizers indirectly reduces the number of greenhouse gases released into the atmosphere. Pulses also promote soil carbon sequestration and, ultimately, reduce soil erosion when included in intercropping farming systems and/or used as cover crops (Anchal et al. 1997). Furthermore, due to their high nutritional value, pulses are also valuable allies in fighting hunger worldwide.

7.16 PSB on Crop Production

Phosphate rock minerals are often too insoluble to provide sufficient P for crop uptake. The use of phosphate solubilizing microorganisms (PSMs) can increase crop yields by up to 70% (Carsky et al. 2001). Combined inoculation of arbuscular mycorrhiza and PSB gives better uptake of both native P from the soil and P coming from the phosphatic rock. Higher crop yields result from solubilization of fixed soil P and applied phosphates by PSB (Cassman et al. 1981a,b). Microorganisms with phosphate solubilizing potential increase the availability of soluble phosphate and enhance plant growth by improving biological nitrogen fixation (Chaturvedi 2006). *Pseudomonas* spp. enhanced the number of nodules, dry weight of nodules, yield components, grain yield, nutrient availability, and uptake in soybean crop (Chauhan et al. 1992; Chela et al. 1993). Phosphate solubilizing bacteria enhanced the seedling length of *Cicer arietinum* while co-inoculation of PSM and plant growth-promoting Rhizobacteria (PGPR) reduced P application by 50% without affecting corn yield. Inoculation with PSB increased sugarcane yield by 12.6% (Chen et al. 2006). Sole application of bacteria increased the biological yield, while the application of the same bacteria along with mycorrhizae achieved the maximum grain weight (Chen et al. 2008). Single and dual inoculation along with P fertilizer was 30–40% better than P fertilizer alone for improving grain yield of wheat, and dual inoculation without P fertilizer improved grain yield up to 20% against sole P fertilization. Mycorrhiza along with *Pseudomonas putida* increased leaf chlorophyll content in barley. Rhizospheric microorganisms can interact positively in promoting plant growth, as well as N and P uptake. Seed yield of green gram was enhanced by 24% following triple inoculation of *Bradyrhizobium* + *Glomus fasciculatum* + *Bacillus subtilis* (Chiezey and Odunze 2012). Growth and phosphorus content in two alpines Carex species increased by inoculation with *Pseudomonas fortini*. Integration of half dose of NP fertilizer with biofertilizer gives crop yield as with full rate of fertilizer, and through reduced use of fertilizers the production cost is minimized and the net return maximized (Crews 1993). Mostly on farm crops using compost may have some limitations. For example, compost could be effectively used during the establishment of the nursery in the field to increase the stock and maintain the health of the plant (Ahmed et al. 2008). Crop nutrient management through an integrated approach is an important aspect provided mainly by bio, organic and chemical fertilizers. However, it is widely accepted that the application of a balanced fertilizer with the effective use of organic sources and beneficial microbes is key to achieving higher crop production and net return (Imran et al. 2016d). Agriculture scientists exploring an alternative source to increase farm products with low input cost and maintain soil health without affecting the soil and plant environment. Plants obtain nutrients from two natural sources, organic matter, and minerals or chemical fertilization. Farmers only rely on chemical fertilizer which not only deteriorates the quality of soil but also increases the cost of production. Soil amendments and bio fertilization is an integrated approach to reduce dependency on chemical fertilizers with the advantage of the low cost of production (Imran 2017a,b). Crop response to biological fertilizers depends largely on crop species/strains of microbe and application method along with climatic conditions. It has been reported that using beneficial microbes reduces the requirement of nitrogen by 50–70% and increases yield by up to 20%. Bacterial inoculation not only provides nitrogen, phosphorus, and growth hormones but also makes the plant healthy and less susceptible to pathogens. Using beneficial microbes along with the optimal dose of fertilizers saves about half the recommended dose of chemical fertilizers (Imran et al. 2016d).

7.17 Mycorrhizas and Crop Production

Mycorrhizas are mutually beneficial associations of fungi with roots. There are broadly two forms – endomycorrhiza (particularly arbuscular types), where fungal hyphae inhabit plant cells, and ectomycorrhiza, where a hyphae sheath forms around the root surface. *Arbuscular mycorrhizas* are crucially important for sustainable cropping. Mycorrhizas increase root access to nutrients, particularly phosphorus, improve water uptake and defend them against invading pathogens.

Commercial formulations are available as soil improvers that enhance amenity and fruit tree and shrub establishment. These are useful for controlling specific replant disorders caused by *Pythium* spp. and other pathogens in top and stone fruit and roses. Mycorrhizal preparations are available from Dragonfly or David Austin Roses.

7.18 Conclusion

To conclude the article, we can say that microbes have a major role in soil fertility. There are many other ways that they can contribute to helping to improve soil fertility besides what has been listed here in this article. Climate change poses serious interlinked challenges in times to come concerning scale and scope, never anticipated in the last century. More or less the most important change in soils expected as a result of these changes would be a gradual improvement in fertility and physical conditions of soils in humid and subhumid climates, change from one major soil-forming process to another in certain fragile tropical soil, and changes in soil property due to poleward retreat of the permafrost boundary. Again changes due to climate change are expected to be relatively well buffered by the mineral composition, the organic matter content, or the structural stability of many soils. The impact of climate change on soil systems should be monitored in different agro-ecological regions on regular basis. Climate change and land degradation are closely linked issues and conservation farming has shown promise in minimizing land degradation. Hence, the potential of conservation agriculture in minimizing the impact of climate change needs thorough investigation. There is a need for harmonization of the database on land degradation keeping in view the productivity and economic losses vis-à-vis climate change effects.

7.19 Summary

Agricultural scientists are looking for an alternative way to increase the low input cost of farm products and maintain soil health without affecting the soil and plant environment. Plants receive nutrients from two natural sources, organic matter, and minerals or chemical fertilizers. Farmers rely solely on chemical fertilizers, which not only impairs soil quality but also increases production costs. Soil modification and bio-fertilization is an integrated method for reducing dependence on chemical fertilizers, which benefits from the lower cost of production. The response of the crop to biological fertilizers depends on the climatic conditions as well as the strain/application of species/microbes. It has been reported that the use of beneficial microbes reduces the need for nitrogen by 50–70% and production by up to 20%. Bacterial inoculation not only causes nitrogen, phosphorus, and growth hormones but also makes the plants healthier and less susceptible to pathogens. Use of beneficial microbes with an optimum dose of fertilizers to save half of the chemical fertilizers. As climate changes, growers face many challenges, including more unexpected rains, soil erosion, and new or different pests and diseases. With increasing temperature and inflammation, the yield of crops decreases rapidly. High temperature reduces soil moisture, crop productivity and also influencing soil nutrients availability and its transportation and uptake. Eliminating drought increases the risk of wasting nutrients from the roots. Under increased temperature, as nutrients are transported to the roots by water, the soil moisture content is reduced, and nutrients such as nitrate, sulfate, Ca, Mg, and Si at longer distances are not available. In drought conditions, root growth and root function impairment reduce the root system's ability to obtain nutrients. In drought conditions, both carbon and oxygen flow to the root nodules, and a decrease in nitrogen accumulation prevents nitrogen fixation in pods. Climate change may change the structure and activity of soil microbial communities such as soil nitrifying bacterial deficiency. Soil organic carbon is the main component of soil organic matter, which gives the soil structure and allows it to store more water for future use in harsh environments. Management of crop nutrients through integrated approaches is an important aspect in which mainly organic, inorganic, and chemical fertilizers are provided. However, it has been widely accepted that the use of balanced fertilizers with the efficient use of organic sources and beneficial microbes is the key to high yields and net returns.

References

Ahmed, M.F., Kennedy, I.R., Choudhury, A.T.M.A. et al. (2008). Phosphorus adsorption in some Australian soils and influence of bacteria on the desorption of phosphorus. *Communications in Soil Science and Plant Analysis* 39: 1269–1294.

Aise, D., Erdal, S., Hasanand, A., and Ahment, M. (2011). Effects of different water, phosphorus and magnesium doses on the quality and yield factors of soybean in Harran plain conditions. *International Journal of Physical Sciences* 6 (6): 1484–1495.

Akpalu, M.M., Siewobr, H., Oppong, D., and Akpalu, S.E. (2014). Phosphorus application and rhizobia inoculation on growth and yield of soybean (*Glycine max* L. Merrill). *American Journal of Experimental Agriculture* 4 (6): 674–685.

Akter, Fahmina, Nurul, I. et al. (2013). Effect of phosphorus and sulphur on growth and yield of soybean (*Glycine max* L.). *International Journal of Bio-resource and Stress Management* 4 (4): 555–560.

Alam, M.N., Jahan, M.S., Ali, M.K. et al. (2007). Effect of vermicompost and chemical fertilizers on growth, yield and yield components of potato in Barind soils of Bangladesh. *Journal of Applied Sciences Research* 3: 1879–1888.

Amanullah, Khan, I., Jan, A. et al. (2015). Compost and nitrogen management influence productivity of spring maize (*Zea mays* L.) under deep and conventional tillage systems in semi-arid regions. *Communications in Soil Science and Plant Analysis* 46: 1566–1578.

Amanullah, M., Muhammad, A., Nawab, K., and Ali, A. (2016). Effect of tillage and phosphorus interaction on yield of mungbean (*Vigna radiata* L., Wilczek) with and without moisture stress condition. *Ponte* 72: 114–139.

Amanullah, Ullah, H., Elshikh, M.S. et al. (2020a). Nitrogen contents in soil, grains, and straw of hybrid rice differ when applied with different organic nitrogen sources. *Agriculture* 10: 386. https://doi.org/10.3390/agriculture10090386.

Amanullah, Inamullah, Alkahtani, J. et al. (2020b). Phosphorus and zinc fertilization improve productivity and profitability under rice-wheat cropping system. *Agronomy* 10: 1085. https://doi.org/10.3390/agronomy10081085.

Amanullah, Inamullah, Alwahibi, M. et al. (2020c). Phosphorus and zinc fertilization improve zinc biofortification in grains and straw of coarse vs. fine rice genotypes. *Agronomy* 10: 1155. https://doi.org/10.3390/agronomy10081155.

Amanullah, Khalid, S., Khalil, F. et al. (2021). Growth and dry matter partitioning response in cereal-legume intercropping under full and limited irrigation regimes. *Scientific Reports (2021)* https://doi.org/10.1038/s41598-021-92022-4.

Anchal, D., Kharwara, P.C., Rana, S.S., and Dass, A. (1997). Response of gram varieties to sowing dates and phosphorus levels under on farm condition. *Himachal Journal of Agricultural Research* 23: 112–115.

Ankomah, A.B., Zapata, F., and Hardarson, G. (1996). Yield, nodulation and N_2 fixation by cowpea cultivars at different phosphorus levels. *Biology and Fertility of Soils* 22: 10–15.

Ansari, A.A. and Sukhraj, K.K. (2010). Effect of vermiwash and vermicompost on soil parameters and productivity of okra in Guyana. *African Journal of Agricultural Research* 5: 1794–1798.

Arancon, N.Q., Edwards, C.A., Bierman, P. et al. (2005). Effects of vermi composts produced from cattle manure, food waste and paper waste on the growth and yield of peppers in the field. *Pedobiologia* 49: 297–306.

Argaw, A. (2012). Evaluation of co-inoculation of *Bradyrhizobium japonicum* and phosphate solubilizing *Pseudomonas* spp. Effect on soybean (*Glycine max* L.) in Assossa Area. *Journal of Agricultural Science and Technology* 14: 213–224.

Assiouty, F.M. and Abo-Sedera, S.A. (2005). Effect of bio and chemical fertilizers on seed production and quality of spinach. *International Journal of Agriculture and Biology* 6: 947–952.

Ayoub, M., Guertin, S., Lussier, S., and Smith, D.L. (1994). Timing and level of nitrogen fertilizer effects on spring wheat yield in eastern Canada. *Crop Science* 34: 748–756.

Azarmi, R., Giglou, M.T., and Taleshmikail, R.D. (2009). Influence of vermicompost on soil chemical and physical properties in tomato field. *African Journal of Biotechnology* 7: 2397–2401.

Aziz, A.L.A., Ahiabor, B.D.K., Opoku, A., and Abaidoo, R.C. (2016). Contributions of Rhizobium inoculants and phosphorus fertilizer to biological nitrogen fixation, growth and grain yield of three soybean varieties on a fluvic luvisol. *American Journal of Experimental Agriculture* 10 (2): 1–11.

Bardan, M.M. (2003). Effect of nitrogenous and phosphetic fertilization on some economical characters of soybean Crawford cultivar under calcareous soil conditions. *Egyptian Journal of Agricultural Research* 81 (2): 433–440.

Begum, A., Islam, M.D.A., Ahmed, Q.M., and Rahman, M. (2015). Effect of nitrogen and phosphorus on the growth and yield performance of soybean. *Research in Agriculture, Livestock and Fisheries* 2 (1): 35–42.

Bekere, W., Endalkachew, W., and Tesfu, K. (2012). Growth and nodulation response of soybean (*Glycine max* L.) to Bradyrhizobium inoculation and phosphorus levels under controlled condition in South Western Ethiopia. *African Journal of Agricultural Research* 7 (30): 4266–4270.

Bellore, S.K. and Mall, L.R. (1975). Chlorophyll content as an ecological index of dry matter production. *Journal of the Indian Botanical Society* 54 (1–2): 75–77.

Berg, R.K. and Lynd, J.Q. (1985). Soil fertility effects on growth, yield, nodulation and nitrogenase activity of Australian winter pea. *Journal of Plant Nutrition* 8: 131–145.

Bhattacharyya, R., Pandey, S.C., Chandra, S. et al. (2010). Fertilization effects on yield sustainability and soil properties under irrigated wheat-soybean rotation of an Indian Himalayan Upper Valley. *Nutrient Cycling in Agroecosystems* 86: 255–268.

Brady, N. (2002). Phosphorus and potassium. In: *The Nature and Properties of Soils*, 352. New Delhi, India: Prentice-Hall of India Pvt. Limited.

Carsky, R.J., Singh, B.B., and Oyewole, R. (2001). Contribution of early–season cowpea to late season maize in the savanna zone of West Africa. *Biological Agriculture and Horticulture* 18: 303–315.

Cassman, K.G., Whitney, A.S., and Fox, R.L. (1981a). Phosphorus requirements of soybean and cowpea as affected by mode of nitrogen nutrition. *Agricultural Journal* 73: 17–22.

Cassman, K.G., Munns, D.N., and Beck, D.P. (1981b). Phosphorus nutrition of *Rhizobium japonicum*: strain differences in phosphate storage and utilization. *Soil Science Society of American Journal* 45: 517–520.

Chaturvedi, I. (2006). Effects of phosphorus levels alone or in combination with phosphate-solubilizing bacteria and farmyard manure on growth, yield and nutrient up-take of wheat (*Triticum aestivum*). *Journal of Agriculture and Social Research* 2 (2): 96–100.

Chauhan, Y.S., Johanson, C., and Venkataratnam (1992). Effect of phosphorus deficiency on phenology and yield components of short duration pigeonpea. *Tropical Agriculture* 69: 235–238.

Chela, G.S., Tiwana, M.S., Thind, I.S. et al. (1993). Effect of bacterial cultures and nitrogen fertility on the yield and quality of maize fodder (*Zea mays* L.). *Annales Biologiques* 9: 83–86.

Chen, Y.P., Rekha, P.D., Arunshen, A.B. et al. (2006). PSB from subtropical soil and their tri-calcium phosphate solubilizing abilities. *Applied Soil Ecology* 34: 33–41.

Chen, Z., Shengwu, M., and Lily, L. (2008). Studies on phosphorus solubilizing activity of a strain of phosphobacteria isolated from chestnut type soil in China. *Bioresource Technology* 99: 6702–6707.

Chiezey, U.F. and Odunze, A.C. (2012). Soybean response to application of poultry manure and phosphorus fertilizer in the Sub-humid Savanna of Nigeria. *Journal of Ecology and the Natural Environment* 1 (2): 025–031.

Crews, T.E. (1993). Phosphorus regulation of nitrogen-fixation in a traditional Mexican agroecosystem. *Biogeochemistry* 21: 141–166.

Devi, K.N., Singh, L.N.K., Devi, T.S. et al. (2012). Response of soybean to sources and levels of phosphorus. *The Journal of Agricultural Science* 4 (6): 44–53.

Devi, K.N., Singh, T.B., Athokpam, H.S. et al. (2013). Influence of inorganic, biological and organic manures on nodulation and yield of soybean and soil properties. *The Australian Journal of Science* 7 (9): 1407–1415.

Diacono, M. and Montemurro, F. (2010). Long-term effects of organic amendments on soil fertility. A review. *Agronomy for Sustainable Development* 30: 401–422.

Diep, C.N., Duong, B.T., Nguyen, and Phan, V.H.L. (2016a). Effects of rhizobia and phosphate-solubilizing bacteria on soybean (*Glycine max* L.) cultivated on Ferralsols of Dak Lak province, Vietnam. *World Journal Pharmacy and Pharmaceutical Science* 5 (4): 318–333.

Diep, C.N., Trung, N.B., and Nhu, V.T.P. (2016b). Isolation and Characterization of rhizospheric bacteria in soybean cultivated on ferralsols of Buonho town, Dak Lak province, Vietnam. *Pharmaceutical Sciences* 5 (2): 34–50.

Drevon, J.J. and Hartwig, U.A. (1997). Phosphorus deficiency increases the argon-induced decline of nodule nitrogenase activity in soybean and alfalfa. *Planta* 201: 463–469.

Dubey, S.K., Balasundaram, V.R., Pant, L.M. et al. (1997). Effect of phosphate dissolving bacteria applied with rock phosphate on nodulation and yield of rainfed soybean under different agro-climatic conditions. *Journal. Indian Society of Soil Science* 45: 503–505.

Elkoca, E., Kantar, F., and Sahin, F. (2008). Influence of nitrogen fixing and phosphorus solubilizing bacteria on the nodulation, plant growth, and yield of chickpea. *Journal of Plant Nutrition* 31: 157–171.

Fageria, N.K., Zimmerman, F.J.P., and Batigar, V.C. (1995). Lime and priterathon on growth and nutrient uptake by upland rice, wheat, common bean and corn in an Oxisol. *Journal of Plant Nutrition* 18: 2519–2532.

FAO (Food and Agriculture Organization of the United Nations) (2014). Soil threats, soil health & sustainability. http://faostat.fao.org/site/567/DesktopDefault.aspx?PageID=567#ancor (accessed 12 May 2015).

Gao, X., Kuyper, T.W., Zou, C. et al. (2007). Mycorrhizal responsiveness of aerobic rice genotypes is negatively correlated with their zinc uptake when nonmycorrhizal. *Plant and Soil* 290: 283–291.

Hasanuzzaman, M. (2019). *Agronomic Crops*. Springer.

Havlin, J., Beaton, J., Tisdale, S., and Nelson, W. (2005). *Soil Fertility and Fertilizers: an Introduction to Nutrient Management*. Upper Saddle River, NJ: Prentice Hall.

Hoomin, L., Bajpai, V.K., Imran, K. et al. (2019). One-pot synthesized nitrogen/phosphorous-doped carbon nanodots diminish Listeria monocytogenes via nano-microbe interaction and surface-linked food wrap in real food matrices. Korean Society of Biological Engineering Conference, pp. 455–455.

Hussain, Z., Marwat, K.B., Munsif, F. et al. (2013). Influence of intercropping in maize on performance of weeds and the associated crops. *Pakistan Journal of Botany* 45: 1729–1734.

Hussain, A., Nabi, G., Ilyas, M. et al. (2020). 19. Effect of zinc and iron on growth, flowering and shelf life of marigold under the agro-climatic conditions of Sawabi. *Pure and Applied Biology* 9: 180–192.

Ibrahim, M. and Khan, A. (2017). Phenology and maize crop stand in response to mulching and nitrogen management. *Sarhad Journal of Agriculture* 33 (3): 426–434.

Ilyas, M., Ayub, G., Imran et al. (2020). Calcium and boron effect on production and quality of autumn potato under chilling temperature. *Communications in Soil Science and Plant Analysis* https://doi.org/10.1080/00103624.2020.1854286.

Imran, K. (2011). Reading assessment techniques among selected secondary school teachers in Pakistan: current trends and practices. *International Journal on New Trends in Education and their Implications* 58.

Imran (2017a). Climate change is a real fact confronting to agricultural productivity. *International Journal of Environmental Sciences & Natural Resources* 3 (3): 555613. https://doi.org/10.19080/IJESNR.2017.03.555613.

Imran, K. (2017b). Compost and nitrogen application influence phenology, growth and biomass yield of spring maize under deep and conventional tillage systems. *Journal of Soils and Crops* 27: 1–6.

Imran (2021). The bioavailability of phosphorus in composite vs. hybrid maize differ with phosphorus and boron fertilization. *Phosphorus, Sulfur, and Silicon and the Related Elements* https://doi.org/10.1080/10426507.2021.1920588.

Imran and Amanullah (2018). Global impact of climate change on water, soil resources and threat towards food security: evidence from Pakistan. *Advances in Plants & Agriculture Research* 8: 350–355.

Imran and Amanullah (2019). Integration of Peach (*Prunus persica* L.) biochar, compost and Peach residues along with beneficial microbes and phosphorus improve agronomic efficiency, phosphorus use efficiency, partial factor productivity and soil P in soybean vs maize crops. *Preprints*. https://doi.org/10.20944/preprints201911.0159.v1.

Imran and Amanullah (2021). Phosphorus and boron application optimizing biofortification of phosphorus and productivity of French Bean (*Phaseolus vulgaris* L.). *Communications in Soil Science and Plant Analysis* https://doi.org/10.1080/00103624.2021.1971689.

Imran, M., Arif, M., Ali, S. et al. (2014). Integration of biochar with organic and inorganic sources of phosphorous for improving maize productivity. *Integration* 4.

Imran, Hussain, I., Khattak, I. et al. (2015a). Roots nodulation, yield and yield contributing parameters of mungbean cultivars as influenced by different phosphorous level in Swat-Pakistan. *Pure and Applied Biology* 4: 557.

Imran, Khan, A.A., Ahmad, F., and Ullah, I. (2015b). Influence of hydrated calcium sulphate ($CaSO_4{\cdot}2H_2O$) and nitrogen levels on water infiltration rate and maize varieties productivity in rainfed area of Swat, Pakistan. *Chemistry and Materials Research* 7: 15–20.

Imran, Asad, A.K., and Fayaz, A. (2015c). Phenology of various rice genotypes as affected by different transplanting dates under cold climatic region of Khyber Pakhtunkhwa-Pakistan. *Journal of Environment and Earth Science*. ISSN 2224-3216; (Paper) ISSN 2225-0948.

Imran, Asad, A.K., Inamullah, and Luqman (2015d). Weeding stages and their effect on yield and yield components of rice in upper Swat, Pakistan. *Pakistan Journal of Weed Science Research* 21 (4): 555–563.

Imran, Inayat, K., Izaz, H. et al. (2015e). Growth and yield of maize hybrids as effected by different sowing dates in Swat Pakistan. *Pure and Applied Biology* 4 (4).

Imran, Inayat, K., Izaz, H. et al. (2015f). Roots nodulation, yield and yield contributing parameters of mungbean cultivars as influenced by different phosphorous level in Swat-Pakistan. *Pure and Applied Biology* 4 (4).

Imran, Hussain, I., Naveed, S. et al. (2016a). Growth and yield of maize hybrids as effected by different sowing dates in Swat Pakistan. *Pure and Applied Biology* 5: 114.

Imran, Asad, A.K., Inamullah, and Ahmad, F. (2016b). Yield and yield attributes of Mungbean (*Vigna radiata* L.) cultivars as affected by phosphorous levels under different tillage systems. *Cogent Food & Agriculture* 2: 1151982.

Imran, K., Naureen, A., Muhammad, Z., and Sobia, N. (2016c). Isolation and identification of heavy metal tolerant bacteria from volta battery effluent, Hattar, Haripur Pakistan. *Pure and Applied Biology* 5: 719–726.

Imran, Khan, A., and Naveed, S. (2016d). Weeds density and late sown maize productivity influenced by compost application and seed rates under temperate environment. *Pakistan Journal of Weed Science Research* 22: 169–181.

Imran, Nasar, J., Ashfaq, A., and Asad, A.K. (2017). Grain yield, yield attributes of wheat and soil physio-chemical characteristics influenced by biochar, compost and inorganic fertilizer application. *Agricultural Research & Technology: Open Access Journal* https://doi.org/10.19080/ARTOAJ.2017.10.555795.

Imran, Amanullah, Rahman, A. et al. (2019a). Organic matter management in cereals based system: symbiosis for improving crop productivity and soil health. In: *Sustainable Agriculture Reviews 29: Sustainable Soil Management: Preventive and Ameliorative Strategies*, vol. 29 (ed. R. Lal and R. Francaviglia), 67. Springer Nature.

Imran, Amanullah, Bari, A., and Ali, R. (2019b). Peach organic sources, phosphorous and beneficial microbes application enhance productivity and profitability of soybean. *Soybean Research* 1: 39.

Imran, Amanullah, Bari, A. et al. (2019c). Climatic variability and agronomic cropping pattern. In: *Agronomic Crops: Volume 1: Production Technologies* (ed. M. Hasanuzzaman), 33–44. Springer Nature (Springer Nature, Agronomic Crops).

Imran, Amanullah, Arif, M. et al. (2020a). Soil application of *Trichoderma* and Peach (*Prunus persica* L.) residues possesses biocontrol potential for weeds and enhances growth and profitability of soybean (*Glycine max*). *Sarhad Journal of Agriculture* 36 (1): 10–20. https://doi.org/10.17582/journal.sja/2020/36.1.10.20.

Imran, Amanullah, and Al-Tawaha, A.R. (2020b). The productivity of subsequent wheat enhanced with residual carbon sources and phosphorus under improved irrigation system. *Communications in Soil Science and Plant Analysis* https://doi.org/10.1080/00103624.2020.1763387.

Imran, Amanullah, Mahmood, T. et al. (2020c). Improvement of rice quality via biofortification of selenium, iron, and zinc and its starring role in human health. In: *Rice Research for Quality Improvement: Genomics and Genetic Engineering*, 699. https://doi.org/10.1007/978-981-15-5337-0_32.

Imran, Amanullah, Khan, A.A. et al. (2020d). Adequate fertilization, application method and sowing techniques improve maize yield and related traits. *Communications in Soil Science and Plant Analysis*.

Imran, Amanullah, Arif, M. et al. (2020e). Integration of Peach (*Prunus persica* L.) residues, beneficial microbes and phosphorus enhance phenology, growth and yield of soybean. *Russian Agricultural Sciences* 46 (3): 223–230. ISSN 1068-3674.

Imran, Amanullah, and Al Tawaha, A.R.M. (2021a). Management of nano-black carbon, phosphorous and bio fertilizer improve soil organic carbon and ensilage biomass of soybean and maize. *Communications in Soil Science and Plant Analysis*. https://doi.org/10.1080/00103624.2021.1966439.

Imran, Amanullah, and Altawaha, A.R. (2021b). Carbon assimilation and dry matter partitioning in soybean ameliorates with the integration of nano-black carbon, along with beneficial microbes and phosphorus fertilization. *Journal of Plant Nutrition*. https://doi.org/10.1080/01904167.2022.2035753.

Imran, Amanullah, Khan, A.A. et al. (2021c). Adequate fertilization, application method and sowing techniques improve maize yield and related traits. *Communications in Soil Science and Plant Analysis* https://doi.org/10.1080/00103624.2021.1925688.

Imran, Amanullah, and Al-Tawaha, A.R. (2021d). Carbon sources application increase wheat yield and soil fertility. *Communications in Soil Science and Plant Analysis* https://doi.org/10.1080/00103624.2020.1865397.

Imran, Amanullah, Ali, A. et al. (2021e). Adequate fertilization, application method and sowing techniques improve maize yield and related traits. *Communications in Soil Science and Plant Analysis* https://doi.org/10.1080/00103624.2021.1925688.

Iqbal, B., Ahmad, B., Ullah, I. et al. (2016). Effect of phosphorus, sulphur and different irrigation levels on phenological traits of Triticale. *Pure and Applied Biology* 5: 303–310. https://doi.org/10.19045/bspab.2016.50039.

Islam, M., Anwar, S., Bashir, S. et al. (2015). Growth and yield components of wheat varieties as affected by dual purpose practices. *Pure and Applied Biology* 4: 491.

Karim, S.A. (2019). Reduction in substrate moisture content reduce final yield of wheat. *Biomedical Journal of Scientific & Technical Research* 15: 11150–11154.

Khalil, S.K., Khan, I., Rab, A. et al. (2016). Impact of rice extract type, concentration and application time on weeds biomass and grain yield of wheat. *Pakistan Journal of Weed Science Research* 22: 555–566.

Khan, A. (2015a). Biochar application and shoot cutting duration (days) influenced growth, yield and yield contributing parameters of *Brassica napus* L. *Journal of Biology, Agriculture and Healthcare* 5: 1–6.

Khan, A. (2015b). Phenological charateristics of *Brassica napus* L. as influenced by biochar application and shoot cutting duration (days). *Civil and Environmental Research* 7: 104–107.

Khan, A.A. (2015c). Weeding stages and their effect on yield and yield components of rice in upper Swat-Pakistan. *Pakistan Journal of Weed Science Research* 21.

Khan, A.A., Khan, M.N., Inamuallah, S.S. et al. (2015a). Effect of potash application on growth, yield and yield components of spring maize hybrids. *Pure and Applied Biology* 4: 195–203.

Khan, A.A., Zada, H., Ahmad, F. et al. (2015b). Yield and yield attributes of rapeseed cultivars as influence by sulfur level under Swat Valley conditions. *Pure and Applied Biology* 4: 296.

Khan, A.Z., Imran, Asim, M. et al. (2016). Impact of fertilizer priming on seed germination behavior and vigor of maize. *Pure and Applied Biology* 5 (4): 744–751. http://dx.doi.org/10.19045/bspab.2016.50093.

Muhammad, N., Khan, D.M., Khattak, M.K., and Imran (2012). A study of Pakistan's various reform policies regarding Islamic seminaries. *Al-Idah* 24: 18–32.

Muhammad, A., Khan, M.Y.A.A., and Ahmad, I. (2017a). *Bioscience Research.*

Muhammad, M., Chattha, M., Hassan, M. et al. (2017b). Response of sunflower to different tillage systems and foliar application of potassium under water deficit conditions. *International Journal of Biology and Biotechnology* 14: 265–270.

Naveed, S., Khan, A.A., and Khattak, I. (2015). Impact of phosphorus levels and seed rates on growth and yield of late sown maize on high elevation in Swat, Pakistan. *Pakistan Journal of Agricultural Research* 28.

Shah, Z. and Khalil, S.K. (2016). Phosphorus and zinc interaction influence leaf area index in fine vs. coarse rice (*Oryza sativa* L.) genotypes in Northwest Pakistan. *Journal of Plant Stress Physiology*.

8

The Declining Trend of Soil Fertility with Climate Change and Its Solution

Imran[1,2], Amanullah[1], and Ibrahim Ortas[3]

[1] *Department of Agronomy, The University of Agriculture, Peshawar, Pakistan*
[2] *Ministry of Agriculture, Agriculture Extension Wing, Merged Area, Khyber Pakhtunkhwa, Pakistan*
[3] *Department of Soil Science and Plant Nutrition, Faculty of Agriculture, University of Çukurova, Adana, Turkey*

CONTENTS

8.1 Introduction

Soils are the support of primary production in terrestrial ecosystems (Imran et al. 2021a,b,c,d,e,f,g) and also provide ecosystem service. At the ecological and human time scales, soil resources are non-renewable. Agricultural production of food and fibre for the human population is strongly dependent on soil resources (Akhtar et al. 2015a; Imran 2018a,b,c,d,e; Khan et al. 2015a,b). Therefore, soil conservation is a critical factor to ensure increasing food demands. Soils can be a carbon (C) source and sink (Lal 2019). Therefore, soils contribute to the regulation of the carbon cycle and its consequent effects on climate change (Amonette and Joseph 2009; Armor et al. 2005; Arshad et al. 2008; Khan et al. 2015a,b; Ayaz et al. 2000; Iqbal et al. 2016a,b). Changes in land

Climate Change and Agriculture: Perspectives, Sustainability and Resilience,
First Edition. Edited by Noureddine Benkeblia.

use constitute the driving force that determines the soil's role as a source or as a sink of C. Tilling of land has led to a loss of organic carbon (OC) and an immediate increase in atmospheric carbon emissions, whereas the reforestation of croplands and covering the surface with plant species causes an increase in carbon sequestration (Iqbal et al. 2015a,b, 2016a,b; Badridze et al. 2009; Bates et al. 1973; Ortas and Yucel 2020). However, sequestration of C under forestry or other non-agricultural land uses often only slowly replaces OC lost through cropping, and such differences in time scale between loss or change of soil properties due to man and their reversal are commonly several orders of magnitude different (Imran et al. 2021a,b,c,d,e,f,g). Furthermore, the properties of soils are sensitive to climate change (Bohnert and Jensen 1996; Bos et al. 2005; Bruinsma 1963; Brussaard et al. 2007; Capell and Doerffling 1993; Chimenti et al. 2006; Ortaş et al. 2016). In mineral soils, the approximate relationship between organic matter (OM) and soil organic C is $1.724 \times \%$ OC $= \%$ OM which is important to estimate soil biological fertility. Temperature increase would cause a higher decomposition rate of the OM. Degradation of OM significantly caused soil fertility degradation and finally increases atmospheric carbon dioxide (CO_2) emission. Increased drought would have the opposite effect (Conde 2011; Cramer et al. 2011; Imran 2018a,b,c,d,e; Imran and Khan 2015a,b,c,d,e, 2017; Duan et al. 2007; Farahbakhsh 2012).

The possible increase in plant productivity due to the fertilizing effect of an increase in atmospheric CO_2 would lead to an increase in OC inputs to the soil, especially with intensive cropping (assuming no hydrological or nutritional limitations). In natural ecosystems where productivity is limited mainly by N, P, and other mineral nutrients (Ortas and Bykova 2020) shortage, increased atmospheric N deposition would increase OC inputs to the soil (Imran 2021). On the contrary, decreased productivity due to the intensification of water stress leads to losses of soil OC. The foreseeable increase in the occurrence of forest fires would cause losses of OC and would increase the risk of erosion (Flowers 2004; Gasco et al. 2012; Goncalves-Alvim and Fernandez 2001; González-Villagra et al. 2018). All over the world, in recent years, significant decreases have been achieved in the OC amounts of the soils with the large forests caused by the increasing drought due to climate changes. Increased soil erosion causes the loss of upper soil horizons richer in OC. In areas in which forest fires are a recurring phenomenon, like in the Mediterranean basin, the production of highly stable forms of OC during the combustion of biomass (charcoal) can contribute to the stabilization of C in the medium term. All these processes do not exclude each other and some of the feedback may be positive, causing a multiplying effect (Heans 1984; Imran 2018a,b,c,d,e; Imran and Khan 2015a,b,c,d,e). OC is intimately linked with the natural fertility and productivity of soils) as a source of macronutrients, especially C, N, and P) concerning the substrate of the soil's microbial activity) humified carbon contributes greatly to the capacity to retain nutrients and pollutants (capacity for cationic and anionic exchange capacity). Humic substances of lower molecular weight (fulvic acids) improve the solubility of some essential micronutrients, and toxic metals (Imran et al. 2015a,b,c,d,e, f,g,h,i,j,k,l, 2016a,b,c, 2017a,b, 2018; ISTA 2005). It is a critical factor in the development and maintenance of soil structure and the stability of soil aggregates and, consequently, of the physical properties that depend on these factors: water infiltration capacity, water holding capacity for plants, aeration, compaction, and erodibility (Ali et al. 2019; Imran and Khan 2015a,b,c,d,e; Imran et al. 2014, 2015a,b,c,d,e,f,g,h,i,j,k,l; Ortaş et al. 2017). Crop and soil management practices have a significant impact on carbon sequestration and aggregate formation (Ortas et al. 2019).

Another process that would probably be affected by climate change is soil salinization. Projected increasing in evapotranspiration and drought would raise the water table, saline intrusion, and accumulation of salts in the rooting soil depth in arid and semiarid lands (Imran et al. 2021a,b,c,d,e,f,g). These processes can be reduced with the application of appropriate farming techniques, tilling, irrigation control, and management of organic amendments, along with the reforestation of barren lands. In short, those measures promoting soil fertility would set off a multiplying effect on the ecosystems' capacity to fix atmospheric carbon in the long term (Imran and Khan 2015a,b,c,d,e; Imran et al. 2015a,b, c,d,e,f,g,h,i,j,k,l; Akhtar et al. 2015b; Khan et al. 2016; Zhan et al. 2011). Of course, land use will almost certainly change as a consequence of climate change, opening opportunities to novel crops, and varieties adapted to the new conditions, including the corresponding changes in cropping systems (Jackson 1962; Jagatheeswari and Ranganathan 2012; Joseph and Jini 2011; Kamara et al. 2014). In this context, the sustainability of nature and agriculture depends on choosing plant species that can adapt to climate changes.

The main potential changes in soil-forming factors (forcing variables) directly resulting from the global change would be in OM supply from biomass, soil temperature regime, and soil hydrology, the latter because of shifts in rainfall zones as well as changes in potential evapotranspiration. Soil changes because of a potential rise in sea level resulting from a net reduction in Antarctic ice cap volume and ocean warming are discussed in Brammer and Brinkman (1990) and are summarized at the end of this chapter (Kammann et al. 2011; Khan et al. 2010; Imran and Amanullah 2018; Imran et al. 2021a,b,c,d,e,f,g; Imran et al. 2015a,b,c,d,e,f,g,h,i,j,k,l, 2019; Lashari et al. 2013; Lehmann and Joseph 2009).

The biggest single change in soils expected as a result of these postulated forcing changes would be a gradual improvement in fertility and physical conditions of soils in humid and sub-humid climates (Imran and Amanullah 2021; Imran et al. 2021a,b,c,d,e,f,g). Another major change would be the poleward retreat of the permafrost boundary, discussed by Goryachkin and Targulian (1990). Other widespread changes would be in degree rather than in kind of Certain tropical soils with low physico chemical activity, such as in the Amazon region, may undergo a radical change from one major soil-forming process to another (Sombroek 1990), as discussed below under processes in soils.

The temperature changes but particularly in rainfall to be expected as a result of global warming are subject to major uncertainties for several reasons. Different global circulation models do not lead to mutually consistent results (an example for Europe is given by Santer 1985), and they are not yet adequately verified. Also, the interaction with changes in location and intensity of major ocean currents and resultant possible modifications in sea surface temperatures is still most uncertain, as well as the interaction with possible major changes in cloudiness and land cover by cower crop or reforestation and the resulting changes in albedo and actual evapotranspiration (Lehmann et al. 2011; Lei et al. 2011; Congming and Vonshak 2002; Maestre et al. 2007; Martinez-Beltran and Manzur 2005; Imran et al. 2015a,b,c,d,e,f,g,h,i,j,k,l, 2017a,b; Imran 2017, 2018a,b,c,d,e).

Indirect effects of climate change on soils through CO_2-induced increases in growth rates or water-use efficiencies, sea-level rise, through climate-induced decrease or increase in vegetative cover, or a change in human influence on soils because of the changes in options for the farmer, for example, may well each be greater than direct effects on soils of higher temperatures or greater rainfall variability and larger or smaller rainfall totals (Mazhar et al. 2016; Miri and Mirjalili 2013; Mittler 2006).

8.2 Possible Changes in Imposing Variables

A gradual, continuing rise in atmospheric CO_2 concentration entailing increased photosynthetic rates and water-use efficiencies of vegetation and crops, hence increases in OM supplies to soils (Figures 8.1–8.3). Minor increases in soil temperatures in the tropics and subtropics; moderate increases and extended periods in which soils are warm enough for microbial activity (warmer than about 5 °C) in temperate and cold climates, parallel to the changes in air temperatures and vegetation zones as summarized by Emanuel et al. (1985).

Minor increases in evapotranspiration in the tropics to major increases in high latitudes are caused both by temperature increase and by extension of the growing period (Palm and Volkenburgh 2012; Palta and Farag 2006; Pant et al. 2015a,b). Increases in amount and variability of rainfall in the tropics; possible decrease in rainfall in a band in the subtropics poleward of the present deserts; minor increases in amount and variability in temperate and cold regions. Peak rainfall intensities could increase in several regions (Islam et al. 2015; Munns

Figure 8.1 Tangible view of field and fertility status as influenced by environmental factors (precipitation, temperature, and light).

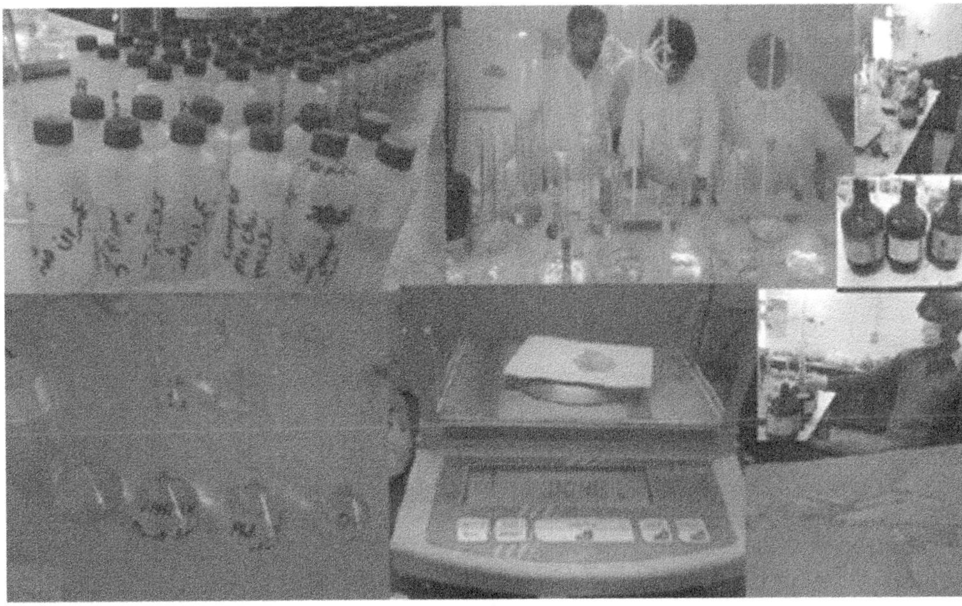

Figure 8.2 Tangible view of soil and water testing laboratory, analyzing soil fertility and nutrients status as influenced by environmental factors (precipitation, temperature, and light).

Figure 8.3 Soil fertility with increased CO_2 concentration and organic amendments influenced crop productivity.

et al. 2006; Munns 2002; Munns and Tester 2008; Nazarli et al. 2011; Neto et al. 2004). Recently mainly in semi-arid climate conditions, there is an increase in the number and amount of sudden precipitation per unit area. The intensive falling precipitation which caused the destruction of plants, the deterioration of the soil and erosion, and a decrease in soil fertility. After rainfall revers are charring tons of OM, soil, and other organic material to sees.

A gradual sea-level rise causes deeper and longer inundation in river and estuary basins and on levee back slope, and brackish-water inundation leading to encroachment of vegetation that accumulates pyrite in soils near the coast (Paz-Ferreiro et al. 2014; Peng et al. 2008; Ji et al. 2007; Prado et al. 2000; Raghothama 1999).

8.3 Processes and Mineralogical Change in Soils

The most rapid processes of physical and manly chemical or mineralogical change under changing external conditions would be loss of salts and nutrient cations where leaching increases, and salinization where net upward water movement occurs because of increased evapotranspiration or decreased rainfall or irrigation water supply (Paz-Ferreiro et al. 2014; Peng et al. 2008; Ji et al. 2007; Prado et al. 2000; Raghothama 1999). The clay mineral composition and the mineralogy of the coarser fractions would generally change little, even over centuries. Exceptions would be the transformation of X-ray amorphous material into the clay mineral halloysite when a volcanic soil previously under perennially moist conditions becomes subject to periodic drying, or the gradual dehydration of goethite to hematite in soils subject to higher temperatures or severe drying, or both. Changes in the surface properties of the clay fraction, while generally slower than salt movement, can take place much faster than changes in bulk composition or crystal structure (Imran et al. 2021a,b,c,d,e,f,g). Such surface changes have a dominant influence on soil's physical and chemical properties (Brinkman 1985, 1990). Clay minerals and fractions (especially high mineral surface area values of 300 and 700 $m^2\ g^{-1}$) are very important to keep organic compounds to increase soil organic material. minerals significantly influence the accumulation and stabilization of OC. Also, clay minerals significantly influence the accumulation and stabilization of soil OC (Singh et al. 2016). Soil OC/material is also very important to mitigate atmospheric CO_2 concentration. The results of Churchman et al. (2020) showed that the addition of clay to soils led to increased OC content. Soil OC is the major source of soil fertility and productivity (Table 8.1).

Changes in the clay mineral surfaces or the bulk composition of the clay fraction of soils are brought about by a small number of transformation processes, listed below (Brinkman 1982). Each of these processes can be accelerated or inhibited by changes in external conditions due to global change (Raza 2012; Rengasamy 2002; Roelofs et al. 2008).

- Hydrolysis by water containing carbon dioxide, which removes silica and basic cations;
- chelation, which dissolves and removes especially aluminium and iron by chelating organic acids;
- ferrolysis, a cyclic process of clay transformation and dissolution mediated by alternating iron reduction and oxidation, which decreases the cation exchange capacity by aluminium interlaying in swelling clay minerals;

Table 8.1 Initial soil inorganic, organic carbon concentration (in 0–15 cm), and effects of organic and inorganic fertilizers application on total soil organic carbon pool.

Treatments	Total soil carbon (%)		Inorganic carbon (%)		SOC (%)		SOC pool (mg ha^{-1})	
Control	4.61	±0.01b	3.80	±0.01a	0.81	±0.00b	17.69	±0.27b
Min. ferti.	4.83	±0.10ab	3.86	±0.07a	0.97	±0.05b	20.99	±1.83ab
Animal manure	5.04	±0.07a	3.74	±0.07a	**1.30**	**±0.11a**	**23.08**	**±2.49a**
Compost	5.02	±0.38a	3.76	±0.20a	1.26	±0.20a	21.40	±3.87ab
Com + Myc.	4.86	±0.05ab	3.75	±0.05a	1.11	±0.10a	21.42	±1.01ab

Mean of three replicates ± SD. Means in the same column followed by the same letter represent significant differences ($p \leq 0.05$) among treatments at the same depth.
Source: Adapted from Ortas et al. (2013).

- dissolution of clay minerals by strong mineral acids, producing acid aluminium salts and amorphous silica;
- reverse weathering, i.e. clay formation and transformation under neutral to strongly alkaline conditions, which may create, e.g. montmorillonite, palygorskite, or analcime.

Hydrolysis and chelation may be accelerated by increased leaching rates. Ferrolysis may occur where soils are subject to reduction and leaching in alternation with oxidation: in a warmer world, this may happen over larger areas than at present, especially in high latitudes and in monsoon climates. Dissolution by strong acids would occur, e.g. where sulphidic materials in coastal plains are oxidized with an improvement of drainage; however, a rise in sea level would reduce the likelihood of this occurring naturally. Reverse weathering could begin in areas drying out during global warming and would continue in most presently arid areas (Rolf et al. 2004; Rubio et al. 2001; Savvides 2015; Naveed et al. 2016; Shamsi and Kobraee 2013; Baqa et al. 2015; Shazma et al. 2016).

These processes would influence the surface properties of the clay fraction only over centuries, even with the changes envisaged as a consequence of global warming. By contrast, direct human action can vastly accelerate some of these processes as is evident, for example, from the severe effects of acid rain on sandy soils in parts of Europe (Van Breemen 1990) or the extremely rapid ferrolysis in soils seasonally inundated by water level fluctuations in the Volta lake in Ghana (Amatekpor 1989).

Not only the speed of soil formation can be accelerated by human action, but also, albeit much more locally, its very nature or direction. In most places, the natural soil-forming processes are not fundamentally changed, but there are certain threshold situations, generally with fragile soils, where even a small change in external conditions may cause a major, and adverse, change from one dominant soil-forming process to another. The four examples summarized below (from Sombroek 1990) illustrate a change from hydrolysis to chelation (Ferralsols to Podzols); irreversible hardening of the subsoil; clay illuviation forming a dense subsoil in originally homogeneous, porous Ferralsols; and salinization (Sidari et al. 2008; Karim and Imran 2019; Tank and Saraf 2010; Tiwari et al. 2011; Tombesi et al. 2018; Samreen et al. 2016).

The yellowish sandy Ferralsols and Ferralic Arenosols of Eastern Amazonia, Kalimantan, and the Zaire basin may rapidly change into Podzols or Albic Arenosols (giant Podzols) with even small increases in total rainfall or stronger seasonality, or increased input of acidic OM. An increase in effective rainfall due to climate change may cause a major increase in the extent of Podzols formed from present-day yellowish sandy Ferralsols where, presently, Podzols occur in patches within the Ferralsols area (Lucas et al. 1987; Dubroeucq and Volkoff 1988). The imperfectly drained loamy Plinthosols on the flat interfluves of Western Amazonia would change into shallow, droughty soils with an irreversibly hardened subsoil if subject to drying out with climate change. Recently very often it has been seen that increasing temperature due to climatic changes and consequently decreasing soil moisture which causes the soil structure degradation.

The deep reddish, porous loamy to clayey Ferralsols of the transition zones between forest and savanna in Eastern Africa, stable under the present vegetation, maybe leached so far that a denser subsoil with washed-in clay is formed below unstable topsoil with little OM, as already observed where the land was cleared several decades ago; the same may happen over more extensive areas under sparser vegetation brought about by a somewhat drier climate (Sidari et al. 2008; Karim and Imran 2019; Tank and Saraf 2010; Tiwari et al. 2011; Tombesi et al. 2018; Samreen et al. 2016).

The silty Fluvisols in the broad river valleys of the Sudano-Sahelian zone of West Africa, such as the interior delta of the Niger river, may become saline or sodic upon even minimal change in precipitation and flooding regimes – as exemplified by current human actions with the same soil-hydrological implications (Sombroek and Zonneveld 1971).

8.4 Effects of Higher CO_2 on Soil Fertility, Physical Conditions, and Productivity

Higher atmospheric CO_2 concentration increases growth rates and water-use efficiency of crops (Figure 8.3) and natural vegetation in so far as other factors do not become limiting (Van Hoorn et al. 2001; Wang et al. 2007; Wang and Xu 2013; Wang and Jin 2007; Wolfe 2007). The higher temperature optima of some plants under increased CO_2 would tend to counteract adverse effects of temperature rise, such as increased nighttime respiration. The shortened growth cycle of a given species because of higher CO_2 and temperature would be compensated for in natural vegetation by adjustments in species composition or dominance (Imran et al. 2020a,b,c,d). In agroecosystems, the choice of longer-duration cultivars or changes in cropping patterns could eliminate unproductive periods that might arise because of the shorter growth cycle of the main crop (Wang and Xu 2013; Wang and Jin 2007; Wolfe 2007).

There will be adequate time to adjust to the changes since these are expected to occur over decades, rather than years or days as in all present experimental situations. The increased productivity is generally accompanied by more litter or crop residues (Imran et al. 2020a,b,c,d; Amanullah et al. 2021; Ilyas et al. 2020a,b,c), a greater total root mass and root exudation, increased mycorrhizal colonization, and activity of other rhizosphere (Ortaş et al. 2019) or soil micro-organisms, including symbiotic and root-zone N, fixers (Figure 8.3). The latter would have a positive effect on N supply to crops or vegetation (Van Hoorn et al. 2001; Wang et al. 2007; Wang and Xu 2013; Wang and Jin 2007; Wolfe 2007). The increased microbial and root activity in the soil would entail higher CO_2 partial pressure in soil air and CO_2 activity in soil water, hence increased rates of plant nutrient release (e.g. K, Mg, micronutrients) from weathering of soil minerals. Similarly, the mycorrhizal activity would lead to better phosphate, zinc, copper, and nitrogen uptake (Ortas 2020). These effects would be in synergy with better nutrient uptake by the more intensive root system due to higher atmospheric CO_2 concentration. There is no *a priori* reason why the degree of synchrony between nutrient release and demand by crops or natural vegetation would be subject to major changes under high CO_2 conditions (Xu et al. 2008; Yeo et al. 1990; Younis et al. 2015; Zakaria et al. 2012; Zargar et al. 2017).

The greater microbial activity tends to increase the number of plant nutrients cycling through soil organisms. The increased production of root material (at similar temperatures) tends to raise soil OM content, which also entails the temporary immobilization and cycling of greater quantities of plant nutrients in the soil. Higher C/N ratios in the litter, reported by some workers under high CO_2 conditions, would entail slower decomposition and slower remobilization of the plant nutrients from the litter and uptake by the root mat and would provide more time for incorporation into the soil by earthworms, termites, etc. Higher soil temperatures would counteract increases in 'stable' soil OM content but would further stimulate microbial activity (Xu et al. 2008; Yeo et al. 1990; Younis et al. 2015; Zakaria et al. 2012; Zargar et al. 2017).

In all experimental situations, whether chamber-type or free-air enrichment, CO_2 increases are rapid or sudden, often to double ambient concentration, sometimes higher. The consequently rapid increases in soil OM dynamics and soil micro-organisms may cause temporary competition for plant nutrients. These temporary effects have on occasion been reported as negative factors affecting plant response to elevated CO_2. However, increased OM dynamics and microbial activity in soils are positive for the soil-plant system when CO_2 concentrations rise gradually over decades, as currently and in the recent past (Tank and Saraf 2010; Tiwari et al. 2011; Tombesi et al. 2018; Samreen et al. 2016; Van Hoorn et al. 2001; Wang et al. 2007, 2020; Wang and Xu 2013). Future experiments could be set up to compensate for the temporary effects caused by the suddenness of the CO_2 increase, for example by artificially higher soil OM contents estimated to be near equilibrium with each stepwise higher CO_2 concentration, in a range between 350 and 600 ppm.

Increased microbial activity due to higher CO_2 concentration and temperature produces greater amounts of polysaccharides and other soil stabilizers (Figure 8.2). Increases in the litter or crop residues, root mass, and OM content tend to stimulate the activity of soil macro-fauna, including earthworms, with consequently improved infiltration rate and bypass flow by the greater number of stable bio pores (Tank and Saraf 2010; Tiwari et al. 2011; Tombesi et al. 2018; Samreen et al. 2016; Van Hoorn et al. 2001; Wang et al. 2007; Wang and Xu 2013). The greater stability and the faster infiltration increase the resilience of the soil against water erosion and consequent loss of soil fertility. The increased proportion of bypass flow also decreases the nutrient loss by leaching during periods with excess rainfall. This refers to the available nutrients in the soil, including well-incorporated fertilizers or manure, but not to fertilizers broadcast on the soil surface. These are subject to loss by runoff or leaching.

- Gradually rising CO_2 as in this century and *transient* global change scenarios.
- Soils with some weatherable minerals at least in the subsoil or substratum within rooting depth.
- Extreme weather events may disrupt some relationships in the figure, so any major increase in their frequency or intensity may counteract the positive effects shown.
- Species composition adjusts, or choices indicated are made to adjust, to the newly attainable biomass or crop production under increased atmospheric CO_2, compensating for shortened growth cycles of existing species or crops. The figure does not include the positive effects of higher temperatures on the length of growing periods in temperate or boreal climates.

These changes increase the resilience of the soil against physical degradation and nutrient loss by increased intensity, seasonality, or variability of rainfall, as well as against some of the unfavourable changes in the rate or direction of soil-forming processes discussed in the next sections (Ji et al. 2007; Prado et al. 2000; Raghothama 1999; Rajalakshmi et al. 2015; Rashid et al. 2003).

If the partial pressure of CO_2 in the soil air would rise, and that of O_2 decreases to levels impairing root function, part of the benefits indicated would not materialize. The improved gas exchange with the atmosphere through increased numbers of stable bio pores would tend to keep CO_2 and O_2 in the soil at 'safe' levels, at least in naturally or artificially well-drained soils. Wetland crops such as rice or jute have their gas exchange mechanisms and would not be affected; neither would natural wetland vegetation (Munns 2002; Munns and Tester 2008; Nazarli et al. 2011; Neto et al. 2004; Palm and Volkenburgh 2012; Palta and Farag 2006).

The positive effect on weathering rate and plant nutrient availability would occur in soils with significant amounts of weatherable minerals, not in very deeply and strongly weathered or otherwise very poor soils.

8.5 Effects of Rainfall and Temperature Changes in Different Climates

In the humid tropics and monsoon climates, increased intensities of rainfall events and increased rainfall totals would increase leaching rates in well-drained soils with high infiltration rates and would cause temporary flooding or water-saturation, hence reduced OM decomposition, in many soils in level or depressional sites (Munns 2002; Munns and Tester 2008; Nazarli et al. 2011; Neto et al. 2004; Palm and Volkenburgh 2012; Palta and Farag. 2006). This may affect a significant proportion of especially the better soils in Sub-Saharan Africa, for example. They would also give rise to greater amounts and frequency of runoff on soils in sloping terrain, with sedimentation downslope and, worse, downstream. Locally, there would be increased chances of the mass movement in the form of landslides or mudflows in certain soft sedimentary materials (such as in the picture . . . in Turkey as well), as discussed below. Soils most resilient against such changes would have adequate cation exchange capacity and anion sorption to minimize nutrient loss during leaching flow and have high structural stability and a strongly heterogeneous system of continuous macropores to maximize infiltration and rapid bypass flow through the soil during high-intensity rainfall (Imran et al. 2019, 2021a,b,c,d,e,f,g, 2015a,b,c,d,e,f,g,h,i,j,k,l; Imran and Amanullah 2018; Lashari et al. 2013).

In subtropical and other subhumid or semi-arid areas, the increased productivity and water-use efficiency due to higher CO_2 would tend to increase ground cover, counteracting the effects of higher temperatures. In recent years, in many part of the water deficiency areas, withdrawals of groundwater are higher than the rainwater recharge to aquifer due to continuous expansion of irrigated agriculture in the areas. In Central Anatolia, Konya-Turkey, in regions where the annual precipitation is 210 mm as a result of too much groundwater withdrawal potholes are formed, and as a result, the plant production cannot be done. For future sustainable water and food security. If there would be locally much less rainfall and increasing intra- and inter-annual variability, these could lead to less dry-matter production and hence, in due course, lower soil OM contents (Imran et al. 2019, 2021a,b,c,d,e,f,g, 2015a,b,c,d,e,f,g,h,i,j,k,l; Imran and Amanullah 2018; Lashari et al. 2013). Periodic leaching during high-intensity rainfall with less standing vegetation could desalinize some soils in well-drained sites, cause increased runoff in others, and lead to soil salinization in depressional sites or where the groundwater table is high. Soils most resilient against the effects of such increasing aridity and rainfall variability would have high structural stability and a strongly heterogeneous system of continuous macropores (the same as in the tropics); hence a rapid infiltration rate, as well as a large available water capacity and a deep groundwater table (Imran 2018a,b,c,d,e; Imran and Khan 2015a,b,c,d,e, 2017; Duan et al. 2007; Farahbakhsh 2012).

Higher temperatures, particularly in arid conditions, entail a higher evaporative demand. Where there is sufficient soil moisture, for example in irrigated areas, this could lead to soil salinization if land or farm water management, or irrigation scheduling or drainage are inadequate. On the other hand, recent experiments by the Salinity Laboratory, Riverside, California, point to increased salt tolerance of crops under high atmospheric CO_2 conditions (E.V. Maas, pers. comm.; Bowman and Strain 1987). Since under abiotic soil conditions, atmospheric CO_2 concentration is increased, which may have negative effects on soil quality and plant production. The application of biological processes such as mycorrhizal fungi and other microbial strains as inoculants provides a cost-effective long-term solution for coping with saline and drought conditions (Ortas et al. 2021).

In temperate climates, minor increases in rainfall totals would be expected to be largely taken up by increased evapotranspiration of vegetation or crops at the expected higher temperatures, so that net hydrologic or chemical effects on the soils might be small. The negative effect on soil OM contents of a temperature rise might be more than compensated by the greater OM supply from vegetation or crops growing more vigorously because of the higher photosynthesis, the greater potential evapotranspiration, and the higher water-use efficiency in a high-CO_2 atmosphere (Imran 2018a,b,c,d,e; Imran and Khan 2015a,b,c,d ,e, 2017; Duan et al. 2007; Farahbakhsh 2012). The temperate zone would thus be likely to have the smallest changes in soils, even in poorly buffered ones, directly caused by the effects of global change. A minor and probably slow, but very visible, change could be a reddening of presently brown soils where increased periods with high summer temperatures would coincide with dry conditions so that the iron oxide haematite would be stable over the presently dominant goethite. This mineralogical change might decrease the intensity and amount of phosphate fixation. An overview of such changes, with emphasis on temperate climate zones, is given by Buol et al. (1990).

In boreal climates, the gradual disappearance of large extents of permafrost and the reduction of frost periods in extensive belts adjoining former permafrost is expected to improve the internal drainage of soils in vast areas, with probable increases in leaching rates. The appreciable increase in the period when the soil temperature is high enough for microbial activity would lead to lower OM contents, probably not fully compensated by increased primary production through somewhat higher net photosynthesis and the longer growing period (Imran 2018a,b,c,d,e; Imran and Khan 2015a,b,c,d,e, 2017; Duan et al. 2007; Farahbakhsh 2012). Depletion of the OM content is one of the major causes underlying the declining agronomic productivity in soils of the semi-arid region (Khresat et al. 2008; Ryan et al. 2008). Especially heavy tillage and breakdown of soil aggregates exacerbate mineralization of SOM and emission of CO_2 (Lal 2010; Larionova et al. 2010). Ander long-term organic fertilizer conditions application of organic fertilizers increased soil aggregation and enhanced C sequestration, and the rate and total amount of soil OC (Ortas et al. 2013).

Paradoxically, the extent of soils subject to periodic reduction could well increase in level areas, despite the greater leaching, because of increased periods when the soils are water-saturated but also sufficiently warm for microbial activity. Soils most resilient against such effects, including the leaching of nutrients and periodic soil reduction, would have similar characteristics as the most resilient ones in other climates: adequate cation exchange capacity and anion sorption to minimize nutrient loss during leaching flow, high structural stability, and a strongly heterogeneous system of continuous macropores to maximize rapid bypass flow during periods with excess meltwater (Armor et al. 2005; Arshad et al. 2008; Khan et al. 2015a,b; Ayaz et al. 2000; Iqbal et al. 2015a,b, 2016a,b; Badridze et al. 2009).

8.6 Some Properties of Clay Surfaces

Soils with naturally high structural stability, for example, Ferralsols and Nitisols, occupy sizeable areas in the tropics. The former is widespread, among others, in northern South America, the latter in East Africa (Armor et al. 2005; Arshad et al. 2008; Khan et al. 2015a,b; Ayaz et al. 2000; Iqbal et al. 2015a,b, 2016a,b; Badridze et al. 2009). The clay fraction in such soils generally has oxidic surfaces: mainly iron(III) and aluminium oxides or hydroxides, while the bulk of the clay fraction may have different compositions. The oxidic surfaces could form from parent materials with moderate or high iron contents under long-continued hydrolysis by water (containing carbon dioxide).

On the other extreme soil with a very low structural stability, or with a severe hazard of failure under load or shock (so-called quick clays). The surfaces of the clay minerals in these soils are generally covered by amorphous, gel-like material with a high silica content

(McKyes et al. 1974). Such material may have originated in earlier periods when the soils were strongly saline and, presumably, subject to processes of reverse weathering (Figure 8.1). Examples are the quick clays of the Champlain Sea sediments in Ontario and Quebec and parts of Scandinavia. Such soils are most likely to generate high proportions of runoff and suspended sediment but are also most liable to mudflows once sloping sites are water-saturated to appreciable depth. Some Andosols are thixotropic and have similarly low stability because of their similar composition, derived from volcanic materials (tuff).

Most soils fall somewhere between these extremes. Vertisols, for example, have moderate or low structural stability, and clay surfaces that are mainly silica, but with generally small amounts of the amorphous coating. In Planosols, if formed by ferrolysis, the clay fraction in the upper, eluvial horizons has been partly decomposed with a residue of amorphous silica, but the remaining smectite or illite has been interlayered with aluminium hydroxide polymers, which has decreased the swell-shrink potential and the cation exchange capacity of the clay fraction. Concurrently, parts of the free iron oxides have been reduced and leached out. The net effect of these changes generally is a decrease in structural stability.

8.7 Resilience Against Physical and Chemical Soil Degradation

As discussed, most soils do not have a high intrinsic resilience against physical soil degradation by, for example, high-intensity rainfall. In natural conditions in humid climates, it is the complete soil cover near ground level combined with the perforating activity of the soil fauna that makes the soil-vegetation system resilient against physical degradation (Armor et al. 2005; Arshad et al. 2008; Khan et al. 2015a,b; Ayaz et al. 2000; Iqbal et al. 2015a,b, 2016a,b; Badridze et al. 2009).

In the Rhine river plain in the Netherlands, for example, most of the originally calcareous alluvial soils have been decalcified within a millennium or so. Only in small areas on the highest levels of that age that have continually remained under forest, soils are still calcareous and even have lime pseudomycelia (filaments) indicative of less humid soil conditions, and abundant vertical macropores produced by earthworms. In these soils, faunal activity is high because of the adequate litter supply: the resulting macropores remain open, protected against rain impact by litter and undergrowth: and excess water from heavy rain passes to the substratum through the macropores without leaching lime from most of the soil mass (Akhtar et al. 2015a; Imran 2018a,b,c,d,e; Imran and Khan 2015a,b,c,d,e; Imran et al. 2015a,b,c,d,e,f,g,h,i,j,k,l; Akhtar et al. 2015b; Khan et al. 2016).

Research to increase the resilience of soils against any adverse effects of climate change could be done in conjunction with that on soil resilience against direct adverse human impacts (Figure 8.4). Until site-specific management procedures have been elaborated, soil and crop (including trees and pasture) management should aim to maintain soil cover and OM supply to soil biota, while minimizing mechanical disturbance by heavy traffic, cultivation, or excessive grazing intensity. Such kind of management may also help to conserve plant nutrients (in soils not flooded for wetland cultivation) since the stable, heterogeneous system of bio pores produced by the soil fauna would favour bypass flow of any excess

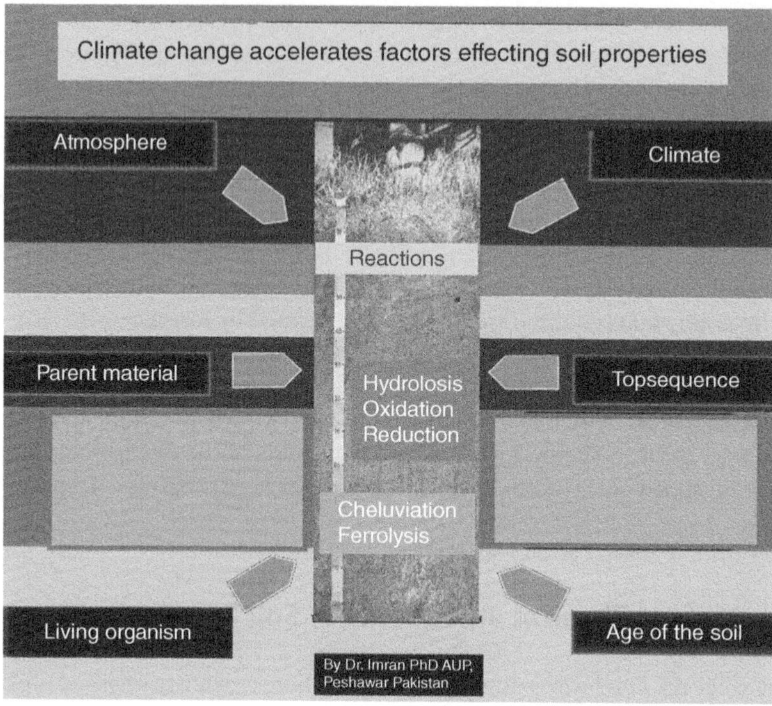

Figure 8.4 Soil properties respond to various climatic factors.

moisture and thus decrease leaching through the soil mass (Akhtar et al. 2015a; Imran 2018a,b,c,d,e; Imran and Khan 2015a,b,c,d,e; Imran et al. 2015a,b,c,d,c,f,g,h,i,j,k,l; Akhtar et al. 2015b; Khan et al. 2016). A single management recipe would not be generally applicable in different conditions.

Soil and crop management may also affect soil properties, such as enhancing the SOC pool, and the development of the soil structure as well. Ortas et al. (2019) reported that managing SOC and soil microorganisms such as mycorrhizal fungi which are a powerful tool to fix more CO_2 from the atmosphere to soil through plant leaves and roots to make the soil more qualitative. Plant roots and mycorrhizal hyphae can physically protect the soil carbon from microbial decomposition through aggregation (Golchin et al. 1994; Jastrow et al. 1998). Root exudates organic compounds such as polysaccharides, carbohydrates, lignin, and lipids have significant effects on aggregation. The aggregation has a significant effect on soil fertility and keeping carbon fixation.

Minimizing damage by certain termite species harming crop performance may necessitate a period without residues on the soil, for example; or crop residues may be needed for feed or fuel. Management methods for wetlands need to be developed that make optimum use of any increased potential productivity while minimizing secondary effects such as increased CH_4 or N_2O emission from the reduced soil. Such factors, and others, should be taken into account in designing an optimum management strategy for any specific natural and cultural environment.

8.8 Resilience Against Soil Reduction (Anoxic Conditions)

Soil reduction, which would limit land suitability for dryland crops, or strong reduction, which would be liable to produce toxins even for wetland crops, may take place once the soil is water-saturated long enough for microbial action to exhaust the oxygen remaining in the soil when water-saturation started. Another necessary condition is the presence of sufficient readily decomposable OM as an energy source for microbial activity (Akhtar et al. 2015a; Imran 2018a,b,c,d,e; Imran and Khan 2015a,b,c,d,e; Imran et al. 2015a,b,c,d,e,f,g,h,i,j,k,l). In most soils, during reduction the redox status is stabilized at an Eh about 100–200 mV near neutrality by the Fe^{2+}–$Fe(OH)_3$ equilibrium, except where the content of readily decomposable OM is very high or the content of free iron(III) oxides very low. In such cases, negative Eh values may occur, and toxic hydrogen sulphide or low-molecular organic compounds – including methane – may be formed.

Resilience against soil reduction in practice depends on the drainage conditions since most soils have sufficient OM for reduction to start within about a week after water saturation. Soils most resilient against a reduction in conditions of increased rainfall variability and incidence of high-intensity rainfall have similar properties as those resilient against the negative effects of other perturbations: high infiltration rate, high structural stability and a permanent heterogeneous system of tubular macropores, good external drainage (Iqbal et al. 2015a,b; Badridze et al. 2009; Bates et al. 1973; Bohnert and Jensen 1996; Bos et al. 2005).

8.9 Soil Reaction (pH)

Most soils would not be subject to rapid pH changes resulting from climate change (Figures 8.1 and 8.2). Exceptions might be found in potential acid sulphate soils, extensive in some coastal plains and estuaries if they become subject to increasingly long dry seasons. Even though most of such soils are clays with moderate or high cation exchange capacity, the amounts of the acid liberated in such soils upon oxidation generally exceed this rapid buffering capacity (Iqbal et al. 2015a,b; Badridze et al. 2009; Bates et al. 1973; Bohnert and Jensen 1996; Bos et al. 2005). Therefore, pH values may temporarily reach 2.5 to 3.5 and a small part of the clay fraction may be decomposed as indicated under *Processes in soils*, above. This then buffers the pH generally between 3.5 and 4 in the long run. Depending on the efficiency with which the excess acid formed can be leached out, the period of extreme acidity and aluminium toxicity may last between less than a year and several decades.

In calcareous soils, soil reaction may range between about 8.5 and 7 depending on the partial pressure of CO_2 in the soil; this range is maintained against the leaching of basic cations by the different soil processes as long as a few percent of finely distributed lime remain. Buffering in non-calcareous soils is less strong but depends on the cation exchange capacity at soil pH. In soils with variable-charge surfaces of the clay fraction, this decreases with acidification (Arshad et al. 2008; Khan et al. 2015a,b; Ayaz et al. 2000; Iqbal et al. 2016a,b).

It should be noted that the simple modelling of accelerated $CaCO_3$ leaching under a doubled atmospheric CO_2 concentration generally does not hold. In most soils, the ongoing decomposition of OM maintains CO_2 concentrations in the soil air far above atmospheric concentration even now, and $CaCO_3$ solubility is determined by the partial pressure of CO_2 in soil air and its activity in soil water, rather than in the atmosphere. Similarly, in the rhizosphere high CO_2 is produced by soil organisms' activity which is change the rhizosphere pH as well. Leaching of lime is thus positively related to the rate of OM decomposition, negatively to gas diffusion rate, and positively to the amount of water percolating through the soil (Arshad et al. 2008; Khan et al. 2015a,b; Ayaz et al. 2000; Iqbal et al. 2016a,b).

In conditions where leaching is accelerated by climate change, it would be possible to find relatively rapid soil acidification after a long period with little apparent change, as has been the case – but after a shorter latent period – in some soils in Europe that have been subject to acid rain for several decades. The soil might be steadily depleted of basic cations, but a pH change may start or may become more rapid, once certain buffering pools are nearly exhausted. Such non-linear and time-delayed effects have been discussed in the context of soil and water pollution by Stigliani (1988); they are also expected to occur in various ways at different times after increased temperatures and changed rainfall patterns will have been operative.

8.10 Effects of a Rising Sea Level on Soils in Coastal Areas

The probable effects on soil characteristics of a gradual eustatic rise in sea level will vary from place to place depending on several local and external factors, and interactions between them (Brammer and Brinkman 1990). In principle, a rising sea level would tend to erode and move back existing coastlines. However, the extent to which this happens will depend on the elevation, the resistance of local coastal materials, the degree to which they are defended by sediments provided by river flow or longshore drift, the strength of longshore currents and storm waves, and on human interventions which might prevent or accelerate erosion (Kammann et al. 2011; Khan et al. 2010; Imran et al. 2019, 2021a,b,c,d,e,f,g, 2015a,b,c,d,e,f,g,h,i,j,k,l; Imran and Amanullah 2018; Lashari et al. 2013; Lehmann and Joseph 2009; Lehmann et al. 2011; Lei et al. 2011).

In major deltas, such as those of the Ganges–Brahmaputra and the major Chinese rivers, sediment supplies delivered to the estuary will generally be sufficient to offset the effects of a rising sea level. Such deltaic aggradation could decrease, however, under three circumstances:

- where human interventions inland, such as large dams or successful soil conservation programs, drastically reduce sediment supply to the delta: e.g., the construction of the Aswan high dam in 1964 has led to coastal erosion and increased flooding of lagoon margins in the Nile delta (Stanley 1988). Similarly, about ten dams built on the Tigris and Euphrates rivers in the Southeast of Turkey play a serious role in preventing erosion as well as water reservoirs;
- where construction of embankments within the delta interrupts sediment supply to adjoining back swamps, exposing them to submergence by a rise in sea level: e.g., embankments along the lower Mississippi River have cut off sediment supplies to adjoining wetlands which formerly offset land subsidence occurring due to compaction of underlying sediments (Day and Templet 1989);
- where land subsidence occurs due to the abstraction of water, natural gas, or oil: e.g., as is presently happening in Bangkok and the northern part of the Netherlands.

In coastal lowlands that are insufficiently defended by sediment supply or embankments, tidal flooding by saline water will tend to penetrate further inland than at present, extending the area of perennially or seasonally saline soils. Where *Rhizophora* mangrove or *Phragmites* vegetation invades the area that would over several decades lead to the formation of potential acid sulphate soils. The impedance of drainage from the land by a higher sea level and by the correspondingly higher levels of adjoining estuarine rivers and their levees will also extend the area of perennially or seasonally reduced soils and increase normal inundation depths and durations in the river and estuary basins and on levee backslope (Imran et al. 2015a,b,c,d,e,f,g,h,i,j,k,l, 2019, 2021a,b,c,d,e,f,g; Imran 2017, 2018a,b,c,d,e; Mazhar et al. 2016; Miri and Mirjalili 2013; Mittler 2006; Kammann et al. 2011; Khan et al. 2010; Imran and Amanullah 2018; Lashari et al. 2013; Lehmann and Joseph 2009; Lehmann et al. 2011; Lei et al. 2011). In sites that become perennially wet, soil OM contents will tend to increase, resulting eventually in peat formation. On the other hand, where coastal erosion removes an existing barrier of mineral soils or mangrove forests, higher storm surges associated with a rising sea level could allow seawater to destroy existing coastal eustatic peat swamps, which might eventually be replaced by freshwater or saltwater lagoons (Kammann et al. 2011; Khan et al. 2010; Imran and Amanullah 2018; Imran et al. 2015a,b,c,d,e,f,g,h,i,j,k,l, 2019, 2021a,b,c,d,e,f,g; Lashari et al. 2013; Lehmann and Joseph 2009; Lehmann et al. 2011; Lei et al. 2011).

The probable response of low-lying coastal areas to a rise in sea level can be estimated in more detail based on the geological and historical evidence of changes that occurred during past periods when sea level was rising eustatically or in response to tectonic or isostatic movements: e.g., around the southern North Sea (Jelgersma 1988); in the Nile delta (Stanley 1988); on the coastal plain of the Guyanas (Brinkman and Pons 1968); in the Musi delta of Sumatera (Brinkman 1987). Contemporary evidence is available in areas where land levels have subsided as a result of the recent abstraction of water, natural gas, or oil from sediments underlying coastal lowlands. Further studies of such contemporary and palaeoenvironments are needed together with location-specific studies to better understand the change processes, identify appropriate responses and assess their technical, ecological, and socio-economic implications (Warrick and Farmer 1990).

8.11 Conclusions

Some major and widespread soil changes expected as a result of any global change are positive, especially the gradual increases in soil fertility and physical qualities consequent on increased atmospheric CO_2. The increased productivity and water-use efficiency of crops and vegetation, and the generally similar or somewhat higher rainfall indicated by several global circulation models, not fully counteracted by higher evapotranspiration, would be expected to lead to widespread increases in ground cover, and consequently better protection against runoff and erosion. Major but less widespread soil changes, including greater biological activity and increased extent of periodic reduction in soils, would be expected where permafrost would disappear. In unprotected low-lying coastal areas, gradual encroachment of *Rhizophora* mangroves or *Phragmites* following more extensive brackish-water inundation may give rise to the formation of potential acid sulphate soil layers after several decades. Deeper and longer-duration flooding of basins and levee back slopes in the adjacent river and estuary plains could lead to more extensive reducing conditions and increased OM contents, and locally to peat formation. Other changes due to climate change (temperature and precipitation) are expected to be relatively well buffered by the mineral composition especially high surface clay minerals, the OM content, or the structural stability of many soils. However, decreases in cover by vegetation or annual or perennial crops, caused by any locally major declines in rainfall not compensated by CO_2 effects, could lead to soil structure degradation and decreased porosity, as well as increased runoff and erosion on sloping sites and by the concomitant more extensive and rapid sedimentation. Changes in options available to land users because of climate change may have similar effects. In certain fragile soils, the nature of the dominant soil-forming process may change for the worse with increased, decreased, or more strongly seasonal rainfall. In most cases, changes in soils by direct human action, on-site or off-site (whether intentional or unintended), are far greater than the direct climate-induced effects. Soil and crop management measures designed to optimize the soil's sustained productive capacity would therefore be generally adequate to counteract any degradation of agricultural land by climate change. Soils of nature areas, or another land with a low intensity of management such as semi-natural forests used for extraction of wood and other products, are less readily protected against the effects of climate change but such soils, too, are threatened less by climate change than by human actions – off-site, such as pollution by acid deposition, or on-site, such as excessive nutrient extraction under very low-input agriculture. Under several climate conditions, low-input ecological soil and crop management using OM, biochar, cover crops, and microorganisms uses are strongly increase soil OC and mitigates atmospheric CO_2 concentration.

To armour the world's soils against any negative effect of climate change, or other extremes in external circumstances such as nutrient depletion or excess (pollution), drought, or high-intensity rains, the best that land users could do, would be:

- to manage their soils to give them maximum physical resilience through a stable, heterogeneous pore system by maintaining a closed ground cover as much as possible;
- to use an integrated plant nutrient management system to balance the input and offtake of nutrients over a cropping cycle or the years, while maintaining soil nutrient levels low enough to minimize losses and high enough to buffer occasional high demands.

Human action and management have been emphasized in these conclusions because most of the world's land is used and, to different degrees, managed rather than under natural conditions.

8.12 Summary

As a result of any global change, some of the larger and wider changes in the soils are progressive, especially the gradual decrease in soil fertility and physical properties after increased environmental temperature. Low productivity of crops and water use efficiency generally similar or somewhat high rainfall can lead to soil degradation and nutrient deprivation. Expect larger but less extensive changes in the soil, including greater biological activity and intermittent reduction in soils. Other changes due to climate change (temperature and rainfall) are expected to contribute relatively well to mineral nutrient composition, OM content, or structural stability of many soils. In some fragile soils, the nature of the dominant soil formation may be inferior, with increasing, decreasing, or more severe seasonal rainfall. In most cases, changes in the soil through direct human action, on-site or off-site, far outweigh the effects of climate change. Soil and crop management measures are designed to improve the soil's productivity and may lead the soil capacity to be sufficient to prevent any reduction of agricultural land through climate change. The land of natural areas, or other lands that have a low intensity of management such as semi-natural forests used for digging wood and other plant products, is less easily protected from the effects of climate change. Such soils are also less vulnerable to climate change than they are to human activities, such as pollution from acid accumulation, or excessive nutrient depletion on site, such as very little input agriculture. To disarm the world's soil against any negative impact of climate change, or external nutrient deficiencies or over-pollution, or drought or high-intensity rains. As a result, soil degradation and soil structure are broken-down and soil OM is mineralized and more carbon is flux into the atmosphere which is caused by climate change. The best that can happen to them is to manage their soil by maintaining maximum physical flexibility by maximizing the off-ground cover through a stable, proportional soaking system. Using an integrated plant nutrient management system to balance nutrient input and offtake during crops or over the years, while maintaining soil nutrient levels substantially and sometimes to minimize losses. Occasionally high demands should be high enough to eliminate them.

References

Akhtar, S.S., Andersen, M.N., and Liu, F. (2015a). Biochar mitigates salinity stress in potato. *Journal of Agronomy and Crop Science* 201: 321–400.

Akhtar, S.S., Andersen, M.N., and Liu, F. (2015b). Residual effects of biochar on improving growth, physiology and yield of wheat under salt stress. *Agricultural Water Management* 158: 61–68.

Ali, I., Khan, A.A., Imran et al. (2019). Humic acids and nitrogen levels optimizing productivity of green gram (*Vigna radiate* L.). *Russian Agriculture Sciences* 45 (1): 43–47.

Amanullah, Khalid, S., Khalil, F. et al. (2021). Growth and dry matter partitioning response in cereal-legume intercropping under full and limited irrigation regimes. *Scientific Reports* https://doi.org/10.1038/s41598-021-92022-4.

Amatekpor, J.K. (1989). The effect of seasonal flooding on the clay mineralogy of a soil series in the Volta lake drawdown area, Ghana. *Land Degradation and Rehabilitation* 1: 89–100.

Amonette, J.E. and Joseph, S. (2009). Characteristics of biochar: microchemical properties. In: *Biochar for Environmental Management: Science and Technology* (ed. J. Lehmann and S. Joseph), 33–52. Earthscan, London: Sterling.

Armor, N.B., Hamed, K.B., Debez, A. et al. (2005). Physiological and antioxidant responses of perennial halophyte *Crithmum maritimum* to salinity. *Plant Science* 4: 889–899.

Arshad, M., Shaharoona, B., and Mahmood, T. (2008). Inoculation with *Pseudomonas* spp. containing ACC-deaminase partially eliminates the effects of drought stress on growth, yield, and ripening of pea (*Pisum sativum* L.). *Pedosphere* 18: 611–620.

Ayaz, F.A., Kadioglu, A., and Turgut, T.R. (2000). Water stress effects on the content of low molecular weight carbohydrates and phenolic acids in *Cienanthe setosa*. *Canadian Journal of Plant Science* 80: 373–378.

Badridze, G., Weidner, A., Asch, F., and Borner, A. (2009). Variation in salt tolerance within Georgian wheat germplasm collection. *Genetic Resources and Crop Evolution* 56: 1125–1130.

Baqa, S., Khan, A.Z., Inamullah et al. (2015). Influence of farm yard manure and phosphorus application on yield and yield components of wheat. *Pure and Applied Biology* 4 (4): 458–464.

Bates, L.S., Waldren, R.P., and Teare, I.D. (1973). Rapid determination of free proline for water stress studies. *Plant and Soil* 39: 205–207.

Bohnert, H.J. and Jensen, R.G. (1996). Strategies for engineering water stress tolerance in plants. *Trends in Biotechnology* 14: 89–97.

Bos, C., Juillet, B., Fouillet, H. et al. (2005). Postprandial metabolic utilization of wheat protein in humans. *The American Journal of Clinical Nutrition* 81: 87–94.

Bowman, W.D. and Strain, B.R. (1987). Interaction between CO_2 enrichment and salinity stress in the C4 non-halophyte *Andropogon glomeratus* (Walter) BSP. *Plant, Cell & Environment* 10: 267–270.

Brammer, H. and Brinkman, R. (1990). Changes in soil resources in response to a gradually rising sea-level. Chapter 12. In: (ed. Scharpenseel et al.), 145–156.

Brinkman, R. (1982). Clay transformations: aspects of equilibrium and kinetics. In: *Soil Chemistry. B. Physicochemical Models. Developments in Soil Science 5B*, 2nde (ed. G.H. Bolt), 433–458. Amsterdam: Elsevier.

Brinkman, R. (1985). Mineralogy and surface properties of the clay fraction affecting soil behavior and management. In: *Soil Physics and Rice* (ed. T. Woodhead), 161–182. Los Baños, Philippines: International Rice Research Institute.

Brinkman, R. (1987). Sediments and soils in the Karang Agung area. In: *Some Aspects of Tidal Swamp Development with Special Reference to the Karang Agung Area, South Sumatra Province* (ed. R. Best, R. Brinkman and J.J. van Roon), 12–22. Mimeo: World Bank.

Brinkman, R. (1990). *Resilience Against Climate Change? Soil Minerals, Transformations and Surface Properties, Eh, pH* (ed. Scharpenseel et al.), 51–60.

Brinkman, R. and Pons, L.J. (1968). *A Pedo-geomorphological Classification and Map of the Holocene Sediments in the Coastal Plain of the Three Guyanas. Soil SurveyPaper No. 4.* Wageningen: Soil Survey Institute (Staring Centre). 40 p, separate map.

Bruinsma, J. (1963). The quantitative analysis of chlorophyll a and b in plant extract. *Photochemistry and Photobiology* 2: 241–249.

Brussaard, L., de Ruiter, P.C., and Brown, G.G. (2007). Soil biodiversity for agricultural sustainability. *Agriculture, Ecosystems and Environment* 121 (3): 233–244. https://doi.org/10.1016/j.agee.2006.12.013.

Buol, S.W., Sanchez, P.A., Kimble, J.M., and Weed, S.B. (1990). Predicted impact of climatic warming on soil properties and use. *American Society of Agronomy Special Publication* 53: 71–82.

Capell, B. and Doerffling, K. (1993). Genotype specific difference in chilling tolerance of maize in relation to chilling induce changes in water status and abscisic acid accumulation. *Physiologia Plantarum* 88: 638–646.

Chimenti, C.A., Marcantonio, M., and Hall, J.A. (2006). Divergent selection for osmotic adjustment results in improved drought tolerance in maize (*Zea mays* L.) in both early growth and flowering phases. *Field Crops Research* 95: 305–315.

Churchman, G.J., Singh, M., Schapel, A. et al. (2020). Clay minerals as the key to the sequestration of carbon in soils. *Clays and Clay Minerals* 68: 135–143.

Conde, A. (2011). Membrane transport, sensing and signaling in plant adaptation to environmental stress (PDF). *Plant & Cell Physiology* 52 (9): 1583–1602. https://doi.org/10.1093/pcp/pcr107. PMID 21828102 – *via Google Scholar*.

Congming, L. and Vonshak, A. (2002). Effects of salinity stress on photosystem II function in cyanobacterial Spirulina platensis cells. *Physiologia Plantarum* 114: 405–413.

Cramer, G.R., Urano, K., Delrot, S. et al. (2011). Effects of abiotic stress on plants: a systems biology perspective. *BMC Plant Biology 11*: 163. https://doi.org/10.1186/1471-2229-11-163. ISSN 1471-2229. PMC 3252258. PMID 22094046.

Day, J.W. and Templet, P.H. (1989). Consequences of sea level rise: implications from the Mississippi delta. *Coastal Management* 17: 241–257.

Duan, D.Y., Li, W.Q., Liu, X.J. et al. (2007). Seed germination and seedling growth of *Suaeda salsa* under salt stress. *Annales Botanici Fennici* 44: 161–169.

Dubroeucq, D. and Volkoff, B. (1988). Evolution des couvertures pédologiques sableuses à podzols géants d'Amazonie (Bassin du haut Rio Negro). *Cahiers ORSTOM, Série Pédologie* 24 (3): 191–214.

Emanuel, W.R., Shugart, H.H., and Stevenson, M.P. (1985). Climatic change and the broad-scale distribution of terrestrial ecosystem complexes. *Climatic Change* 7: 29–43.

Farahbakhsh, H. (2012). Germination and seedling growth in un-primed and primed seeds of fennel as affected by reduced water potential induced by NaCl. *International Research Journal of Applied and Basic Sciences* 3: 737–744.

Flowers, T.J. (2004). Improving crop salt tolerance. *Journal of Experimental Botany* 55: 307–319.

Gasco, G., Paz-Ferreiro, J., and Mendez, A. (2012). Thermal analysis of soil amended with sewage sludge and biochar from sewage sludge pyrolysis. *Journal of Thermal Analysis and Calorimetry* 108: 769–775.

Golchin, A., Oades, J.M., Skjemstad, J.O., and Clarke, P. (1994). Soil-structure and carbon cycling. *Australian Journal of Soil Research* 32: 1043–1068.

Goncalves-Alvim, S.J. and Fernandez, G.W. (2001). Biodiversity of galling insects: historical, community and habitat effects in four neotropical savannas. *Biodiversity and Conservation* 10: 79–98. https://doi.org/10.1023/a:1016602213305.

González-Villagra, J., Rodrigues-Salvador, A., Nunes-Nesi, A. et al. (2018). Age-related mechanism and its relationship with secondary metabolism and abscisic acid in *Aristotelia chilensis* plants subjected to drought stress. *Plant Physiology and Biochemistry* 124: 136–145. https://doi.org/10.1016/j.plaphy.2018.01.010. ISSN 0981-9428. PMID 29360623.

Goryachkin, S.V. and Targulian, V.O. (1990). *Climate-induced Changes of the Boreal and Subpolar Soils* (ed. Scharpenseel et al.), 191–209. Moscow: USSR.

Heans, D.L. (1984). Determination of total organic carbon in soils by an improved chromic acid digestion and spectrophotometric procedure. *Communications in Soil Science and Plant Analysis* 15: 1191–1213.

Ilyas, M., Ayub, G., Imran et al. (2020a). Calcium and boron effect on production and quality of autumn potato under chilling temperature. *Communications in Soil Science and Plant Analysis*. https://doi.org/10.1080/00103624.2020.1854286.

Ilyas, M., Ayub, G., Imran et al. (2020b). Effect of calcium and boron on production and quality of autumn potato under chilling temperature. *Communication in Soil Science and Plant Analysis*. https://doi.org/10.1080/00103624.2020.1854286.

Ilyas, M., Ayub, G., Imran et al. (2020c). Calcium and boron effect on production and quality of autumn potato under chilling temperature. *Communication in Soil Science and Plant Analysis*. https://doi.org/10.1080/00103624.2020.1854286.

Imran (2017). Climate change is a real fact confronting to agricultural productivity. *International Journal of Environmental Sciences & Natural Resources* 3 (3): 555613. https://doi.org/10.19080/IJESNR.2017.03.555613.

Imran (2018a). Climate change is threat toward agronomy (base of food, fiber system), and food security. *Food and Nutrition Journal* 3: 160. https://doi.org/10.29011/2575- 7091.100060.

Imran (2018b). Phosphorous fertilization influenced weeds attributes and phenological characteristics of mungbean cultivars (Vigna radiata L.). *Russian Agricultural Sciences* 44 (3): 229–234.

Imran (2018c). Physiological and morphological traits of agronomic crops influenced by climate change. *Modern Concepts & Developments in Agronomy* 1 (4). MCDA.000524.

Imran (2018d). Ecological environmental variability influence growth and yield potential of rice under northern climatic scenario. *Russian Agricultural Sciences* 44 (1): 18–24.

Imran (2018e). Organic matter amendments improve soil health, productivity and profitability of maize and soybean. *Annals of Reviews and Research* 1 (3): 555564.

Imran (2021). The bioavailability of phosphorus in composite vs. hybrid maize differ with phosphorus and boron fertilization. *Phosphorus, Sulfur, and Silicon and the Related Elements*. https://doi.org/10.1080/10426507.2021.1920588.

Imran and Amanullah (2018). Review. Global impact of climate change on water, soil resources and threat towards food security: evidence from Pakistan. *Advances in Plants & Agriculture Research* 8 (5): 350–355. https://doi.org/10.15406/apar.2018.08.00349.

Imran and Amanullah (2021). Phosphorus and boron application optimizing biofortification of phosphorus and productivity of French bean (*Phaseolus Vulgaris* L.). *Communication in Soil Science and Plant Analysis*. https://doi.org/10.1080/00103624.2021.1971689.

Imran and Khan, A.A. (2015a). Grain yield and phenology of maize cultivars influenced by various phosphorus sources. *Food Science and Quality Management* 37. ISSN 2224-6088 (Paper) ISSN 2225-0557.

Imran and Khan, A.A. (2015b). Biochar application and shoot cutting duration (days) influenced growth, yield and yield contributing parameters of *Brassica napus* L. *Journal of Biology, Agriculture and Healthcare* 5 (5): ISSN 2224-3208 (Paper) ISSN 2225-093X (Online).

Imran and Khan, A.A. (2015c). Influence of compost application and seed rates on production potential of late sown maize on high elevation in Swat, Pakistan. *Journal of Environmental Studies and Sciences* ISSN 2224-3216 (Paper) ISSN 2225-0948.

Imran and Khan, A.A. (2015d). Phenological characteristics of *Brassica napus* L. as influenced by biochar application and shoot cutting duration (days). *Civil and Environmental Research* ISSN 2224-5790 (Paper) ISSN 2225-0514 (Online).

Imran and Khan, A.A. (2015e). Effect of transplanting dates on yield and yield components of various rice genotypes in hilly area cold climatic region of Khyber Pakhtunlhwa-Pakistan. *Journal of Biology, Agriculture and Healthcare* ISSN 2224-3208 (paper) ISSN 2225-093x (online).

Imran and Khan, A.A. (2017). Canola yield and quality enhanced with sulphur fertilization. *Russian Agricultural Sciences* 43: 113. https://doi.org/10.3103/S1068367417020100.

Imran, Khan, A.A., Irfanullah, and Ahmad, F. (2014). Production potential of rapeseed (*Brassica napus* L.) As influenced by different nitrogen levels and decapitation stress under the rainfed agro-climatic condition of swat Pakistan. *Journal of Global Innovations in Agricultural and Social Sciences*. ISSN (Online): 2311-3839; ISSN (Print): 2312-5225.

Imran, Khan, A.A., Ahmad, F., and Irfanullah (2015a). Nitrogen levels, tillage practices and irrigation timing influenced yeild, yeild components and oil contents of Canola. *Civil and Environmental Research*. ISSN 2224-5790 (Paper) ISSN 2225-0514 (Online).

Imran, Khan, A.A., Ullah, I. et al. (2015b). Yield and yield attributes of rapeseed cultivars as influence by sulfur level under swat valley conditions. *Pure & Applied Biology* 4 (3): 296–301.

Imran, Khan, A.A., Inamullah, and Luqman (2015c). Weeding stages and their effect on yield and yield components of rice in upper Swat, Pakistan. *Pakistan Journal of Weed Science Research* 21 (4): 555–563.

Imran, Khan, A.A., and Ahmad, F. (2015d). Phenology of various rice genotypes as affected by different transplanting dates under cold climatic region of Khyber Pakhtunkhwa-Pakistan. *Journal of Environmental Studies and Sciences*. ISSN 2224-3216 (Paper). ISSN 2225-0948.

Imran, Khan, A.A., Ahmad, F., and Irfanullah (2015e). Influence of hydrated calcium sulphate ($CaSO_4 \cdot 2H_2O$) and nitrogen levels on water infiltration rate and maize varieties productivity in rainfed area of Swat, Pakistan. *Chemistry and Materials Research*. ISSN 2224-3224 (Print) ISSN 2225-0956 (Online).

Imran, Khan, A.A., Zada, H. et al. (2015f). Graine yield and yield components of wheat cultivar "Siran 2010" as affected by phosphorous levels under rain fed condition. *International Journal of Research in Science*. ISSN 2224-3186 (Paper) ISSN 2225-0921 (Online).

Imran, Khan, A.A., Akhtar, K. et al. (2015g). Rice seedling characteristics of various genotypes influenced by different sowing dates in Swat-Pakistan. *Journal of Environmental Studies and Sciences*. ISSN 2224-3216 (Paper) ISSN 2225-0948.

Imran, Maula, F., Uzair, M., and Zada, H. (2015h). Farmers income enhancement through off-season vegetables production under natural environment in Swat-Pakistan. *Journal of Environmental Studies and Sciences*. ISSN 2224-3216 (Paper) ISSN 2225-0948.

Imran, I., Hussain, I., Khattak, A.R. et al. (2015i). Roots nodulation, yield and yield contributing parameters of mungbean cultivars as influenced by different phosphorous level in swat-Pakistan. *Pure and Applied Biology* 4 (4).

Imran, Khattak, I., Hussain, I. et al. (2015j). Growth and yield of maize hybrids as effected by different sowing dates in Swat Pakistan. *Pure and Applied Biology* 4 (4).

Imran, Uzair, M., Maula, F. et al. (2015k). Introduction and promotion of off-season vegetables production under natural environment in hilly area of upper swat-Pakistan. *Journal of Biology, Agriculture and Healthcare* 5 (11).

Imran, Naveed, S., Khan, A.A., and Khattak, I. (2015l). Impact of phosphorus levels and seed rates on growth and yield of late sown maize on high elevation in Swat, Pakistan. *Pakistan Journal of Agricultural Research* 28 (4): 406–413.

Imran, Khan, A.A., Khan, I.U., and Naveed, S. (2016a). Weeds density and late sown maize productivity influenced by compost application and seed rates under temperate environment. *Pakistan Journal of Weed Science Research* 22 (1): 169–181.

Imran, Khan, A.A., Inam, I., and Ahmad, F. (2016b). Yield and yield attributes of Mungbean (*Vigna radiata* L.) cultivars as affected by phosphorous levels under different tillage systems. *Cogent Food & Agriculture* 2: 1151982.

Imran, Zada, H., Naveed, S. et al. (2016c). Variable rates of phosphorous application influenced phenological traits of green gram (*Vignaradiata* L.). *Journal of Agricultural Research* 1 (3).

Imran, Bari, A., Ali, R. et al. (2017a). Traditional rice farming accelerate CH_4 & N_2O emissions functioning as a stronger contributors of climate change. *International Journal of Environmental Sciences & Natural Resources*. https://doi.org/10.19080/ARTOAJ.2017.09.555765.

Imran, Jamal, N., Alam, A., and Khan, A.A. (2017b). Grain yield, yield attributes of wheat and soil physio-chemical characteristics influenced by biochar, compost and inorganic fertilizer application. *Agri Res & Tech: Open Access J*. https://doi.org/10.19080/ARTOAJ.2017.10.555795.

Imran, Amanullah, Bari, A., and Ali, R. (2018). Peach sources, phosphorous and beneficial microbes enhance productivity of soybean. *Soybean Research* 16 (2): 39–48.

Imran, Amanullah, Khan, A.A. et al. (2019). Production statistics and modern technology of maize cultivation in Khyber Pakhtunkhwa Pakistan. *Plant Science Archives* 2 (2): 1–12.

Imran, Amanullah, and Al-Tawaha, A.R. (2020a). The productivity of subsequent wheat enhanced with residual carbon sources and phosphorus under improved irrigation system. *Communication in Soil Science and Plant Analysis*. https://doi.org/10.1080/00103624.2020.1763387.

Imran, Amanullah, Arif, M. et al. (2020b). Soil application of *Trichoderma* and Peach (*Prunus persica* L.) residues possesses biocontrol potential for weeds and enhances growth and profitability of soybean (Glycine max). *Sarhad Journal of Agriculture* 36 (1): 10–20. https://doi.org/10.17582/journal.sja/2020/36.1.10.20.

Imran, Amanullah, Khan, A.A. et al. (2020c). Adequate fertilization, application method and sowing techniques improve maize yield and related traits. *Communication in Soil Science and Plant Analysis* https://doi.org/10.1080/00103624.2021.1925688.

Imran, Amanullah, Arif, M. et al. (2020d). Integration of Peach (*Prunus persica* L.) residues, beneficial microbes and phosphorus enhance phenology, growth and yield of soybean. *Russian Agricultural Sciences* 46 (3): 223–230. ISSN 1068-3674.

Imran, Amanullah, and Al Tawaha, A.R.M. (2021a). Management of nano-black carbon, phosphorous and bio fertilizer improve soil organic carbon and ensilage biomass of soybean and maize. *Communication in Soil Science and Plant Analysis*. https://doi.org/10.1080/00103624.2021.1966439.

Imran, Amanullah, and Al-Tawaha, A.R. (2021b). Carbon assimilation and dry matter partitioning in soybean ameliorates with the integration of nano-black carbon, along with beneficial microbes and phosphorus fertilization. *Journal of Plant Nutrition*. https://doi.org/10.1080/01904167.2022.2035753.

Imran, Amanullah, Khan, A.A. et al. (2021c). Adequate fertilization, application method and sowing techniques improve maize yield and related traits. *Communication in Soil Science and Plant Analysis*. https://doi.org/10.1080/00103624.2021.1925688.

Imran, Amanullah, and Al-Tawaha, A.R. (2021d). Carbon sources application increase wheat yield and soil fertility. *Communication in Soil Science and Plant Analysis*. https://doi.org/10.1080/00103624.2020.1865397.

Imran, Amanullah, and Al Tawaha, A.R.M. (2021e). Management of nano-black carbon, phosphorous and bio fertilizer improve Soil organic carbon and ensilage biomass of soybean and maize. *Communications in Soil Science and Plant Analysis*. https://doi.org/10.1080/00103624.2021.1966439.

Imran, Amanullah, and Altawaha, A.R. (2021f). Carbon assimilation and dry matter partitioning in soybean ameliorates with the integration of nano-black carbon, along with beneficial microbes and phosphorus fertilization. *Journal of Plant Nutrition*.

Imran, Amanullah, Khan, A.A. et al. (2021g). Adequate fertilization, application method and sowing techniques improve maize yield and related traits. *Communications in Soil Science and Plant Analysis*. https://doi.org/10.1080/00103624.2021.1925688.

Iqbal, B., Jan, M.T., Inamullah et al. (2015a). Integrated management of compost type and fertilizer-N in maize. *Pure and Applied Biology* 4 (4). https://doi.org/10.19045/bspab.2015.44001.

Iqbal, B., Jan, M.T., Imran et al. (2015b). Growth and phenology of maize as affected by integrated management of compost and fertilizers nitrogen. *Pure and Applied Biology* 4 (4).

Iqbal, B., Ahmad, B., Ullah, I. et al. (2016a). Effect of phosphorus, sulphur and different irrigation levels on phenological traits of Triticale. *Pure and Applied Biology* 5 (2): 303–310. https://doi.org/10.19045/bspab.2016.50039.

Iqbal, B., Jan, M.T., Muhammad, Z. et al. (2016b). Phenological traits of maize influenced by integrated management of compost and fertilizer nitrogen. *Pure and Applied Biology* 5 (1): 58–63.

Islam, M., Anwar, S., Bashir, S. et al. (2015). Growth and yield components of wheat varieties as affected by dual purpose practices. *Pure and Applied Biology* https://doi.org/10.19045/bspab.2015.44006.

ISTA (2005). *International Rules for Seed Testing*. Zurich, Switzerland: International Seed Testing Association.

Jackson, M. (1962). *Soil Chemical Analysis*, vol. 496. London: Constable and Company Limited.

Jagatheeswari, D. and Ranganathan, P. (2012). Influence of mercuric chloride on seeds germination, seedling growth and biochemical analysis of green gram (*Vigna radiata* (L.) Wilczek. Var. Vamban-3). *International Journal of Pharmaceutical and Biological Archives* 3: 291–295.

Jastrow, J.D., Miller, R.M., and Lussenhop, J. (1998). Contributions of interacting biological mechanisms to soil aggregate stabilization in restored prairie. *Soil Biology & Biochemistry* 30: 905–916.

Jelgersma, S. (1988). A future sea-level rise: its impacts on coastal lowlands. In: *Geology and Urban Development. Atlas of Urban Geology*, vol. 1, 61–81. UN-ESCAP.

Ji, P., Ron, G.K., and M. (2007). Understanding abiotic stress tolerance mechanisms: recent studies on stress response in rice. *Journal of Integrative Plant Biology* 49 (6): 742–750. https://doi.org/10.1111/j.1744-7909.2007.00495.x.

Joseph, B. and Jini, D. (2011). Development of salt stress-tolerant plants by gene manipulation of antioxidant enzymes. *Asian Journal of Agricultural Research* 5: 17–27.

Kamara, A., Kamara, A., Mansaray, M.M., and Sawyerr, P.A. (2014). Effects of biochar derived from maize stover and rice straw on the germination of their seeds. *American Journal of Agriculture and Forestry* 2: 246–249.

Kammann, C.I., Linsel, S., Gobling, J.W., and Koyro, H. (2011). Influence of biochar on drought tolerance of *Chenopodium quinoa* willd and on soil-plant relation. *Plant and Soil* 345: 195–210.

Karim, S.A. and Imran (2019). Reduction in substrate moisture content reduce final yield of wheat. *Biomedical Journal of Scientific & Technical Research* 1–5. https://doi.org/10.26717/BJSTR.2019.15.002656.

Khan, M.A., Islam, E., Shirazi, M.U. et al. (2010). Physiological responses of various wheat genotypes to salinity. *Pakistan Journal of Botany* 42: 3497–3505.

Khan, A., Khan, M.N., Uallah, I. et al. (2015a). Effect of potash application on growth, yield and yield components of spring maize hybrids. *Pure and Applied Biology* 4 (2): 195–203.

Khan, A.A., Imran, Inamullah, F.A.S. et al. (2015b). Phenological traits of rice as influenced by seedling age and number of seedling per hill under temperate region. *Journal of Biology, Agriculture and Healthcare*. ISSN 2224-3208 (Paper) ISSN 2225-093X (Online).

Khan, A.Z., Imran, Muhammad, A. et al. (2016). Impact of fertilizer priming on seed germination behavior and vigor of maize. *Pure and Applied Biology* 5 (4): 744–751. https://doi.org/10.19045/bspab.2016.50093.

Khresat, S., Al-Bakri, J., and Al-Tahhan, R. (2008). Impacts of land use/cover change on soil properties in the Mediterranean region of northwestern Jordan. *Land Degradation & Development* 19: 397–407.

Lal, R. (2010). Managing soils and ecosystems for mitigating anthropogenic carbon emissions and advancing global food security. *Bioscience* 60: 708–721.

Lal, R. (2019). Carbon cycling in global drylands. *Current Climate Change Reports* 5: 221–232.

Larionova, A.A., Kurganova, I.N., de Gerenyu, V.O.L. et al. (2010). Carbon dioxide emissions from agrogray soils under climate changes. *Eurasian Soil Science* 43: 168–176.

Lashari, M.S., Liu, Y., Li, W. et al. (2013). Effect of amendment of biochar manure compost in conjunction with pyroligneous solution on soil quality and wheat yield of a salt stressed cropeland from central China Great Plain. *Food Crop Research* 144: 113–118.

Lehmann, J. and Joseph, S. (2009). Biochar for environmental management: an introduction. In: *Biochar for Environmental Management: Science and Technology* (ed. J. Lehmann and S. Joseph), 1–12. London: Earthscan.

Lehmann, J., Rillig, M.C., Thies, J. et al. (2011). Biochar effects on soil biota. *Soil Biology and Biochemistry* 43: 1812–1836.

Lei, G., Shen, M., Li, Z.-G. et al. (2011). EIN2 regulates salt stress response and interacts with a MA3 domain-containing protein ECIP1 in Arabidopsis. *Plant, Cell & Environment* 34 (10): 1678–1692. https://doi.org/10.1111/j.13653040.2011.02363.x. ISSN 1365-3040.

Lucas, Y., Boulet, R., Chauvel, A., and Veillon, L. (1987). Systèmes sols ferrallitiques-podzols en région amazonienne. In: *Podzols et podzolisation* (ed. D. Righi and A. Chauvel), 53–68. Paris: AFES-INRA.

Maestre, F.T., Cortina, J., and Bautista, S. (2007). Mechanisms underlying the interaction between Pinus halepensis and the native late-successional shrub Pistacia lentiscus in a semi-arid plantation. *Ecography* 27 (6): 776–786. https://doi.org/10.1111/j.0906-7590.2004.03990.x.

Martinez-Beltran, J. and Manzur, C.L. (2005). Overview of salinity problems in the world and FAO strategies to address the problem. *Proceedings of the International Salinity Forum, Riverside, California* 2005 (April): 311–313.

Mazhar, R., Ilyas, N., Saeed, M. et al. (2016). Biocontrol and salinity tolerance potential of *Azospirillum lipoferum* and its inoculation effect in wheat crop. *International Journal of Agriculture and Biology* 3: 494–500.

McKyes, E., Sethi, A., and Yong, R.N. (1974). Amorphous coatings on particles of sensitive clay soils. *Clays and Clay Minerals* 22: 427–433.

Miri, Y. and Mirjalili, S.A. (2013). Effect of salinity stress on seed germination and some physiological traits in primary stages of growth in purple coneflower (*Echinacea purpurea*). *International Journal of Agronomy and Plant Production* 4: 142–146.

Mittler, R. (2006). *Abiotic stress, the field environment and stress combination. Trends in Plant Science* 11 (1): 15–19. https://doi.org/10.1016/j.tplants.2005.11.002. PMID 16359910.

Munns, R. (2002). Comparative physiology of salt and water stress. *Plant, Cell & Environment* 25: 239–250.

Munns, R. and Tester, M. (2008). Mechanisms of salinity tolerance. *Annual Review of Plant Biology* 59: 651–681.

Munns, R.R., James, A., and Lauchli, A. (2006). Approaches to increasing the salt tolerance of wheat and other cereals. *Journal of Experimental Botany* 57: 1025–1043.

Naveed, S., Ibrar, M., Khattak, I. et al. (2016). Anthelmintic, antilice, insecticidal, cytotoxic and phytotoxic potential of ethanolic extracts of two wild medicinal plants *Iphiona grantioides* and *Plucheaarguta*. *Woulfenia Journal* 23 (11): 13–25.

Nazarli, H., Faraji, F., and Zardashti, M.R. (2011). Effect of drought stress and polymer on osmotic adjustment and photosynthetic pigments of sunflower. *Carcetriari Agronomica in Maldova* 1: 45.

Neto, A.D.A., Prisco, J.T., Enéas-Filho, J. et al. (2004). Effects of salt stress on plant growth, stomatal response and solute accumulation of different maize genotypes. *Brazilian Journal of Plant Physiology* 16 (1): 31–38. https://doi.org/10.1590/S1677-04202004000100005. ISSN 1677-0420.

Ortas, I. (2020). Mycorrhizas in fruit nutrition: important breakthroughs. In: *Fruit Crops*. Elsevier.

Ortas, I. and Bykova, A. (2020). Effects of long-term phosphorus fertilizer applications on soil carbon and CO(2)flux. *Communications in Soil Science and Plant Analysis* 51: 2270–2279.

Ortas, I. and Yucel, C. (2020). Do mycorrhizae influence cover crop biomass production? *Acta Agriculturae Scandinavica Section B Soil and Plant Science* 70: 657–666.

Ortas, I., Akpinar, C., and Lal, R. (2013). Long-term impacts of organic and inorganic fertilizers on carbon sequestration in aggregates of an Entisol in Mediterranean Turkey. *Soil Science* 178: 12–23.

Ortaş, I., Razzaghi, S., and Rafique, M. (2016). Arbuscular mycorrhizae: effect of rhizosphere and relation with carbon nutrition. In: *Plant-Microbe Interaction: An Approach to Sustainable Agriculture* (ed. D. Choudhary, A. Varma and N. Tuteja), 125–152. Singapore: Springer.

Ortaş, I., Lal, R., and Kapur, S. (2017). Carbon sequestration and mycorrhizae in Turkish soils. In: *Carbon Management, Technologies, and Trends in Mediterranean Ecosystems*, 139, 149. Springer.

Ortas, I., Bulutekin, B., and Öztürk, F. (2019). Rhizosphere and non-rhizosphere soil structure development under phosphorus fertilization application. *Фундаментальные концепции физики почв: развитие, современные приложения и перспективы* 369–379.

Ortaş, I., Rafique, M., and Iqbal, M.T. (2019). Mycorrhizae Resource Allocation in Root Development and Root Morphology. In: *Plant Microbe Interface*, 1–26. Springer.

Ortas, I., Rafique, M., and Çekiç, F. (2021). Do mycorrhizal fungi enable plants to cope with abiotic stresses by overcoming the detrimental effects of salinity and improving drought tolerance? In: *Symbiotic Soil Microorganisms*, 391–428. Springer.

Palm, B. and Volkenburgh, V. (2012). Serpentine tolerance in Mimuslus guttatus does not rely on exclusion of magnesium. *Functional Plant Biology* https://doi.org/10.1071/FP12059.

Palta, J.P. and Farag, K. (2006). Methods for enhancing plant health, protecting plants from biotic and abiotic stress related injuries and enhancing the recovery of plants injured as a result of such stresses. US Patent 7,101,828.

Pant, B.-D., Burgos, A., Pant, P. et al. (2015a). The transcription factor PHR1 regulates lipid remodeling and triacylglycerol accumulation in Arabidopsis thaliana during phosphorus starvation. *Journal of Experimental Botany* 66 (7): 1907–1918. https://doi.org/10.1093/jxb/eru535. ISSN 0022-0957. PMC 4378627. PMID 25680792.

Pant, B.-D., Pant, P., Erban, A. et al. (2015b). Identification of primary and secondary metabolites with phosphorus status-dependent abundance in Arabidopsis, and of the transcription factor PHR1 as a major regulator of metabolic changes during phosphorus limitation. *Plant, Cell & Environment* 38 (1): 172–187. https://doi.org/10.1111/pce.12378. ISSN 1365-3040. PMID 24894834.

Paz-Ferreiro, J., Fu, S., Mendez, A., and Gasco, G. (2014). Interactive effects of biochar and the earthworm *Pontoscolex corethrurus* on plant productivity and soil enzymes activities. *Journal of Soils and Sediments* 14: 483–494.

Peng, Y., Gao, Z., Gao, Y. et al. (2008). Ecophysiological characteristics of alfalfa seedlings in response to various mixed salt-alkaline stresses. *Journal of Integrative Plant Biology* 50: 29–39.

Prado, F.E., Boero, C., Gallarodo, M., and Gonzalez, A.J. (2000). Effect of NaCl on germination, growth and soluble sugar content in *Chenopodium quinoa* wil seeds. *Botanical Bulletin - Academia Sinica Taipei* 14: 27–34.

Raghothama, K.G. (1999). *Phosphate acquisition. Annual Review of Plant Physiology and Plant Molecular Biology* 50 (1): 665–693. https://doi.org/10.1146/annurev.arplant.50.1.665. PMID 15012223.

Rajalakshmi, A.G., Kumar, S.K., Bharathi, C.D. et al. (2015). Effect of biochar in seed germination: *in-vitro* study. *International Journal of Biosciences and Nanosciences* 2: 132–136.

Rashid, A., Saleem, Q., Nazir, A., and Kazım, H.S. (2003). Yield potential and stability of nine wheat varieties under water stress conditions. *International Journal of Agriculture and Biology* 5: 7–9.

Raza, A. (2012). Grain and field annual 2012 gain report. In: *Global Agricultural Information Network*, 1–3. USDA Foreign Agricultural Services.

Rengasamy, P. (2002). Transient salinity and subsoil constraints to dryland farming in Australian sodic soils: an overview. *Australian Journal of Experimental Agriculture* 42: 351–361.

Roelofs, D., Jiwan, P., and kwangy I. (2008). *Functional ecological genomics to demonstrate general and specific responses to abiotic stress. Functional Ecology* 22: 8–18. https://doi.org/10.1111/j.1365-2435.2007.01312.x.

Rolf, D., Vinebrooke, K., and Mittler, J. (2004). *Impacts of multiple stressors on biodiversity and ecosystem functioning: the role of species co-tolerance. Oikos* 104 (3): 451–457. https://doi.org/10.1111/j.0030-1299.2004.13255.x.

Rubio, V., Linhares, F., Solano, R. et al. (2001). A conserved MYB transcription factor involved in phosphate starvation signaling both in vascular plants and in unicellular algae. *Genes & Development* 15 (16): 2122–2133. https://doi.org/10.1101/gad.204401. ISSN 0890-9369. PMC 312755. PMID 11511543.

Ryan, J., Masri, S., Ibrikci, H. et al. (2008). Implications of cereal-based crop rotations, nitrogen fertilization, and stubble grazing on soil organic matter in a Mediterranean-type environment. *Turkish Journal of Agriculture and Forestry* 32: 289–297.

Samreen, U., Ibrar, M., Lalbadshah et al. (2016). Ethnobotanical study of subtropical hills of Darazinda, Takht-e-Suleman range F.R D.I. Khan, Pakistan. *Pure and Applied Biology* 5 (1): 149–164. https://doi.org/10.19045/bspab.2016.50020.

Santer, B. (1985). The use of general circulation models in climate impact analysis - a preliminary study of the impacts of a CO_2-induced climatic change on West European agriculture. *Climatic Change* 7: 71–93.

Savvides, A. (2015). Chemical priming of plants against multiple abiotic stresses: mission possible? *Trends in Plant Science* 21 (4): 329–340. https://doi.org/10.1016/j.tplants.2015.11.003. hdl:10754/596020.

Shamsi, K. and Kobraee, S. (2013). Biochemical and physiological responses of three wheat cultivars (*Triticum aestivum* L.) to salinity stress. *Annals of Biological Research* 4: 180–185.

Shazma, A., Israeel, Iqbal, B. et al. (2016). Nitrogen and phosphorus fertilization of improved varieties for enhancing phenological traits of wheat. *Pure and Applied Biology* 5 (3): 511–519. https://doi.org/10.19045/bspab.2016.50065.

Sidari, M., Mallamaci, C., and Muscolo, A. (2008). Drought, salinity and heat differently affect seed germination of *Pinus pinea*. *Journal of Forest Research* 13: 326–330.

Singh, M., Sarkar, B., Biswas, B. et al. (2016). Adsorption-desorption behavior of dissolved organic carbon by soil clay fractions of varying mineralogy. *Geoderma* 280: 47–56.

Sombroek, W.G. (1990). *Soils on a Warmer Earth: the Tropical Regions* (ed. H.W. Scharpenseel, M. Schomaker and A. Ayoub), 157–174. UNEP 1990.

Sombroek, W.G. and Zonneveld, L.S. (1971). Ancient dune fields and fluviatile deposits in the Rima-Sokoto river basin (N.W. Nigeria). Stiboka (Staring Centre). Soil Survey Paper No. 5. Wageningen. 109 p.

Stanley, D.J. (1988). Subsidence in the Northeastern Nile delta: rapid rates, possible causes, and consequences. *Science* 240: 497–500.

Stigliani, W.M. (1988). Changes in valued 'capacities' of soils and sediments as indicators of nonlinear and time-delayed environmental effects. *Environmental Monitoring and Assessment* 10: 245–307.

Tank, N. and Saraf, M. (2010). Salinity-resistant plant growth promoting rhizobacteria ameliorates sodium chloride stress on tomato plants. *Journal of Plant Interactions* 5: 51–58.

Tiwari, S., Singh, P., Tiwari, R. et al. (2011). Salt-tolerant rhizobacteria-mediated induced tolerance in wheat (*Triticum aestivum*) and chemical diversity in rhizosphere enhance plant growth. *Biological Fertile Soils* 47: 907–916.

Tombesi, S., Frioni, T., Poni, S., and Palliotti, A. (2018). Effect of water stress "memory" on plant behavior during subsequent drought stress. *Environmental and Experimental Botany* 150: 106–114. https://doi.org/10.1016/j.envexpbot.2018.03.009. ISSN 0098-8472.

Van Breemen, N. (1990). Impact of anthropogenic atmospheric pollution on soils. In: *Climate Change and Soil Processes*, 137–144. Nairobi: UNEP.

Van Hoorn, J.W., Nader, K., Hamdy, A., and Mastrorilli, M. (2001). Effect of salinity on yield and nitrogen uptake of four grain legumes and on biological nitrogen contribution from the soil. *Agriculture Water Management* 51: 87–98.

Wang, H. and Jin, J. (2007). Effects of zinc deficiency and drought on plant growth and metabolism of reactive oxygen species in maize (*Zea mays* L.). *Agricultural Sciences in China* 6: 988–995.

Wang, G. and Xu, Z. (2013). The effects of biochar on germination and growth of wheat in different saline-alkali soil. *Asian Agricultural Research* 5: 116–119.

Wang, W., Vinocur, B., and Altman, A. (2007). Plant responses to drought, salinity and extreme temperatures towards genetic engineering for stress tolerance. *Planta* 218 (1): 1–14. https://doi.org/10.1007/s00425-003-1105-5. PMID 14513379.

Wang, X.Y., He, T.H., Gen, S.Y. et al. (2020). Soil properties and agricultural practices shape microbial communities in flooded and rainfed croplands. *Applied Soil Ecology* 147.

Warrick, R. and Farmer, G. (1990). The greenhouse effect, climatic change and rising sea level: implications for development. *Transactions of the Institute of British Geographers* 15: 5–20.

Wolfe, A. (2007). *Patterns of Biodiversity*. Ohio State University. https://doi.org/10.1016/j.agee.2006.12.013.

Xu, D., Huang, J., Guo, S. et al. (2008). Over-expression of a TFIII-type zinc finger protein gene ZFP252 enhances drought and salt tolerance in rice (*Oryza sativa* L.). *Federation of European Biochemical Societies Letters* 582: 1037–1043.

Yeo, A.R., Yeo, M.E., Flowers, S.A., and Flowers, T.J. (1990). Screening of rice (*Oryza sativa* L.) genotypes for physiological characters contributing to salinity resistance, and their relationship to overall performance. *Theoretical and Applied Genetics* 79: 377–384.

Younis, U., Athar, M., Malik, S.A. et al. (2015). Biochar impact on physiological and biochemical attributes of spinach *Spinacia oleracea* (L.) in nickel contaminated soil. *Global Journal Environment Science Management* 1: 245–254.

Zakaria, M., Daniel, S., Murphy, V., and Abbott, L.K. (2012). Biochars influence seed germination and early growth of seedlings. *Plant and Soil* 353: 273–287.

Zargar, Majeed, S., Nagar, P. et al. (2017). Aquaporins as potential drought tolerance inducing proteins: towards instigating stress tolerance. *Journal of Proteomics* 169: 233–238. https://doi.org/10.1016/j.jprot.2017.04.010. ISSN 1874-3919.

Zhan, H.X., Chang, Z.J., Wei, A.L. et al. (2011). Impact of drought to wheat physiological index. *Journal of Shanxi Agriculture Sciences* 39: 1049–1051.

9

Nano-Black Carbon Is an Organic Tool for the Alleviation of Abiotic Stresses and Its Certain Damages Under Changing Climate

Imran[1,2], Amanullah[1], Muhammad Arif[1], Abdel Rahman Mohammad Said Al-Tawaha[3], and Tariq Mahmood[4]

[1] *Department of Agronomy, The University of Agriculture, Peshawar, Pakistan*
[2] *Ministry of Agriculture, Agriculture Extension Wing, Merged Area, Khyber Pakhtunkhwa, Pakistan*
[3] *Department of Biological Sciences, Al-Hussein Bin Talal University, Ma'an, Jordan*
[4] *Department of Bioinformatics/Agriculture, Hazara University, Mansehra, Pakistan*

CONTENTS

9.1 Introduction

The adverse impact of environmental factors on a living organism in specific circumstances is known as abiotic stress (Amanullah et al. 2020a,b,c; Imran et al. 2020a,b,c,d ,e,f; Rolf et al. 2004). The non-living variable must influence the environment beyond its normal range of variation to adversely affect the population performance or individual physiology of the organism in a significant way (Akhtar et al. 2015b; Ping et al. 2007; Khan et al. 2016; Zhan et al. 2011). Abiotic stress is essentially unavoidable. Abiotic stress effects on plants are especially dependent, if not solely dependent, on environmental factors, so it is particularly constraining. Abiotic stress is the most

Climate Change and Agriculture: Perspectives, Sustainability and Resilience,
First Edition. Edited by Noureddine Benkeblia.

harmful factor concerning the growth and productivity of crops worldwide. Mittler (2006) stated that abiotic stressors are most harmful when they occur together, in combinations of abiotic stress factors. *These findings were concerning those of Roelofs et al. (2008) who revealed that salinity and drought stress combine reduced the grain yield of maize.*

Abiotic stress comes in many forms (Amonette and Joseph 2009; Armor et al. 2005; Arshad et al. 2008). The most common of the stressors are the easiest to identify, but there are many other, less recognizable abiotic stress factors that affect environments constantly (Ayaz et al. 2000). The most basic Abiotic stressors include High winds, Extreme temperatures, Drought, Flood, and other natural disasters, such as tornadoes and wildfires (Iqbal et al. 2016a,b; Wang et al. 2007). Lesser-known stressors generally occur on a smaller scale. They include poor edaphic conditions like rock content and pH levels, high radiation, compaction, contamination, and other, highly specific conditions like rapid rehydration during seed germination (Bruinsma 1963; Brussaard et al. 2007). Abiotic stress, as a natural part of every ecosystem, will affect the plant in a variety of ways (Bohnert and Jensen 1996; Bos et al. 2005). Although these effects may be either beneficial or detrimental, the location of the area is crucial in determining the extent of the impact that abiotic stress will have (Khan et al. 2015; Akhtar et al. 2015a; Imran 2018a,b,c; Imran et al. 2015a,b,c,d,e). The higher the latitude of the area affected, the greater the impact of abiotic stress will be on that area (Capell and Doerffling 1993; Chimenti et al. 2006). So boreal forest is at the mercy of whatever abiotic stress factors may come along, while tropical zones are much less susceptible to such stressors (Cramer et al. 2011; Iqbal et al. 2015a,b, 2016a,b).

A plant's first line of defence against abiotic stress is in its roots (Imran et al.2021a,b,c, d,c,f,g). If the soil holding the plant is healthy and biologically diverse, the plant will have a higher chance of surviving stressful conditions (Conde 2011; Cramer et al. 2011; Imran 2018a,b,c; Imran and Khan 2017; Duan et al. 2007; Farahbakhsh 2012). The plant responses to stress are dependent on the tissue or organ affected by the stress. For example, transcriptional responses to stress are tissue or cell-specific in roots and are quite different depending on the stress involved (Maestre et al. 2007). Various publications report a generally positive effect of Nano-black carbon soil amendment on field crops under stress conditions. It has been reported that a single application of $20\,t\,ha^{-1}$ Nano-black carbon to a Colombian savanna soil resulted in an increase in maize yield by 28–140% as compared with the unamended control under the saline stress condition (Badridze et al. 2009; Bates et al. 1973). Nano-black carbon acts synergistically to improve crop performance under stress conditions. Specific mechanisms underlying the contribution of Nano-black carbon to plant response are poorly understood (Flowers 2004; Gasco et al. 2012; Goncalves-Alvim and Fernandez 2001; Gonzalez-Villagra et al. 2018). Regional conditions including climate, soil chemistry, and soil condition all influence Nano-black carbon agronomic benefits (Imran et al. 2015a,b,c,d,e, 2017). Evidence is mounting that Nano-black carbon in soil has significant effects on soil microorganisms (Lehmann et al. 2011; Imran et al. 2016a,b).

The majority of the studies assessed that microbial biomass increased in Nano-black carbon amended soils (Imran et al. 2018) which act as a biocontrol agent and mitigate abiotic stress induce damages (Heans 1984; Imran 2018a,b,c).

9.2 Salinity Stress Mitigation with Nano-Black Carbon

Nano-black carbon application in agriculture possesses the potential to tolerate abiotic stress (Khan et al. 2010; Imran et al. 2019) such as high salinity is the disruption of the Na^+/K^+ ratio in the cytoplasm of the plant cell (Jackson 1962; Jagatheeswari and Ranganathan 2012). High concentrations of Na+, for example, can decrease the capacity of the plant to take up water and also alter enzyme and transporter functions (Joseph and Jini 2011; Kamara et al. 2014). Soil salinization, the accumulation of water-soluble salts to levels that negatively impact plant production, is a global phenomenon affecting approximately 831 million hectares of land (Martinez-Beltran and Manzur 2005; Kammann et al. 2011; Lashari et al. 2013; Lehmann and Joseph 2009). More specifically, the phenomenon threatens 19.5% of the world's irrigated agricultural land and 2.1% of the world's non-irrigated (dry-land) agricultural lands (Lei et al. 2011; Maestre et al. 2007; Martinez-Beltran and Manzur 2005; Imran et al. 2017). High soil salinity content can be harmful to plants because water-soluble salts can alter osmotic potential gradients and consequently inhibit many cellular functions (Raghothama 1999). For example, high soil salinity content can inhibit the process of photosynthesis by limiting a plant's water uptake; high levels of water-soluble salts in the soil can decrease the osmotic potential of the soil and consequently decrease the difference in water potential between the soil and the plant's roots, thereby limiting electron flow from H_2O to P680 in Photosystem II's reaction centre (Rubio et al. 2001; Lehmann et al. 2011; Lei et al. 2011; Congming and Vonshak 2002).

The enzyme activation in plants with Nano-black carbon application has been proposed to increase tolerance to abiotic stresses in crop plants which may increase a stress-inducing chemical in the plant so that the plant begins preparing defence mechanisms (Paz-Ferreiro et al. 2014; Peng et al. 2008; Ping et al. 2007; Prado et al. 2000; Raghothama 1999). Thus, when abiotic stress occurs, the plant has already prepared defence mechanisms that can be activated faster and increase tolerance (Congming and Vonshak 2002). Over generations, many plants have mutated and built different mechanisms to counter salinity effects (Pant et al. 2015a,b; Rajalakshmi et al. 2015; Rashid et al. 2003; Imran 2018a,b,c; Iqbal et al. 2017). A good combatant of salinity in plants is the Nano-black carbon amendment. Nano-black carbon is known for regulating plant growth and development and dealing with stress conditions (Palm and Volkenburgh 2012; Palta and Farag 2006; Pant et al. 2015a,b). With Nano-black carbon application under stress conditions, plants adapt very differently from one another, even from a plant living in the same area (Imran et al. 2021a,b,c,d,e,f,g).

When a group of different plant species was prompted by a variety of different stress signals, such as drought or cold, each plant responded uniquely. Hardly any of the responses were similar, even though the plants had become accustomed to the same home environment (Neto et al. 2004; Rolf et al. 2004; Rubio et al. 2001; Savvides 2015; Shahida et al. 2016; Shamsi and Kobraee 2013; Shamsuddin et al. 2015; Shazma et al. 2016).

9.3 Plant Wilting Stress Mitigation with Nano-Black Carbon Amendments

Plant wilting occurs when water is scarce (Imran et al. 2021a,b,c,d,e,f,g). Drought stress is one of the main causes of crop loss in the agricultural world. This is due to water's necessity in so many fundamental processes in plant growth. It has become especially important in recent years to find a way to combat drought stress (Sidari et al. 2008; Karim and Imran 2019; Tank and Saraf 2010; Tiwari et al. 2011; Tombesi et al. 2018; Ulfat et al. 2016). A decrease in precipitation and a subsequent increase in drought is extremely likely in the future due to an increase in global warming. (Zargar et al. 2017; Van Hoorn et al. 2001; Wang et al. 2007; Wang and Xu 2013; Wang and Jin 2007; Wolfe 2007). Plants have come up with many mechanisms and adaptations to try and deal with drought stress. One of the leading ways that plants combat drought stress is by increasing water infiltration rate and soil porosity (Xu et al. 2008; Yeo et al. 1990; Younis et al. 2015; Zakaria et al. 2012; Zargar et al. 2017). Nano-black carbon application can enhance soil porosity and increase water infiltration in various types of soil. It has been reported that Nano-black carbon application regulates hormone secretion that maintains stomatal opening and closing (Imran et al. 2021a,b,c,d,e,f,g). The hormone which is regulated by Nano-black carbon application is abscisic acid (ABA). Synthesis of ABA causes the ABA to bind to receptors. This binding then affects the opening of ion channels thereby decreasing turgor pressure in the stomata and causing them to close. Recent studies showed that how ABA levels increased in drought-stressed plants (Van Hoorn et al. 2001; Wang et al. 2007; Wang and Xu 2013; Wang and Jin 2007). They showed that when plants were placed in a stressful situation they produced more ABA to try and conserve any water they had in their leaves (Goncalves-Alvim and Fernandez 2001; Rolf et al. 2004; Rubio et al. 2001; Savvides 2015; Shahida et al. 2016; Shamsi and Kobraee 2013). Nano-black carbon application can regulate another extremely important factor dealing with drought stress and uptake of water is aquaporins (AQPs). AQPs are integral membrane proteins that makeup channels. This channel's main job is the transport of water and other necessary solutes. AQPs are both transcriptionally and post-transcriptionally regulated by many different factors such as ABA, GA3, pH, and Ca^{2+} and the specific levels of AQPs in certain parts of the plant, such as roots or leaves, help to draw as much water into the plant as possible (Rubio et al. 2001; Savvides 2015; Shahida et al. 2016; Shamsi and Kobraee 2013; Shamsuddin et al. 2015).

9.4 Relationship of Soil Microbiota with Nano-Black Carbon

Evidence is mounting that Nano-black carbon in soil has significant effects on soil microorganisms, as reviewed recently by Lehmann et al. (2011) and Imran et al. (2020a). In the majority of the studies assessed in that review, microbial biomass increased in Nano-black carbon-amended soils (Palm and Volkenburgh 2012; Palta and Farag 2006; Pant et al. 2015a,b; Paz-Ferreiro et al. 2014; Peng et al. 2008; Ping et al. 2007; Prado et al. 2000; Raghothama 1999). Nano-black carbon addition also caused significant changes in microbial community composition and enzyme activities in both bulk soil and the rhizosphere (Lehmann et al. 2011; Lei et al. 2011; Congming and Vonshak 2002). For instance, the Nano-black carbon amendment was generally characterized by an increase in the relative abundance of members of the Actinobacteria and Bacteriodetes phyla. While little is understood regarding the mechanisms by which Nano-black carbon affects the microbial abundance and community structure, it is well-known that soil microorganisms can have a tremendous impact on plant productivity (Imran et al. 2020a,b,c,d,e,f). Therefore, Nano-black carbon-induced changes in soil microorganisms may certainly play a role in 'The Nano-black carbon Effect' (Lehmann et al. 2011; Lei et al. 2011; Congming and Vonshak 2002).

9.5 Nutrients Stress Mitigation with Nano-Black Carbon

In the agriculture sector, very little attention has been given to Nano-black carbon utilization (Imran 2021), because crop response and soil health benefits have yet to be quantified, and because the mechanisms by which it improves soil fertility are poorly understood (Bohnert and Jensen 1996; Bos et al. 2005; Bruinsma 1963; Brussaard et al. 2007; Capell and Doerffling 1993; Chimenti et al. 2006). The positive effects of Nano-black carbon on crop productivity under conditions of extensive agriculture are frequently attributed to direct effects of Nano-black carbon (Imran et al. 2021a,b,c,d,e,f,g; Ilyas et al. 2020a,b) supplied nutrients and several other indirect effects, including increased water and nutrient retention (Amanullah et al. 2021), improvements in soil pH, increased soil cation exchange capacity, effects on P and S transformations and turnover, neutralization of phytotoxic compounds in the soil, improved soil physical properties (Imran 2021), promotion of mycorrhizal fungi, and alteration of soil microbial populations and functions (Izhar et al. 2019; Imran et al. 2015a,b,c,d,e; Jackson 1962; Jagatheeswari and Ranganathan 2012; Joseph and Jini 2011; Kamara et al. 2014). Yet, the Nano-black carbon effect is also evident under conditions of intensive production where many of these parameters are not limited (Gasco et al. 2012; Goncalves-Alvim and Fernandez 2001). Nano-black carbon addition to soil alters microbial populations in the rhizosphere, albeit via mechanisms not yet understood, and may cause a shift towards beneficial microorganism populations that promote plant growth and resistance to biotic stresses. In addition to some scant evidence for Nano-black carbon-induced plant protection against soil-borne diseases, the induction of systemic resistance towards several foliar pathogens in three crop systems has been demonstrated (Khan et al. 2010; Imran et al. 2015a,b,c,d,e, 2019; Imran and Amanullah 2018; Lashari et al. 2013; Lehmann and Joseph 2009). There are indications that Nano-black carbon induces responses along with both systemic acquired resistance (SAR) and induced systemic resistance (ISR) pathways, resulting in a broad spectrum controlling capacity in the canopy (Lei et al. 2011; Congming and Vonshak 2002). This chapter examines the effects of Nano-black carbon soil amendment on the different soil–plant–microbe interactions that may have a role in plant health. Improvement of plant responses to disease can be one of the benefits gained from applying Nano-black carbon to soil (Paz-Ferreiro et al. 2014; Peng et al. 2008; Ping et al. 2007; Prado et al. 2000; Raghothama 1999; Rajalakshmi et al. 2015; Rashid et al. 2003; Imran 2018a,b,c; Iqbal et al. 2017; Raza 2012; Rengasamy 2002; Roelofs et al. 2008).

9.6 Plant's Development of Nano-Black Carbon Amendments

Under field and greenhouse conditions, Nano-black carbon has a positive effect on plant growth and development. Early studies reported that charcoal added to soil increased the yield of moong, soybean, and pea (Paz-Ferreiro et al. 2014; Peng et al. 2008; Ping et al. 2007; Prado et al. 2000; Raghothama 1999; Rajalakshmi et al. 2015; Rashid et al. 2003; Imran 2018a,b,c; Iqbal et al. 2017; Raza 2012; Rengasamy 2002; Roelofs et al. 2008) and soybean. Shoot and root biomass of birch and pine were greater in charcoal amended soil (Rashid et al. 2003; Imran 2018a,b,c; Iqbal et al. 2017). Similarly, five years following the soil application of charcoal, biomass production of sugi trees (*Cryptomeria japonica*) was substantially increased (Sidari et al. 2008; Karim and Imran 2019). A single application of 20 t ha^{-1} Nano-black carbon to a Colombian savanna soil resulted in an increase in maize yield by 28–140% as compared with the unamended control in the 2nd to 4th years after application (Peng et al. 2008; Ping et al. 2007; Prado et al. 2000). With the addition of Nano-black carbon (at 90 g kg^{-1}) to a tropical, low-fertility Ferralsol, the proportion of N fixed by bean plants (*Phaseolus vulgaris*) increased from 50% (without Nano-black carbon) to 72%, and biomass production and bean yield were significantly improved (Van Hoorn et al. 2001; Wang et al. 2007; Wang and Xu 2013; Wang and Jin 2007; Wolfe 2007; Xu et al. 2008; Yeo et al. 1990; Younis et al. 2015). On the same type of soil, total N recovery in soil, crop residues, and grains were significantly higher on compost (16.5%), charcoal (18.1%), and charcoal plus compost treatments (17.4%) in comparison to mineral-fertilized plots (10.9%). Nano-black carbon soil application resulted in higher upland rice (*Oryza sativa*) grain yields at sites in northern Laos with low P availability and improved the response to N and NP chemical fertilizer treatments (Munns et al. 2006; Munns 2002; Munns and Tester 2008; Nazarli et al. 2011; Neto et al. 2004). Large vol-ume applications of Nano-black carbon (30 and 60 t ha^{-1}) in the Mediterranean basin increased durum wheat (*Triticum durum*) biomass and yield by up to 30%, an effect which was sustained for two consecutive seasons (Nazarli et al. 2011; Neto et al. 2004). Overall, these results demonstrate the potential of Nano-black carbon application to improve plant productivity. The means by which Nano-black carbon improves crop response can be attributed to direct effects via bio-char-supplied nutrients (Munns et al. 2006; Munns 2002; Munns and Tester 2008), and to several other indirect effects, including increased nutrient retention (Rajalakshmi et al. 2015; Rashid et al. 2003; Imran 2018a,b,c; Iqbal et al. 2017; Raza 2012; Rengasamy 2002; Roelofs et al. 2008); improvements in soil pH; increased soil cation exchange capacity (Rashid et al. 2003; Imran 2018a,b,c; Iqbal et al. 2017; Raza 2012; Rengasamy 2002; Roelofs et al. 2008); effects on P and S transformations and turnover (Iqbal et al. 2017; Raza 2012; Rengasamy 2002; Roelofs et al. 2008); neutralization of phytotoxic compounds in the soil (Rajalakshmi et al. 2015; Raza 2012; Rengasamy 2002; Roelofs et al. 2008); improved soil physical properties including water retention (Rajalakshmi et al. 2015; Rashid et al. 2003; Imran 2018a,b,c; Iqbal et al. 2017; Raza 2012; Rengasamy 2002; Roelofs et al. 2008); promotion of mycorrhizal fungi (Rajalakshmi et al. 2015; Rashid et al. 2003; Imran 2018a,b,c; Iqbal et al. 2017; Raza 2012; Rengasamy 2002; Roelofs et al. 2008); and alteration of soil microbial populations and functions (Rubio et al. 2001; Savvides 2015; Shahida et al. 2016; Shamsi and Kobraee 2013; Shamsuddin et al. 2015). Many of these effects are interrelated and may act synergistically to improve crop performance (Rubio et al. 2001; Savvides 2015; Shahida et al. 2016; Shamsi and Kobraee 2013; Shamsuddin et al. 2015). Specific mechanisms underlying the contribution of Nano-black carbon to plant response are poorly understood. Regional conditions including climate, soil chemistry, and soil condition all influence Nano-black carbon agronomic benefits. Besides, dissimilar biomass feedstocks and pyrolysis conditions create Nano-black carbon with different physical and chemical properties (Savvides 2015; Shahida et al. 2016; Shamsi and Kobraee 2013; Shamsuddin et al. 2015), resulting in different impacts on plant response (Shahida et al. 2016; Shamsi and Kobraee 2013; Shamsuddin et al. 2015). Given that the Nano-black carbon–soil–plant–water–environment is highly complex, it is difficult to isolate those factors which play an instrumental role in the 'Nano-black carbon Effect'. To reduce the number of potential factors involved, Savvides (2015) tested whether Nano-black carbon addition could impact plant growth when nutritional and soil physical aspects of Nano-black carbon amendment were eliminated. This was achieved by examining the impact of a nutrient-poor, wood-derived Nano-black carbon on tomato.

9.7 Nano-Black Carbon Application and Phosphate Starvation in Plants

Many researchers reported that with no P fertilization, plant growth and development significantly reduced and affect the plant's dry biomass and productivity which ultimately led to food insurgency (Imran et al. 2020a,b,c,d,e,f; Imran et al. 2019; Ali et al. 2019). It has been reported that P is an important plant metabolism component and makes 0.2% of the plant's dry weight (Karim and Imran 2019; Imran et al. 2018; Imran 2018a,b,c). Therefore, P fertilization is needed as it is an important major plant nutrient after N. In limited P application or no P fertilization, plants cannot grow as it is a key component of nucleic acids, phospholipids, and ATP, and without these components, the plant cannot survive (Imran 2018a,b,c). Phosphorus is involved in controlling and regulating several enzymatic functions and reactions in the metabolic pathway and thus regulating plant growth and development and ultimately plant productivity and yield (Imran and Amanullah 2018; Imran. 2018e; Imran and Khan 2017). Other researchers reported that P is very rarely available in soil or only available in the soil out of the rhizosphere which cannot support the plant for high growth rate and development (Amanullah et al. 2020a,b,c; Imran et al. 2020a,b,c,d,e,f; Imran et al. 2017). Phosphorus application is necessary for agriculture systems to ensure crop productivity and food security (Imran et al. 2017; Imran et al. 2015a,b,c,d,e). Limited P supply in bulk soil affect P uptake by plants as it is only moving by the process of diffusion and do not uptake or very low amount is uptake by the crop plants which is an alarming threat for low productivity (Imran et al. 2015a,b,c,d,e; Imran et al. 2019). Phosphorus (P) is an essential macronutrient required for plant growth and development, but most of the world's soil is limited in this important plant nutrient (Palta and Farag 2006; Pant et al. 2015a,b; Paz-Ferreiro et al. 2014; Peng et al. 2008; Ping et al. 2007; Prado et al. 2000). Plants can utilize P mainly in the form of soluble inorganic phosphate (Pi) but are subjected to abiotic stress of P-limitation when there are not sufficient soluble PO_4 available in the soil. Phosphorus forms insoluble complexes with Ca and Mg in alkaline soils and Al and Fe in acidic soils that makes it unavailable for plant roots (Imran 2018a,b,c; Iqbal et al. 2017; Raza 2012; Rengasamy 2002; Roelofs et al. 2008; Rolf et al. 2004; Rubio et al. 2001; Savvides 2015). When there is limited bioavailable P in the soil, plants show extensive abiotic stress phenotype such as short primary roots and more lateral roots and root hairs to make more surface available for Pi absorption, exudation of organic acids, and phosphatase to release Pi from complex P containing molecules and make it available for growing plants organs (Imran 2018a,b,c; Iqbal et al. 2017; Raza 2012; Rengasamy 2002; Roelofs et al. 2008; Rolf et al. 2004; Rubio et al. 2001; Savvides 2015). It has been shown that Nano-black carbon application is a master regulator of P-starvation response in plants (Gonzalez-Villagra et al. 2018). Nano-black carbon amended soil has been shown to regulate extensive secondary metabolites during phosphorus limitation stress (Bos et al. 2005; Bruinsma 1963; Brussaard et al. 2007; Capell and Doerffling 1993).

9.8 Conclusion

Nano-black carbon is an activated carbon soil conditioner that can alleviate the negative impacts of salinity, drought, heat and cold, water-logging, and heavy metal stresses. Drought and salt stress negatively affect soil fertility and plant growth. The application of Nano-black carbon ameliorates the negative effects of drought and salt stress on plants. The Nano-black carbon application increased the plant growth, biomass, and yield under either drought or salt stress and also increased photosynthesis, nutrient uptake, and modified gas exchange characteristics in drought and salt-stressed plants. Under drought stress, Nano-black carbon increased the water holding capacity of the soil and improved the physical and biological properties of soils. Under salt stress, Nano-black carbon decreased Na^+ uptake, while increased K^+ uptake by plants. Nano-black carbon-mediated increase in salt tolerance of plants is primarily associated with improvement in soil properties, thus increasing plant water status, reduction of Na^+ uptake, increasing uptake of minerals, and regulation of stomatal conductance and phytohormones. This chapter highlights both the potential of Nano-black carbon in alleviating abiotic stresses in plants and the prospect of the role of Nano-black carbon under abiotic stresses in plants.

9.9 Summary

Organic biomass thermal degradation in limited oxygen forms a solid coproduct known as Nano-black carbon (biochar). There is a growing interest worldwide in Nano-black carbon application in agriculture and farm produce. The main reasons for Nano-black carbon application in agricultural land and cultural practices might be due to (i) pyrolysis can be a source of renewable biofuels. (ii) Many biomass waste materials can be treated by pyrolysis and thus converted into fuel sources. (iii) The long-term sequence of carbon dioxide that can be introduced into the atmosphere by adding Nano-black carbon to the soil. Nano-black carbon amendment of soil aids in improving soil fertility and crop productivity. To meet global food security, crop production faces several real challenges, including soil fertility and organic matter depletion, climate change, unmanaged rainfall patterns, and pathogens. Abiotic pressures such as heat and cold, drought, salinity, water storage, and heavy metal pressure are emerging as important factors in reducing crop production globally. Nano-black carbon is a high carbonate modification created by the absence of oxygen or limited supply of a wide variety of feedstocks. Nano-black carbon typically has a high level of the area, capacity for cation exchange, reactive functional groups, and highly reactive carbon. Due to these characteristics, Nano-black carbon has been widely associated with multiple abiotic stress reductions to enhance crop growth, development, production, soil fertility, and soil quality.

References

Akhtar, S.S., Andersen, M.N., and Liu, F. (2015a). Nano-black carbon mitigates salinity stress in potato. *Journal of Agronomy and Crop Science* 201: 321–400.

Akhtar, S.S., Andersen, M.N., and Liu, F. (2015b). Residual effects of nano-black carbon on improving growth, physiology and yield of wheat under salt stress. *Agricultural Water Management* 158: 61–68.

Izhar, A., Khan, A.A., Imran et al. (2019). Humic acids and nitrogen levels optimizing productivity of green gram (*Vigna radiate* L.). *Russian Agriculture Sciences* 45 (1): 43–47.

Amanullah, Inamullah, Alkahtani, J. et al. (2020a). Phosphorus and zinc fertilization improve productivity and profitability under rice-wheat cropping system. *Agronomy 2020* (10): 1085. https://doi.org/10.3390/agronomy10081085.

Amanullah, Inamullah, Alwahibi, M. et al. (2020b). Phosphorus and zinc fertilization improve zinc biofortification in grains and straw of coarse vs. fine rice genotypes. *Agronomy 2020* 10: 1155. https://doi.org/10.3390/agronomy10081155.

Amanullah, Ullah, H., Elshikh, M.S. et al. (2020c). Nitrogen contents in soil, grains, and straw of hybrid rice differ when applied with different organic nitrogen sources. *Agriculture 2020* (10): 386. https://doi.org/10.3390/agriculture10090386.

Amanullah, Khalid, S., Khalil, F. et al. (2021). Growth and dry matter partitioning response in cereal-legume intercropping under full and limited irrigation regimes. *Scientific Reports (2021)*. https://doi.org/10.1038/s41598-021-92022-4.

Amonette, J.E. and Joseph, S. (2009). Characteristics of nano-black carbon: microchemical properties. In: *Nano-black Carbon for Environmental Management: Science and Technology* (ed. J. Lehmann and S. Joseph), 33–52. Earthscan, London: Sterling.

Shazma, A., Israeel, Iqbal, B. et al. (2016). Nitrogen and phosphorus fertilization of improved varieties for enhancing phenological traits of wheat. *Pure and Applied Biology* 5 (3): 511–519. https://doi.org/10.19045/bspab.2016.50065.

Armor, N.B., Hamed, K.B., Debez, A., and Grignonand Crabbedly, C. (2005). Physiological and antioxidant responses of perennial halophyte *Crithmum maritimum* to salinity. *Plant Science* 4: 889–899.

Arshad, M., Shaharoona, B., and Mahmood, T. (2008). Inoculation with *Pseudomonas* spp. containing ACC-deaminase partially eliminates the effects of drought stress on growth, yield, and ripening of pea (*Pisum sativum* L.). *Pedosphere* 18: 611–620.

Ayaz, F.A., Kadioglu, A., and Turgut, T.R. (2000). Water stress effects on the content of low molecular weight carbohydrates and phenolic acids in *Cienanthe setosa*. *Candian Journal of Plant Sciences* 80: 373–378.

Badridze, G., Weidner, A., Asch, F., and Borner, A. (2009). Variation in salt tolerance within Georgian wheat germplasm collection. *Genetic Resources and Crop Evolution* 56: 1125–1130.

Shamsuddin, B., Khan, A.Z., Inamullah et al. (2015). Influence of farm yard manure and phosphorus application on yield and yield components of wheat. *Pure and Applied Biology* 4 (4): 458–464.

Bates, L.S., Waldren, R.P., and Teare, I.D. (1973). Rapid determination of free proline for water stress studies. *Plant and Soil* 39: 205–207.

Bohnert, H.J. and Jensen, R.G. (1996). Strategies for engineering water stress tolerance in plants. *Trends in Biotechnology* 14: 89–97.

Bos, C., Juillet, B., Fouillet, H. et al. (2005). Postprandial metabolic utilization of wheat protein in humans. *The American Journal of Clinical Nutrition* 81: 87–94.

Bruinsma, J. (1963). The quantitative analysis of chlorophyll a and b in plant extract. *Photochemistry and Photobiology* 2: 241–249.

Brussaard, L., de Ruiter, P.C., and Brown, G.G. (2007). *Soil biodiversity for agricultural sustainability. Agriculture, Ecosystems and Environment* 121 (3): 233–244. https://doi.org/10.1016/j.agee.2006.12.013.

Capell, B. and Doerffling, K. (1993). Genotype specific difference in chilling tolerance of maize in relation to chilling induce changes in water status and abscisic acid accumulation. *Physiologia Plantarum* 88: 638–646.

Chimenti, C.A., Marcantonio, M., and Hall, J.A. (2006). Divergent selection for osmotic adjustment results in improved drought tolerance in maize (*Zea mays* L.) in both early growth and flowering phases. *Field Crops Research* 95: 305–315.

Conde, A. (2011). Membrane transport, sensing and signaling in plant adaptation to environmental stress. *Plant & Cell Physiology* 52 (9): 1583–1602. https://doi.org/10.1093/pcp/pcr107. PMID 21828102 – *via Google Scholar*.

Congming, L. and Vonshak, A. (2002). Effects of salinity stress on photosystem II function in cyanobacterial Spirulina platensis cells. *Physiologia Plantarum* 114: 405–413.

Cramer, G.R., Urano, K., Delrot, S. et al. (2011). Effects of abiotic stress on plants: a systems biology perspective. *BMC Plant Biology* 11 (163). https://doi.org/10.1186/1471-2229-11-163. ISSN 1471-2229. PMC 3252258. PMID 22094046.

Duan, D.Y., Li, W.Q., Liu, X.J. et al. (2007). Seed germination and seedling growth of *Suaeda salsa* under salt stress. *Annales Botanici Fennici* 44: 161–169.

Farahbakhsh, H. (2012). Germination and seedling growth in un-primed and primed seeds of fennel as affected by reduced water potential induced by NaCl. *International Research Journal of Applied and Basic Sciences* 3: 737–744.

Flowers, T.J. (2004). Improving crop salt tolerance. *Journal of Experimental Botany* 55: 307–319.

Gasco, G., Paz-Ferreiro, J., and Mendez, A. (2012). Thermal analysis of soil amended with sewage sludge and Nano-black carbon from sewage sludge pyrolysis. *Journal of Thermal Analysis and Calorimetry* 108: 769–775.

Goncalves-Alvim, S.J. and Fernandez, G.W. (2001). *Biodiversity of galling insects: historical, community and habitat effects in four neotropical savannas. Biodiversity and Conservation* 10: 79–98. https://doi.org/10.1023/a:1016602213305.

Gonzalez-Villagra, J., Rodrigues-Salvador, A., Nunes-Nesi, A. et al. (2018). Age-related mechanism and its relationship with secondary metabolism and abscisic acid in *Aristotelia chilensis* plants subjected to drought stress. *Plant Physiology and Biochemistry* 124: 136–145. https://doi.org/10.1016/j.plaphy.2018.01.010. ISSN 0981-9428. PMID 29360623.

Heans, D.L. (1984). Determination of total organic carbon in soils by an improved chromic acid digestion and spectrophotometric procedure. *Communications in Soil Science and Plant Analysis* 15: 1191–1213.

Ilyas, M., Ayub, G., Imran et al. (2020a). Effect of calcium and boron on production and quality of autumn potato under chilling temperature. *Communications in Soil Science and Plant Analysis*. https://doi.org/10.1080/00103624.2020.1854286.

Ilyas, M., Ayub, G., Imran et al. (2020b). Calcium and boron effect on production and quality of autumn potato under chilling temperature. *Communications in Soil Science and Plant Analysis*. https://doi.org/10.1080/00103624.2020.1854286.

Imran (2018a). Ecological environmental variability influence growth and yield potential of rice under northern climatic scenario. *Russian Agricultural Sciences* 44 (1): 18–24.

Imran (2018b). Phosphorous fertilization influenced weeds attributes and phenological characteristics of mungbean cultivars (*Vigna radiata* L.). *Russian Agricultural Sciences* 44 (3): 229–234.

Imran (2018c). Organic matter amendments improve soil health, productivity and profitability of maize and soybean. *Annals of Review Research* 1 (3): 555564.

Imran (2021). The bioavailability of phosphorus in composite vs. hybrid maize differ with phosphorus and boron fertilization. *Phosphorus, Sulfur, and Silicon and the Related Elements*. https://doi.org/10.1080/10426507.2021.1920588.

Imran and Amanullah (2018). Review. Global impact of climate change on water, soil resources and threat towards food security: evidence from Pakistan. *Advances in Plants & Agriculture Research* 8 (5): 350–355. https://doi.org/10.15406/apar.2018.08.00349.

Imran and Khan, A.A. (2017). Canola yield and quality enhanced with sulphur fertilization. *Russian Agricultural Sciences* 43: 113. https://doi.org/10.3103/S1068367417020100.

Imran, Khan, A.A., Inamullah, and Luqman (2015a). Weeding stages and their effect on yield and yield components of rice in upper Swat, Pakistan. *Pakistan Journal of Weed Science Research* 21 (4): 555–563.

Imran, Naveed, S., Khan, A.A., and Khattak, I. (2015b). Impact of phosphorus levels and seed rates on growth and yield of late sown maize on high elevation in Swat, Pakistan. *Pakistan Journal of Agricultural Research* 28 (4): 406–413.

Imran, A., Khan, A., Ullah, I. et al. (2015c). Yield and yield attributes of rapeseed cultivars as influence by sulfur level under swat valley conditions. *Pure and Applied Biology* 4 (3): 296–301.

Imran, I., Hussain, I., Khattak, A. et al. (2015d). Roots nodulation, yield and yield contributing parameters of mungbean cultivars as influenced by different phosphorous level in Swat-Pakistan. *Pure and Applied Biology* 4 (4): 557–567.

Imran, Khattak, I., Hussain, I. et al. (2015e). Growth and yield of maize hybrids as effected by different sowing dates in Swat Pakistan. *Pure and Applied Biology* 4 (4). https://doi.org/10.19045/bspab.2016.50015.

Imran, Khan, A.A., Khan, I.U., and Naveed, S. (2016a). Weeds density and late sown maize productivity influenced by compost application and seed rates under temperate environment. *Pakistan Journal of Weed Science Research* 22 (1): 169–181.

Imran, Khan, A.A., Inam, I., and Ahmad, F. (2016b). Yield and yield attributes of Mungbean (*Vigna radiata* L.) cultivars as affected by phosphorous levels under different tillage systems. *Cogent Food & Agriculture* 2 (2016): 1151982.

Imran, Bari, A., Ali, R. et al. (2017). Traditional rice farming accelerate CH_4 & N_2O emissions functioning as a stronger contributors of climate change. *International Journal of Environmental Sciences & Natural Resources* 2017. https://doi.org/10.19080/ARTOAJ.2017.09.555765.

Imran, Amanullah, Bari, A., and Ali, R. (2018). Peach sources, phosphorous and beneficial microbes enhance productivity of soybean. *Soybean Research* 16 (2): 39–48.

Imran, Amanullah, Khan, A.A. et al. (2019). Production statistics and modern technology of maize cultivation in Khyber Pakhtunkhwa Pakistan. *Plant Science Archives* 2 (2): 1–12.

Imran, Amanullah, Arif, M. et al. (2020a). Soil application of *Trichoderma* and Peach (*Prunus persica* L.) residues possesses biocontrol potential for weeds and enhances growth and profitability of soybean (Glycine max). *Sarhad Journal of Agriculture* 36 (1): 10–20. https://doi.org/10.17582/journal.sja/2020/36.1.10.20.

Imran, Amanullah, and Al-Tawaha, A.R. (2020b). The productivity of subsequent wheat enhanced with residual carbon sources and phosphorus under improved irrigation system. *Communications in Soil Science and Plant Analysis*. https://doi.org/10.1080/00103624.2020.1763387.

Imran, Amanullah, and Al-Tawaha, A.R. (2020c). The productivity of subsequent wheat enhanced with residual carbon sources and phosphorus under improved irrigation system. *Communications in Soil Science and Plant Analysis*. https://doi.org/10.1080/00103624.2020.1763387.

Imran, Amanullah, Khan, A.A. et al. (2020d). Adequate fertilization, application method and sowing techniques improve maize yield and related traits. *Communications in Soil Science and Plant Analysis* 52: 2318–2330.

Imran, Amanullah, Arif, M. et al. (2020e). Integration of peach (*Prunus persica* L.) residues, beneficial microbes and phosphorus enhance phenology, growth and yield of soybean. *Russian Agricultural Sciences* 46 (3): 223–230. ISSN 1068-3674.

Imran, Amanullah, Arif, M. et al. (2020f). Soil application of *Trichoderma* and Peach (*Prunus persica* L.) residues possesses biocontrol potential for weeds and enhances growth and profitability of soybean (Glycine max). *Sarhad Journal of Agriculture* 36 (1): 10–20. https://doi.org/10.17582/journal.sja/2020/36.1.10.20.

Imran, Amanullah, and Al Tawaha, A.R.M. (2021a). Management of nano-black carbon, phosphorous and bio fertilizer improve soil organic carbon and ensilage biomass of soybean and maize. *Communications in Soil Science and Plant Analysis*. https://doi.org/10.1080/00103624.2021.1966439.

Imran, Amanullah, and Altawaha, A.R. (2021b). Carbon assimilation and dry matter partitioning in soybean ameliorates with the integration of nano-black carbon, along with beneficial microbes and phosphorus fertilization. *Journal of Plant Nutrition*. https://doi.org/10.1080/01904167.2022.2035753.

Imran, Amanullah, Khan, A.A. et al. (2021c). Adequate fertilization, application method and sowing techniques improve maize yield and related traits. *Communications in Soil Science and Plant Analysis*. https://doi.org/10.1080/00103624.2021.1925688.

Imran, Amanullah, and Al-Tawaha, A.R. (2021d). Carbon sources application increase wheat yield and soil fertility. *Communications in Soil Science and Plant Analysis*. https://doi.org/10.1080/00103624.2020.1865397.

Imran, Amanullah, Al Tawaha, A.R. et al. (2021e). *Use of Biochar in Agriculture: An Inspiring Way in Existing Scenario. Organic Farming for Sustainable Development*. CRC Press/Taylor & Francis.

Imran, Amanullah, Al Tawaha, A.R. et al. (2021f). *Organic Amendments: Right Way to Soil Sustainability. Organic Farming for Sustainable Development*. CRC Press/Taylor & Francis.

Imran, Amanullah, and Al-Tawaha, A.R. (2021g). Carbon sources application increase wheat yield and soil fertility. *Communications in Soil Science and Plant Analysis*. https://doi.org/10.1080/00103624.2020.1865397.

Iqbal, B., Jan, M.T., Inamullah et al. (2015a). Integrated management of compost type and fertilizer-N in maize. *Pure and Applied Biology* 4 (4): 453–457.

Iqbal, B., Jan, M.T., Imran et al. (2015b). Growth and phenology of maize as affected by integrated management of compost and fertilizers nitrogen. *Pure and Applied Biology* 4 (4).

Iqbal, B., Ahmad, B., Ullah, I. et al. (2016a). Effect of phosphorus, sulphur and different irrigation levels on phenological traits of triticale. *Pure and Applied Biology* 5 (2): 303–310. https://doi.org/10.19045/bspab.2016.50039.

Iqbal, B., Jan, M.T., Muhammad, Z. et al. (2016b). Phenological traits of maize influenced by integrated management of compost and fertilizer nitrogen. *Pure and Applied Biology* 5 (1): 58–63.

Iqbal, A., Amanullah, A.A., Iqbal, M. et al. (2017). Integrated use of phosphorus and organic matter improve fodder yield of moth bean (*Vigna aconitifolia* (Jacq.)) under irrigated and dryland conditions of Pakistan. *Journal of AgriSearch* 4 (1): 10–15.

Jackson, M. (1962). *Soil Chemical Analysis*, vol. 496. London: Constable and Company Limited.

Jagatheeswari, D. and Ranganathan, P. (2012). Influence of mercuric chloride on seeds germination, seedling growth and biochemical analysis of green gram (*Vigna radiata* (L.) Wilczek. Var. Vamban-3). *International Journal of Pharmaceutical and Biological Archives* 3: 291–295.

Ping, J., Geo, K., and Ron, M. (2007). *Understanding abiotic stress tolerance mechanisms: recent studies on stress response in rice. Journal of Integrative Plant Biology* 49 (6): 742–750. https://doi.org/10.1111/j.1744-7909.2007.00495.x.

Joseph, B. and Jini, D. (2011). Development of salt stress-tolerant plants by gene manipulation of antioxidant enzymes. *Asian Journal of Agricultural Research* 5: 17–27.

Kamara, A., Kamara, A., Mansaray, M.M., and Sawyerr, P.A. (2014). Effects of nano-black carbon derived from maize stover and rice straw on the germination of their seeds. *American Journal of Agriculture and Forestry* 2: 246–249.

Kammann, C.I., Linsel, S., Gobling, J.W., and Koyro, H. (2011). Influence of nano-black carbon ondrought tolerance of *Chenopodium quinoa* willd and on soil-plant relation. *Plant and Soil* 345: 195–210.

Karim, S.A. and Imran (2019). Reduction in substrate moisture content reduce final yield of wheat. *Biomedical Journal of Scientific & Technical Research* 1–5. https://doi.org/10.26717/BJSTR.2019.15.002656.

Khan, M.A., Islam, E., Shirazi, M.U. et al. (2010). Physiological responses of various wheat genotypes to salinity. *Pakistan Journal of Botany* 42: 3497–3505.

Khan, A.A., Inamullah, I.F.A.S., and Khan, L.Z.M.N.M.N. (2015). Phenological traits of rice as influenced by seedling age and number of seedling per hill under temperate region.

Journal of Biology, Agriculture and Healthcare. ISSN 2224-3208 (Paper) ISSN 2225-093X (Online).

Khan, A.Z., Imran, A.M., Khalil, A. et al. (2016). Impact of fertilizer priming on seed germination behavior and vigor of maize. *Pure and Applied Biology* 5 (4): 744–751. https://doi.org/10.19045/bspab.2016.50093.

Lashari, M.S., Liu, Y., Li, W. et al. (2013). Effect of amendment of nano-black carbon manure compost in conjunction with pyroligneous solution on soil quality and wheat yield of a salt stressed cropeland from central China Great Plain. *Food Crop Research* 144: 113–118.

Lehmann, J. and Joseph, S. (2009). Nano-black carbon for environmental management: An introduction. In: *Nano-black Carbon for Environmental Management: Science and Technology* (ed. J. Lehmann and S. Joseph), 1–12. London: Earthscan.

Lehmann, J., Rillig, M.C., Thies, J. et al. (2011). Nano-black carbon effects on soil biota. *Soil Biology and Biochemistry* 43: 1812–1836.

Lei, G., Shen, M., Li, Z.-G. et al. (2011). *EIN2 regulates salt stress response and interacts with a MA3 domain-containing protein ECIP1 in Arabidopsis. Plant, Cell & Environment* 34 (10): 1678–1692. https://doi.org/10.1111/j.13653040.2011.02363.x. ISSN 1365-3040.

Maestre, F.T., Cortina, J., and Bautista, S. (2007). *Mechanisms underlying the interaction between Pinus halepensis and the native late-successional shrub Pistacia lentiscus in a semi-arid plantation. Ecography* 27 (6): 776–786. https://doi.org/10.1111/j.0906-7590.2004.03990.x.

Martinez-Beltran, J. and Manzur, C.L. (2005). Overview of salinity problems in the world and FAO strategies to address the problem. *Proceedings of the International Salinity Forum*, Riverside, California, April 2005, 311–313.

Mittler, R. (2006). *Abiotic stress, the field environment and stress combination. Trends in Plant Science* 11 (1): 15–19. https://doi.org/10.1016/j.tplants.2005.11.002. PMID 16359910.

Munns, R. (2002). Comparative physiology of salt and water stress. *Plant, Cell and Environment* 25: 239–250.

Munns, R. and Tester, M. (2008). Mechanisms of salinity tolerance. *Annual Review of Plant Biology* 59: 651–681.

Munns, R.R., James, A., and Lauchli, A. (2006). Approaches to increasing the salt tolerance of wheat and other cereals. *Journal of Experimental Botany* 57: 1025–1043.

Shahida, N., Ibrar, M., Khattak, I. et al. (2016). Anthelmintic, antilice, insecticidal, cytotoxic and phytotoxic potential of ethanolic extracts of two wild medicinal plants *Iphiona Grantioides* and *Plucheaarguta*. *Woulfenia Journal* 23 (11): 13–25.

Nazarli, H., Faraji, F., and Zardashti, M.R. (2011). Effect of drought stress and polymer on osmotic adjustment and photosynthetic pigments of sunflower. *Carcetriari Agronomica in Maldova* 1: 45.

Neto, A., De, A.D., Prisco, J.T. et al. (2004). Effects of salt stress on plant growth, stomatal response and solute accumulation of different maize genotypes. *Brazilian Journal of Plant Physiology* 16 (1): 31–38. https://doi.org/10.1590/S1677-04202004000100005. ISSN 1677-0420.

Palm, B. and Volkenburgh, V. (2012). Serpentine tolerance in *Mimuslus guttatus* does not rely on exclusion of magnesium. *Functional Plant Biology*.

Palta, J.P. and Farag, K. (2006). Methohasds for enhancing plant health, protecting plants from biotic and abiotic stress related injuries and enhancing the recovery of plants injured as a result of such stresses. US Patent 7,101,828.

Pant, B.D., Burgos, A., Pant, P. et al. (2015a). The transcription factor PHR1 regulates lipid remodeling and triacylglycerol accumulation in *Arabidopsis thaliana* during phosphorus starvation. *Journal of Experimental Botany* 66 (7): 1907–1918. https://doi.org/10.1093/jxb/eru535. ISSN 0022-0957. PMC 4378627. PMID 25680792.

Pant, B.-D., Pant, P., Erban, A. et al. (2015b). *Identification of primary and secondary metabolites with phosphorus status-dependent abundance in Arabidopsis, and of the transcription factor PHR1 as a major regulator of metabolic changes during phosphorus limitation. Plant, Cell & Environment* 38 (1): 172–187. https://doi.org/10.1111/pce.12378. ISSN 1365-3040. PMID 24894834.

Paz-Ferreiro, J., Fu, S., Mendez, A., and Gasco, G. (2014). Interactive effects of nano-black carbon and the earhworm *Pontoscolex corethrurus* on plant productivity and soil enzymes activities. *Journal of Soils and Sediments* 14: 483–494.

Peng, Y., Gao, Z., Gao, Y. et al. (2008). Ecophysiological characteristics of alfalfa seedlings in response to various mixed salt-alkaline stresses. *Journal of Integrative Plant Biology* 50: 29–39.

Prado, F.E., Boero, C., Gallarodo, M., and Gonzalez, A.J. (2000). Effect of NaCl on germination, growth and soluble sugar content in *Chenopodium quinoa* wil seeds. *Botanical Bulletin – Academia Sinica Taipei* 14: 27–34.

Raghothama, K.G. (1999). *Phosphate Acquisition. Annual Review of Plant Physiology and Plant Molecular Biology* 50 (1): 665–693. https://doi.org/10.1146/annurev.arplant.50.1.665. PMID 15012223.

Rajalakshmi, A.G., Kumar, S.K., Bharathi, C.D. et al. (2015). Effect of nano-black carbon in seed germination: *in-vitro* study. *International Journal of Biosciences and Nanosciences* 2: 132–136.

Rashid, A., Saleem, Q., Nazir, A., and Kazım, H.S. (2003). Yield potential and stability of nine wheat varieties under water stress conditions. *International Journal of Agriculture and Biology* 5: 7–9.

Raza, A. (2012). Grain and field annual 2012 gain report. Global agricultural information network. *USDA Foreign Agrilcultural Services* 1–3.

Rengasamy, P. (2002). Transient salinity and subsoil constraints to dryland farming in Australian sodic soils: an overview. *Australian Journal of Experimental Agriculture* 42: 351–361.

Roelofs, D., Jiwan, P., and Kwangy, I. (2008). *Functional ecological genomics to demonstrate general and specific responses to abiotic stress. Functional Ecology* 22: 8–18. https://doi.org/10.1111/j.1365-2435.2007.01312.x.

Rolf, D., Vinebrooke, K., and Mittler, J. (2004). *Impacts of multiple stressors on biodiversity and ecosystem functioning: the role of species co-tolerance. Oikos* 104 (3): 451–457. https://doi.org/10.1111/j.0030-1299.2004.13255.x.

Rubio, V., Linhares, F., Solano, R. et al. (2001). A conserved MYB transcription factor involved in phosphate starvation signaling both in vascular plants and in unicellular algae. *Genes & Development* 15 (16): 2122–2133. https://doi.org/10.1101/gad.204401. ISSN 0890-9369. PMC 312755. PMID 11511543.

Ulfat, S., Ibrar, M., Lalbadshah, S.N. et al. (2016). Ethnobotanical study of subtropical hills of Darazinda, Takht-e-Suleman range F.R D.I. Khan, Pakistan. *Pure and Applied Biology* 5 (1): 149–164. https://doi.org/10.19045/bspab.2016.50020.

Savvides, A. (2015). Chemical priming of plants against multiple abiotic stresses: mission possible? *Trends in Plant Science* 21 (4): 329–340. https://doi.org/10.1016/j.tplants.2015.11.003. hdl:10754/596020.

Shamsi, K. and Kobraee, S. (2013). Biochemical and physiological responses of three wheat cultivars (*Triticum aestivum* L.) to salinity stress. *Annals of Biological Research* 4: 180–185.

Sidari, M., Mallamaci, C., and Muscolo, A. (2008). Drought, salinity and heat differently affect seed germination of Pinus pinea. *Journal of Forest Research* 13: 326–330.

Tank, N. and Saraf, M. (2010). Salinity-resistant plant growth promoting rhizobacteria ameliorates sodium chloride stress on tomato plants. *Journal of Plant Interactions* 5: 51–58.

Tiwari, S., Singh, P., Tiwari, R. et al. (2011). Salt-tolerant rhizobacteria-mediated induced tolerance in wheat (*Triticum aestivum*) and chemical diversity in rhizosphere enhance plant growth. *Biological Fertile Soils* 47: 907–916.

Tombesi, S., Frioni, T., Poni, S., and Palliotti, A. (2018). *Effect of water stress "memory" on plant behavior during subsequent drought stress. Environmental and Experimental Botany* 150: 106–114. https://doi.org/10.1016/j.envexpbot.2018.03.009. ISSN 0098-8472.

Van Hoorn, J.W., Nader, K., Hamdy, A., and Mastrorilli, M. (2001). Effect of salinity on yield and nitrogen uptake of four grain legumes and on biological nitrogen contribution from the soil. *Agriculture Water Management* 51: 87–98.

Wang, H. and Jin, J. (2007). Effects of zinc deficiency and drought on plant growth and metabolism of reactive oxygen species in maize (*Zea mays* L.). *Agricultural Sciences in China* 988–995.

Wang, G. and Xu, Z. (2013). The effects of Nano-black carbon on germination and growth of wheat in different saline-alkali soil. *Asian Agricultural Research* 5: 116–119.

Wang, W., Vinocur, B., and Altman, A. (2007). *Plant responses to drought, salinity and extreme temperatures towards genetic engineering for stress tolerance. Planta* 218 (1): 1–14. https://doi.org/10.1007/s00425-003-1105-5. PMID 14513379.

Wolfe, A. (2007). *Patterns of Biodiversity*. Ohio State University. https://doi.org/10.1016/j.agee.2006.12.013.

Xu, D., Huang, J., Guo, S. et al. (2008). Over-expression of a TFIII-type zinc finger protein gene ZFP252 enhances drought and salt tolerance in rice (*Oryza sativa* L.). *Federation of European Biochemical Societies Letters* 582: 1037–1043.

Yeo, A.R., Yeo, M.E., Flowers, S.A., and Flowers, T.J. (1990). Screening of rice (*Oryza sativa* L.) genotypes for physiological characters contributing to salinity resistance, and their relationship to overall performance. *Theoretical and Applied Genetics* 79: 377–384.

Younis, U., Athar, M., Malik, S.A. et al. (2015). Nano-black carbon impact on physiological and biochemical attributes of spinach *Spinacia oleracea* (L.) in nickel contaminated soil. *Global Journal Environment Science Management* 1: 245–254.

Zakaria, M., Daniel, S., Murphy, V., and Abbott, L.K. (2012). Nano-black carbons influence seed germination and early growth of seedlings. *Plant and Soil* 353: 273–287.

Zargar, S.M., Nagar, P., Deshmukh, R. et al. (2017). Aquaporins as potential drought tolerance inducing proteins: towards instigating stress tolerance. *Journal of Proteomics* 169: 233–238. https://doi.org/10.1016/j.jprot.2017.04.010. ISSN 1874-3919.

Zhan, H.X., Chang, Z.J., Wei, A.L. et al. (2011). Impact of drought to wheat physiological index. *Journal of Shanxi Agriculture Sciences* 39: 1049–1051.

10

Biological Nitrogen Fixation in Nonlegume Plants and Changing Climate

Abdel Rahman Mohammad Said Al-Tawaha[1], Pratibha Vyas[2], Arun Karnwal[3], Noureddine Benkeblia[4], Swapnil Ganesh Sanmukh[5], Eduard Torrents Serra[5], Imran[6,7], Amanullah[6], Shah Khalid[6], Abdel Razzaq Al-Tawaha[8], Abhijit Dey[9], Nujoud Alimad[10], Devarajan Thangadurai[11], Jeyabalan Sangeetha[12], Saher Islam[13], and Mohammad Shatnawi[14]

[1] *Department of Biological Sciences, Al-Hussein Bin Talal University, Ma'an, Jordan*
[2] *Department of Microbiology, College of Basic Sciences and Humanity, Punjab Agricultural University, Ludhiana, Punjab, India*
[3] *Department of Microbiology, School of Bioengineering and Biosciences, Lovely Professional University, Phagwara, Punjab, India*
[4] *Department of Life Sciences, The University of the West Indies, Kingston, Jamaica*
[5] *Institute for Bioengineering of Catalonia (IBEC), The Institute of Science and Technology, Barcelona, Spain*
[6] *Department of Agronomy, The University of Agriculture, Peshawar, Pakistan*
[7] *Ministry of Agriculture, Agriculture Extension Wing, Merged Area, Khyber Pakhtunkhwa, Pakistan*
[8] *Department of Crop Science, Faculty of Agriculture, University Putra Malaysia, Serdang, Selangor, Malaysia*
[9] *Department of Life Sciences, Presidency University, Kolkata, West Bengal, India*
[10] *Department of Plant Protection, Faculty of Agriculture, Damascus University, Damascus, Syria*
[11] *Department of Botany, Karnatak University, Dharwad, Karnataka, India*
[12] *Department of Environmental Science, Central University of Kerala, Kasaragod, Kerala, India*
[13] *Department of Biochemistry and Biotechnology, Institute of Biochemistry and Biotechnology, University of Veterinary and Animal Sciences, Lahore, Pakistan*
[14] *Department of Biotechnology, Faculty of Agricultural Technology, Al Balqa Applied University, Salt, Jordan*

CONTENTS

Climate Change and Agriculture: Perspectives, Sustainability and Resilience, First Edition. Edited by Noureddine Benkeblia.

10.1 Introduction

One major challenge for agriculture in the twenty-first century is the production of agricultural crops in an environmentally sound and sustainable way. An enhanced production of crops is necessary to provide sufficient food for the increasing human population. To get higher productivity, chemical fertilizers and pesticides are extensively used, which impose an extensive burden on the agriculture. Mineral nutrients such as nitrogen and phosphorus are regarded as critical factors limiting plant biomass and productivity in a wide variety of ecosystems (Abebe et al. 2005; Abera et al. 2005; Al-Ajlouni et al. 2009; Al-Juthery et al. 2018; Al-Kiyyam et al. 2008; Al-Tawaha et al. 2005, 2010; Nikus et al. 2004a,b; Tawaha and Turk 2004; Turk and Tawaha 2004). Nitrogen, the most important nutrient, is a limiting factor in plant growth (Tawaha and Turk 2002; Tawaha et al. 2003; Turk and Tawaha 2001, 2002a,b; Turk et al. 2003a,b,c). Nitrogen cycle, a complex biogeochemical cycle containing four microbial processes including nitrogen fixation, nitrification, denitrification, and nitrogen mineralization, has a huge impact on maintaining soil fertility (Jetten 2008). Large amounts of nitrogenous fertilizers are used to increase agriculture productivity to fulfill the demands of ever-increasing human populations. Among the plant nutrients, nitrogenous fertilizers have the lowest input efficiency, thereby contributing substantially to environmental pollution. The application of nitrogenous fertilizers has serious implications for causing pollution as the excess reduced nitrogen (ammonium) in agricultural ecosystem is acidified through the process of nitrification (Santi et al. 2013; Tchan et al. 1991).

For sustainability in agriculture, careful utilization and management of the internal resources are sought. Nitrogen-fixing plants present an 'eco-friendly' and economical alternatives to improve internal resources and reduce the utilization of the external inputs. Biological nitrogen fixation (BNF), a process of the reduction of nitrogen to ammonia by microorganisms collectively called as diazotrophs, helps in maintaining soil nitrogen reserves, thereby, decreasing the utilization of fertilizers to attain higher crop production (Roper and Gupta 2016; Santi et al. 2013).

Microorganisms with the ability to fix nitrogen fall into two genera with 87 species of archaebacteria, 38 bacterial genera, and 20 cyanobacterial genera (Zahran 2001). Diazotrophs are generally classified as free-living, associative, and symbiotic, though distinction between the former, i.e. free-living and associative is not well defined. Nitrogen-fixing bacteria include free-living aerobic – *Azotobacter*, *Beijernickia*, *Klebsiella,* and *Paenibacillus*; free-living anaerobic – *Clostridium* and *Desulfovibrio*; symbiotic – rhizobia and *Frankia*; and associative – *Azospirillum* (Choudhury and Kennedy 2004; Bellenger et al. 2014). Symbiotic bacteria including rhizobia and *Frankia* fix atmospheric nitrogen in the nodules, whereas the other two systems fix nitrogen while living freely with the utilization of carbohydrates either excreted from the plants' roots or from the degradation of soil organic matter (Yoneyama et al. 2019). However, for fixing nitrogen, these microorganisms have to face a competition with high number of non-nitrogen-fixing microorganisms in the rhizosphere for acquiring the carbohydrates (Yoneyama et al. 2017). Rhizobia-legume symbiosis is considered as the main supplier of the fixed nitrogen; however, other nitrogen-fixing systems are equally important (Fox and Papanicolaou 2007). It is difficult to study BNF in nonleguminous systems due to the diverse sites of nitrogen fixation as compared to leguminous systems (Yoneyama et al. 2019). The techniques of molecular biology have provided insight to increase our understanding of N_2-fixing microbial systems and the molecular mechanisms used for nitrogen fixation in nonleguminous plants. The knowledge is useful while developing new plant-diazotrophic combinations for different environmental conditions in order to exploit these nitrogen-fixing microorganisms as inoculants for sustainable agriculture.

10.2 Nitrogen Fixation and Sustainable Agriculture

Basic principles for managing soil required for sustainable agriculture as suggested by Greenland (1975) include replacing chemical nutrients utilized by crops, maintaining physical conditions of soil, nonoccurrence of pests, diseases or weeds, nonincrease in toxic elements or soil acidity, and management of soil erosion. In majority of the systems, most of the principles are not being adhered to. The use of large amounts of nitrogenous fertilizers in agriculture is a global concern in view of their high cost and environmental pollution. Therefore, new and novel solutions are sought to maintain the balance between productivity and sustainability of agriculture. To lessen the burden imposed on our natural resources and environment, ecologically safe, efficient, and cost-effective alternatives are required.

BNF is suggested as an alternative to nitrogenous chemical fertilizers for improving internal resources and reducing external inputs. A large number of microorganisms fix atmospheric nitrogen including (i) symbiotic such as *Rhizobium*, and *Frankia* (ii) free-living aerobic bacteria such as *Azotobacter*, *Bacillus*, and *Pseudomonas* (iii) associative bacteria such as *Azospirillum*. In most of the cropping systems, symbiotic nitrogen fixation offers a major source of nitrogen, while in rice crops, *Anabaena* and *Azolla* are of particular value. Likewise, nitrogen-fixing free-living and associative microorganisms are also equally important in different systems.

10.3 Key Features in Achieving N_2-Fixing Activity

10.3.1 The Nitrogenase Reaction

BNF, the conversion of N_2 to ammonia catalyzed by the enzyme nitrogenase, is a very complex process requiring an energy source in the form of ATP and a reducing potential. In addition, it also requires a mechanism for the utilization of ammonia which ultimately will neutralize the generated alkalinity. The overall reaction of BNF can be represented as:

$$\begin{aligned} &N_2 + 8H^+ + 8Fd^- + 16MgATP^{2-} + 18H_2O \rightarrow 2NH_4^+ + 2OH^- + H_2 \\ &+ 8Fd + 16MgADP^- + 16H_2PO_4^- \end{aligned}$$

During symbiotic nitrogen fixation, nitrogenase reductase and dinitrogenase – the two proteins present in the nodules of plants under low redox conditions – reduce N_2 to NH_3 with the use of NADH and energy in the form of ATP. However, in other systems, nitrogen fixation takes place by the utilization of energy and carbon sources by bacteria from the environment and after the death and lysis of the bacterial cells, fixed nitrogen is released (James 2000; White et al. 2012). However, an important question arises about the efficiency of the reductant and energy sources supplied to the diazotrophs.

10.3.2 The Energy Requirement

BNF requires energy similar to that of nitrate assimilation which is the key nitrogen source in crops except paddy (Rosenblueth et al. 2018). The reaction is energy demanding and requires 16 ATP molecules to fix one molecule of nitrogen (Hubbell and Kidder 2009). However, the total energy required for the reaction is about 25 ATP molecules per molecule of N_2 fixed, as some more energy is necessary for lifting the electrons to a large redox potential for nitrogenase. The rate of BNF is high in carbon-rich environments. The energy requirement estimates are, in general, based on theoretical as well as on the results of many experiments.

10.3.3 Sensitivity to Oxygen

The sensitivity of nitrogenase to oxygen is an important factor in BNF (Dalton and Kramer 2007). Oxidative phosphorylation responsible for producing ATP for fuelling nitrogen fixation requires oxygen but the enzyme nitrogenase is inactivated by oxygen. Nitrogenase contains two proteins namely the Mo–Fe protein with molecular weight in the range of 220 000–270 000 and the Fe protein of molecular weight ranging from 55 000 to 66 800. Since, both the components of the nitrogenase are inactivated by oxygen, it is obligatory to maintain low levels of oxygen so that nitrogen fixation is carried out optimally (Pienkos et al. 1983). Diazotrophic organisms have evolved with the ability to overcome the nitrogenase sensitivity along with the energy-intensive nature of nitrogen fixation.

10.4 Evolution of N_2-Fixing Symbiosis

Nitrogen fixation by microorganisms is an ancient process, as evident from the concept that a nitrogen crisis occurred somewhere around 3.5 Ga due to a decrease in CO_2 concentration and simultaneous decrease in nitrogen oxidation (Fani et al. 2000; Kasting and Siefert 2001). However, a data-driven attempt to address the evolution of BNF with molecular dating techniques recently assumes its origin around 1.5–2.2 Ga (Boyd et al. 2011). During this period, fixed nitrogen levels became limited, O_2 concentrations and soluble Mo concentrations started increasing, leading to a consequence of producing oxygen by growing populations of phototrophs (Anbar et al. 2007; Anbar 2008; Canfield 2005). Increase in the oxygen concentration might have allowed the diversification of global biome, increasing a nitrogen pool demand and ultimately evolving a biological nitrogen-fixing symbiosis to increase the nitrogen pool locally (Boyd et al. 2011).

The nitrogenase enzyme is limited to bacteria and archaebacteria, absent in eukarya and within archaebacteria also, and present only in methanogens (Boyd et al. 2011; Dos Santos et al. 2012). Various aerobes and facultative anaerobes have developed different mechanisms for nitrogen fixation in oxygenic conditions, as nitrogenase is highly oxygen sensitive. Nitrogen fixation in strict anaerobes takes place under anaerobic conditions whereas in facultative anaerobes during their anaerobic growth period. Haemoglobin, a high affinity oxygen-binding protein, provides a microaerobic environment for symbiotic nitrogen fixation and allows the diazotroph to carry out aerobic respiration along with nitrogen fixation (Ott et al. 2005). Likewise, in obligate aerobes, high respiration rates lead to oxygen consumption at the cell membrane known as the cytochrome-dependent respiratory protection mechanism, thus, maintaining low intracellular oxygen concentration, protecting the nitrogen fixation apparatus (Poole and Hill 1997). Due to the highly complex *nif* gene clusters in these microorganisms able to fix nitrogen in aerobic environment, it is proposed that they have evolved later in the evolutionary history (Boyd et al. 2011). However, identifying the most ancient nitrogen-fixing systems has been a challenge.

10.5 Association of Diazotrophs with Nonlegumes

Diazotrophic organisms are generally classified as symbiotic, free-living, and associative; however, many organisms fall into more than one group as the differentiation of these groups is not clearly marked. The nitrogen-fixing bacteria include free-living aerobic *Azotobacter*, *Beijernickia*, *Klebsiella*, and *Paenibacillus*, free-living anaerobic *Clostridium* and *Desulfovibrio*, symbiotic rhizobia and *Frankia*, and associative *Azospirillum*. *Acetobacter*-sugarcane system is becoming a very important and effective model for experiments (Pedraza 2008). Similarly, the endophytic bacterium *Azoarcus* spp. able to colonize grasses including Kallar grass and rice have been found to fix nitrogen (Hurek and Reinhold-Hurek 2003). *Burkholderia* comprising 67 validly published species, with nitrogen-fixing *Burkholderia kururiensis*, *B. phynatum*, *B. tuberum*, and *B. vietnamiensis* (Coenye and Vandamme 2003). *Herbaspirillum*, an endophyte capable of colonizing maize, rice, sorghum, and sugarcane has been found to fix about 31–45% of total plant nitrogen in rice (James et al. 2002). In addition, many species of the family *Enterobacteriaceae* are diazotrophs.

In symbiotic associations, the plant provides sugars as energy source to nitrogen fixers required for fixing nitrogen and in return, nitrogen is provided to the host plant by the microbial partner for its growth. An important nonleguminous association is between the actinomycete *Frankia* with trees and shrubs. Another such example is the symbiosis of *Azolla* with a cyanobacterium *Anabaena azollae*, an important association in rice. Likewise, associative nitrogen-fixing organisms also exchange nutrients between their plant host without forming any differentiated structures, whereby the association is just more casual than cooperative (Franche et al. 2009). In strains of *Azospirillum*, aerobic heterotrophs grow extensively in the rhizosphere of Gramineae family members fix nitrogen under microaerobic conditions (Dobbelaere et al. 2001).

10.6 Application of Microbial Inoculations

Most of the nitrogen absorbed by the plants mainly includes nitrate (NO^{3-}) and to some extent ammonium (NH^{4+}). Apart from this, the information about the uptake of larger (or heavier) molecules with organically bound nitrogen is limited and conditional

(Werner 2007). The optimum growth and rice yield are dependent on the nitrogen supplements which are later absorbed through soil by the rice plant. Presently, most of this nitrogen supply is through nitrogen fertilizers, which provide the required amount of nitrogen supply, but at its consequences, as most of this nitrogen supply is lost through denitrification, leaching, and other similar factors causing environmental issues and leads to bio-magnification (Kecskés et al. 2016). It is also reported that following bacterial inoculations there is a significant increase in the uptake of nitrogen and from the soil and reduces the application of nitrogen through fertilizers (Choudhury et al. 2014).

The effect of nitrogen fixation and indole-3-acetic acid (IAA) such as phytohormones are important for the promotion of growth in rice seedlings (Keyeo et al. 2011). Also, not just nitrogen fixation but the production of phytohormones affect the growth of paddy crops significantly. In presence of diazotrophic bacterial inoculations, which differ in both the rate of nitrogen fixation and production of phytohormones, it was observed that those bacterial inoculations which produce more amount of phytohormones like IAA, yield reduction in paddy growth as compared with those which produce such hormones in lesser quantity (Keyeo et al. 2011).

The growth of rice plants, yield as well as diseases are reported to be regulated by the bacterial strains available in the environment around the fields (Yasmin et al. 2016). Hence, it must be understood that the microbial communities available around the rice rhizosphere can affect significantly the growth of the rice plant, yield as well as disease occurrence considerably along with the reduction in the use of chemical fertilizers further providing an economic benefit to the paddy crop cultivators.

In another study, where the effect of zinc solubilizing bacterial inoculation on growth promotion through zinc uptake was assessed reveal that such bacterial inoculations significantly enhance total Zinc uptake as well as methionine concentration in plants (Vaid et al. 2014). Also, lower uptake of Zinc renders shunted rice plant growth and reduced yield. Furthermore, consumption of such a low Zinc diet in the developing world results in poor growth affects brain function and development, as well as immunity (Benton 2008; Graham 2008; Vaid et al. 2014).

It is a well-established concept that crop productivity can be considerably increased by avoiding various environmental issues by using plant growth-promoting rhizobacteria (PGPR). They provide various benefits like increased nitrogen fixation, siderophore, and phytohormone production and also in controlling plant pathogens (Vaid et al. 2014). Due to their importance *Rhizobium*, *Azotobacter*, *Enterobacter*, *Azospirillum*, *Bacillus*, *Pseudomonas*, and other agriculturally important species are extensively used as bacterial inoculations as bio-fertilizers (Vaid et al. 2014; Yasmin et al. 2016). Also, they are reported to promote growth in different rice genotypes (Sharma et al. 2014).

In a recent study, commercial microbial inoculant Bacillus subtilis C-3102 (Thervelics®-with carrier material) was used to study its effect on the rice (*Oryza sativa* cv. Milky princess) and barley (*Hordeum vulgare* cv. Sachiho Golden) plant species to observe how it affects the growth, phytohormone as well as protease production (Jamily et al. 2019). It was concluded that, amongst several strains of *Bacillus subtilis*, the highest protease activity and IAA production was from NBRC109335 and *Bacillus subtilis* C-3102 (commercial strain) clearly indicating that phosphate solubilizing activity, as well as siderophore production, has no effect on growth-promoting activity and was confirmed from control as well as inoculation studies through soil experiments. However, the protease activity was positively regulating plant growth (Kumar et al. 2012). The positive and negative effects of PGPR and deleterious rhizobacteria (DRB) respectively was also supported by this study (Jamily et al. 2019; Suarez et al. 2014). Also, combined or consortium inoculation is reported to be more effective rather than a single inoculation of one strain (García-de-Salamone et al. 2012).

The alarming situation of environmental pollution by the use of chemical fertilizers and pesticides is increasing every day, which affects not just the environment but also the plants and animals alike. Considering microbial fertilizers, PGPB, plant growth-promoting bacteria; PGPM, plant growth promotion microorganisms seem to be the best alternative to provide required nutrients, promote growth and yield as well as to avoid serious health conditions within consuming populations (Stamenkovic et al. 2018). Due to this farming with microbes and microbial seed treatment must be promoted for agricultural applications for improving crop quality (Quarles 2018) and human health (Alori and Babalola 2018).

10.7 Conclusion

Nitrogen is a restrictive factor in plant growth and it is the most important nutrient. The nitrogen cycle, a complex biogeochemical cycle that includes four microbial processes such as nitrogen fixation, nitrification, denitrification, and nitrogen mineralization, has a significant impact on soil fertility. Because inorganic nitrogen fertilizers are both environmentally and economically costly, interest in alternative nitrogen fertilizer sources has been renewed in order to boost confidence in sustainable agriculture. BNF has been shown to have numerous agronomic, economic, and environmental advantages for nonlegume farming systems.

10.8 Summary

One major challenge for agriculture in the twenty-first century is the production of agricultural crops in an environmentally sound and sustainable way. An enhanced production of crops is necessary to provide sufficient food for the increasing human population. To get higher productivity, chemical fertilizers and pesticides are extensively used which impose an extensive burden on the agriculture. Nitrogen, the most important nutrient is a limiting factor in plant growth. Nitrogen cycle, a complex biogeochemical cycle containing four microbial processes including nitrogen fixation, nitrification, denitrification, and nitrogen mineralization has a huge impact on maintaining soil fertility. Inorganic nitrogen fertilizers being environmentally and economically costly, attention to the use of alternative sources of nitrogen fertilizer has been renewed to give confidence for sustainable agriculture. BNF has been demonstrated to have many agronomic, economic, and environmental benefits for the nonlegume farming system. In this chapter, we discuss traditional and new aspects of BNF of rice-based farming system with examples primarily chosen from developing countries.

References

Abebe, G., Hattar, B., and Al-Tawaha, A. (2005). Nutrient availability as affected by manure application to cowpea (*Vigna unguiculata* L. Walp.) on calacarious soils. *Journal of Agriculture & Social Sciences* 1: 1–6.

Abera, T., Feyisa, D., Yusuf, H. et al. (2005). Grain yield of maize as affected by biogass slurry and NP fertilizer rate at Bako, Western Oromiya. *Ethopia Bioscience Research* 2: 31–38.

Al-Ajlouni, M.M., Al-Ghzawi, A.A., and Al-Tawaha, A.M. (2009). Crop rotation and fertilization effect on barley yield grown in arid conditions. *Journal of Food, Agriculture and Environment* 88: 869–872.

Al-Juthery, H.W., Habeeb, K.H., Altaee, F.J.K. et al. (2018). Effect of foliar application of different sources of nano-fertilizers on growth and yield of wheat. *Bioscience Research* 15: 3976–3985.

Al-Kiyyam, M.A., Turk, M., Al-Mahmoud, M., and Al-Tawaha, A.R. (2008). Effect of plant density and nitrogen rate on herbage yields of marjoram under Mediterranean conditions. *Journal American-Eurasian Journal of Agricultural and Environmental Science* 3: 153–158.

Alori, E.T. and Babalola, O.O. (2018). Microbial inoculants for improving crop quality and human health in Africa. *Frontiers in Microbiology* 9: 2213.

Al-Tawaha, A.M., Turk, M.A., Lee, K.D. et al. (2005). Impact of fertilizer and herbicide application on performance of ten barley genotypes grown in northeastern part of Jordan. *International Journal of Agriculture and Biology* 7: 162–166.

Al-Tawaha, A.M., Yadav, S.S., Turk, M. et al. (2010). Crop production and management technologies for drought prone environments. In: *Climate Change and Drought Management in Cool Season Grain Legume Crops*. Springer.

Anbar, A.D. (2008). Elements and evolution. *Science* 322: 1481–1483. https://doi.org/10.1126/science.1163100.

Anbar, A.D., Duan, Y., Lyons, T.W. et al. (2007). A whiff of oxygen before the great oxidation event. *Science* 317: 1903–1906. https://doi.org/10.1126/science.1140325.

Bellenger, J.P., Xu, Y., Zhang, X. et al. (2014). Possible contribution of alternative nitrogenases to nitrogen fixation by asymbiotic N2-fixing bacteria in soils. *Soil Biology and Biochemistry* 69: 413–420.

Benton, D. (2008). Micronutrient status, cognition and behavioral problems in childhood. *European Journal of Nutrition* 47: 38–50.

Boyd, E.S., Anbar, A.D., Miller, S. et al. (2011). A late methanogen origin for molybdenum-dependent nitrogenase. *Geobiology* 9: 221–232. https://doi.org/10.1111/j.1472-4669.2011.00278.x.

Canfield, D.E. (2005). The early history of atmospheric oxygen: homage to Robert, M. Garrels. *Annual Review of Earth and Planetary Sciences* 33: 1–36. https://doi.org/10.1146/annurev.earth.33.092203.122711.

Choudhury, A.T.M.A. and Kennedy, I.R. (2004). Prospects and potentials for systems of biological nitrogen fixation in sustainable rice production. *Biology and Fertility of Soils* 39: 219–227.

Choudhury, A.T., Kecskés, M.L., and Kennedy, I.R. (2014). Utilization of BNF technology supplementing urea N for sustainable rice production. *Journal of Plant Nutrition 37*, vol. 10: 1627–1647.

Coenye, T. and Vandamme, P. (2003). Diversity and significance of *Burkholderia* species occupying diverse ecological niches. *Environmental Microbiology* 5: 719–729.

Dalton, D.A. and Kramer, S. (2007). Nitrogen-fixing bacteria in non-legumes. In: *Plant-Associated Bacteria*, 105–130. Dordrecht: Springer.

Dobbelaere, S., Croonenborghs, A., Thys, A. et al. (2001). Responses of agronomically important crops to inoculation with Azospirillum. *Functional Plant Biology* 28: 871–879.

Dos Santos, P, Fang, Z, Mason, S, Setubal, J, & Dixon, R 2012, Distribution of nitrogen fixation and nitrogenase-like sequences amongst microbial genomes, *BMC Genomics*, vol.13, pp. 162. https://doi.org/10.1186/1471-2164-13-162

Fani, R., Gallo, R., and Liò, P. (2000). Molecular evolution of nitrogen fixation: the evolutionary history of the *nifD*, *nifK*, *nifE*, and *nifN* genes. *Journal of Molecular Evolution* 51: 1–11.

Fox, J.F. and Papanicolaou, A.N. (2007). The use of carbon and nitrogen isotopes to study watershed erosion processes, 1. *JAWRA Journal of the American Water Resources Association* 43: 1047–1064.

Franche, C., Lindström, K., and Elmerich, C. (2009). Nitrogen-fixing bacteria associated with leguminous and non-leguminous plants. *Plant and Soil* 321: 35–59.

García-de-Salamone, I.E., Funes, J.M., Di Salvo, L.P. et al. (2012). Inoculation of paddy rice with *Azospirillum brasilense* and *Pseudomonas fluorescens*: impact of plant genotypes on rhizosphere microbial communities and field crop production. *Applied Soil Ecology* 61: 196–194.

Graham, R.D. (2008). Micronutrient deficiencies in crops and their global significance. In: *Micronutrient Deficiencies in Global Crop Production* (ed. B.J. Alloway), 41–61. New York: Springer.

Greenland, D.J. (1975). Bringing the green revolution to the shifting cultivator. *Science* 190: 841–844.

Hubbell, D.H. and Kidder, G. (2009). *Biological nitrogen fixation*, 1–4. University of Florida IFAS Extension Publication SL16.

Hurek, T. and Reinhold-Hurek, B. (2003). *Azoarcus* sp. strain BH72 as a model for nitrogen-fixing grass endophytes. *Journal of Biotechnology* 106: 169–178.

James, E.K. (2000). Nitrogen fixation in endophytic and associative symbiosis. *Field Crops Research* 65: 197–199.

James, E.K., Gyaneshwar, P., Mathan, N. et al. (2002). Infection and colonization of rice seedlings by the plant growth-promoting bacterium *Herbaspirillum seropedicae* Z67. *Molecular Plant–Microbe Interactions* 15: 894–896.

Jamily, A.S., Koyama, Y., Win, T. et al. (2019). Effects of inoculation with a commercial microbial inoculant Bacillus subtilis C-3102 mixture on rice and barley growth and its possible mechanism in the plant growth stimulatory effect. *Journal of Plant Protection Research* 59 (2): 193–195.

Jetten, M.S. (2008). The microbial nitrogen cycle. *Environmental Microbiology* 10: 2903–2909.

Kasting, J.F. and Siefert, J.L. (2001). Biogeochemistry: the nitrogen fix. *Nature* 412: 26–27. https://doi.org/10.1038/35083660.

Kecskés, M.L., Choudhury, A.T.M.A., Casteriano, A.V. et al. (2016). Effects of bacterial inoculant biofertilizers on growth, yield and nutrition of rice in Australia. *Journal of Plant Nutrition* 39: 377–388.

Keyeo, F., Ai'shah, O.N., and Amir, H.G. (2011). The effects of nitrogen fixation activity and phytohormone production of diazotroph in promoting growth of rice seedlings. *Biotechnology* 10: 267–273.

Kumar, P., Dubey, R.C., and Maheshwari, D.K. (2012). Bacillus strains isolated from rhizosphere showed plant growth promoting and antagonistic activity against phytopathogens. *Micro-biological Research* 167: 493–499.

Nikus, O., Al-Tawaha, M., and Turk, M.A. (2004a). Effect of manure supplemented with phosphate fertilizer on the fodder yield and quality of two sorghum cultivars (*Sorghum bicolor* L.). *Bioscience Research* 1: 1–7.

Nikus, O., Turk, M.A., and Al-Tawaha, A.M. (2004b). Yield response of sorghum (*Sorghum bicolor* L.) to manure supplemented with phosphate fertilizer under semi-arid Mediterranean conditions. *International Journal of Agriculture and Biology* 6: 889–893.

Ott, T., Van Dongen, J.T., Günther, C. et al. (2005). Symbiotic leghemoglobins are crucial for nitrogen fixation in legume root nodules but not for general plant growth and development. *Current Biology* 15: 531–535. https://doi.org/10.1016/j.cub.2005.01.042.

Pedraza, R.O. (2008). Recent advances in nitrogen-fixing acetic acid bacteria. *International Journal of Food Microbiology* 125: 25–35.

Pienkos, P.T., Bodmer, S., and Tabita, F.R. (1983). Oxygen inactivation and recovery of nitrogenase activity in cyanobacteria. *Journal of Bacteriology* 153: 182–190.

Poole, R.K. and Hill, S. (1997). Respiratory protection of nitrogenase activity in *Azotobacter vinelandii*–roles of the terminal oxidases. *Bioscience Repository* 17: 303–317. https://doi.org/10.1023/A:1027336712748.

Quarles, W. (2018). Farming with microbes: microbial seed treatments. *IPM Practitioner* 36: 1–8.

Roper, M.M. and Gupta, V.V.S.R. (2016). Enhancing non-symbiotic N2 fixation in Agriculture. *The Open Agriculture Journal* 10: pp.

Rosenblueth, M., Ormeño-Orrillo, E., López-López, A. et al. (2018). Nitrogen fixation in cereals. *Frontiers in Microbiology* 9: 1794.

Santi, C., Bogusz, D., and Franche, C. (2013). Biological nitrogen fixation in non-legume plants. *Annals of Botany* 111: 743–767.

Sharma, A., Shankhdhar, D., Sharma, A., and Shankhdhar, S.C. (2014). Growth promotion of rice genotypes by pgprs isolated from rice rhizospheres. *Journal of Soil Science and Plant Nutrition* 14: 505–517.

Stamenkovic, S., Beskoski, V., Karabegovic, I. et al. (2018). Microbial fertilizers: a comprehensive review of current findings and future perspectives. *Spanish Journal of Agricultural Research* 16: 1–18.

Suarez, D.E.C., Gigon, A., Freitas, R.P. et al. (2014). Combined effects of earthworms and IAA-producing rhizobacteria on plant growth and development. *Applied Soil Ecology* 80: 100–107.

Tawaha, A.M. and Turk, M.A. (2002). Lentil (*Lens culinaris* Medic.) productivity as influenced by rate and method of phosphate placement in a Mediterranean environment. *Acta Agronomica Hungarica* 50: 197–191.

Tawaha, A.M. and Turk, M.A. (2004). Field pea seeding management for semi-arid Mediterranean conditions. *Journal of Agronomy and Crop Science* 190: 86–92.

Tawaha, A.M., Singh, V.P., and Turk, M.A. (2003). A review on growth, yield components and yield of barley as influenced by genotypes, herbicides and fertilizer application. *Research on Crop* 4: 1–9.

Tchan, Y.T., Zeman, A.M.M., and Kennedy, I.R. (1991). Nitrogen fixation in para-nodules of wheat roots by introduced free-living diazotrophs. *Plant and Soil* 137: 43–47.

Turk, M.A. and Tawaha, A.M. (2001). Common vetch (*Vicia sativa* L.) productivity as influenced by rate and method of phosphate fertilization in a Mediterranean environment. *Agriculture Mediterranean* 131: 108–111.

Turk, M.A. and Tawaha, A.M. (2002a). Onion (*Allium cepa* L.) as influenced by rate and method of phosphorus placement. *Crop Research* 23: 105–107.

Turk, M.A. and Tawaha, A.M. (2002b). Impact of seeding rate, seeding date and method of phosphorous application in faba bean (*Vicia faba L. minor*) in the absence of moisture stress. *Biotechnology, Agronomy, Society and Environment* 6: 171–178.

Turk, M. and Tawaha, A.R. (2004). Effect of variable sowing ratios and sowing rates of bitter vetch on the herbage yield of oat-bitter vetch mixed cropping. *Proceedings 7th International Oat Conference/PirjoPeltonen-Sainio and Mari Topi-Hulmi (eds.).* MTT.

Turk, M.A., Tawaha, A.M., and Shatnawi, M. (2003a). Lentil (*Lens culinaris* Medik) Response to plant density, sowing date, phosphorus fertilization and Ethephon application in the absence of moisture stress. *Journal of Agronomy and Crop Science* 189: 1–6.

Turk, M.A., Tawaha, A.M., and Samara, N. (2003b). Effects of seeding rate and date, and phosphorus application on growth and yield of narbon vetch (*Vicia narbonensis*). *Agronomie* 23: 1–4.

Turk, M.A., Hameed, K.M., Aqeel, A.M., and Tawaha, A.M. (2003c). Nutritional status of durum wheat grown in soil supplemented with olive mill by-products. *Agrochimica* 47: 209–219.

Vaid, S.K., Kumar, B., Sharma, A. et al. (2014). Effect of zinc solubilizing bacteria on growth promotion and zinc nutrition of rice. *Journal of Soil Science and Plant Nutrition* 14: 889–810.

Werner, W. (2007). *Fertilizers*, Chapter 9, 98–111. Federal Republic of Germany: Agrikulturchemisches Institut der Universitat Bonn, Bonn. https://application.wiley-vch.de/books/sample/3527316043_c01.pdf.

White, J.F., Crawford, H., Torres, M.S. et al. (2012). A proposed mechanism for nitrogen acquisition by grass seedlings through oxidation of symbiotic bacteria. *Symbiosis* 57: 161–171.

Yasmin, S., Zaka, A., Imran, A. et al. (2016). Plant growth promotion and suppression of bacterial leaf blight in rice by inoculated bacteria. *PLoS ONE* 11: 1–19.

Yoneyama, T., Terakado-Toonooka, J., and Minamisawa, K. (2017). Exploration of bacterial N2-fixation systems in association with soil-grown sugarcane, sweet potato, and paddy rice: a review and synthesis. *Soil Science and Plant Nutrition* 63: 578–590.

Yoneyama, T., Terakado-Tonooka, J., Bao, Z., and Minamisawa, K. (2019). Molecular analyses of the distribution and function of diazotrophic rhizobia and methanotrophs in the tissues and rhizosphere of non-leguminous plants. *Plants* 8: 408.

Zahran, H.H. (2001). Rhizobia from wild legumes: diversity, taxonomy, ecology, nitrogen fixation and biotechnology. *Journal of Biotechnology* 91: 143–153.

11

Role of Phosphorus in Imparting Abiotic Stress Tolerance to Plants

Bhaswatee Baroowa, Sreyashi Paul, and Nirmali Gogoi

Department of Environmental Science, Tezpur University, Tezpur, Assam, India

CONTENTS

Climate Change and Agriculture: Perspectives, Sustainability and Resilience, First Edition. Edited by Noureddine Benkeblia.

11.1 Introduction

Plant nutrients play a greater role in improving the abiotic stress tolerance in crop plants. Phosphorus (P), next to nitrogen, is a crucial macronutrient for plants. The primary role of P in a plant is to store and transfer energy produced via photosynthesis for use in growth and reproductive processes. Adequate P levels promote root growth and winter hardiness, stimulate tillering, and hasten maturity. P is used in many aspects of plant and cell biology such as photophosphorylation, genetic transfer, the use of adenosine triphosphate (ATP) as energy exchange currency, the transportation of nutrients, and phospholipid cell membranes (Hernández and Munné-Bosch 2015).

Adverse environmental conditions such as low and high temperatures, osmotic shock and water deficit, excess soil moisture, elevated carbon dioxide (CO2), and salinity are composed of a complex set of phenomena and is an integral factor of climate change, which exerts erratic impacts on the environment. Prolonged exposure to these abiotic stresses transforms metabolism and induces impairment of biomolecules. Plants being sessile organisms are more intensely affected by abiotic stress. Thus, they evolve various defence mechanisms to cope with these stresses, e.g. accumulation of osmolytes, osmoprotectants, and enzymatic and non-enzymatic antioxidants.

The survival and productivity of crop plants exposed to environmental stresses are dependent on their ability to develop adaptive mechanisms to avoid or tolerate stress. Abiotic stress management practices involve control at the biochemical and molecular levels. These include the ability and promptness at which a stress signal is perceived and communicated to produce stress signal molecules and trigger stress-protective mechanisms. Abiotic stresses can also be managed by proper plant nutrition depending upon the type of soil and stress involved. Efficient management of macro and micronutrients is proved to be effective for developing abiotic stress resistance in plants as better plant nutrition can effectively alleviate the adverse effects of such unfavourable conditions. Application of nutrients such as nitrogen (N), phosphorus (P), potassium (K), calcium (Ca), and magnesium (Mg) reduces the toxicity of reactive oxygen species (ROS) by triggering the antioxidant enzymes such as superoxide dismutase (SOD), catalase (CAT), and peroxidize (POD) in the plant cells. These antioxidants scavenge the ROS and reduce the photo-oxidation and maintain the integrity of the chloroplast membrane and thereby the rate of photosynthesis in plants. Both nitrogen and phosphorus nutrition can increase crop water use, thereby reducing soil evaporation. For instance, P fertilizers increased stomatal conductance, photosynthesis, membrane stability, leaf water potential, and root growth during drought to improve osmotic adjustment in beans and sorghum (Al-Karaki et al. 1996).

11.2 Importance of Phosphorus (P) in Plant Metabolism

P is a crucial macronutrient that constitutes around 0.2% of plant mass, a structural constituent of nucleic acids, phospholipids, and ATP (Kuwahara et al. 2016). After nitrogen, phosphorous is the second growth-limiting nutrient in plants (Schachtman et al. 1998). It functions as a vital element of numerous biochemical reactions and metabolic cycles, comprising glycolysis and several stages of C3 and C4 cycles. In nature, P occurs in two forms – organic and inorganic (Pi). The organic P comprises of un-decomposed animal and plant residues, soil microbes, and organic matter; while the available aluminium, iron, and calcium compounds of varying solubility are the sources of inorganic P to plants. Production of photoassimilate depends on the P availability in the cytoplasm and is linked to the transport of trioses from chloroplasts to the cytoplasm and the process of sucrose synthesis (Hendrickson et al. 2004). At the time of photosynthesis, chloroplasts obtain Pi and release Triose-P to the cytoplasm. The flow between carbon and Pi is facilitated by the Triose-P transporter, which transports the antiport system of Triose-P/Pi across the chloroplast membrane (Flügge et al. 2003). Furthermore, the assimilated partitioning between starch and sucrose is dependent on a three-factor interaction: the concentration of cytoplasmic Pi, which regulates the export of Triose-P. Being a crucial mineral nutrient, phosphorus is essential in moderately large amounts to conserve and transfer energy for cellular metabolism (Müller et al. 2015). It is a component of significant molecules such as nucleic acids, phospholipids, and ATP (Stigter and Plaxton 2015). P is a relatively stable element in plants where deficiency causes a decrease in seedling vigour, plant establishment, and development of roots (Peng and Li 2005). Plants take up P in the form of $H_2PO_4^-$ or HPO_4^{2-} ions depending on the soil pH. Under normal conditions, there is a need for energized transport of inorganic phosphorus (Pi) across the plasma membrane to the plant cell from the soil as the concentration of Pi in the cytoplasm is relatively high (Schachtman et al. 1998).

11.3 Phosphorus (P) Stress and Plant Growth

Phosphorus availability in acidic or calcareous soil may be limited owing to the low total phosphorus or/and due to the production of insoluble complex (Vance et al. 2003). In response to P deficiency, higher phosphatase activity in the rhizosphere has been reported in various plant species and has been found greater in P-deficient soils. Richardson and Simpson (2011), in a study, reported that exhaustion of organic P in soil was related to a substantial increase in the activity of both mono- and di-ester phosphatases. Phosphorus stress can be mitigated by the application of a large amount of phosphate-containing fertilizers. Nonetheless, this might result in soil pollution which is not economically sustainable (Withers et al. 2001). It has been reported that phosphate solubilizing bacteria (PSB) are the most promising bacteria among the plant growth-promoting rhizobacteria (PGPR); which may be used as biofertilizers for plant growth and nutrient use efficiency. Thus, the application of PSB in agricultural fields is much more beneficial as it enhances nitrogen phosphorus potassium (NPK) uptake and minimizes the harmful effect of agrochemicals, thereby promoting the sustainable production of agricultural crops.

Under P deficiency, plants tend to have a poor concentration of phospholipids while higher concentration of galactolipids (mainly di-galactosyl-diacyl-glycerol (DGDG)) compared to those growing under P sufficiency (Benning and Ohta 2005). The low concentration of phospholipids makes it useful for other important cellular functions and makes the lipid moiety diacylglycerol (DAG) available for the biosynthesis of galactolipid. It has been found that the decrease in phospholipid content is due to the involvement of the phospholipid-hydrolyzing enzymes, such as phospholipase D (PLD), and phospholipase

C (PLC). In P-starved Arabidopsis (*Arabidopsis thaliana*), the expression of one PC-hydrolysing PLC was induced greatly while the gene knockout of that PLC has no seeming consequence on the lipid composition under normal and phosphorus-limited growth conditions (Nakamura et al. 2005).

11.4 Phosphorus (P) Availability in Soil

Phosphate is produced by soil microorganisms and these microorganisms when cultured in artificial media are able to utilize P from wide forms of organic P present in soil. Among these, a predominant form of organic P known as inositol phosphates is identified in many soils (Richardson and Simpson 2011). These organic P substrates when applied to soil are hydrolyzed quickly (Macklon et al. 1997). On the contrary, in the treatment of soil suspensions with excess of phosphatase activity, considerable amounts of orthophosphate can be released (George et al. 2007). A study by Bünemann (2008) revealed that hydrolysis of 60% of the total organic P is carried out by phosphatases with highest quantities being released by phytases (monoester phosphatases active against phytate). The release of orthophosphate from soil is efficiently carried out by both plant and microbial phosphatases indicating the efficiency of microbial enzymes for P release (Tarafdar et al. 2001). At present, the extent of independently occurring P mineralization is not known. Thus, a competition occurs between plants and microorganisms for the uptake of nutrients which might either result in net P mineralization or P immobilization (Cheng 2009) and also subject to microbial turnover times and thus enhances P availability for plant uptake. Therefore, microorganisms are vital for the soil phosphorus cycle, which aids in mediating the availability of P to plants (Figure 11.1).

11.5 P Accumulation and Use Efficiency

Soil microorganisms hold a significant amount of total soil P which is almost equal to or greater than what plant biomass hold. In soil bulk, microbial P accounts for 2 to 10% of total soil P which varies with stage of soil development, and in the litter layer, this can be as much as 50% (Achat et al. 2010). Essentially, microbial P is a highly dynamic soil P pool, which may undergo considerable change in response to environmental stresses such as availability of soil moisture, and carbon and temperature fluctuations. Organic amendments applied to soil are decomposed by microorganisms which also mineralized the organic P. For example, the addition of crop residue enhances the quick incorporation of P into microbial biomass with greater recovery of P over short period of time (Damon et al. 2014). Several fold increase in microbial P in relation to added C is linked to decline of orthophosphate (or dissolved organic P) in soil solution. This takes place in soil either amended with P or without P (Alori et al. 2017). Thus, it is obvious that microorganisms can efficiently compete with plants for available soil orthophosphate which also characterize a pool of immobilized soil P, temporarily not available to plants. Nonetheless, over time, entire microbial P is potentially available to plants and this is why it is suggested that immobilization of P within the biomass is crucial for regulating the P supply in soil solution (Seeling and Zasoski 1993). Radioactive tracer studies showed a considerable release

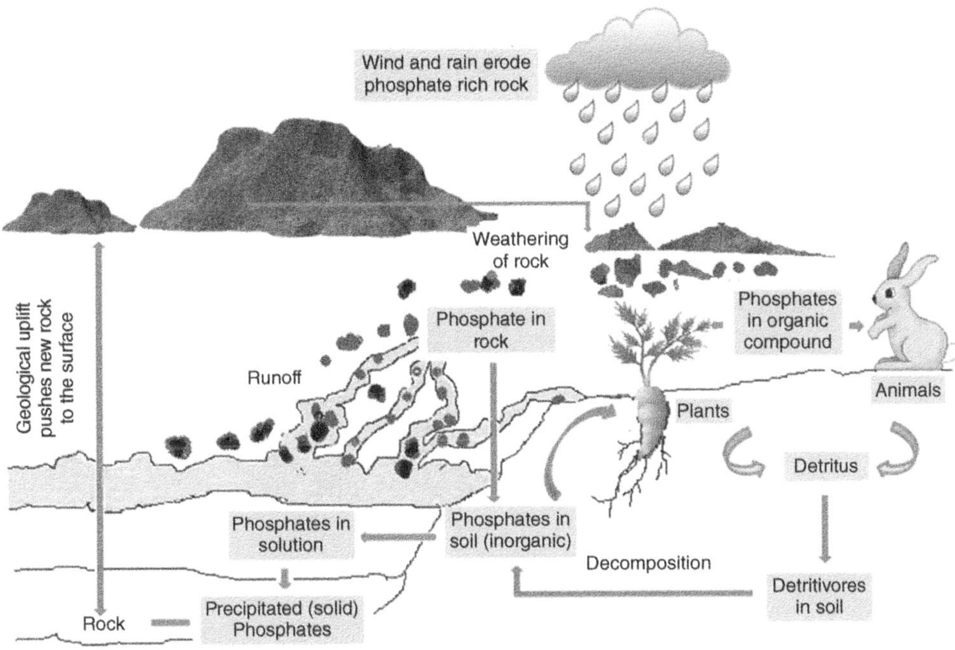

Figure 11.1 Bio-geochemical cycle of phosphorus (P).

of orthophosphate from microbial turnover compared to basal mineralization of soil. To be available for plants, this orthophosphate must diffuse through the rhizosphere (Richardson and Simpson 2011). The rate of diffusion basically depends on its immobilization by microbes and performance during the competition for direct nutrient uptake. Therefore, the rate of P release from microorganisms or the turnover time from the microbial biomass within the rhizosphere will have major inferences on plant P availability.

Several mechanisms have been proposed through which microorganisms can increase the plant's P uptake capacity from soil; they are (i) enlarged growth of root through either an extension of prevailing root systems or by hormonal stimulation of root growth, branching, or development of root hair (Hayat et al. 2010); (ii) modification of sorption balances that may cause increased net transfer of orthophosphate ions into soil solution or enable the movement of organic P either directly or indirectly through microbial turnover (Seeling and Zasoski 1993); and (iii) through initiation of metabolic processes that are vital in direct solubilization and mineralization of P from sparingly existing forms of soil inorganic and organic P (Richardson et al. 2009).

11.6 Phosphorus and Drought Stress

Being an important stress factor, drought hampers plants' growth and development causing significant losses in productivity worldwide. In addition to the physiological and biochemical impact of drought stress on plants, imbalances in mineral nutrition also occur

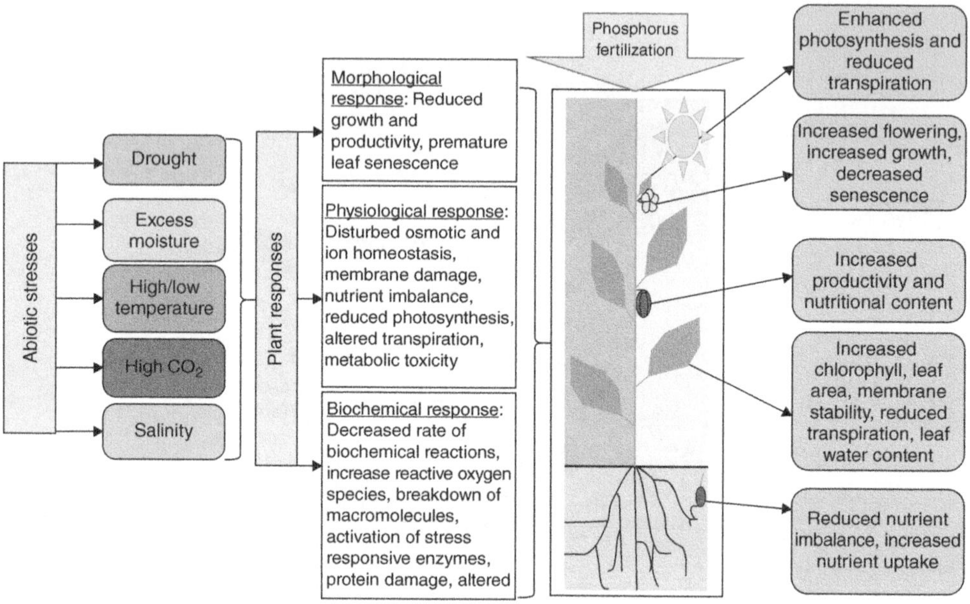

Figure 11.2 Abiotic stress responses in plants and alleviation by phosphorus (P) application.

leading to secondary effects (Waqas et al. 2019). Under drought stress, nutrient transport from the root to the shoot is reduced by reducing transpiration rates and altering the cell transporter functions (Figure 11.2).

A number of mechanisms have been projected that describe the positive effects of P fertilization on plants such as increased root growth, leaf area and photosynthesis (Singh and Sale 2000), improved cell-membrane stability and water relations (Abid et al. 2018), stomatal conductance, and nitrate reductase activity (Dinh et al. 2017). Researches indicate that in a number of species, drought tolerances and water use efficiency (WUE) can be enhanced by improved P nutrition (Waraich et al. 2011). In a study by Kang et al. (2014), it was reported that when P fertilizer was applied deep into the soils of semi-arid regions the grain yield and WUE in winter wheat increased due to stimulation of deep root development. Apart from P nutrition, drought tolerance is also dependent on the type of soil a particular plant is growing (Graciano et al. 2005). For example, P fertilization was found to be more effective in clay soils, even under moderate drought in *Eucalyptus grandis* seedlings, as compared to sandy soils where P nutrition proved effective only under well-watered conditions (Graciano et al. 2005). A systematic understanding of the effects of drought stress on the mineral nutrition of plants will help to formulate the suitable strategies for implementation to reduce the amount of drought-induced damages and the subsequent nutrient deficiency (da Silva et al. 2011).

11.6.1 P Availability in Soil Under Drought

Sufficient water supply is required for the availability and absorption of P by plants as the movement of phosphate ions in soil takes place through diffusion and with reduced soil water, the water-filled pores decrease their radii with increased soil porosity and reduced P mobility (Faye et al. 2006).

Thus, drought stress leads to decline in P absorption and transport in plants. A drop in available P forms in soil and an increase in unavailable P decreases the P uptake and subsequently reduced foliar P content in plant (Sardans and Peñuelas 2007). In a study by Fisher (1980), it was observed that drought stress significantly reduced the concentration of P in both stems and leaves of Townsville stylo (*Stylosanthes humilis*) and this reduction was more prominent in stems as compared to the leaves. In both the parts, P level returned to the same quantity as the control after stress recovery. This study proves the unreliability of the chemical analysis performed to detect the P status of the soil unless the plants have not grown under stress for at least 6 weeks. A decrease in P content in soil has been reported in *Spartinaal terniflora* (roots and shoot) due to dry soil (Brown et al. 2006). According to Sardans and Peñuelas (2007), 40% decrease in aboveground plant P accumulation was observed when grown under 22% reduced soil moisture due to lower increase in aerial biomass. Thus, water is a crucial element to maintain the P availability of the soil and plant.

11.6.2 Drought Tolerance by P Fertilization

P fertilization can increase drought tolerance in many plant species through increased WUE (Payne et al. 1992), and shoot dry matter (Rodriguez et al. 1996). According to Jones et al. (2005), sufficient P nutrition can mitigate the negative effects of drought on the yield and quality of malting barley (Table 11.1). It has been reported that adequate P nutrition in wheat enhances the rate of leaf appearance, final number of leaves, and leaf area per plant under drought conditions (Gutiérrez-Boem and Thomas 1998). Similarly, total P uptake and P translocation to seeds have been found to be alleviated by P fertilization in two soybean cultivars and thereby restricting yield loss (Jin et al. 2006). P fertilization has also been found to offset the harmful effect of drought on the leaf water potential of two pearl millet cultivars (Faye et al. 2006). In an experiment, drought stress has been reported to lower crop yield in plants without P nutrition while plants supplemented with P under water deficiency resulted in higher leaf area and dry weight of shoot, root, tiller, etc. indicating the positive involvement of P under drought (Kuwahara et al. 2016). Thus, underwater deficiency P nutrition can mitigate the damaging effects of water stress on plants (Table 11.2).

11.7 Phosphorus and Excess Moisture Stress

One of the major abiotic stresses is waterlogging in soils which greatly affects growth and productivity of numerous plant species around the world (Jackson and Colmer 2005). This condition is brought about by a high rainfall and poor drainage conditions (Caetano and Dias-Filho 2008). As a result, there is severe oxygen deficiency in the soil owing to the lower gas diffusion to root in addition to microbial respiration (George et al. 2012). This reduced oxygen condition of soil results in decreased aerobic root respiration, energy production, and nutrient uptake in plants (Kulichikhin et al. 2008).

Table 11.1 Increase in grain yield as a result of P fertilization under various abiotic stress.

Stress	Crop	Dosage of P	Yield increase (%)	References
Drought	Rapeseed	$150\,kg\,ha^{-1}$	25.4	Mamnabi et al. (2020)
Drought	Wheat	$120\,kg\,ha^{-1}$	87	Mumtaz et al. (2014)
High temperature	Rice	2 g/12 kg soil	7	Fahad et al. (2016a,b)
Salinity	chickpea	$200\,kg\,ha^{-1}$ of P_2O_5	30	Sadji-Ait Kaci et al. (2017)
Salinity	Okra	$30\,kg\,ha^{-1}$	10.5	Singh et al. (2017)
		$60\,kg\,ha^{-1}$	18.9	
		$90\,kg\,ha^{-1}$	26.8	

11.7.1 P Availability in Soil Under Excess Moisture

Under wet soil conditions, P availability can be dramatically reduced. Decreased chemical transformation and microbial activity in saturated soils reduce phosphorus release from the organic and mineral complex (Alori et al. 2017). Moreover, under waterlogged conditions the beneficial fungi, mycorrhizae, which generally improve the absorption of P and other nutrients, are frequently decreased (Bonfante and Genre 2010). Thus there is limited access of plants to soil P due to the limited volume of soil being explored by the plant root system. This decline in soil P under waterlogged condition indicates the O_2 deficiency. Under waterlogging, average leaf P concentration is reduced by about fourfold (0.06%, Singh and Setter 2017). Contrastingly, some reports say that waterlogging of soil may increase P availability (Oesterheld and Lavado 1997; Wang et al. 2017). This increase might have resulted from the mechanisms, namely – (i) change of ferric phosphates into more soluble ferrous compounds due to variations in redox potential. Though the soil is deficient in iron phosphates (Chacon et al. 2006), solubilization of that small fraction by waterlogging causes the availability of P. (ii) P availability might increase by a flush of P from the organic pool as the organic fraction of P principally predominates over the inorganic forms (>80% of total P, Jiao et al. 2007). P release from soil micro-organisms killed under anaerobic conditions or solubilization of soil organic P fractions might lead to this flush of P. It has been earlier reported that P release can be brought about by momentary declines in microbial populations (Sharma et al. 2013). Irrespective of the significance of these mechanisms, the parallel impact of waterlogging in fertilization treatments suggests that native P, instead of added P, was the fraction whose availability increased under waterlogged condition. Lastly, there is sufficient indication that the actual diffusion coefficient of P rises with increased soil water content (Mariano et al. 2002). These mechanisms explain that waterlogged conditions provided higher P than non-waterlogged soil. Therefore, this mechanism can be useful for the greater P uptake efficiency.

11.7.2 Excess Moisture Tolerance by P Fertilization

The higher availability of P nutrient in waterlogged soil imparted tolerance in a variety of grass (*Brachiaria ruziziensis*) (Jiménez et al. 2015). This is due to the formation of aerenchyma to maintain the root aerobic respiration (Colmer 2003; Visser et al. 2000) allowing

Table 11.2 Crop traits are improved by phosphorus application under different environmental stress.

Environmental stress	Crop	Growth stage	Dosage of P	Traits	References
Drought	*Alnus cremastogyne*	2 months after growing at an interval of 30 days	129.3 mg P (NaH_2PO_4, 25.5% P) in the form of mixed in 200 ml of water	Improved growth and leaf relative water content, photosynthetic rate, and maximal quantum efficiency of PSII	Tariq et al. (2018)
Drought	Field pea (*Pisum sativum*)	Flowering stage	60 mg P kg^{-1}	Increased root length distribution in deeper soil layers	Jin et al. (2015)
Drought	Gold Phoebe (*Phoebe zhennan*)	2 months after growing	129.3 mg P (NaH_2PO_4, 25.5% P) mixed in 200 ml of water	Increased root biomass, improved leaf relative water content, net photosynthetic rate, and maximal quantum efficiency of PSII	Tariq et al. (2017)
Drought	Soybean [*Glycine max* (L.) Merr.]	Podding stage	30 mg P kg^{-1} (added as KH_2PO_4)	Enhanced the concentration and accumulation of nitrogen (N) and P in shoots and seeds, improved grain yield and quality	Jin et al. (2006)
Drought	*Phaseolus vulgaris* L.	Pre-flowering stage	Pi foliar spray (10 and 20 g Pi l^{-1})	Lower reduction in pod number and seed dry weight per plant	Santos et al. (2004)
Drought	Rapeseed	Entire life cycle	150 kg ha^{-1}	Increased the antioxidant enzymes activities, soluble sugars, chlorophyll content, leaf water content, membrane stability index, and stomatal conductance	Mamnabi et al. (2020)
Elevated CO_2	Field pea (*Pisum sativum*)	Flowering stage	60 mg P kg^{-1}	Increased concentrations of soluble sugars, increased root length distribution in deeper soil layers	Jin et al. (2014)
High temperature	Rice (*Oryza sativa*)	3 weeks after sowing	2 g/12 kg soil as triple super phosphate [$Ca(H_2PO_4)_2{\cdot}2H_2O$]	Increase grain size and water use efficiency	Fahad et al. (2016a,b)

(*Continued*)

Table 11.2 (Continued)

Environmental stress	Crop	Growth stage	Dosage of P	Traits	References
Low temperature	Winter wheat (*Triticum aestivum*)	–	150 mg P_2O_5 kg^{-1} soil	Increased seed yield, soluble sugar, water-soluble protein	Nie et al. (2015)
Salinity	*Cenostigma pyramidale*	After 6 months of development	Foliar spray of monoammonium phosphate solution in leaves (10 g Pi l^{-1})	Increased stomatal conductance, photosynthesis, and biomass	Frosi et al. (2018)
Salinity	Maize (*Zea mays* L.)	60 days after germination	160 kg P ha^{-1}	Stem diameter, leaf number, and area, leaves chlorophyll index, and shoot and root dry weights	Soltani et al. (2015)
Salinity	Sorghum	Early tillering, upstream	60 kg P ha^{-1}	Increase in stem height, dry material yield, nitrogen and phosphorus uptake, proline accumulation	Belouchrani et al. (2019)

the plant to adjust under waterlogged condition. In a study conducted by Ayub et al. (2002), the application of phosphorus together with nitrogen increased the grain yield of fodder maize (Table 11.1).

11.8 Phosphorus and Temperature Stress (Low/Chilling Temperature Stress, High Temperature Stress)

Phosphorus present in soil organic fractions is released by mineralization engaging soil microorganisms and is highly influenced by soil temperature.

11.8.1 P Availability in Soil Under High/Low Rhizosphere Temperature

Both high and low temperature is an important factor in the interface of sediment and water as it raises the water column and hampers the P release from sediments (Wang et al. 2015). It has been found that in the sediments, the concentration of labile P increases with the increased temperature leading to faster P release (Vincent et al. 2014). Environmental factor such as soil temperature greatly changes the dynamic pool of microbial P (Richardson and Simpson 2011) and thus significantly affect the mechanisms involved in P uptake.

Under field conditions, phosphate solubilizing microorganisms (PMS) need to endure a variety of abiotic stresses depending on the agro-climatic conditions together with seasonal changes. High or low temperatures greatly influence the efficacy of different PMS and the isolation of numerous PSMs under such situations proves their resistance to stress (Chang and Yang 2009). P solubilization by a number of Streptomyces, Bacillus, and Aspergillus strains had been documented at 50 °C which could aid composting (Chaiharn and Lumyong 2008). A number of PMS are also reported that solubilize P at low-temperature conditions (Batool and Iqbal 2019; Mujahid et al. 2015).

Pseudomonas fluorescens mutant was found to solubilize mineral P to a similar degree at both 10 and 25 °C (Trivedi and Sa 2008). Again, mutants of *Pseudomonas corrugate* were isolated with the ability of P solubilization at 4 °C as well as at 25 °C (Zhu et al. 2011). Bacteria isolated from cold conditions of the Himalaya established good P solubilizing capacity at low temperature. P-solubilizing halophilic bacteria (*Kushneria* sp. YCWA18) isolated from the coast of the Yellow Sea of China, could grow quickly at 28 °C at 6% (w/v) NaCl concentration (Singh et al. 2004). Thus, numerous PSMs can endure many stress conditions but very few can retain the phenotype under adverse environmental conditions.

11.8.2 Physiological Changes in Plants at Membrane Level Due to Temperature Stress

It is commonly known that plant cell membranes are among the prime targets under abiotic stresses and maintenance of cellular integrity and function is an indicator of stress tolerance (Bajji et al. 2002). Under conditions of extreme temperature, fluidity of membranes changes and lipid peroxidation increases and membrane permeability is often hampered (Dias et al. 2010; Su et al. 2007). A number of cellular alterations resulting from either high or low temperature enhance the production of toxic compounds, particularly

ROS leading to oxidative stress (Mittler 2002) (Figure 11.2). Being toxic, ROS can induce protein oxidation, DNA impairment, membrane lipid peroxidation, and pigment damage (Hasanuzzaman et al. 2013). Plants, however, have evolved several mechanisms for extreme temperatures that help in reducing damage and imparting cellular homeostasis (Kotak et al. 2007). As a measure of electrolyte leakage from tissues, study on cell membrane thermostability (CMT) is used a selection of various crop species for heat stress tolerance (Ismail and Hall 1999; Srinivasan et al. 1996; Tang et al. 2007). In wheat plants, heat-induced damage enhanced the lipid peroxidation and decreased the thermostability of the membrane by 28 and 54% respectively leading to remarkable increase in electrolyte leakage (Savicka and Škute 2010). Again, in rice plants, heat stress increased the leaf temperature which declined the activity of antioxidant enzymes and consequently increase malondialdehyde (MDA) content (Hurkman et al. 2009). High temperature stress-triggered membrane lipid peroxidation and intensified membrane injury were also detected in cotton, sorghum, and soybean (Mohammed and Tarpley 2010; Tan et al. 2011; Young et al. 2004). Under chilling conditions, plant cell membranes go through alterations in lipid and fatty acid compositions so as to conserve the function of chloroplast at low temperature (Liu et al. 2018). These changes might be the result of an increased concentration of unsaturated fatty acids in galactolipids, such as linolenic acid, during low-temperature acclimation. In *Coffea dewevrei*, an electrolyte leakage was confirmed by the production of MDA under cold conditions which indicated lipid peroxidation (Campos et al. 2003).

Under extreme temperature, plants have evolved a variety of responses which is often related to enhanced activities of antioxidative defence enzymes that imparts stress tolerance to plants (Almeselmani et al. 2009). Recurrent heat stress results in the accumulation of ROS at the outer surface of the plasma membrane which might cause depolarization of the membrane (Qi et al. 2011), stimulation of Ca-induced RBOHD (the ROS-producing enzyme RBOHD located at the plasma membrane), etc. Thus, under temperature stress, the physiology of the cellular membrane is significantly hampered.

11.8.3 Temperature Tolerance by P Fertilization

The nutritional status of the plant is greatly affected due to its ability to alleviate temperature stress (Waraich et al. 2012). As higher light intensity is also associated with high temperature, it disturbs the uptake of mineral nutrients in plants and negatively affects plant growth (Ma et al. 2016). Under heat stress, P fertilization of rice plants resulted in higher grain yield production with better quality due to improved photosynthesis, WUE, and grain size, which compensated for the adversities of high temperature stress (Fahad et al. 2016a,b) (Table 11.2). During hot and dry summers, the application of foliar KH_2PO_4 has been found to delay leaf senescence with higher yield in winter wheat (Amanullah et al. 2016). Under high temperature, foliar P treatments proved beneficial in enhancing cell sugar content and protection of cell membranes (Bita and Gerats 2013).

Low soil temperature has been reported to inhibit plant growth due to restricted root growth (Sebastian et al. 2016). P also plays a crucial role in imparting tolerance against cold stress. Foliar application of P in the form of KH_2PO_4 enhanced the grain weight of winter wheat under low-temperature conditions (Mosali et al. 2006). Under such conditions, P application aided in vital early growth resulting in a larger root system, increased WUE, and higher yields (Table 11.2) (Shen et al. 2013) (Table 11.2). Thus, at low temperatures, P application at early growth stages is considered to be an effective strategy to alleviate the harmful effects of temperature stress by stimulating root development and establishing a nearly ideal root architecture (Table 11.2).

11.9 Phosphorus and High CO_2

At present times, CO_2 concentration is expected to rise by 443–541 ppm in 2050 and going to be two times the present concentration by 2100 (Zhou et al. 2020). It has been reported that under elevated CO_2, P requirement of a plant would be higher compared to those grown at ambient CO_2 as the rate of photosynthesis increases in response to CO_2 especially in C3 plants. This evidently shows that owing to the heavy metabolic demand for P acquisition, this limiting nutrient would be reactive to future climatic conditions (Pandey et al. 2015). Thus, with respect to the rising CO_2 concentrations, P is an exclusive nutrient among all the essential plant nutrients.

11.9.1 Impact of P Nutrition Under Excess CO_2

Influence of CO_2 on growth and yield of various crops in distinct under regimes of P availability and thereby controlling various physiological processes. These physiological processes comprise dry matter production and partitioning, photosynthesis, metabolite synthesis, activities of enzymes such as acid phosphatase, acid invertase, sucrose synthase, and ribulose-1,5-bisphosphate carboxylase oxygenase (RuBisCO), growth parameters, root exudation, arbuscular-mycorrhizal colonization, and nutrient uptake and use efficiency (Pandey et al. 2015). The interactive effect of CO_2 and P nutrition on plant response suggests that the requirement of P of a plant grown under high CO_2 would be higher than those grown under normal ambient conditions due to increases in photosynthetic rate in response to higher CO_2 especially in C3 plants (Pandey et al. 2015). Therefore, this evidently shows that because of the heavy metabolic demand for P acquisition, this limiting nutrient would be responsive to future climatic conditions. For enhanced plant productivity; increased usage of P fertilizers and the constant rise in CO_2 has led to imbalances in P (Peñuelas et al. 2013). Under high CO_2, plant growth is enhanced but P deficiency leads to growth retardation especially in the shoots. Nonetheless, when both these factors are individually applied, the root growth and exudation, as well as increased mycorrhizal association, are increased (Püschel et al. 2016).

Even though the specific effects of high CO_2 or P stress have been studied widely but their combined impact on plant growth is yet to be understood. The harmful effects induced by low P could be moderately overcome by high CO_2 at least during the early growth stage. This is primarily due to the influence of high CO_2 on RuBisCO activation, RuBP regeneration capacity, and reduction in resistance to CO_2 diffusion across stomata and mesophyll. Contrariwise, the stimulation of growth at high CO_2 again raises the P demand. Thus, longstanding exposure of plants to increased CO_2 causing high photosynthetic acclimation is overcome by augmented P supply. This is because assimilation of extra carbon (C) at high CO_2 involves more energy in the form of ATP, also the energy-rich compounds modulate RuBisCO activity

(Xu et al. 2016). Under high CO_2 and P starvation, alteration in below-ground processes comprise enhanced root growth with increased secretion of ATPase enzymes and organic acids and increased colonization with mycorrhizal fungi (Roy-Bolduc and Hijri 2011). These modifications might result in improved uptake of P under high CO_2 as long as adequate P is accessible in the soil. Under P stress, internal P recycling, scavenging, and remobilization cause more efficient P utilization (van de Wiel et al. 2016). High CO_2 might support these processes as several P stress-responsive genes such as PEP-associated proteins (PAPs) and reactive nitrogen species (RNS), involved in P recycling are up-regulated under high $CO_{2.}$

It is therefore of utmost importance to know how elevated levels of CO_2 can impact crop yields under P-limiting conditions. Likely requirements for higher inputs of P fertilizers under elevated CO_2 conditions could even accelerate the depletion of rock P reserves (Pandey et al. 2015).

11.9.2 P Availability in Soil Under High CO_2

P acquisition and utilization efficacy of soil should be enhanced in order to sustain food production particularly in P deficient soils (López-Arredondo et al. 2014). P availability of plants is increased by extensive root developments and is thus determined by the plant C status while the P status of the plant impacts plant growth rate and photosynthesis resulting in multiple C–P interactions. Earlier reports on elevated CO_2 and P interactions showed that increased CO_2 aid in the growth of C3 grasses even under P starved condition (Newbery et al. 1995; Pandey et al. 2015) while in case of legumes which are also C3 plants, limited growth is observed under low-P conditions (Lam et al. 2012, Singh et al. 2014). These variations in responses with plant groups depend on C partitioning and P uptake efficiencies by their root system (Jin et al. 2015). C limitation during AM (*Arbuscular mycorrhiza)* development both in the soil and within the roots can be alleviated at high CO_2 concentration and lead to improved mycorrhiza formation. This might consecutively lead to mycorrhizal P uptake and enable the growth of legumes grown under elevated CO_2 in P deficient soil.

Increased CO_2 levels affect the ability of the plants to uptake P from the soil. Under elevated CO_2, the inflow of C from plants to soil is enhanced by the increased discharge of root exudates (Lin et al. 2000; Song et al. 2014). Functional molecules present in these exudates improve the solubility of rhizosphere P and thus increase in plant P nutrition (Richardson et al. 2009). Moreover, change in rhizosphere pH and microbial activity is brought about by root exudates (Shen et al. 2011) and these impacts of high CO_2 can alter P availability in the rhizosphere and therefore affect plant P acquisition (de Graaff et al. 2006).

11.10 Phosphorus and Salinity Stress

Due to the rising sea levels, intrusion of saltwater or salinization is becoming a concern around the world (Wilson et al. 2019). As a result, there is extreme build-up of Na^+ and Cl^- ions in plant cells. Salinity is closely linked with the imbalance of mineral nutrients. Crops are extremely sensitive to salinity throughout their ontogeny. As a result, salinity decreases the productivity of almost all crops in saturated soil when exceeded the threshold value. The threshold of salinity in most of the crops is low (ranging from 1 to 2.5 dSm^{-1})

(Singh and Singh 2017). Among other macronutrients, P is the most crucial for crops in the vegetative as well as reproductive stage (Ye et al. 2019). Plant growth is affected by salinity due to their concentration in soil solution which reduces the osmotic potential thereby decreasing the water availability. Also, excessive ion accumulation may also cause toxicity which finally results in plant death (Flowers and Flowers 2005; Wilson et al. 2006).

11.10.1 P Availability in Saline Soil

P plays a crucial role in the early vegetative growth phase as it aids in tillering and the development of roots (Peltonen-Sainio et al. 2006; Slaton et al. 2002). As a result of low total P and elevated soil P fixation, the availability of P in arid soil is much less than the humid regions (Shen et al. 2011). Majority of the water-soluble P which is applied to the soil as fertilizers are readily converted into unavailable forms (Rivaie et al. 2008). In alkaline soils, the occurrence of lime further aggravates its availability while in calcareous soil, the reaction between lime and P of the soil solution results in the formation of calcium phosphate compounds (Chhabra and Thakur 2000). Thus, low P availability in calcareous soil leads to low P fertilizer efficiency. Plants grown under such conditions can be stunted due to reduced internodes and poor root systems. Most of the soil P in alkaline soil is present in the form of soluble sodium phosphate compounds (NaH_2PO_4). The uptake of excess N with P fertilizers causes a decline in plant growth and yield (Mahmood et al. 2013).

11.10.2 Salinity Tolerance by P Fertilization

A number of mitigation strategies to combat the deleterious effects of salinity on crops have been reported. However, nutrient management practices to alleviate salinity have stood out as the most promising ones (Hellal and Abdelhamid 2013). P is one of the most vital elements for plants as it aids root growth, maturation of fruit, grain, fibre and fruit formation, and also in plant vigour. Salinity and P interaction affects plant growth and increasing P results in increasing salinity tolerance as reported by Bargaz et al. (2016). On the other hand, it has been documented in *Oriza sativa* and *Pistacia vera* that only a moderate P supply is required to enhance salinity tolerance while higher P input resulted in excess P accumulation thereby declining overall plant growth and productivity (Aslam et al. 1996; Shahriaripour et al. 2011). Again, legumes are extremely P-dependent crops, and thus P deficiency affects plant growth and nitrogen fixation capacity (Lrsquo et al. 2012). Higher P addition to crops limits the risk of nutrient deficiency thereby reducing the chance of further damage to crops (Table 11.2) (Soares et al. 2019).

11.11 Conclusion and Future Work

Being a macronutrient, phosphorus is indispensable for many aspects of plant and cell biology such as photophosphorylation, genetic transfer, the use of ATP as energy exchange currency, the transportation of nutrients, and phospholipid cell membranes. Environmental factors such as low/high temperature, deficiency in soil water, excess soil moisture, salinity, and elevated CO_2. lead to decline in P absorption and transport in plants which negatively affects the rate of nutrient transport from the root to the shoot by reducing transpiration rates and altering the cell transporter functions thereby hampering growth, development and yield processes of crop plants. P fertilization under drought has been found to enhance tissue proline levels to regulate the cellular osmotic potential which is a kind of osmotic adjustment to combat water scarcity. The higher availability of P in soil allows the plant to adjust during the waterlogged condition by the formation of aerenchyma – to maintain the root aerobic respiration. During the heat and cold stress, P fertilization promotes larger root system, increased WUE and higher grain yield production with better quality due to improved photosynthesis. Phosphorus application can significantly increase stomatal conductance, photosynthesis, and biomass production in many crop species under saline conditions. The interaction of elevated CO_2 and P nutrition is greatly beneficial for plants as the inflow of C from plants to soil is enhanced by the increased discharge of root exudates. Functional molecules present in these exudates improve the solubility of rhizosphere P and which facilitate in plant P nutrition. With respect to the above discussed abiotic stresses, it can be concluded that P is an exclusive nutrient, and optimizing P application may become one strategy that will help in minimizing the impact of climate change in the near future.

11.12 Summary

Abiotic stresses such as drought, excess moisture, salinity, temperature (heat and cold), and elevated CO_2, are critically affecting the agro-ecosystem. This is leading to a decline in global crop productivity by more than 50% every year. Adverse environmental conditions lower the rate of nutrient transport from the root to the shoot thereby hampering the growth, development, and yield processes of crop plants. Phosphorus (P) is one of the most abundant macronutrients in plant tissues constituting around 0.2% of plant mass and is required for many aspects of plant biology. A comprehensive understanding of the impact of drought, excess soil moisture, high/low temperature, salinity, and elevated CO_2 on crop growth and productivity will be critical in evaluating the impact of climate change and climate variability on crop production. Also, assessing the role of externally applied P in tolerating these unfavourable conditions will give an insight into management practices against these adverse conditions. Therefore, this chapter aims to highlight the latest facets of our understanding of how crop nutrient management and mineral status can impart resistance against various abiotic stresses especially through the external application of P.

References

Abid, M., Ali, S., Qi, L.K. et al. (2018). Physiological and biochemical changes during drought and recovery periods at tillering and jointing stages in wheat (*Triticum aestivum* L.). *Scientific Reports* 8: 1–15.

Achat, D.L., Morel, C., Bakker, M.R. et al. (2010). Assessing turnover of microbial biomass phosphorus: combination of an isotopic dilution method with a mass balance model. *Soil Biology and Biochemistry* 42: 2231–2240.

Al-Karaki, G.N., Clark, R.B., and Sullivan, C.Y. (1996). Phosphorous nutrition and water stress effects on proline accumulation in sorghum and bean. *Journal of Plant Physiology* 148: 745–751.

Almeselmani, M., Deshmukh, P.S., and Sairam, R.K. (2009). High temperature stress tolerance in wheat genotypes: role of antioxidant defence enzymes. *Acta Agronomica Hungarica* 57: 1–14.

Alori, E.T., Glick, B.R., and Babalola, O.O. (2017). Microbial phosphorus solubilization and its potential for use in sustainable agriculture. *Frontiers in Microbiology* 8: 971.

Amanullah, S.A., Iqbal, A., and Fahad, S. (2016). Foliar phosphorus and zinc application improve growth and productivity of maize (*Zea mays* L.) under moisture stress conditions in semi-arid climates. *Journal of Microbial Biochemical Technology* 8: 433–439.

Aslam, M., Flowers, T.J., Qureshi, R.H., and Yeo, A.R. (1996). Interaction of phosphate and salinity on the growth and yield of rice (*Oryza sativa* L.). *Journal of Agronomy and Crop Science* 176: 249–258.

Ayub, M., Nadeem, M.A., Sharar, M.S., and Mahmood, N. (2002). Response of maize (*Zea mays* L.) fodder to different levels of nitrogen and phosphorus. *Asian Journal of Plant Sciences* 1: 352–354.

Bajji, M., Kinet, J.M., and Lutts, S. (2002). The use of the electrolyte leakage method for assessing cell membrane stability as a water stress tolerance test in durum wheat. *Plant Growth Regulation* 36: 61–70.

Bargaz, A., Nassar, R.M., Rady, M.M. et al. (2016). Improved salinity tolerance by phosphorus fertilizer in two *Phaseolus vulgaris* recombinant inbred lines contrasting in their P-efficiency. *Journal of Agronomy and Crop Science* 202: 497–507.

Batool, S. and Iqbal, A. (2019). Phosphate solubilizing rhizobacteria as alternative of chemical fertilizer for growth and yield of *Triticum aestivum* (Var. Galaxy 2013). *Saudi Journal of Biological Sciences* 26: 1400–1410.

Belouchrani, A.S., Latati, M., Ounane, S.M. et al. (2019). Study of the interaction salinity: phosphorus fertilization on sorghum. *Journal of Plant Growth Regulation* 9: 1–6.

Benning, C. and Ohta, H. (2005). Three enzyme systems for galactoglycerolipid biosynthesis are coordinately regulated in plants. *Journal of Biological Chemistry* 280: 2397–2400.

Bita, C. and Gerats, T. (2013). Plant tolerance to high temperature in a changing environment: scientific fundamentals and production of heat stress-tolerant crops. *Frontiers in Plant Science* 4: 273.

Bonfante, P. and Genre, A. (2010). Mechanisms underlying beneficial plant–fungus interactions in mycorrhizal symbiosis. *Nature Communications* 1: 48.

Brown, C.E., Pezeshki, S.R., and DeLaune, R.D. (2006). The effects of salinity and soil drying on nutrient uptake and growth of *Spartina alterniflora* in a simulated tidal system. *Environmental and Experimental Botany* 58: 140–148.

Bünemann, E.K. (2008). Enzyme additions as a tool to assess the potential bioavailability of organically bound nutrients. *Soil Biology and Biochemistry* 40: 2116–2129.

Caetano, L.P.S. and Dias-Filho, M.B. (2008). Responses of six *Brachiaria* spp. accessions to root zone flooding. *Revista Brasileira de Zootecnia* 37: 795–801.

Campos, P.S., nia Quartin, V., chicho Ramalho, J., and Nunes, M.A. (2003). Electrolyte leakage and lipid degradation account for cold sensitivity in leaves of *Coffea* sp. plants. *Journal of Plant Physiology* 160: 283–292.

Chacon, N., Silver, W.L., Dubinsky, E.A., and Cusack, D.F. (2006). Iron reduction and soil phosphorus solubilization in humid tropical forests soils: the roles of labile carbon pools and an electron shuttle compound. *Biogeochemistry* 78: 67–84.

Chaiharn, M. and Lumyong, S. (2008). Phosphate solubilisation potential and stress tolerance of rhizobacteria from rice soil in Northern Thailand. *World Journal of Microbiology and Biotechnology* 25: 305–314.

Chang, C.H. and Yang, S.S. (2009). Thermo-tolerant phosphate solubilising microbes for multi-functional biofertilizer preparation. *Bioresource Technology* 100: 1648–1658.

Cheng, W. (2009). Rhizosphere priming effect: its functional relationships with microbial turnover, evapotranspiration, and C–N budgets. *Soil Biology and Biochemistry* 41: 1795–1801.

Chhabra, R. and Thakur, N.P. (2000). Long-term study on phosphorus fertilization for rice-wheat cropping system in an alkali soils in the Indo-Gangetic plains. In: *Long-term Soil Fertility Experiments in Rice-Wheat Cropping Systems* (Rice–Wheat Consortium) Paper Series 6, Rice Wheat Consortium for the Indo-Gangetic Plains, New Delhi (ed. I.P. Abrol, K.F. Bronson, J.M. Duxbury and R.K. Gupta), 31–39.

Colmer, T.D. (2003). Long-distance transport of gases in plants: a perspective on internal aeration and radial oxygen loss from roots. *Plant, Cell & Environment* 26: 17–36.

Damon, P.M., Bowden, B., Rose, T., and Rengel, Z. (2014). Crop residue contributions to phosphorus pools in agricultural soils: a review. *Soil Biology and Biochemistry* 74: 127–137.

Dias, A.S., Barreiro, M.G., Campos, P.S. et al. (2010). Wheat cellular membrane thermotolerance under heat stress. *Journal of Agronomy and Crop Science* 196: 100–108.

Dinh, T.H., Watanabe, K., Takaragawa, H. et al. (2017). Photosynthetic response and nitrogen use efficiency of sugarcane under drought stress conditions with different nitrogen application levels. *Plant Production Science* 20: 412–422.

Fahad, S., Hussain, S., Saud, S. et al. (2016a). Exogenously applied plant growth regulators enhance the morpho-physiological growth and yield of rice under high temperature. *Frontiers in Plant Science* 7: 1250.

Fahad, S., Hussain, S., Saud, S. et al. (2016b). A combined application of biochar and phosphorus alleviates heat-induced adversities on physiological, agronomical and quality attributes of rice. *Plant Physiology and Biochemistry* 103: 191–198.

Faye, I., Diouf, O., Guissé, A. et al. (2006). Characterizing root responses to low phosphorus in pearl millet [*Pennisetumglaucum* (L.) R. Br.]. *Agronomy Journal* 98: 1187–1194.

Fisher, M.J. (1980). The influence of water stress on nitrogen and phosphorus uptake and concentrations in *Townsville stylo* (*Stylosanthes humilis*). *Australian Journal of Experimental Agriculture* 20: 175–180.

Flowers, T.J. and Flowers, S.A. (2005). Why does salinity pose such a difficult problem for plant breeders? *Agricultural Water Management* 78: 15–24.

Flügge, U.I., Häusler, R.E., Ludewig, F., and Fischer, K. (2003). Functional genomics of phosphate antiport systems of plastids. *Physiologia Plantarum* 118: 475–482.

Frosi, G., Barros, V.A., Oliveira, M.T. et al. (2018). Arbuscular mycorrhizal fungi and foliar phosphorus inorganic supply alleviate salt stress effects in physiological attributes, but only arbuscular mycorrhizal fungi increase biomass in woody species of a semiarid environment. *Tree Physiology* 38 (1): 25–36.

George, T.S., Simpson, R.J., Hadobas, P.A. et al. (2007). Accumulation and phosphatase-lability of organic phosphorus in fertilised pasture soils. *Australian Journal of Agricultural Research* 58: 47–55.

George, E., Horst, W.J., and Neumann, E. (2012). Adaptation of plants to adverse chemical soil conditions. In: *Marschner's Mineral Nutrition of Higher Plants*, 409–472. Petra Marschner: Academic Press.

de Graaff, M.A., van Groenigen, K.J., Six, J. et al. (2006). Interactions between plant growth and soil nutrient cycling under elevated CO_2: a meta-analysis. *Global Change Biology* 12: 2077–2091.

Graciano, C., Guiamet, J.J., and Goya, J.F. (2005). Impact of nitrogen and phosphorus fertilization on drought responses in *Eucalyptus grandis* seedlings. *Forest Ecology and Management* 212: 40–49.

Gutiérrez-Boem, F.H. and Thomas, G.W. (1998). Phosphorus nutrition affects wheat response to water deficit. *Agronomy Journal* 90: 166–171.

Hasanuzzaman, M., Nahar, K., Alam, M. et al. (2013). Physiological, biochemical, and molecular mechanisms of heat stress tolerance in plants. *International Journal of Molecular Sciences* 14: 9643–9684.

Hayat, R., Ali, S., Amara, U. et al. (2010). Soil beneficial bacteria and their role in plant growth promotion: a review. *Annals of Microbiology* 60: 579–598.

Hellal, F.A. and Abdelhamid, M.T. (2013). Nutrient management practices for enhancing soybean (*Glycine max* L.) production. *Acta Biológica Colombiana* 18: 239–250.

Hendrickson, L., Chow, W.S., and Furbank, R.T. (2004). Low temperature effects on grapevine photosynthesis: the role of inorganic phosphate. *Functional Plant Biology* 31: 789–801.

Hernández, I. and Munné-Bosch, S. (2015). Linking phosphorus availability with photo-oxidative stress in plants. *Journal of Experimental Botany* 66: 2889–2900.

Hurkman, W.J., Vensel, W.H., Tanaka, C.K. et al. (2009). Effect of high temperature on albumin and globulin accumulation in the endosperm proteome of the developing wheat grain. *Journal of Cereal Science* 49: 12–23.

Ismail, A.M. and Hall, A.E. (1999). Reproductive-stage heat tolerance, leaf membrane thermostability and plant morphology in cowpea. *Crop Science* 39: 1762–1768.

Jackson, M.B. and Colmer, T.D. (2005). Response and adaptation by plants to flooding stress. *Annals of Botany* 96: 501–505.

Jiao, Y., Whalen, J.K., and Hendershot, W.H. (2007). Phosphate sorption and release in a sandy-loam soil as influenced by fertilizer sources. *Soil Science Society of America Journal* 71: 118–124.

Jiménez, J.D.L.C., Cardoso, J.A., Arango-Londoño, D. et al. (2015). Influence of soil fertility on waterlogging tolerance of two *Brachiaria* grasses. *Agronomía Colombiana* 33: 20–28.

Jin, J., Wang, G., Liu, X. et al. (2006). Interaction between phosphorus nutrition and drought on grain yield, and assimilation of phosphorus and nitrogen in two soybean cultivars differing in protein concentration in grains. *Journal of Plant Nutrition* 29: 1433–1449.

Jin, J., Lauricella, D., Armstrong, R. et al. (2014). Phosphorus application and elevated CO_2 enhance drought tolerance in field pea grown in a phosphorus-deficient vertisol. *Annals of Botany* 116: 975–985.

Jin, J., Tang, C., and Sale, P. (2015). The impact of elevated carbon dioxide on the phosphorus nutrition of plants: a review. *Annals of Botany* 116: 987–999.

Jones, C.A., Jacobsen, J.S., and Wraith1, J.M. (2005). Response of malt barley to phosphorus fertilization under drought conditions. *Journal of Plant Nutrition* 28: 1605–1617.

Kang, L.Y., Yue, S.C., and Li, S.Q. (2014). Effects of phosphorus application in different soil layers on root growth, yield, and water-use efficiency of winter wheat grown under semi-arid conditions. *Journal of Integrative Agriculture* 13: 2028–2039.

Kotak, S., Larkindale, J., Lee, U. et al. (2007). Complexity of the heat stress response in plants. *Current Opinion in Plant Biology* 10: 310–316.

Kulichikhin, K.Y., Chirkova, T.V., and Fagerstedt, K.V. (2008). Intracellular pH in rice and wheat root tips under hypoxic and anoxic conditions. *Plant Signaling & Behavior* 3: 240–242.

Kuwahara, F.A., Souza, G.M., Guidorizi, K.A. et al. (2016). Phosphorus as a mitigator of the effects of water stress on the growth and photosynthetic capacity of tropical C_4 grasses. *Acta Scientiarum Agronomy* 38: 363–370.

Lam, S.K., Chen, D., Norton, R., and Armstrong, R. (2012). Does phosphorus stimulate the effect of elevated [CO_2] on growth and symbiotic nitrogen fixation of grain and pasture legumes? *Crop & Pasture Science* 63: 53–62.

Lin, W., Zhang, F., and Bai, K. (2000). Response of plant rhizosphere to atmospheric CO_2 enrichment. *Chinese Science Bulletin* 45: 97–100.

Liu, X., Zhou, Y., Xiao, J., and Bao, F. (2018). Effects of chilling on the structure, function and development of chloroplasts. *Frontiers in Plant Science* 9: 1715.

López-Arredondo, D.L., Leyva-González, M.A., González-Morales, S.I. et al. (2014). Phosphate nutrition: improving low-phosphate tolerance in crops. *Annual Review of Plant Biology* 65: 95–123.

Lrsquo, B., Sifi, B., Zaman-Allah, M. et al. (2012). Genotypic variability for tolerance to salinity and phosphorus deficiency among N2-dependent recombinant inbred lines of Common Bean (*Phaseolus vulgaris*). *African Journal of Microbiology Research* 6 (20): 4205–4213.

Ma, Q., Cao, X., Wu, L. et al. (2016). Light intensity affects the uptake and metabolism of glycine by pakchoi (*Brassica chinensis* L.). *Scientific Reports* 6: 21200.

Macklon, A.E.S., Grayston, S.J., Shand, C.A. et al. (1997). Uptake and transport of phosphorus by *Agrostis capillaris* seedlings from rapidly hydrolysed organic sources extracted from 32 P-labelled bacterial cultures. *Plant and Soil* 190 (1): 163–167.

Mahmood, I.A., Arshad, A., Aslam, M. et al. (2013). Phosphorus availability in different salt-affected soils as influenced by crop residue incorporation. *International Journal of Agriculture and Biology* 15 (3): 472–478.

Mamnabi, S., Nasrollahzadeh, S., Ghassemi-Golezani, K., and Raei, Y. (2020). Improving yield-related physiological characteristics of spring rapeseed by integrated fertilizer management under water deficit conditions. *Saudi Journal of Biological Sciences* 27: 797–804.

Mariano, I.O.D.S., Fernandes, L.A., Faquin, V., and Andrade, A.T.D. (2002). Phosphorus critical levels and availability in lowland soils cultivated with flooded rice. *Scientia Agricola* 59: 113–120.

Mittler, R. (2002). Oxidative stress, antioxidants and stress tolerance. *Trends in Plant Science* 7: 405–410.

Mohammed, A.R. and Tarpley, L. (2010). Effects of high night temperature and spikelet position on yield-related parameters of rice (*Oryza sativa* L.) plants. *European Journal of Agronomy* 33: 117–123.

Mosali, J., Desta, K., Teal, R.K. et al. (2006). Effect of foliar application of phosphorus on winter wheat grain yield, phosphorus uptake, and use efficiency. *Journal of Plant Nutrition* 29: 2147–2163.

Mujahid, T.Y., Subhan, S.A., Wahab, A. et al. (2015). Effects of different physical and chemical parameters on phosphate solubilization activity of plant growth promoting bacteria isolated from indigenous soil. *Journal of Pharmacy and Nutrition Sciences* 5: 64–70.

Müller, J., Gödde, V., Niehaus, K., and Zörb, C. (2015). Metabolic adaptations of white lupin roots and shoots under phosphorus deficiency. *Frontiers in Plant Science* 6: 1014.

Mumtaz, M.Z., Aslam, M., Jamil, M., and Ahmad, M. (2014). Effect of different phosphorus levels on growth and yield of wheat under water stress conditions. *Journal of Environment and Earth Science* 4: 23–30.

Nakamura, Y., Awai, K., Masuda, T. et al. (2005). A novel phosphatidylcholine-hydrolyzing phospholipase C induced by phosphate starvation in Arabidopsis. *Journal of Biological Chemistry* 280: 7469–7476.

Newbery, R.M., Wolfenden, J., Mansfield, T.A., and Harrison, A.F. (1995). Nitrogen, phosphorus and potassium uptake and demand in *Agrostis capillaris* – the influence of elevated CO_2 and nutrient supply. *New Phytologist* 130: 565–574.

Nie, Z., Li, S., Hu, C. et al. (2015). Effects of molybdenum and phosphorus fertilizers on cold resistance in winter wheat. *Journal of Plant Nutrition* 38: 808–820.

Oesterheld, G.R.M. and Lavado, C.R.A.R.S. (1997). Mechanisms for the increase in phosphorus uptake of waterlogged plants: soil phosphorus availability. *Oecologia* 1: 150–155.

Pandey, R., Zinta, G., Abd Elgawad, H. et al. (2015). Physiological and molecular alterations in plants exposed to high [CO_2] under phosphorus stress. *Biotechnology Advances* 33: 303–316.

Payne, W.A., Malcolm, D.C., Hossner, L.R. et al. (1992). Soil phosphorus availability and pearl millet water-use efficiency. *Crop Science* 32: 1010–1015.

Peltonen-Sainio, P., Kontturi, M., and Peltonen, J. (2006). Phosphorus seed coating enhancement on early growth and yield components in oat. *Agronomy Journal* 98: 206–211.

Peng, Z. and Li, C. (2005). Transport and partitioning of phosphorus in wheat as affected by P withdrawal during flag-leaf expansion. *Plant and Soil* 268: 1–11.

Peñuelas, J., Poulter, B., Sardans, J. et al. (2013). Human-induced nitrogen–phosphorus imbalances alter natural and managed ecosystems across the globe. *Nature Communications* 4: 1–10.

Püschel, D., Janoušková, M., Hujslová, M. et al. (2016). Plant–fungus competition for nitrogen erases mycorrhizal growth benefits of *Andropogon gerardii* under limited nitrogen supply. *Ecology and Evolution* 6: 4332–4346.

Qi, Y., Wang, H., Zou, Y. et al. (2011). Over-expression of mitochondrial heat shock protein 70 suppresses programmed cell death in rice. *FEBS Letters* 585: 231–239.

Richardson, A.E. and Simpson, R.J. (2011). Soil microorganisms mediating phosphorus availability update on microbial phosphorus. *Plant Physiology* 156: 989–996.

Richardson, A.E., Hocking, P.J., Simpson, R.J., and George, T.S. (2009). Plant mechanisms to optimise access to soil phosphorus. *Crop & Pasture Science* 60: 124–143.

Rivaie, A.A., Loganathan, P., Graham, J.D. et al. (2008). Effect of phosphate rock and triple superphosphate on soil phosphorus fractions and their plant-availability and downward movement in two volcanic as soils under *Pinus radiata* plantations in New Zealand. *Nutrient Cycling in Agroecosystems* 82: 75–88.

Rodriguez, D., Goudriaan, J., Oyarzabal, M., and Pomar, M.C. (1996). Phosphorus nutrition and water stress tolerance in wheat plants. *Journal of Plant Nutrition* 19: 29–39.

Roy-Bolduc, A. and Hijri, M. (2011). The use of mycorrhizae to enhance phosphorus uptake: a way out the phosphorus crisis. *Journal of Biofertilizers and Biopesticides* 2: 1–5.

Sadji-Ait Kaci, H., Chaker-Haddadj, A., and Aid, F. (2017). Interactive effects of salinity and two phosphorus fertilizers on growth and grain yield of *Cicer arietinum* L. *Acta Agriculturae Scandinavica Section B Soil and Plant Science* 67: 208–216.

Santos, M.G.D., Ribeiro, R.V., Oliveira, R.F.D., and Pimentel, C. (2004). Gas exchange and yield response to foliar phosphorus application in *Phaseolus vulgaris* L. under drought. *Brazilian Journal of Plant Physiology* 16: 171–179.

Sardans, J. and Peñuelas, J. (2007). Drought changes phosphorus and potassium accumulation patterns in an evergreen Mediterranean forest. *Functional Ecology* 21: 191–201.

Savicka, M. and Škute, N. (2010). Effects of high temperature on malondialdehyde content, superoxide production and growth changes in wheat seedlings (*Triticum aestivum* L.). *Ekologija* 56: 26–33.

Schachtman, D.P., Reid, R.J., and Ayling, S.M. (1998). Phosphorus uptake by plants: from soil to cell. *Plant Physiology* 116: 447–453.

Sebastian, N., Erika, H., and Christian, K. (2016). Critically low soil temperatures for root growth and root morphology in three alpine plant species. *Alpine Botany* 126: 11–21.

Seeling, B. and Zasoski, R.J. (1993). Microbial effects in maintaining organic and inorganic solution phosphorus concentrations in a grassland topsoil. *Plant and Soil* 148: 277–284.

Shahriaripour, R., Tajabadi Pour, A., and Mozaffari, V. (2011). Effects of salinity and soil phosphorus application on growth and chemical composition of pistachio seedlings. *Communications in Soil Science and Plant Analysis* 42: 144–158.

Sharma, S.B., Sayyed, R.Z., Trivedi, M.H., and Gobi, T.A. (2013). Phosphate solubilizing microbes: sustainable approach for managing phosphorus deficiency in agricultural soils. *Springer Plus* 2: 587.

Shen, J., Yuan, L., Zhang, J. et al. (2011). Phosphorus dynamics: from soil to plant. *Plant Physiology* 156: 997–1005.

Shen, J., Li, C., Mi, G. et al. (2013). Maximizing root/rhizosphere efficiency to improve crop productivity and nutrient use efficiency in intensive agriculture of China. *Journal of Experimental Botany* 64: 1181–1192.

da Silva, E.C., Nogueira, R.J.M.C., da Silva, M.A., and de Albuquerque, M.B. (2011). Drought stress and plant nutrition. *Plant Stress* 5 (Special Issue 1): 32–41.

Singh, D.K. and Sale, P.W.G. (2000). Growth and potential conductivity of white clover roots in dry soil with increasing phosphorus supply and defoliation frequency. *Agronomy Journal* 92: 868–874.

Singh, S.P. and Setter, T.L. (2017). Effect of waterlogging on element concentrations, growth and yield of wheat varieties under farmer's sodic field conditions. *Proceedings of the National Academy of Sciences, India, Section B: Biological Sciences* 87: 513–520.

Singh, A.P. and Singh, M. (2017). Effect of phosphorus and zinc on yield and nutrient uptake by okra (*Abelmoschus esculentus* L.) under different salinity conditions. *International Journal of Chemical Studies* 5: 284–288.

Singh, B.K., Millard, P., Whiteley, A.S., and Colin Murrell, J. (2004). Unravelling rhizosphere–microbial interactions: opportunities and limitations. *Trends in Microbiology* 12: 386–393.

Singh, S.K., Reddy, V.R., Fleisher, D.H., and Timlin, D.J. (2014). Growth, nutrient dynamics, and efficiency responses to carbon dioxide and phosphorus nutrition in soybean. *Journal of Plant Interactions* 9: 838–849.

Singh, M., Singh, A., Prasad, S.M., and Singh, R.K. (2017). Regulation of plants metabolism in response to salt stress: an omics approach. *Acta Physiologiae Plantarum* 39 (2): 1–17.

Slaton, N.A., Wilson, C.E., Norman, R.J. et al. (2002). Rice response to phosphorus fertilizer application rate and timing on alkaline soils in Arkansas. *Agronomy Journal* 94: 1393–1399.

Soares, J.C., Santos, C.S., Carvalho, S.M. et al. (2019). Preserving the nutritional quality of crop plants under a changing climate: importance and strategies. *Plant and Soil* 443: 1–26.

Soltani, S.S., Hashemimajd, K., and Najafi, N. (2015). The effect of diammonium phosphate fertilization on salinity tolerance of maize (*Zea mays* L.). *Journal of Soil Environment* 1: 18–27.

Song, C., Ballantyne, F.I.V., and Smith, V.H. (2014). Enhanced dissolved organic carbon production in aquatic ecosystems in response to elevated atmospheric CO_2. *Biogeochemistry* 118: 49–60.

Srinivasan, A., Takeda, H., and Senboku, T. (1996). Heat tolerance in food legumes as evaluated by cell membrane thermostability and chlorophyll fluorescence techniques. *Euphytica* 88: 35–45.

Stigter, K.A. and Plaxton, W.C. (2015). Molecular mechanisms of phosphorus metabolism and transport during leaf senescence. *Plants* 4: 773–798.

Su, K., Bremer, D.J., and Keeley, S.J. (2007). & Fry JD, Effects of high temperature and drought on a hybrid bluegrass compared with Kentucky bluegrass and tall fescue. *Crop Science* 47: 2152–2161.

Tan, W., Meng, Q.W., Brestic, M. et al. (2011). Photosynthesis is improved by exogenous calcium in heat-stressed tobacco plants. *Journal of Plant Physiology* 168: 2063–2071.

Tang, Y.I., Wen, X., Lu, Q. et al. (2007). Heat stress induces an aggregation of the light-harvesting complex of photosystem II in spinach plants. *Plant Physiology* 143: 629–638.

Tarafdar, J.C., Yadav, R.S., and Meena, S.C. (2001). Comparative efficiency of acid phosphatase originated from plant and fungal sources. *Journal of Plant Nutrition and Soil Science* 164: 279–282.

Tariq, A., Pan, K., Olatunji, O.A. et al. (2017). Phosphorous application improves drought tolerance of *Phoebe zhennan*. *Frontiers in Plant Science* 8: 1561.

Tariq, A., Pan, K., Olatunji, O.A. et al. (2018). Phosphorous fertilization alleviates drought effects on *Alnus cremastogyne* by regulating its antioxidant and osmotic potential. *Scientific Reports* 8: 5644.

Trivedi, P. and Sa, T. (2008). *Pseudomonas corrugate* (NRRL B-30409) mutants increased phosphate solubilisation, organic acid production and plant growth at lower temperatures. *Current Microbiology* 56: 140–144.

Vance, C.P., Uhde-Stone, C., and Allan, D.L. (2003). Phosphorus acquisition and use: critical adaptations by plants for securing a non-renewable resource. *New Phytologist* 157: 423–447.

Vincent, A.G., Sundqvist, M.K., Wardle, D.A., and Giesler, R. (2014). Bioavailable soil phosphorus decreases with increasing elevation in a subarctic tundra landscape. *PLoS One* 9 (3): e92942.

Visser, E.J.W., Colmer, T.D., Blom, C.W.P.M., and Voesenek, L.A.C.J. (2000). Changes in growth, porosity, and radial oxygen loss from adventitious roots of selected mono-and dicotyledonous wetland species with contrasting types of aerenchyma. *Plant, Cell & Environment* 23: 1237–1245.

Wang, Z., Rahman, A.B.M., Wang, G. et al. (2015). Hormonal interactions during cluster-root development in phosphate-deficient white lupin (*Lupinusalbus* L.). *Journal of Plant Physiology* 177: 74–82.

Wang, X., Deng, Z., Zhang, W. et al. (2017). Effect of waterlogging duration at different growth stages on the growth, yield and quality of cotton. *PLoS One* 12: e0169029.

Waqas, M.A., Kaya, C., Riaz, A., and Li, Y.E. (2019). Potential mechanisms of abiotic stress tolerance in crop plants induced by Thiourea. *Frontiers in Plant Science* 10: 1336.

Waraich, E.A., Ahmad, R., Ashraf, M.Y. et al. (2011). Improving agricultural water use efficiency by nutrient management in crop plants. *Acta Agriculturae Scandinavica Section B Soil and Plant Science* 61: 291–304.

Waraich, E.A., Ahmad, R., Halim, A., and Aziz, T. (2012). Alleviation of temperature stress by nutrient management in crop plants: a review. *Journal of Soil Science and Plant Nutrition* 12: 221–244.

van de Wiel, C.C., van der Linden, C.G., and Scholten, O.E. (2016). Improving phosphorus use efficiency in agriculture: opportunities for breeding. *Euphytica* 207: 1–22.

Wilson, C., Liu, X., Lesch, S.M., and Suarez, D.L. (2006). Growth response of major USA cowpea cultivars: II. Effect of salinity on leaf gas exchange. *Plant Science* 170: 1095–1101.

Wilson, B.J., Servais, S., Charles, S.P. et al. (2019). Phosphorus alleviation of salinity stress: effects of saltwater intrusion on an Everglades freshwater peat marsh. *Ecology* 100: e02672.

Withers, P.J.A., Edwards, A.C., and Foy, R.H. (2001). Phosphorus cycling in UK agriculture and implications for phosphorus loss from soil. *Soil Use and Management* 17: 139–149.

Xu, Z., Jiang, Y., Jia, B., and Zhou, G. (2016). Elevated-CO_2 response of stomata and its dependence on environmental factors. *Frontiers in Plant Science* 7: 657.

Ye, T., Li, Y., Zhang, J. et al. (2019). Nitrogen, phosphorus, and potassium fertilization affects the flowering time of rice (*Oryza sativa* L.). *Global Ecology and Conservation* 20: e00753.

Young, L.W., Wilen, R.W., and Bonham-Smith, P.C. (2004). High temperature stress of *Brassica napus* during flowering reduces micro- and megagametophyte fertility, induces fruit abortion, and disrupts seed production. *Journal of Experimental Botany* 55: 485–495.

Zhou, R., Yu, X., Wen, J. et al. (2020). Interactive effects of elevated CO_2 concentration and combined heat and drought stress on tomato photosynthesis. *BMC Plant Biology* 20: 260.

Zhu, F., Qu, L., Hong, X., and Sun, X. (2011). Isolation and characterization of a phosphate-solubilizing halophilic bacterium *Kushneria* sp. YCWA18 from Daqiao Saltern on the coast of Yellow Sea of China. *Evidence-based Complementary and Alternative Medicine*, special issue 2011: 1–6.

12

Climate Change and Cereal Production

Abdel Rahman Mohammad Said Al-Tawaha[1], *Javeid Ahmed Dar*[2], *Amreena Sultan*[2], *Noureddine Benkeblia*[3], *Amanullah*[4], *Imran*[4,5], *Shah Khalid*[4], *Abdel Razzaq Al-Tawaha*[6], *Nidal Odat*[7], and *Hiba Alatrash*[8]

[1] *Department of Biological Sciences, Al-Hussein Bin Talal University, Ma'an, Jordan*
[2] *SKUAST-K, Shalimar, Srinagar, Jammu and Kashmir, India*
[3] *Department of Life Sciences, The University of the West Indies, Kingston, Jamaica*
[4] *Department of Agronomy, The University of Agriculture, Peshawar, Pakistan*
[5] *Ministry of Agriculture, Agriculture Extension Wing, Merged Area, Khyber Pakhtunkhwa, Pakistan*
[6] *Department of Crop Science, Faculty of Agriculture, Universiti Putra Malaysia, Serdang, Selangor, Malaysia*
[7] *Department of Medical Laboratories, Faculty of Science, Al-Balqa Applied University, Al Salt, Jordan*
[8] *Department of Field Crops, General Commission for Scientific Agricultural Research, Aleppo, Syria*

CONTENTS

Climate Change and Agriculture: Perspectives, Sustainability and Resilience,
First Edition. Edited by Noureddine Benkeblia.

12.1 Introduction

Global climate change and uncertainty are the main worries of human life. Although the continual dry period and floods influence severely the necessities of billions of human beings that are dependent on soil for most of their required demand (Al-Tawaha et al. 2010), the world's financial resources are critically affected in many instances due to undesirable events such as floods, droughts, landslides, and thermal waves. The ingenuous catastrophes such as tremors and stellar eruption are nevertheless associated with climatic hazards that will alter the atmospheric configuration and cause relevant calamities of the atmospheric conditions. The rise in the deleterious atmospheric substances/impurities is due to the discharge of emissions like CO_2 due to the burning of nonrenewable fuel sources, chlorofluorocarbons, sulfur hexafluoride, per fluorocarbons, and hydrofluorocarbons. Ozone layer depletion and filtered ultraviolet radiation, volcanic eruptions, humans' involvement in the permanent removal of forest trees, the appearance of fires of the forest plantation, and the damage to the coverage of marsh lands and wetlands are the causative means for extreme weather patterns.

The destruction of the forest ecosystem persistently expropriates the heavy rainstorm and then permits it to be assimilated by the uppermost layer of soil, thus creating drizzle that comes toward the earth's catastrophic upper layer of soil, which in turn causes droughts and floods. On the contrary, inadequacy of trees such as conifer, evergreen, pinus, salix, and populus can induce water stress in droughty periods and make the agricultural land rapidly drier. Among the various atmospheric gases, carbon dioxide is the dominant gas in the atmosphere, and it plays important role in global warming. Atmospheric warming is warming of the atmospheric surface due to the discharge of various greenhouse gases (GHGs), thus raising the atmospheric temperature of the world over a prolonged time. These alterations on the surface of temperature of air and subsequent detrimental effects on rainfall over a prolonged time are called as climate uncertainty. However, when the aforementioned criteria appear every year, it is called as climate diversity. Climate change is the alteration that could be assigned instantly that changes the configuration of the world's atmospheric conditions. Although experts frequently utilize the idea for any change in the weather conditions, originating either naturally or from human sources. The IPPC defined climate change as the alterations in the state of the climate that can be described by the variations that continue for a prolonged duration.

12.2 Weather and Climate

Weather encompasses the forecasting situations comparatively – temperature, sunshine, precipitation, storm, mist, and snowflakes at specific intervals with reference to area and day durations. The concept *climate* represents the altogether continued features of the weather accomplished at a given place. The rural life sustainability of an area is an important parameter that depends on climatic variables. Hence climactic variability in nutshell depends on the elements of a particular area. Thus, the changes and the statistical data related to the storms, hurricanes, or abnormally warmer periods are a part of climate change. Global climate has changed enormously over a prolonged period, as depicted by the evidence of earth's science that glaciers and the level of oceans have changed, as evident by the history of humankind during several 100 years. Though the effects of variations during the previous periods are not usually obvious but are predominantly well studied and are to be associated with variations in the marine current turbine, eruption of Mount Etna, and common causes. The temperatures of the ground surface have been increased immensely during the prolonged period of time. There is distinct proof that rise in the mean ocean temperatures results in global climate (air) causing extensive melting of snowpack and icicle, thus elevating the mean oceanic level. The fourth evaluation statement accomplishes that the greenhouse warming is unambiguous. Global climate change has already had an observable impact on Pacific Ocean. Different experts have proven that the GHGs of atmosphere exhibit as a mantle that in turn absorbs entering sun energy and retains the global surface hotter than otherwise it could be and that can enhance the emission of GHGs into the atmosphere and that could cause extra atmospheric heating.

12.3 Distribution and Production of Cereals

Cereal crops are cultivated in abundance and supply greater energy globally than any other types of crops, and hence they are regarded as staple crops in traditional diets. They are also the main source of proteins and carbohydrates. In many developing countries, grains in the form of corn or rice form practically the entire diet. In developed countries, cereal crop utilization is both more ordinary and changed but still considerable. Growing of different cereal crops is almost similar to most annual plants and accordingly one planting gives one harvest. Triticale, rye, wheat, barley, and oats are the crops of the winter (Odat 2018). These crops are hardy that grow in hot weather at 30 °C, but this changes from species and variety. Although the warm cultivated cereal crops are tender, they thrive in hot weather. Maize, rice, wheat, and oat are the four cereal crops produced world over; however, wheat, maize, and rice suffice about half of the global requirement of calories and protein. Rice, maize, and wheat form about 87% of total grain production in the world and 43% of all food calories in 2008 (Food and Agriculture Organization (FAO) 2011). Maize is considered as the essential food of Australia, New Zealand, and North America. However, the production of maize in 2018 was recorded as 1148 million metric tons (MMT); in case of rice the worldwide production was recorded as 506 MMT in 2018 and 500 MMT in 2019/2020. In the case of wheat, the production was 734 MMT (2018) and 762 MMT (2019/2020). In the case of oats, the production was recorded as 20 MT (2018) and 23 MT (2019/2020). Oat is popular worldwide as an early meal food and cattle feed. In human utilization, oats can be taken in the form of oatmeal.

12.4 Effects of Climate Change on Cereals

The variations in the global climate may cause major threats in the production of foodstuffs. The temperatures will increase and the amount of precipitation (rain) will lower and rainfall declines for others. Besides this, the profitable land for agriculture will be decreased due to coastal flooding. Primarily, the cereals are susceptible to climate uncertainty (Turk and Tawaha 2001a,b, 2002a,b,c, 2003a,b; Tawaha et al. 2002). These changes that influence moisture levels of soils and the temperature indicate the diversity of useful and harmful microorganisms. Because of the greater calamities covering the world climatic uncertainties, there are projections of agricultural land being decreased that differ from 10 to 50%; however, it is abundantly understood that high global warming has the potential to change the production yield of *Triticum sativum* (wheat), *Zea mays* (corn), *Oryza sativa* (rice), and *Glycine max* (soybeans), which are essential food for billions of people. Although the uncertainty of world climate may have the worldwide effects on sustainable farming, geographical changes may be dominant next in 2030. As per one assumption the percentage of CO_2 would be two times the pre-industrialization levels, and the various additional emissions of GHGs may be increased considerably and temperature in Africa and Northern America may increase by 2 °C. However, if these variations take place, in North America the projected rainfall may be 10% lower than the current rainfall. It may be 10% greater though more rains influence the assurance of enhancing the productivity of agriculture and the raised warming can obviate this benefit by lowering moisture of the top soil. Therefore, drier areas of the agriculture crops will proceed to endure insufficient supplies of H_2O.

12.4.1 Crops and Temperature

Several unproven hypotheses exist subsequent to the undertaking of the global warming project's potential effects on crops. When the cereal crops are at blooming stage they are severely susceptible to change in moisture and warmer conditions; in the subsequent period of crop growth, the cereal crops could withstand hotter, unfavorable conditions and the periodical rainfall varies intervalley, area to area/zone to zone. Atmospheric temperature and rainfall alterations influenced by climate change will interact with different atmospheric gases, pesticides, fertilizers, insects, plant pathogens, weeds, and the organic matter of soil to generate unanticipated responses (Abebe et al. 2005). Despite these uncertainties, average global temperature rise of slightly more than one-half degree centigrade would lengthen the frost-free growing season in the corn belt by two weeks. However, if temperatures continue to increase beyond a specific threshold, a crop's productive summer growing season could become shorter, thus reducing the yield.

12.4.2 UV Radiation

The CFS discharge has critically exhausted atmosphere's protective ozone layer. In general each 1% reduction in the ozone layer causes a 2% increase in the amount of ultraviolet radiation that reaches the earth. In a recent study two thirds of the 300 species and cultivars examined appeared susceptible to ultraviolet radiation damage. This suggests that a 25% depletion in the ozone layer could reduce soybean yield by 20%. Unlike soybeans, some crops may be more tolerant of ultraviolet radiation. However, such crops also may be more susceptible to disease. Despite the fact that wheat appears to withstand ultraviolet radiation, "Red Hard" infection rates increased from 9 to 20% when experimental ultraviolet radiation was raised to 16 percent over ambient levels. Disease rates in rice also have increased when rice is exposed to higher ultraviolet radiations than normal.

12.4.3 Carbon Dioxide

Carbon dioxide is an essential compound for photosynthesis. It also increases water use efficiency in plants. Therefore, increasing carbon dioxide levels in the atmosphere should improve the rates of growth and water use among many crops. Decrease in carbon dioxide leads to effects of climate change in cereals.

12.4.4 Pest Attack

If global warming raises the temperature by 2 °C in the United States and slightly less in Africa, insects will multiply and prosper. During a growing season, some insects produce 500 offsprings per female every two weeks. Rising temperatures will lengthen the breeding season and increase the reproductive rate that in turn will raise the total number of insects attacking a crop and subsequently increase crop losses.

12.5 Introduction and Climate Change Scenario in Cereals

Recently, the Asia-Pacific region is likely to face the worst impacts on cereal crop yields. Loss in yields of wheat, rice, and maize are estimated to be approximately 50, 17, and 6%, respectively. This loss of yield will threaten the food security of at least 1.6 billion people in South Asia. The projected rise in temperature of 0.5 to 1.2 °C will be the major cause of grain yield reduction in most areas of South Asia. India is considered to be the second largest producer of wheat and the national productivity of wheat is about 2708 kg ha^{-1}. The Northern Indian states such as Uttar Pradesh, Punjab, Haryana, Uttaranchal, and Himachal Pradesh are some of the major wheat producing states. Here the impact of climate change would be profound, and only a 1 °C rise in temperature could reduce wheat yield in Uttar Pradesh, Punjab, and Haryana. In Haryana, night temperatures during February and March in 2003–2004 were recorded to be 3 °C above normal, and subsequently wheat production declined from 4106 to 3937 kg ha^{-1} in this period. An assessment of the impact of climate change on wheat production states that the country's annual wheat output could plunge by 6 million tons with every 1 °C rise in temperature. However, by utilizing adaptation strategies such as changing the planting dates and using different varieties, it is possible to moderate the losses. By adapting certain agronomic strategies, it was estimated that at a 1 °C rise, 3 million tons could be restored. The assessment also found that the impact of climate change on wheat production varies significantly by region. North India and other areas with higher potential productivity were less impacted by a rise in temperature than the low-productivity regions. If there is no mechanism or strategy to cope with rainfall variability, then rained crops will be more heavily impacted than irrigated ones. Climate change impacts cereal production mainly through heat and water stress, but it is also associated with water logging, frost, and disease and pest dynamics (Porter et al. 2014; Amanullah et al. 2019; Khalid et al. 2019). Under climate change scenarios, and without adaptation, yields of wheat, maize, and rice are predicted to decrease in both tropical and temperate regions (Challinor et al. 2014). For example, wheat yields a real-ready slowing in most locations, and model ensembles show that for each 1 °C increase in global mean temperature, grain yields may decline by 6% (Porteret al 2014). Not only are temperatures predicted to increase generally (Battisti and Naylor 2009), short periods of extreme heat are expected to become more frequent. Some models estimate that by the end of the twenty-first century, a current 1-in-20 years hottest day will become a 1-in-10 years event or even occur annually or biannually in many regions (Jagadish et al. 2007). Frequent short episodes of high temperature stress could have adverse effects on seed number, seed weight, and thus cereal yields (Jagadish et al. 2010). This may have negative repercussions on crop production as found in Russia in the year 2010, when the summer periods exhibited 30% reduction in grain yield (Wegren 2011).

The five main cereal crops viz., rice, wheat, maize, pearl, sorghum, and millet make about 44% of the calories and is used per capita globally, due to the mentioned fact that "the statistic increasing up to nearly 55% in least well developed countries." Imperative experimentation associations over these crop cereals and domains possess such capability to stimulate biotechnological/molecular and genetic perceptive of prominent attributes. These participations could enhance chances to improve digenetic gains, and the various difficulties faced by crop cereals since a long period of time are encountered by having very ordinary physiological basis. Moreover, novel tools in genetics are performing an abruptly significant part in enhancing the yields of cereal crops (Odat et al. 2015). Highlighting these research experimentations and synergies across crops would aid to enhance their adoption, development, as well as upliftment of boundaries of genetic bases of adjustment of crop (Valluru et al. 2014). The consultative group on international agricultural centers have a conspicuous usefulness in operating investigations for the development of adaptability on climatic conditions in cereals because of their considerable selection of genetic crop requirements, comprehensive dry spell phenotyping cassettes grid throughout the globe, and magnificent scientific research associations that are in connection private and public things across the world. "National programs of action in selected zones, besides have the features of access to different ecologies, providing ideal opportunities for screening and phenotyping in target environments, as well as in-depth knowledge of local adaptive needs." Developed scientific institutes furnish earlier knowledge on adjustment of plants and modern innovations particularly in the area of structural genomics. International cooperation between these various kinds of agencies, as well as the private sectors, will be important in establishing and assessing germplasm for enhanced adaptation to abiotic and biotic stresses.

12.5.1 Rice (*Oryza sativa*)

Research carried out by Indian Agricultural Research Institute (IARI) proved that the cereal yield of *O. sativa* (rice) is not affected by the increase in temperature by >1 °C, although the rise of 1 to 4 °C the yield of grain will be decreased by 10%. Therefore, relatively increased temperatures along with climate uncertainty can impact the global production of *O. sativa* inducing probability of deficit. We have recognized that the basmati varieties of rice are particularly sensitive to higher temperature, which in turn results in induced pollen abortion and thus the formation of grain declines. The pattern of rainfall is the dominant determining part for *O. sativa* production yield in arid conditions. Greater diversity in allocation and especially lowering in rainfall can severely affect *O. sativa* production yield and a possibility of 100% crop loss if drastic dry spell happens throughout the reproductive periods. In highland farms, if *O. sativa* crop receives up to 200 mm of rain in one day and for the next 20 days no rain is received, the wetness stress will critically affect final production of *O. sativa*. Forecasting reduces rice production in tropical regions, but in India it increases. This movement creates a burden to India because it will lead to hunger crisis on a large scale due to lower rice yield. Investigations on increase in rice yields due to rise in night temperature reveals that hot nights have a severe effect on *O. sativa* yield. There is 10% decline in yield for every 1 °C rise in night temperature. India (eastern region) has vast agro-climatic zones that promote genetic resources. According to an investigation done by the IARI, the effect of climatic uncertainty causes increased temperature and decreased radiation in turn decreasing *O. sativa* productivity in the northeastern region.

12.5.1.1 Effect of Climate Change on Rice

Rice is a fundamental staple meal in majority of countries, both developed and developing, and more significantly, 90% of the population in South Asian countries depend on rice. Climate change, increasing urbanization, drought, and desertification have all contributed to a substantial decline in rice output, prompting researchers to create new kinds with improved yield. Thus, new cultivars with an enhanced nutritional profile may provide a plethora of advantages and may represent a significant step toward eliminating malnutrition in less developed countries (Thangadurai et al. 2020; Sangeetha et al. 2020). A favourable outcome for rice production in a number of regions despite the unpredictability of the climate. For example, the world's temperature rise may cause increased rice production in the northern regions of countries such as China (cultivating two rice crops in one year). The main effect of climate change on *O. sativa* is likely to be negative (Imran et al. 2020). The International Food Policy Research Institute (IFPRI) report on the effect of climate uncertainty on agriculture and costs of adaption predicts that prices of rice by 2050 will increase by 32 and 37%. In South Asia productivity of rice will be reduced by 14%, 10% in East Asia and the pacific, and 15% in sub-Saharan Africa.

12.5.1.2 Sea Level Rise

Scientists forecasted that increasing temperatures cause melting of polar ice caps and glaciers that in turn cause rise in seawater levels on an average of about 1 m by the end of twenty-first century. The rise in sea level also affects the cultivation of rice in low lying areas of Asia (Vietnam). More than half of the rice production in Vietnam grown in the Mekong River delta is impacted by sea level rise. Hence forecasting precisely the rise in sea level in advance would result in the prevention of such huge losses.

12.5.1.3 Salinity

Oryzae sativa cereal crop can withstand to some extent salty conditions and production might be hindered when the level of ocean increases and abundant quantities of saline impurities are present. Though the impacts of saline conditions can saturate all across the continent and intrinsically alter the systems of hydrology and plant growth and development (Lee et al. 2005; Othman et al. 2006; Odat et al. 2015; Al-Tawaha and Al-Ghzawi 2013; Hasan et al. 2018; Abu Obaid et al. 2018; Al-Tawaha et al. 2018; AL-Issa et al. 2020; Osman et al. 2021).

12.5.1.4 Flooding

Oryza sativa is a separate cereal crop that can grow in moist conditions where other crops are unable to grow. Severe flooding is a serious threat for *O. sativa* growing areas; however, *O. sativa* will never survive under severe flooded conditions for longer intervals. The increase in the level of ocean in coastal regions, enhanced flooding caused by sea level rise in coastal areas, and the severity of tropical storms with climatic uncertainty resulted in reducing *O. sativa* production. About 20 Mha of global *O. sativa* cultivating area is at peril due to intermittent flooding to the level of submergence, especially in the principal *O. sativa* (rice) producing nations viz., Bangladesh and India. The main flooding regions are likely to enhance in frequency of climatic variations and *O. sativa* cultivating regions, previously not subjected to flooding.

12.5.1.5 Increased CO_2 Levels and Higher Temperatures

The rise in the levels of CO_2 and temperature can alter the yield of *O. sativa*. The higher CO_2 levels usually results in increase of biomass production but not compulsory yield. Rice yields can be decreased by higher temperatures because they can fabricate *O. sativa* flowers unfruitful, that means grain formation does not occur. Excessive respirational losses associated with increased temperatures make the *O. sativa* cereal less productive. Although there are various forecasting for upraised CO_2 levels, temperature, and humidity variations, the association of discussed circumstances makes the predictions of future *O. sativa* rice production yields difficult. Research conducted by the International Rice Research Institute reported that rise of 1 °C in night temperatures can reduce *O. sativa* yields by 10%.

12.5.1.6 Water Scarcity

Rice cereal needs adequate H_2O for rising. Stress conditions for 5 days in highland *O. sativa* farming regions and for two 14 days in shallow low lying *O. sativa* cultivating regions can considerably diminish rice production. The average rice yield decline in unirrigated and stress-affected areas varies from 17 to 40%. The losses of rice yield production and scarcity of food has been reported in severe drought areas. With the beginning of climatic uncertainties, the intensity of severe droughts are to enhance in rainfed *O. sativa* cultivating regions and stresses would additionally increase H_2O scarcity in watered regions. Water shortage impacts more than 23 million hectares of rainless *O. sativa* production regions in east South Asia. The severe periods of drought impact production of rice in the United States, Australia, China, and other rice producing nations.

12.5.1.7 Diseases, Pests, and Weeds

While pest and disease control has contributed to tripling food production over the past 40 years, diseases continue to affect 10–16% of the world crop (Basit et al. 2021; Imran et al. 2021a). Field supervisions in 100 of growers' farms during the past 10 years reveal that pathogenic diseases and insects of *O. sativa* are severely altered by climatic uncertainties. The non-uniform showers of rainfall, deficit of H_2O, and associated water stresses as well as blast and brown spot of rice. However, new environmental circumstances and conversion in the practices of production that the growers might acquire to manage with climate variations may lead to minimizations of diseases such as sheath blight and insects such as cutworms or whorl maggots. Comparatively, new crop health are appearing progressively. Rice cereal weed competition is expected to increase and threaten the productivity of *O. sativa*. In addition, severe weather incidents newly led to surprising *Hamsters cricetinae* (rodent) population disruption in Asian countries due to nonseasonal and contemporary cropping.

12.5.1.8 Methods to Cope with Climate Change on Rice

Establishment and advancement of profitable systems of irrigation, managed with H_2O saving methods can aid to build the magnificent usage of fixed H_2O. Altered modifications of cropping system, enhanced fertilizer supplies, and good fertilizer management plans should be adopted to accessible H_2O sources, soil improvement, and leveling in the occasions of stress and severe drought periods. The proper seed bed practices, proper seed direct sowing, and an optimum nutrient supplement help in making rice seedlings healthier, taller, and lesser susceptible to flood.

12.5.1.9 Inhibiting Emissions from Rice

The weather, day length, and growing degree days (GDDs) influence the growth and development of cereal crops, allowing them to be more sensitive to the particular settings in which they are grown in various seasons. The global temperature is increasing as a result of human activities such as the combustion of fossil fuels and the clearing of forest area for the construction of new buildings. The increase in temperature has a negative impact on crop growth and development. Disturbance is primarily responsible for a change in the phenological development of crops, which has an impact on their economic output. Scientists and farmers respond to these phenological changes in part by altering sowing times and cultivar shifts, which may either increase or reduce the length of time crops are in growth. Combinative H_2O economizing fertilizer management mechanizations help to sustain production yields and inhibit the global warming gases discharged from the *O. sativa* fields, although contemporaneously inhibiting prices and conserving the profitable inputs and elegant management residual effects on rice and inhibiting the time cooking of *O. sativa* to decline the effect of rice on climatic uncertainty.

12.5.1.10 Reducing Methane Emissions

Oryza sativa cereal crop is usually cultivated under anaerobic soil conditions in flooded regions that discharge CH_4 in the soil as organic matter disintegrates. CH_4 is a GHG that is nearly 25 times more powerful than CO_2. Methane (CH_4) constitutes about 20% of the increased greenhouse effect and rice accounts about 10%. International Rice Research Institute reveals that cultivating rice in flooded conditions has been observed to be tremendously feasible practice where the health of soil would be protected. Under water conditions have been a fundamental part of production of *O. sativa* in Asian countries for centuries, and they are accountable for 75% of world production of *O. sativa*. Water saving method helps to reduce the emission of methane from rice fields.

12.5.1.11 Reducing Nitrous Oxide Emissions

The different cultural practices like as alternate wetting and drying (AWD) decrease the CH_4 discharges. However, they can enhance formation of N_2O, the one more global warming gas that is 300 times more powerful than CO_2. The massive use of nitrogen in acidic soils, associated with highly unsaturated farms generates N_2O (nitrous oxide). To reduce N_2O, technologies of H_2O saving should be paralleled to precision nutrient management.

12.5.1.12 Rice Residues

Because of the strong need for food, agriculture productivity is rising as the world's population grows. This, in turn, results in a significant quantity of agricultural waste being generated. The pyrolysis of agricultural wastes is one of the most appropriate methods for producing biochar, which may also be used as a fuel source. The burning of rice straw (3 Mt $year^{-1}$) results in GHG emissions of about 10 953 (CO_2e) Gg $year^{-1}$, while the combustion of biochar generated from rice straw at various temperatures results in emissions ranging from 2131 to 3239 (CO_2e) Gg $year^{-1}$, depending on the temperature used (Sakhiya et al. 2021). On the other hand, the burning of *O. sativa* residues and adding the obtained carbon black to rice fields rather than incorporating untreated harvest residues of rice that may suppress field methane discharge by nearly 80%.

12.5.2 Maize (*Zea mays* L.)

Maize is the third utmost essential crop in India and has a magnificent importance in food safety particularly in drought areas. *Z. mays* cultivation in arid and semi-arid tropical zones are usually susceptible to atmospheric conditions, mainly rainfall. Consequently changes in the precipitation as well as the highest and lowest temperatures over the north-east and west-south monsoon period will have contradictory effect on maize cereal crops. In Tamil Nadu, evaluations stipulate that there will be an inhibition in yield up to 3.0, 9.3, and 18.3% in 2021, 2050, and 2080 from current yields. "The higher Gangetic Plain described by lower cold temperature, *Zea mays* production can increase up to a 2.7 °C rise in temperature" Higher increasing temperatures result in significant function in impacting *Z. mays* yield in contrast to precipitation, which does not have a great effect on cold season production yields as the cereal in well-irrigated belts "*Zea mays* yield throughout the monsoon could be inhibited up to 35% in several Southern Plateau regions and 55% in Mid Indo-Gangetic Plains, whereas the higher Indo-Gangetic Plain is predicted to comparatively unaffected."

In March 2004 high temperature severely troubled the crops such as wheat, mustard, rapeseed, linseed, vegetables, *Pisum sativum*, and *Campellina sinensis* all over India. In the region of Himachal Pradesh, production decrease was calculated to be about 20 and 60% in consonance with the type of grain. *Z. mays* and *Solanum tubersoum* harvest was 15 to 20 days and the period of flowering of *Malus domestica* was earliest by 15 days.

Nelson et al. (2014) planned a procedure to differentiate impacts of a set of nine climates cereal and profitable forecasting models undergoing the outline of representative concentration pathway 8.5 "World annual green house discharges continuing to increase all over the twenty-first century" without constituting for carbon dioxide, nutrients of agronomic cereals. The writers differentiate impacts to the shocks of extrinsic climatic uncertainties on the production of four crops viz., coarse rice, oil seeds, coarse grains, and wheat that contribute 70% of the worlds harvested cropping area. The main biophysique impact of the climatic variability disturbance is 17% decrease in production. The economizing classic models transfer the shocking impact on the apparent variants. The manufacturers acquiesce to the price increases combined with the shock by accentuating mitigating principles that facilitate the ultimate yield variations to 11% and thus increasing the cropping area by a mean of 11%. Expenditure reduces slightly, with a mean decline of 3% variations in fair trade shares cancels out across regions but the share of world trade production increases by 1% on average. The medium prices producer enhances by 20%. The magnitude of responsiveness is prevalent in all the models, but the direction of responsiveness changes notably throughout the models and cereal crops. Although the mean utilization reduces comparatively, less prices enhance the nature of inelasticity the world demands and probably enhanced costs of food substantially for poor stricken growers. Variations to trade shares throughout the regions cancel out and the share of world trade enhances by 1%.

A 1 °C increase in temperature had no dominant impact on production potential. In northern India but under rainfed and irrigated conditions, the production enhanced in several regions. "The potential wheat yields has been reduced in various regions when an increase of 2 °C rise in temperature The impact on rainfed and irrigated productivity differed with location. The natural climatic uncertainty also had substantial impact on the magnitude of response to the climatic variability." The evaporation and transpiration was decreased in rainfed and under irrigated conditions.

Gangadhar Rao and Sinha (1994) studied the effect of global warming on *Triticum aestivum* in India and revealed that *T. aestivum* cereal yield declines occur by deleterious impacts of temperature throughout the grain filling and the ripening stages of growth and development. Overall, their research shows that the crop's qualities, such as how sensitive it is to temperature during the filling grain period, play a big role in figuring out how changes in the climate affect the crop's yield. The level of carbon dioxide is 425 ppm and temperature rise options of 0, 1, and 2 °C was presumed. Carbon dioxide concentration at 425 ppm and rise in temperature is zero, and enormously enhanced production of grain at all levels of yield and at all places. On the other hand, irrigated yields revealed a slight increase in various regions where current yields and was found to be more than 3.5 tons ha^{-1}. It has been seen that in peninsular and central India, where irrigated yields are between 2 and 4 tons ha^{-1}, the effects ranged from a large decrease to a large increase. However, the dryland production significantly declined. However, the capacity is very less at several regions with less productivity potential. It has been seen that at few locations there was a slight increase. There was a little decline in production capability i.e. 1.5 to 5.8% in subtropical regions above 23°N and 17–18% decline in tropical regions.

Rupa Kumar et al. (1992) under upcoming situations of inclined GHG concentration on cereals crops revealed that pronounced increase in both temperature and rainfall in the twenty-first century would remarkably become "ostentatious" after the 2040s in India. Throughout the region south of 25°N like as Khajuraho and Varanasi the maximum temperature will rise up to 4 °C in 2050s. In northern regions the temperature will exceed by 4 °C. This investigation also reveals that the rise in the slight temperature up to 4 °C throughout the nation may exceed in northeast Asia (India), southern peninsula. However, there is generally a decline in the number of rainy days throughout the main part of the country. This decrease is seen more in central and western part "by more than 15 days" in foothills of Himalayas i.e. Uttaranchal state and the number of rainy days may increase by 5–10 in northeast India. However, increase in GHG will lead to a rise in the rainy days severity by 1–4 mm per day.

12.6 Effects of Climate Change on Cereal Production

Cereal grains, such as wheat, maize, and rice, are referred to as main crops since they are staple meals for the majority of the world's population. The average world temperature is expected to rise by the end of this century as a result of the increased emission of GHGs into the atmosphere. As a consequence of this shift in climate, agricultural production may be reduced, increasing the likelihood of food insecurity. Agriculture, on the other hand, is one of the most significant contributors to GHG emissions, and poorer yields may prompt more activity to satisfy the need for food, resulting in greater amounts of GHGs being emitted into the environment (Imran et al. 2020, 2021b; Hussain et al. 2021; Saranraj et al. 2021; Tawaha et al. 2021; Amanullah et al. 2021). Grain cereals comprising of maize, wheat, and rice are regarded as elementary cereal crops as they are considered to be essential food to most of the population throughout the world. Subsequently in 2050, 70–100% increase in food cereal supply is essential to supply foodstuff to the expected 9.8 billion people of global population. For example, in the growing of *T. sativum*, the rise in temperature by up to 1 °C would inhibit the yields in the range of 3 to 10%. "In a globe where by population size is ascending and reaching unacceptable levels an intense inhibition in agriculture production yields of main agronomic cereal crops which can considerably inhibit ascertaining the food security." To tackle this issue, the research community should find substitute cereal crops that can be cultivated even though there is global warming circumstance. Simultaneously these crops could release lesser GHGs and should be less economic intensive and be abundant in utmost major and minor nutrients imperative for our wellness. Millets are the cereal crops that may have several advantageous properties mentioned earlier and when differentiated to other main cereal crops such as paddy, wheat, and maize. Pearl millets are vigorous species that have inadequate growing phases that are generally 60 to 100 days depending on the type of variety. Millets crops withstand drought that can be cultivated under a diverse range of climatic conditions. In addition, scientists reported that millets discharge less GHGs that would be profitable in hindering the contributions of agriculture and food industry to global warmings. In a situation where the foods and H_2O catastrophes are imitating in the face millets could become a dominant nourishment for performing and achieving the goal of nutritional and food security. Most experts are sure that climate warming is invariable within a shorter period of time and needs world policy alterations and sustainable agricultural principles for a longer period of time to manage the enormous environmental damage.

12.7 Climate Change Impact on Crops Yield and Food Security

12.7.1 Paddy (*Oryza sativa*)

Oryza sativa L. is cultivated in more than 100 countries in the world (Saleh et al. 2021). Several investigations revealed that temperature, irrigation, and nutrient fertilizer application plays a vital role in the total yield of *Orzya sativa* (Al-Tawaha et al. 2020). The investigation found yields of *Oryza sativa* decreased enormously under the substitute drying and wetting irrigation system, when differentiated to yields under continuous surface flooding. However, the similar investigation also assessed the controlled discharge of nitrogenous fertilizer using the substitute AWD irrigation system. In the AWD, the irrigation supply of H_2O was reduced to 28 and 42% in two years of studies differentiated to the flooding system. Even though at reduced consumption, this method accomplished higher *O. sativa* production yield and biomass of the plant such as shoot, root, and panicle. Moreover different experts have revealed that the climatic warming has a reduced impact on the yields of *O. sativa* found all over the globe. Over the past century the world temperatures enhanced by 0.5 to 0.6 °C. The rise in temperature due to enhanced respiration in the plant followed by enhancement in carbon metabolism and inhibited *O. sativa* yield. However, the *O. sativa* flowers become sterile when the temperature increases and ultimately it resulted in disrupting the reproduction process. IFPRI studied that climate uncertainty would decrease the *O. sativa* production yield in the range of 10 to 15% that would result in decrease in the market price by up to 32 to 37%. By using ancient records from 1999 to 2007, the crop growth simulation model for rice (ORYZA model) 2000 forecasted 2 °C rise in temperature would decrease *O. sativa* production yields by up to 0.36 tons ha^{-1} that results in massive economic loss in Malaysia. Matthew et al. studied the effect of rise in atmospheric CO_2 in *O. sativa* productivity in many regions of Asia and showed that the production may be reduced by up to 4%. However, the several regions where production may increase by rise in world temperatures usually in the zones with cooler climatic conditions resulted in multiple cropping patterns, this could not sufficiently countervail the loss in the final production in different regions of Asia that cultivate bulk of *O. sativa* throughout the globe; also investigation revealed that utilizing cultivars with short maturing periods helped to manage the production yields. The various investigations which indicate that the climatic warms will enhance the production of *O. sativa* (2.9–34%) be contingent on immensely bright frame work essentially due to the heat in the night and produce moves into the regions which have cold climatic conditions that could warm up due to climatic uncertainties in the 2050s and 2080s. Although the discussed investigations do not consequently examine the social and economic impacts in areas that are presently cultivating *O. sativa* and the convenient approaches for these growers to manage the production loss of *O. sativa*, several investigations assure that in the long term the production had enormously declined as the world temperature rises at the end of the century. However, if no useful managements to combat decreased production are applied as early as possible, the climatic warming can result in scarcity of foods thereby causing food insecurity for coming generations.

12.7.2 Maize (*Zea mays* L.)

Modeling of maize yields shows that *Z. mays* production yield in Malawi will reduce 14% by mid-century and 33% by the end of century as result of climate uncertainty. Likewise in northeast zones of China, the modeling global warming predicted that the utmost temperature increase "where temperature increases by 1.32 °C could result to reduce in *Zea mays* production up to 35% in 2030 in comparison to productivity in 2008." In the regions of the United States the climatic warming has depicted 2.5% reduction in *Z. mays* production from 1970 to 1999, and the rainfall modeling described that by "the 2050s corn yields are projected to reduce further by 20 to 50% depending on the present discharge scenarios." H_2O provision is also a crucial component for *Z. mays* yield. Investigation indicated that from 1999 to 2002 the production of *Z. mays* cultivated under water is "12.44 tons ha^{-1} was 16.5% greater than when cultivated under non-irrigated condition 10.68 tons ha^{-1}." Production of *Z. mays* cultivated under two water methods especially drip water method and raised bed were differentiated. The outcomes revealed that when compared with the drip method, the raised bed irrigation system was significantly useful for increasing the biological, plant, and grain yield by 5, 1, and 21% approximately. Scientists investigated the variations in the *Z. mays* production yields in Africa with a 2 °C increase in temperature and a 20% decrease in rainfall and found that production declined by 10% under such circumstances. The historical record used by Lobell et al. explains the importance of moisture on the total production yields of *Z. mays*. They found that loss of 1 and 2% was noticed in the final yield production of *Z. mays* for every day the crop spends at a temperature 30 °C stress and rainfed conditions. They found that wetness plays the dominant importance in *Z. mays* capability to withstand the increased temperatures.

12.7.3 Wheat (*Triticum aestivum*)

World climate uncertainties and the severe weather episodes will have a massive effect on the *T. aestivum* production; it is the most important and extensively used cereal. In France, winter *T. aestivum* production had examined using a model technique depend on ancient production and network available collected data from 1950 to 2015. The model forecasted that climate uncertainties on the *T. aestivum* production could be decreased by 3.5–12.9% in the average term from 2037 to 2065 and also expected that cold *T. aestivum* yield productions could decline by 14.6–17.2% at the end of century. Scientists in China found that the yield of *T. aestivum* rates could be decreased by 3 to 10% by 1 °C rise in temperature during the growing periods found from the data between 1970 and 2000. The rise in temperature over the past 2–10 years may have caused a 4.5 percent loss in crop productivity or a decline in the usage of resources like fertilizers, water, etc. Investigational observations on the *T. sativum* production yields by 1981 to 2009 across

China revealed extensive changes in the *T. aestivum* productivity in several climactic zones. They showed that the yields of *T. aestivum* in Northern China rose by 1 to 13% while a decrease of 1 to 10% was seen in Southern China. In Turkey *T. aestivum* production yield forecasted model with prolonged prediction to 2100 was to estimate production based on zonal data and the outcomes revealed that *T. aestivum* yield could reduce by 8 to 23% at the end of 2100. Also these models forecasted that the temperature rises on the surface of earth, the land surface becomes drier resulting in lesser production. The important factor for proper growth, balanced development, and higher yields of *T. sativum* is H_2O. The two winter wheat varieties such as 14SAWYT306 and Baviaan had grown with three irrigation schedules with the reduced system having no supply of water. The three schedules were supply of water for the development of the extension stem, supply of water for extension stem to the ripening, and supply of water all over the growth of the cereal crop. However, the observations revealed that supply of water treatments enormously enhanced grain yield in comparison to without irrigation treatment for both varieties in comparison to without supply of water all over the growth period enhanced the number of grains per spike, number of tillers, grain production, maturity index, and grain protein in the range of 20.58, 26.07, 42.72, 16.71, and 3.31% similarly. Thus we conclude that irrigation is very important for wheat production under semi-arid conditions (Turk et al. 2003c).

Mall and Singh (2000) reported slight variations in the cultivating growing temperatures throughout the years and showed the crucial form of climate impacting the production yearly in *T. aestivum* variations. The minimum and maximum temperatures in Delhi exhibit increasing trend both in summer or *kharif* cultivating season as well as winter or *rabi* cultivating season that starts after the summer monsoon. It has been observed that there was slight reduction drift in solar rays during *kharif* and *rabi* growing seasons after 1989. These alterations show that the heating trend of solar radiation is related to cereal growth, and any decline in this can enormously decrease the production of agriculture.

Lal et al. (1998) studied the susceptibility of *T. aestivum* and *O. sativa* crops in India (northwest) to climate uncertainty with the susceptibility investigations with CERES *T. aestivum* and CERES-*O. sativa* models and observed that under elevated carbon dioxide levels the yields of *O. sativa* and *T. sativum* increased considerably between15 and 28% for a doubling of carbon dioxide. A 2 °C increase in temperature canceled out the positive impact of elevated carbon dioxide on *T. sativum* and *O. sativa*. The collective impact of increased carbon dioxide and dictated thermal stress on the *T. sativum* (*O. sativa*) crop is 21% (4%) enhancement in yield for the irrigation program currently accomplished in the region. While as the unfavorable effects of such as water scarcity on *T. aestivum* cereal crops could be reduced to a definite considerable extent under elevated carbon dioxide levels, they could greatly be sustained for the *O. sativa* crops results in about 20% net decrease in the yields of *O. sativa*.

12.8 Potential Strategies to Increase Cereal Crops Production

Uncertainty about the climate has the potential to affect wheat yield directly through water stress and heat, as well as indirectly through fertilizer supplies and pests. Experts are sure that the global warming due to the discharge of CO_2 is the irreversible trend all over a period of 100 years and could have long-term repercussion on the planet. By now, enhancing the yields of different cereal crops and decrease world food losses are compulsory to keep up with the population growth and to cope with the effect of climate variability. The feasible practical answers for promising the enhanced yields of crops and countering the impacts of climate changes are given as next.

12.8.1 Breeding

The moderate temperatures throughout the world are predicted to rise by 2.5 to 4.5 °C at the "end of this century." The sustenance of convenient temperatures throughout the growing period is essential for the cereals in various periods of growing because most of the cereals are extremely susceptible to climatic and environmental variations (e.g. higher temperatures can inhibit the pollen viability), which in turn results in less kernel numbers and eventually the yield throughout the flowering and heading periods (Odat 2018). Conventionally breeding helps to enhance crop yields instead of high temperature tolerance. Although, there is enhancing obligation for recent cultivars of cereals that can tolerate harsh climatic incidents and rising growing temperatures with calm yield rates. Investigations are conducted on utilizing and prolonging prevailing genetic vulnerability to produce new high-yielding and heat-tolerant crop genotypes. For example, new varieties of *Vigna unguiculata* (Cowpea) have been formed that had greater total yield grains even when cultivated in extremely high temperatures. This was an improvement from the conventional variety of *V. unguiculata* whose yield decreased enormously where the night temperature had risen higher than 20 °C. As well as the recurrent selection technique using progenitor of "*T. tauschii* as a donor gene, has been effectively used to enhance the grain filling and to produce greater grains in BC1F6 *Triticum aestivum* plants." Moreover, genes associated to temperature with standness in *S. tubersoum* had been chosen through three-cycle recurrent selection method breeding, which in turn leads to enhance the productivity up to 37%. The strategy used a method that had applied to create cultivars that could give superb yield in the climates of tropical regions putting *S. tubersoum* cultivars indigenous to Europe. In more investigations the overexpression system of "glycerol-3-phosphate acyltransferase was considered to result in enhanced profusion of the thylakoid membrane lipids in transgenic *Nicotiana tabacum* plants. These plants exhibit quick renaissance rates after heat stress differentiated to wild plants." This exhibited that glycerol-3-phosphate acyltransferase plays a vital role in enhancing the immunity of plants to heat. It attains overexpression of enzymes alike to glycerol-3-phosphate acyltransferase at high levels in cereals such as maize and wheat through biotechnological techniques, which then enhances tolerance to heat.

12.8.2 Irrigation and Fertilizer Efficiency

The use of chemical fertilizer and H_2O plays an important role in ascertaining the production yields of different cereal crops throughout the various stages of farming (Turk and Tawaha 2001a,b;. Turk et al. 2006; Ananthi et al. 2017; Al-Juthery et al. 2018; Hayyawi et al. 2018; Al-Ghzawi et al. 2019; Shatnawi et al. 2021). The crop yields enhanced within a determined rise in H_2O and application of chemical fertilizers. Few developing nations such as India, China, Egypt, Iran, Mexico, Pakistan, and Algeria are tackling concerns with H_2O deficit due to variations of climatic conditions, enough agricultural activities, industry usage, and population explosion. For example, in "China the wetlands have shrunk up to 9%, like as 340 000 km^2 for 2003". Hence enhancing the efficiency of irrigation should be regarded as a preference as a factor of approach for the improvement of feasible agricultural practices and enhancing the production of several cereal crops. The IFPRI described that enhancing the efficiency of irrigation would increase the crop yields by 21%. The assessment of the water use efficiency in California for various growing strategies found that efficiency of H_2O use was 60 to 85% in flood surface irrigation, 88 to 90% for drip irrigation, and 70 to 90% for sprinkler irrigation. Using drip water method would excellently reduce usage of H_2O and helps to enhance yields and decrease the water usage quantity by greater than 50%.

Moreover the application of fertilizer is one more important factor that directly effects the production of crop. Over the past few decades the calculated 30 to 60% increase in crop production was through the application of chemical fertilizers. Presently the consumption of chemical fertilizer is increasing in both developed as well as developing countries. The FAO described that the chemical fertilizer application reached 44.93 million tons in 2014 and up by 2% in comparison to 2013. The global need for chemical nutrients is reported to "be growing at 1.8% per annum from 2014 to 2019." Although massive usage of chemical fertilizers has been shown to have adverse impacts on the soil and on the attributes of cereals, various investigations exhibited that the excessive use of chemical fertilizers can lead to acidification of soil. Moreover, nitrogen leaching and NH_4 "volatilization also accounted to environmental threats in the areas where the NH_3 and nitrogen fertilizers are applied." Furthermore, an investigation revealed that the pH of soil reduced to 4.3 after 13 years of yield in Shouguang and Shandong regions of China, which eventually would lead to decreased crop yields and quality. However, with the application of chemical fertilizers at proper levels, it can save all the resources.

12.8.3 Increasing the Cultivation Area of Tolerant Cereal Crops

Pennisetum glaucum are regarded as the most important drought-tolerant crop among the main cereals such as *O. sativa*, *T. sativum*, and *Z. mays*; it has been reported that growing of *O. sativa* will move to the locations that are warming-up, i.e. conventionally cold regions as the world temperatures rise in few nations of Asia such as China and India. To countervail the damage of *O. sativa* crop, *P. glaucum* would be cultivated in the rainfed locations as a result of their powerful root system that helps them to tolerate water stresses. Furthermore, pearl millets need no chemical fertilizers because pearl millet is a C4 crop that means it has the capability to fix carbon at a lower transpirational rate in comparison to other cereal crops.

12.9 Conclusion

The cultivating needs of various cereal crops and effect of climatic uncertainty variations on the production of several cereals throughout the globe. The methods of breeding and techniques of irrigation had exhibited as the finding solutions to tackle the discussed threats and difficulties caused by climatic warming. Climatic variations and mainly the rise at ordinary temperatures may decrease the production of main agronomic cereals. To achieve food safety, we must emphasize the use and production of cereal grains that can endure climate changes as an alternative to the current major cereals, especially in dry land and waterless places that are in danger of food poverty. Millet is one such crop that is highly resistant to the higher temperatures and this is the only crop that has the capability to cultivate in insufficient quality soils. Although pearl millet yield can assist to manage climate uncertainties since it releases lesser GHGs than other cereals and has lesser environmental impact since it needs little chemical fertilizers and H_2O while growing on marginal land. Therefore, efforts should be made to encourage the growers to grow these crops and to increase their market value, especially in developing countries. Experts and growers should develop diversification strategies based on local conditions. This should be part of the approach for accomplishing food security within the context of climate change and a rapidly increasing population.

12.10 Summary

Cereal (wheat, corn, rice) production is highly sensitive to climate change. It is affected by long-term trends in average rainfall and temperature and interannual of climate variability. In this chapter we discuss the following: (i) the impacts of climate change on cereal production – the need for the intensification and diversification of sustainable production, (ii) climate change impact on cereal production – need for adaptation to climate change, (iii) climate variability impact on wheat production – adaptation and mitigation strategies, (iv) climate variability impact on corn production – adaptation and mitigation strategies, and (v) climate variability impact on rice production – adaptation and mitigation strategies.

References

Abebe, G., Sahile, G., and Al-Tawaha, A.R.M. (2005). Evaluation of potential trap crops on Orobanche soil seed bank and tomato yield in the central rift valley of Ethiopia. *World Journal of Agricultural Sciences* 1: 148–151.

Abu Obaid, A.M., Melnyk, A.V., Onichko, V.I. et al. (2018). Evaluation of six sunflower cultivar for forage productivity under salinity condition. *Advances in Environmental Biology* 12: 13–15.

Al-Ghzawi, A.L.A., Al Khateeb, W., Rjoub, A. et al. (2019). Lead toxicity affects growth and biochemical content in various genotypes of barley (*Hordeum vulgare* L.). *Bulgarian Journal of Agricultural Science* 25: 55–61.

Al-Issa, R.M.S., Odat, N., Qrunfleh, I. et al. (2020). The impact of NaCl on different genotypes of tomato (*Solanum Lycopersicon* Mil) on germination, some physiology characteristics and gene expression. *Eurasian Journal of Biosciences* 14: 4467–4470.

Al-Juthery, H.W., Habeeb, K.H., Altaee, F.J.K. et al. (2018). Effect of foliar application of different sources of nano-fertilizers on growth and yield of wheat. *Bioscience Research* 15: 3976–3985.

Al-Tawaha, A.M. and Al-Ghzawi, A. (2013). Effect of Chitosan coating on seed germination and salt-tolerance of lentil (*Lens culinaris* L.). *Research on Crops* 14: 489–491.

Al-Tawaha, A.M., Shyam, S.Y., Turk, M. et al. (2010). Crop production and management technologies for drought prone environments. In: *Climate Change and Drought Management in Cool Season Grain Legume Crops*, 325–350. Springer.

Al-Tawaha, A.R., Turk, M.A., Al-Tawaha, A.R.M. et al. (2018). Using chitosan to improve growth of maize cultivars under salinity conditions. *Bulgarian Journal of Agricultural Science* 24: 437–442.

Al-Tawaha, A.R.M.S., Singh, S., Singh, V. et al. (2020). Improving water use efficiency and nitrogen use efficiency in rice through breeding and genomics approaches. In: *Rice Research for Quality Improvement: Genomics and Genetic Engineering* (ed. A. Roychoudhury), 307–337. Singapore: Springer.

Amanullah, Khan, N., Khan, M.I. et al. (2019). Wheat biomass and harvest index increases with integrated use of phosphorus, zinc and beneficial microbes under semiarid climates. *Journal of Microbiology, Biotechnology and Food Sciences* 9: 242–247.

Amanullah, Yar, M., Khalid, S. et al. (2021). Phenology, growth, productivity, and profitability of mungbean as affected by potassium and organic matter under water stress vs. no water stress conditions. *Journal of Plant Nutrition* 45 (5): 1–22.

Ananthi, T., Amanullah, M.M., and Al-Tawaha, A.R.M.S. (2017). A review on maize-legume intercropping for enhancing the productivity and soil fertility for sustainable agriculture in India. *Advances in Environmental Biology* 11: 49–64.

Basit, A., Farhan, M., Abbas, M. et al. (2021). Do microbial protein elicitors PeaT1 obtained from *Alternaria tenuissima* and PeBL1 from *Brevibacillus laterosporus* enhance defense response against tomato aphid (*Myzus persicae*)? *Saudi Journal of Biological Sciences* 28: 3242–3248.

Battisti, D.S. and Naylor, R.L. (2009). Historical warnings of future food insecurity with un precedented seasonal heat. *Science* 323: 240–244.

Challinor, A.J., Watson, J., Lobell, D.B. et al. (2014). A meta-analysis of crop yield under climate change and adaptation. *Nature Climate Change* 27: 1–5.

Food and Agriculture Organization (FAO) (2011). Production Stat FAOSTAT. Retrieved 26 August 2008.

Gangadhar Rao, D. and Sinha, S.K. (1994). Impact of climate change on simulated wheat production in India. In: *Implications of Climate Change for International Agriculture: Crop Modelling Study* (ed. C. Rosenzweig and I. Iglesias), 1–17. Washington, DC: USEPA230-B-94-003. USEPA.

Hasan, M., Odat, N., Qrunfleh, I. et al. (2018). Microsatellite analysis of genetic diversity and population structure of Jordanian barley (*Hordeum vulgare* L.) reveals genetic polymorphism and divergence associated with inflorescence type. *Research on Crops* 19: 86–96.

Hayyawi, W.A., Habeeb, K.H., Jawad, F. et al. (2018). Effect of foliar application of different sources of nano-fertilizers on growth and yield of wheat. *Bioscience Research* 15: 3988–3997.

Hussain, I., Ali, I., Ullah, S. et al. (2021). Agricultural soil reclamation and restoration of soil organic matter and nutrients via application of organic, inorganic and bio fertilization

(Mini review). In: *IOP Conference Series: Earth and Environmental Science*, vol. 788, No. 1, 012165. IOP Publishing.

Imran, Amanullah, Mahmood, T. et al. (2020). Improvement of rice quality via biofortification of selenium, iron, and zinc and its starring role in human health. In: *Rice Research for Quality Improvement: Genomics and Genetic Engineering* (ed. A. Roychoudhury). Singapore: Springer https://doi.org/10.1007/978-981-15-5337-0_32.

Imran, Amanullah, and Al Tawaha, A.R.M. (2021a). Efficacy of pre and post emergence herbicides alone and in combination for effective weeds control without effecting growth and development of maize (*Zea mays* L.). *Russian Agricultural Sciences* 47: 261–269.

Imran, Amanullah, and Al-Tawaha, A.R.M. (2021b). Carbon sources application increase wheat yield and soil fertility. *Communications in Soil Science and Plant Analysis* 52: 695–703.

Jagadish, S.V.K., Craufurd, P.Q., and Wheeler, T.R. (2007). High temperature stress and spikelet fertility in rice (*Oryza sativa* L.). *Journal of Experimental Botany* 58: 1627–1635.

Jagadish, S.V.K., Muthurajan, R., Oane, R. et al. (2010). Physiological and proteomic approaches to address heat tolerance during anthesis in rice (*Oryza sativa* L.). *Journal of Experimental Botany* 61: 143–156.

Khalid, S., Khan, H.A., Arif, M. et al. (2019). Organic matter management in cereals based system: symbiosis for improving crop productivity and soil health. In: *Sustainable Agriculture Reviews*, vol. 29, 67–92. (ed. R. Lal, R. Francaviglia). Cham: Springer.

Lal, M., Singh, K.K., Srinivasan, G. et al. (1998). Vulnerability of rice and wheat yields in NW-India to future change in climate. *Agricultural and Forest Meteorology* 89: 101–114.

Lee, K.D., Tawaha, A.R.M., and Supanjani (2005). Antioxidant status, stomatal resistance and mineral composition of hot pepper under salinity and boron stress. *Bioscience Research* 2: 148–154.

Mall, R.K. and Singh, K.K. (2000). Climate variability and wheat yield progress in Punjab using the CERES-wheat and WTGROWS models. *Yayu Mandal* 30: 35–41.

Nelson, G.C., Valin, H., Sands, R.D. et al. (2014). Climate change effects on agriculture: economic responses to biophysical shocks. *Proceedings of the National Academy of Sciences* 111: 3274–3279.

Odat, N. (2018). Molecular and biochemical responses of barley (*Hordeum vulgare* L.) to NaCl salinity stress and salicylic acid. *Research on Crops* 19: 101–106.

Odat, N., Hasan, M., Obeidat, M. et al. (2015). Identifying selection signatures related to domestication process in barley (*Hordeum vulgare* L.) landraces of Jordan using microsatellite markers. *Jordan Journal of Biological Sciences* 8: 307–303.

Osman, N.N.A.E., Shatnawi, M., Shibli, R.S.R. et al. (2021). Salts induced salinity and in vitro multiplication of *Paronychia argentea*. *Ecological Engineering & Environmental Technology* 22: 55–64.

Othman, Y., Al-Karaki, G., Al-Tawaha, A.R., and Al-Horani, A. (2006). Variation in germination and ion uptake in barley genotypes under salinity conditions. *World Journal of Agricultural Sciences* 2: 11–15.

Porter, J.R., Xie, L., Challinor, A.J. et al. (2014). Food Security and Food Production Systems. In: *Climate Change 2014: Impacts, Adaptation, and Vulnerability. Part A: Global and Sectoral Aspects. Contribution of Working Group II to the Fifth Assessment Report of the*

Intergovernmental Panel on Climate Change, vol. 40 (ed. C.B. Field, V.R. Barros, D.J. Dokken, et al.), 485–433. Cambridge, UK and New York, NY, USA: Cambridge University Press.

Rupa Kumar, K., Pant, G.B., Parthasarathy, B., and Sontakke, N.A. (1992). Spatial and subseasonal patterns of the long-term trends of Indian summer monsoon rainfall. *International Journal of Climatology* 12: 257–268.

Sakhiya, A.K., Anand, A., Aier, I. et al. (2021). Sustainable utilization of rice straw to mitigate climate change: a bioenergy approach. *Materials Today: Proceedings* 46: 5366–5371.

Saleh, M.M., Elabd, A.B., and Al Tawaha, A.R.M. (2021). Principle component analysis among exotic and Egyptian rice genotypes. *Advances in Environmental Biology* 15: 16–23.

Sangeetha, J., Al-Tawaha, A.R.M.S., Thangadurai, D. et al. (2020). Genetic engineering for developing herbicide resistance in rice crops. In: *Rice Research for Quality Improvement: Genomics and Genetic Engineering*, 209–232. (ed. A. Roychoudhury). Singapore: Springer.

Saranraj, P., Sivasakthivelan, P., Al-Tawaha, A.R.M. et al. (2021). Macronutrient management for the cultivation of soybean (*Glycine max* L.): a review. In: *IOP Conference Series: Earth and Environmental Science*, vol. 788, No. 1, 012055. IOP Publishing.

Shatnawi, M., Osman, N.A.E., Shibli, R. et al. (2021). Effect of heavy metal on the in vitro growth of *Paronchia argentea* and its antimicrobial activity. *Ecological Engineering & Environmental Technology* 3: 142–151.

Tawaha, A.M., Turk, M.A., and Maghaereh, G.A. (2002). Response of barley to herbicide versus mechanical weed control under semiarid conditions. *Journal of Agronomy and Crop Science* 188: 106–112.

Tawaha, A.R.M., Khanum, S., Al Tawaha, A.R. et al. (2021). Use of mycorrhiza in organic farming. In: *IOP Conference Series: Earth and Environmental Science*, vol. 788, No. 1, 012167. IOP Publishing.

Thangadurai, D., Kordrostami, M., Islam, S. et al. (2020). Genetic enhancement of nutritional traits in rice grains through marker-assisted selection and quantitative trait loci. In: *Rice Research for Quality Improvement: Genomics and Genetic Engineering* (ed. A. Roychoudhury). Singapore: Springer https://doi.org/10.1007/978-981-15-5337-0_21.

Turk, M.A. and Tawaha, A.M. (2001a). Common vetch (*Vicia sativa* L.) productivity as influenced by rate and method of phosphate fertilization in a Mediterranean environment. *Agricoltura Mediterranea* 131: 108–111.

Turk, M.A. and Tawaha, A.M. (2001b). Wheat response to 2,4-D application at two growths stages under semiarid conditions. *Acta Agronomica Hungarica* 49: 387–391.

Turk, A.M. and Tawaha, A.M. (2002a). Effect of sowing rates and weed control methods on winter wheat under Mediterranean environment. *Pakistan Journal of Agronomy* 1: 25–27.

Turk, A.M. and Tawaha, A.M. (2002b). Response of winter wheat to seeding rate with or without Ethrel spray under irrigation. *Bulgarian Journal of Agricultural Science* 8: 37–42.

Turk, A.M. and Tawaha, A.M. (2002c). Response of sorghum genotypes to weed management under Mediterranean conditions. *Pakistan Journal of Agronomy* 1: 31–33.

Turk, M.A. and Tawaha, A.M. (2003a). Weed control in cereals in Jordan. *Crop Protection* 22: 239–246.

Turk, M.A. and Tawaha, A.M. (2003b). The response of wild oats (*Avena fatua* L.) to sowing rate and herbicide application. *African Journal of Range and Forage Science* 20: 239–242.

Turk, M.A., Hameed, K.M., Aqeel, A.M., and Tawaha, A.M. (2003c). Nutritional status of durum wheat grown in soil supplemented with olive mills by products. *Agrochimica* 5: 209–219.

Turk, M.A., Assaf, T.A., Hameed, K.M., and Al-Tawaha, A.M. (2006). Significance of mycorrhizae. *World Journal of Agricultural Sciences* 2: 16–20.

Valluru, R., Reynolds, M.P., and Salse, J. (2014). Genetic and molecular bases of yield-associated traits: a translational biology approach between rice and wheat. *Theoretical and Applied Genetics* 127: 1463–1489.

Wegren, S. (2011). Foods Security and Russia's 2010 drought. *Eurasian Geography and Economics* 52: 140–156.

13

Impact of Climate Change on Tea Cultivation and Adaptation Strategies: Special Emphasis on Tea Pests in North East India

Azariah Babu[1], Somnath Roy[2], Rupanjali Deb Baruah[3], Bhabesh Deka[1], Kamruza Z. Ahmed[3], Sourajit Bayen[2], and Suman Sarkar[1]

[1] *Tea Research Association, North Bengal Regional R & D Centre, Nagrakata, West Bengal, India*
[2] *Department of Entomology, Tocklai Tea Research Institute, Tea Research Association, Jorhat, Assam, India*
[3] *Climate and GIS Laboratory, Soils Department, Tocklai Tea Research Institute, Tea Research Association, Jorhat, Assam, India*

CONTENTS

13.1 Introduction

Climate change, a result of global warming, has begun to have global consequences. Drought (Figure 13.1), flood (Figure 13.2), storm (Figure 13.3), landslides (Figure 13.4), and hail are some of the negative effects of climate change on tea plantations (Figure 13.5). The agricultural sector is one of the most affected, as the nature and qualities of vegetation and crops are determined by the environmental conditions of a region/country. Food production

Climate Change and Agriculture: Perspectives, Sustainability and Resilience, First Edition. Edited by Noureddine Benkeblia.

Figure 13.1 Drought effect due to long term dry spell and high temperature.

Figure 13.2 Flood in tea plantation due to heavy rainfall in short duration.

systems are particularly sensitive to variations in temperature and precipitation, which can result in pest and disease outbreaks, lowering harvest and threatening the country's food security. A rise in the mean temperature above a certain threshold can shorten the growing season of many crops, resulting in lower agricultural production. As the temperature rises, more water becomes scarce, and the demand for irrigation water rises. Drought severity will be amplified in warmer parts of the world. In terms of soil temperature, while

Figure 13.3 Damage to tea plantation due to storm.

Figure 13.4 Landslides in tea sections.

Figure 13.5 Damage to tea plant due to hail.

nutrient availability will rise in the short term, increased soil temperature will lead to a decrease in organic matter content, resulting in a decrease in soil fertility in the long run. Bug pests will become more common as the temperature rises, as higher temperatures favour insect development and winter survival. Plant disease epidemiology will also be affected. In times of rapidly changing climate and inclement weather, forecasting disease outbreaks will be more challenging. Increased prevalence of harsh weather and environmental instability may impair the effectiveness of insecticides on targeted pests or result in more injury to non-targeted creatures.

Climate change is increasingly being felt in India's agricultural systems, not just by policymakers, but also by tea growers. Tea (*Camellia sinensis* L.) is grown as a monoculture in more than 45 nations across the globe, except in North America, in the latitudinal region of 45°N to 34°S. Tea, along with coffee and cocoa, is one of the world's top three non-alcoholic beverages with therapeutic and social benefits. Tea was grown in Argentina, Bangladesh, China, India, Indonesia, Iran, Japan, Kenya, Malawi, Myanmar, Nepal, Russia, South Africa, South Korea, Sri Lanka, Thailand, Turkey, Uganda, Viet Nam, and Zimbabwe until 2018. China, India, Kenya, and Sri Lanka are among the world's leading tea producers. Tea is now regarded as a highly profitable agricultural plant when compared to other crops in terms of poverty reduction and income production for tea growers. India is the world's second-largest producer of tea, accounting for 24% of global production. The best-known tea-growing regions in India are in the north-eastern part of the country, near the Himalayan foothills. Dooars, Terai, Darjeeling, and Assam are among these areas. In the far north, such as Himachal Pradesh, there are just a few small, isolated tea gardens. Along the mountain range that runs north-south along South India's west coast, there are some notable tea-growing regions. The Nilgiris, Annamalais in Tamil

Nadu, and the high and mid ranges of Kerala are the tea-growing regions of South India. Approximately 75% of India's tea is produced in Assam and North Bengal. Tea can thrive in a variety of temperatures from subtropical to tropical, but it requires a lot of rain and humidity during the growing season. Tea yield, crop distribution, and quality are all influenced by the weather. Tea in north-eastern India, despite its economic importance, is facing issues as a result of climate change, raising questions about its long-term survival. Tea growers are already dealing with decreased and unpredictable rainfall distribution, as well as rising temperatures, which have a significant impact on yields and productivity. Tea production is governed by three key factors: genotype, environment, and management tactics. The interaction (G×E×M) of these three parameters affects the system's long-term viability and production. When it comes to the adaptation to changing environment, the adaptability of a genotype (G) or variations is considered as a crucial factor by researchers. However, whether these are evaluated for future climatic circumstances, which of these clones would confront, remains to be seen. If research efforts are focused using modern scientific instruments, this factor 'G' can be adjusted to some extent to suit future climatic abrasions. The element 'E,'which refers to current weather (climate in a broader sense) conditions, is difficult to manage; nevertheless, microclimate creation can be attempted to some extent, for example, increasing humidity in a garden can delay the onset of drought. The most important element that can be managed is 'M,' which can be addressed with both preventative and adaptive strategies.

With the changing climate, shrubs are becoming more stressed, pest populations/disease incidences are increasing, and monthly crop cycles are varying unpredictably. The pest environment has changed as a result of rising temperatures and changing rainfall patterns, as well as poor practices such as indiscriminate application of synthetic pesticides, which is posing a serious threat to tea growers. In such circumstances, effective methods must be established and implemented in order to maintain the crop, which provides the country with precious foreign cash.

13.2 China: Impact of Climate Change in Tea Plantations and Possible Adaptation Strategies

China is the world's largest producer of tea, and to determine climate change trends over a 60-year period, four typical cities were studied: Haikou (an island city in south China that produces black tea), Kunming (a plateau in south China that produces Puer and black tea), Hangzhou (in east China that produces green tea), and Jinan (in north China that produces green tea in nearby coastal areas). In the last 50 years, the sub-tropical city Hangzhou and the plateau city Kunming both saw a 1.6°C increase in annual mean temperature, contrasted to Haikou, the tropical city, and Jinan, the temperate city, which both experienced 1.0 and 1.1°C increases in annual mean temperature. Despite no change in annual rainfall, the number of wet days in Haikou, Kunming, Hangzhou, and Jinan decreased by 13.1, 13.2, 12.0, and 6.7 days, respectively, in the last 50 years, associated with a decrease in atmospheric relative humidity in all four cities. Annual sunshine time declined in all four cities, and if current trends continue, tea yield will suffer as the mean and percentage of sunshine time decreases.

In the previous few years, extreme climate fluctuations have become common. In 2009, Kunming city faced a drought with only 565.8 mm of rainfall, resulting in a 20% decline in tea yield and the complete drying of tea plants covering around 3300 acres. Since 2003, tea farmers in Hangzhou have experienced late spring cold practically every year, resulting in billion-dollar losses. Hangzhou had its warmest and driest summer since 1950 in 2013. Temperature rises may cause tea-growing regions to relocate to higher altitudes, extending the tea-growing season and plucking season. Tea is supposed to benefit from CO_2 enrichment because it improves photosynthesis and respiration. Some of the negative effects of climate change in China include a decrease in the number of rainy days, increased relative humidity, increased insect and disease incidence, and deterioration of soil health.

Pests, illnesses, and weeds are becoming more common as the climate changes (Lal 2005). The temperature has a positive correlation with disease incidence, but sunshine time has a negative correlation (Li et al. 2010). Tea blister blight, which is frequent in tropical tea-growing countries, has also become an issue in the subtropics, according to studies. Anthracnose disease (*Colletotrichum gloeosporioides*) was also shown to increase dramatically when carbon dioxide levels in the atmosphere increased according to reports (Li et al. 2016). Pests and illnesses are projected to cost the agricultural industry 20-25% of its gross output value (Wang et al. 2006).

Winter is a good time to minimize the number of pests and pathogens, but shifting climatic trends and rising temperatures cause them to multiply, resulting in an increase in the number of yearly generations and the rate of reproduction of some pests, extending the destructive period. Tea geometrid (*Ectropis oblique*) in Hangzhou, which has 6 annual generations under normal weather circumstances, has been recorded to grow to seven generations if the mean temperature in October exceeds 20°C (Wang and Jin 2010). The combination of a rise in ambient temperature and a drop in relative humidity favours particular bugs. In Hangzhou district, the number of days with daily mean temperatures above normal is connected with the amount of tea green leafhoppers and their reproductive rate (Yang 2005).

In China, there are three stages of adaptation and mitigation strategies: (i) government policy; (ii) technology and technique development; and (iii) community development. Establishing worldwide and national networks to analyze the impact of climate change, incorporating international and interdisciplinary research, as well as R&D activities to fine-tune new technologies, are examples of government programmes. The government is also in favour of enhancing field infrastructure, including drainage and irrigation systems, as well as promoting organic agriculture and good agricultural practices (GAP) in tea production, all of which can help to mitigate climate change. Innovative agriculture techniques and technologies are critical in combating climate change by lowering negative environmental impacts. The use of slow-release nitrogen (N) fertilizer and methods that improve N use efficiency may help to reduce greenhouse gas emissions. Increased soil organic content can improve C sequestration, which has an impact on soil structure, biodiversity, and water retention capacity. Under the current climate change scenario, the development of novel resistant cultivars with great tolerance to drought, heat, and cold is critical and necessary. New cultivars with high net photosynthesis, high N usage efficiency, and good CO_2 responsiveness are urgently needed. Tea growers and the local community should have access to more information on the effects of climate change and how to implement mitigation and adaptation techniques.

13.3 Kenya: Impact of Climate Change in Tea Plantations and Possible Adaptation Strategies

The tea business in Kenya is one of the most important contributions to the country's socio-economic progress. Tea is grown in 18 counties that are divided into two zones, Zones I (humid) and II (dry) (sub-humid). Climate change in Kenya is manifested by rising temperatures, unpredictable rainfall patterns, and an increase in the frequency of extreme weather events such as hail storms, droughts, and frost. Bore et al. (2013) and Cheserek et al. (2015) conducted a study in Timbilil Tea Estate, Kericho, Magura Tea Estate, Sotik, and Kangaita Tea farm, Kirinyaga, found increasing temperature trends with annual rise of 0.02 °C at Timbilil, 0.22 °C in Sotik-Magura, and 0.01 °C in Kangaita. In the tea-growing districts of Kericho, Sotik, and Nandi Hills, hail is responsible for an estimated loss of over 2,000 tons each year. Rainfall that is unevenly distributed and unexpected, with significant rain in some months followed by a protracted dry spell, has resulted in a decrease in soil moisture, leading to osmotic stress. Frost occurrence was previously rare but has recently increased, resulting in significant losses. One such example is the January 2012 frost, which resulted in a 30 percent yield loss and forced the crops to be harvested for three months. Climate change trends may reduce the number and area of acceptable tea-growing regions in Kenya.

Adopting drought-tolerant clones, applying water conservation tactics, and field management that is successful against climate change are some of the measures Kenya is taking to mitigate the effects of climate change. It is vital to replenish main nutrients such as nitrogen, phosphorus, and potassium that are reduced during harvesting by applying the appropriate amount of fertilizer based on the clone's output potential. Keep in mind that using too much of this type of fertilizer can deplete soil moisture and cause death in drought-tolerant clones. Shade trees protect tea plants from frost, limit evapotranspiration (ET), and act as windbreakers, so controlling climate to some extent. Because they do not compete for resources with tea bush, *Grevillea robusta* (Silver Oak), *Hakea saligna* (Willow Hakea), *Millettia dura,* and *Sesbania sesbans* (Egyptian pea) are recommended. Tea cultivation in a changing environment requires the identification of sensitive clones and the development of new varieties/clones that are drought and cold tolerant. Such clones should be tested before being launched for commercialization if they perform well. TRFK 301/4, TRFK 301/5, TRFK 430/90, TRFK 371/3 (drought tolerant), and TRFK 306/1 are some of the clones developed by the Tea Research Institute (drought, frost, pest, and disease resistant).

13.4 Sri Lanka: Impact of Climate Change in Tea Plantations and Possible Adaptation Strategies

Tea was initially introduced to Sri Lanka in 1839, and it has since risen to become the world's fourth-largest tea producer, accounting for over 7% of global tea production. Tea is being grown on 205,000 acres of land in 14 administrative districts across Sri Lanka. In Sri Lanka, there are three main climatic zones based on yearly rainfall: Wet zone (W),

Intermediate zone (I), and Dry zone (D). Low country (300m) (L), mid-country (300-900m) (M), and upcountry (>900m) (U) regions can be classified based on elevation. Sri Lanka may be split into 46 agro-ecological regions using a combination of the two classifications (AER). By analyzing meteorological data and soil characteristics from 1961 to 2010, these AERs were identified as extremely vulnerable (WL1a, WL1b, WL2a, WM2a, WM2b, WM3a, IM2b, IM3a, and IM3c), vulnerable (WM1a, WM1b, WM3b, IM1a, IM2a, IU3a, IU3d, and IU3e) (Wijeratne and Chandrapala 2013). Poor soil is linked to the effects of climate change, according to this study, and can make tea more vulnerable in different tea-growing regions. When comparing tea plantations cultivated at low and mid-elevations to tea grown at high elevations, trends reveal that low and mid-elevation tea plantations are more affected. The negative effects are primarily due to rising ambient temperatures and dry weather conditions, which may cause tea-growing places to relocate. Sri Lanka has experienced low rainfall in all agro-ecological zones for the previous 20 years. From 1976 to 2013, the area suitable for tea growing and tea production dropped (Thushara 2015). Root growth and development are influenced by soil structure, chemical composition, pH, and nutrient availability. Productivity is hampered by a lack of moisture in the soil, acidic soil, and a lack of organic matter in the soil. In Sri Lanka, a lower-elevation tea production has been reported due to a soil moisture shortage (Wijeratne and Fordham 1996).

It was suggested that climate change adaptation be accomplished by land classification and mapping in conjunction with a crop development programme. The most appropriate adaptation being conducted in all major tea growing regions across the world is the development of diverse resistant and tolerant clones that perform better in heat and drought stress, and are potent against pest and disease incidence as a result of changing climate. Other major variables that can help tea estates mitigate the negative effects of climate change are crop diversification, soil development, and irrigation.

13.5 India: Impact of Climate Change in Tea Plantations and Possible Adaptation Strategies

13.5.1 Agro-Climatic Conditions Ideal for Tea Cultivation in North East India

Different agro-ecosystem environments influence the interactions between biotic elements such as agricultural plants, weeds, insect pests, diseases, nematodes, and abiotic factors such as temperature, humidity, rainfall, and soil parameters. Tea, as a perennial monoculture crop, requires a climate that is somewhat hot and humid for maximum growth. Unfavourable conditions have been reported, including a long dry spell and alternate waves of warm and cool winds (Willson and Clifford 1992; Barua 1989; Goswami 2011). High humidity, heavy dew, and morning fog promote the growth and development of tea bushes' new leaves. Temperatures between 20 and 30 °C are optimal for tea production, and temperatures above 35 °C or below 10 °C are toxic or harmful to the tea bushes, affecting proper growth. Tea crops require between 1500 and 3000 mm of annual rainfall, which should be evenly spread throughout the year (Goswami 2011). Increased day length to 13 hours during the winter months by supplementary weak artificial illumination was reported to enhance shoot growth, hastened bud break, and inhibited flowering of tea bushes under irrigated conditions (Barua and Barua 1969), and thus day length influences the growth and dormancy of tea bushes. Dormancy in tea plants occurs after six weeks of less than 11-hour day duration on a worldwide scale (Goswami 2011). Around 18° North and South latitudes, seasonal dormancy appears. Due to the combined impact of short days and cold temperatures, tea bushes in North East India (25°–27°N latitude) are dormant for around three months during the winter season (Barua 1989, Goswami 2011). As a result, because tea is a shade-loving plant, shade trees are typically planted before spreading tea seeds, and tea grows more quickly when planted alongside shade trees. Other variables that help to the optimal growth of tea plants include well-drained fertile acidic soil (pH range of 4.5–5.5) on hill slopes and neighbouring plains.

Due to its reliance on relatively stable temperatures and consistent rainfall patterns, the Indian tea industry is significantly impacted by climate change; however, rising temperatures and changing rainfall patterns pose a threat to Indian tea plantations, affecting both the quantity and quality of tea produced. Since canalization and impoundment disrupted the flow of water in many rivers' tributaries, intensive irrigation is essential for tea and other agricultural crops to survive. Various issues, such as illegal deforestation and encroachment, have resulted in the loss of greenery by diminishing the density of forest cover. This is one of the main reasons for the entry of sylvan pests into tea plantations.

13.5.2 Influence of Climate Change and Pesticides on Biotic Community Associated with Tea

The primary elements balancing the life processes of the biotic communities that exist in the tea environment are climate and soil. Tea insect and mite pests such as the tea mosquito bug, looper caterpillars, and red spider mites have altered their growth and survival methods in order to cope with the environment and adapt to climate change. Changes in the micro and macro environment, such as (i) rising ambient temperatures and reduced or changing rainfall patterns due to climate change; (ii) deforestation and planting of alternate hosts or shade trees; and (iii) pest management using synthetic or chemical pesticides, can all be linked to the appearance of these pests. Chemical pesticides have been used widely and indiscriminately in recent years, resulting in resurgence of main pests, secondary pest outbreaks, environmental contamination, resistance development, and unwanted residues in processed tea. Due to behavioural and biochemical changes in pest populations, chemical pesticide control of tea pests is becoming more difficult. Pests like the tea mosquito bug change their egg-laying patterns in response to pesticides by ovipositing eggs in safer places, such as the lower tier of tea shoots, even when only a small amount of organochloride or pyrethroid pesticides are present, which can be considered a form of behavioural adaptation. The Red Slug Caterpillars of *Eterusia magnifica* show another example of adaptive behaviour and avoidance techniques in the form of temporary pest relocation. Pests develop resistance through defence enzymes, cuticular sclerotization with dark colour development, and increased lipid accumulation in the body as the main mechanisms to withstand the high toxicity of pesticides as a result of indiscriminate application of chemical pesticides for an extended period of time. According to studies, increased levels of general esterases (GE), glutathione S transferases (GST), and cytochrome P 450 (CYP) in the tea mosquito bug and red spider mite confer resistance or tolerance to chemical pesticides. The rising expense of chemical pesticides and their application is draining developing countries' economies. Tea growers have traditionally relied on synthetic chemical insecticides to control tea pests. Even when massive chemicals are available on the market, pesticide choices are often limited, which is why the Food Safety and Standards Authority of India (FSSAI), European Union (EU), FAO, Codex Alimentarius Commission, and Environmental Protection Agency (EPA of USA) have notified Maximum Residue Limits (MRLs) in tea, along with the introduction of the Plant Protection Code (PPC) by the TGA. Conservation and release of natural enemies; microbial bio-control agents (bacteria, virus, nematodes, fungi); use of botanical extracts as herbal pesticides; and cultural practices are examples of non-chemical pest management strategies. Many published reports on Integrated Pest Management (IPM) packages for tea are currently available in both North East and South India. In addition to the use of other eco-friendly pest management measures, IPM frequently incorporates the prudent use of chemical pesticides as a last resort or strategy. Regardless of the availability of various non-chemical pest management options, tea growers require a robust strategy and its effective implementation because it is a challenge for them to combat the pest situation that is causing significant economic loss and damage to tea plants as a result of climate change, changes in agronomic practises, and natural changes in pest dynamics. To deal with inconvenient and unstable conditions like these, new tactics must be explored and innovated, as well as existing pest management strategies must be modified or fine-tuned.

13.5.3 Climate Change Trends in Major Tea Growing Regions of North East India

Tea is a perennial crop that can last for 40–50 years before needing to be removed and replanted. Thus, the plantation is stressed at some point as a result of the continual intake of nutrients from the soil as well as pest attacks caused by climate variables, among other things. Climate conditions, particularly rainfall distribution, are one of the most critical elements in determining tea yield (Mitra 1974). A desirable climatic element for optimum tea cultivation is considered to be well-distributed rainfall. Tea is reliant on natural precipitation, but it is being harmed as the rainy season shortens. Flooding in Assam caused by the Brahmaputra river overflowing produces water logging in tea-growing zones, which is an abiotic stress for tea plants, impacting their development and survival. The annual rainfall transmits minimal information; however, the rainfall distribution is more essential. Monsoonal rainfall in North Eastern India is often high (May to September). Rainfall is either insufficient or missing in a variety of regions, including near Golaghat, Nowgong, sections of the North Bank, Cachar, the Dooars, the Terai, and so on. Tea growing locations in North Eastern India can be separated into three main regions: the Assam plains' Brahmaputra and Barak Valleys, the Dooars and Terai region of North Bengal, and the Darjeeling Hills of North Bengal. The south-north monsoon brings copious rain to the Brahmaputra valley throughout the summer months, causing problems with water disposal, while less rain falls during the winter and early spring, leaving shrubs thirsty. The annual rainfall ranges from 2100 to 2900 mm. Annual rainfall in the Barak Valley, Dooars, and Terai ranges from 3000 to 5000 mm; however, the distribution is irregular. The rainfall in Darjeeling ranges from 2500 to 5000 mm (Jain 1999). It is widely assumed that the rainfall pattern in North Eastern India has changed dramatically in recent years. Extreme event frequency has increased dramatically in recent years, necessitating a shift in climate research efforts to focus on long-term climate analyses and trends.

For the two climatic factors of rainfall and temperature, a study of long meteorological data was conducted. The Climate Research Centre of the Tocklai Tea Research Institute, Jorhat, Assam, India, has archived rainfall and temperature datasets for seven locations (Tocklai, Jorhat, South Bank, Assam; Dikom, Upper Assam; Silcoorie, Cachar, Assam; Thakurbari, North Bank, Assam; Darjeeling, North Bengal; Bengdubi, Terai, North Bengal; Nagrakata, Dooars). This study uses data from two locations: Tocklai, South Bank, Assam, and Dooars, Nagrakata, North Bengal, for the sake of brevity. Figures 13.6 and 13.7 show the annual total rainfall in the South Bank region of Assam and Dooars, Nagrakata region of North Bengal, respectively.

In general, rainfall in Assam's tea-growing regions has been dropping, with rainfall in Tocklai (South bank), Assam alone having declined by more than 200 mm. There are yearly changes in rainfall, with some years being rainy and others being dry, but the frequency of extreme rainfall events has increased in the previous 40 years, with more droughts and floods than in the past. The irregular nature of rainfall poses a threat to tea growers in certain tea-growing regions in that, although they must deal with deluges and waterlogging difficulties due to excessive rain during one part of the growing season, severe drought may threaten the tea fields at other times. Not only that, but the rainfall distribution has been adversely impacted, with individual year analyses revealing significant fluctuations in rainfall and an increase in concentration during monsoonal seasons. The rainfall datasets also revealed that rainfall decreases during both the productive (April–October) and dormant (November–March) seasons. During the winter, evaporation far outnumbers rainfall,

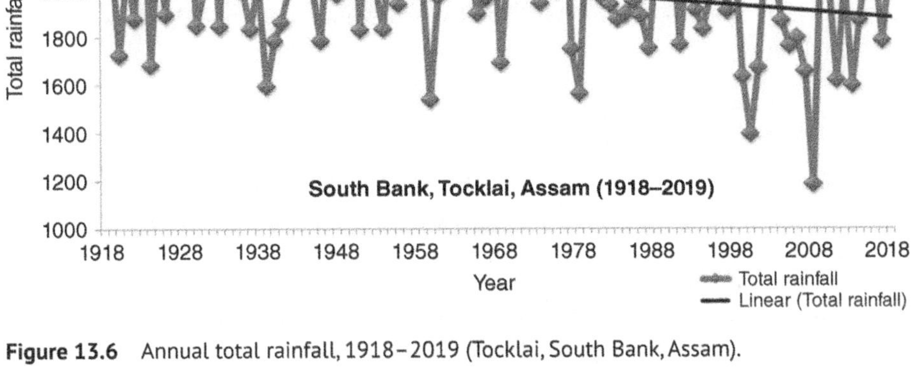

Figure 13.6 Annual total rainfall, 1918–2019 (Tocklai, South Bank, Assam).

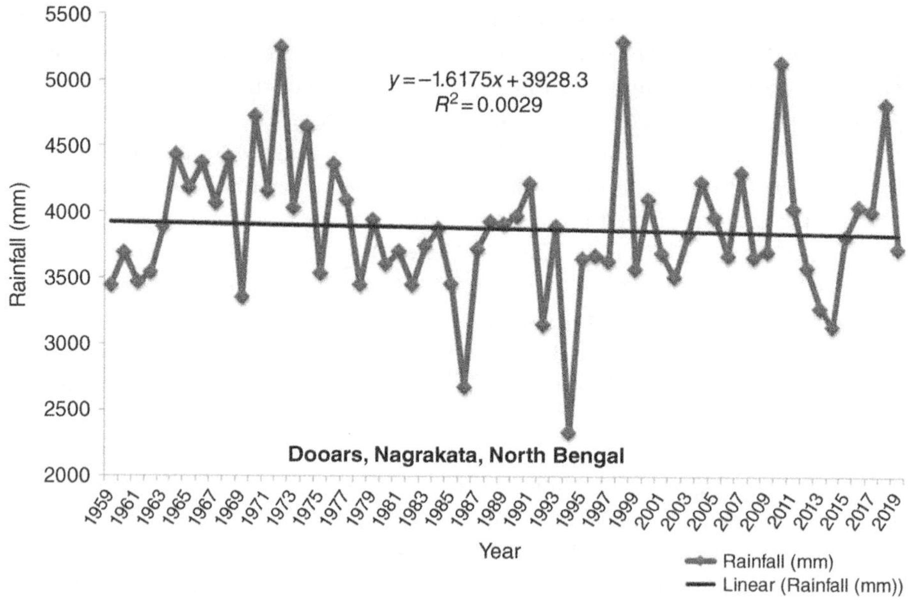

Figure 13.7 Annual total rainfall, 1959–2019 (Dooars, Nagrakata, North Bengal).

resulting in water shortages, which are detrimental to unpruned teas and early flush teas. This necessitates irrigation in Assam teas, especially during the winter, which was previously primarily rain-fed. In terms of monthly rainfall, there has been a decrease, in practically every month. Tea, which is predominantly cultivated as rain-fed crop in Assam, may be severely harmed by such shortfalls in rain.

To detect historical patterns, the yearly average minimum lowest temperature for the archived datasets available was plotted and is displayed in Figures 13.8 and 13.9 for Tocklai, South Bank of Assam and Dooars, Nagrakata of North Bengal area, respectively.

The minimum temperature in several tea-growing regions in North East India has been noted to be rising in general. However, there has been a steep increase in Tocklai, South Bank region, with an increase of 1.4 °C over 96 years, while there has been no notable increase in Dooars, North Bengal region. Furthermore, there is a rise in minimum temperature in North Bengal for several individual years, which appears to be concealed in long-term averages.

Tea, like all plants, has a preferred temperature range. Tea is said to grow best between the temperatures of 13 and 30 °C (Barua 2008); however, a review of long-term temperature data found that not only the number of days above 30 °C is increasing, but the number of days above 35 °C is also increasing. Figures 13.10 and 13.11 depict statistics for the number of days with temperatures above 35 °C in the South Bank region of Assam and the Dooars, Nagrakata region of North Bengal, respectively.

A closer examination of Figure 13.10 revealed that, in comparison to the rest of the eras, the frequency of days with temperatures more than 35 °C has increased significantly in the previous 40 years. This raises various problems, such as why tea continues to grow well

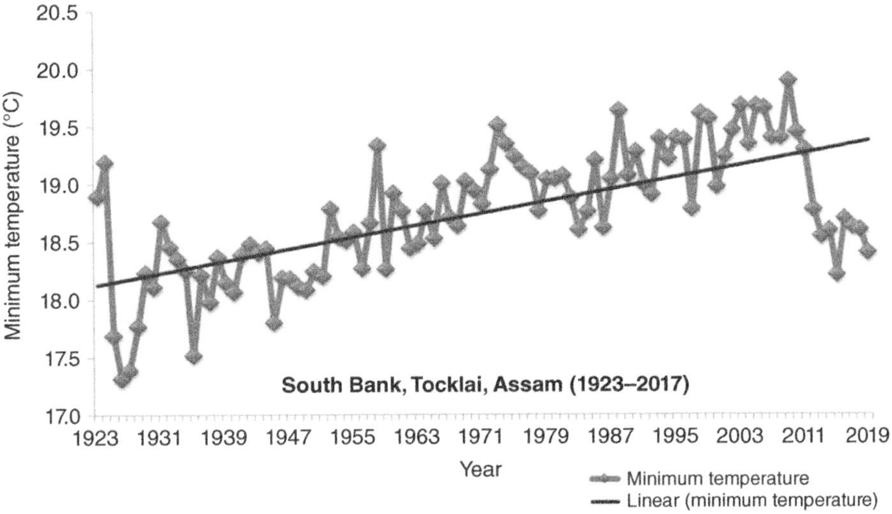

Figure 13.8 Annual average minimum temperature, 1923–2019 (Tocklai, South Bank, Assam).

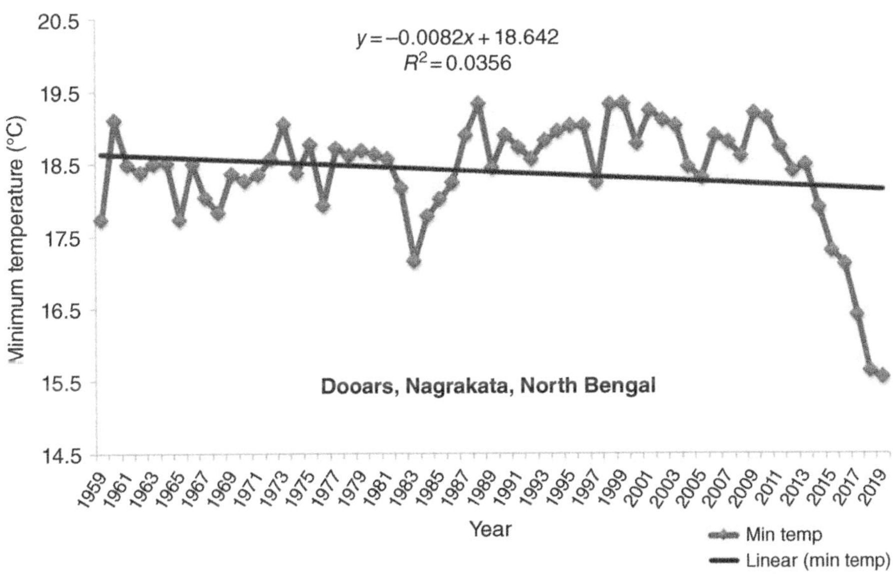

Figure 13.9 Annual average minimum temperature, 1959–2019 (Dooars, Nagrakata, North Bengal).

even when the limitations are exceeded, implying that the crop is self-adapting to fluctuating temperatures. However, it is yet unknown how long such self-adaptation can last at a given temperature. Temperatures in these locations have risen over the last century, and as they continue to rise in the twenty-first century, the chance of its tea-friendly environment envelope shifting or disintegrating grows.

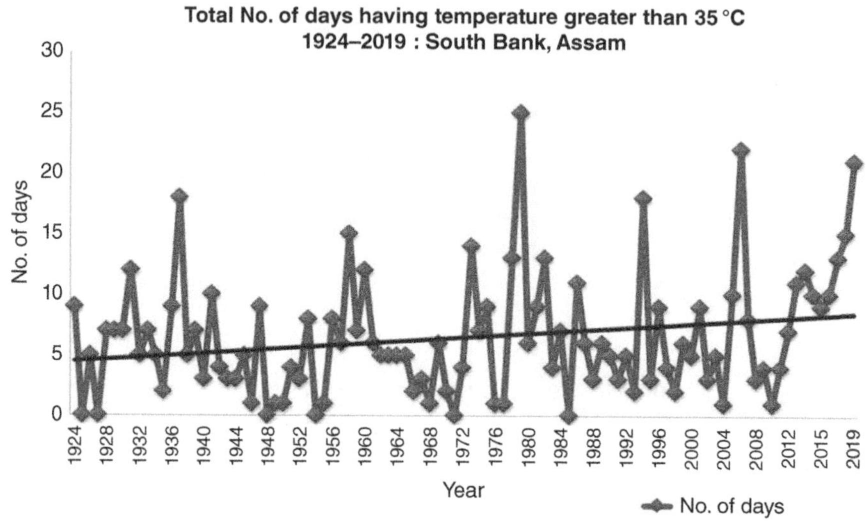

Figure 13.10 Number of days with temperature greater than 35 °C, 1924–2019 (Tocklai, South Bank, Assam).

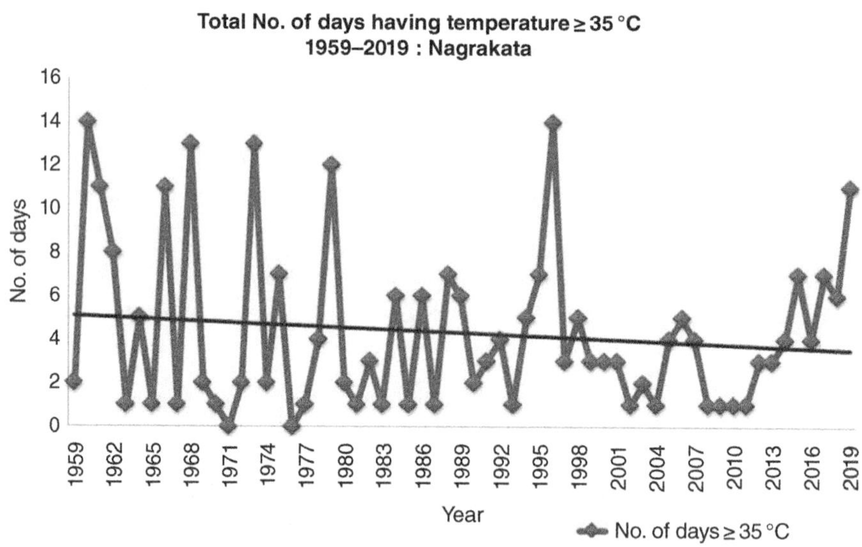

Figure 13.11 Number of days with temperature greater than 35 °C, 1959–2019 (Dooars, Nagrakata, North Bengal).

13.6 Impact of Temperature and Carbon on Insect Physiology and Abundance

Although increased temperature and other climatic conditions affect the physiology and life history of insects, such changes also affect the host plant, which influences the life cycle of insects (Willson and Clifford 1992; Hare 1992; Gutierrez et al. 2008; Musser and Shelton 2005). The temperature has a considerable impact on the reproductive potential, feeding rate, distribution pattern, duration of development stages, number of annual generations, and movement of insect and mite pests. When tea pests such as geometrid moths, *Hyposidra infixaria*, and *Ectropis* sp., tea mosquito bug, *Helopeltis theivora*, and red spider mite, *Oligonychus coffeae*, were exposed to varying temperatures, their developmental periods changed.

They complete their life-cycle faster during the summer months due to the higher accumulation of degree-days than projected throughout the winter (Prasad and Mukhopadhyay 2013; Das 2015; Saren et al. 2015). According to a degree-day study on tea pests, rising temperatures can lead to increased fecundity, the production of extra broods, and a faster pace of development, all of which can lead to increased population and crop damage.

Rainfall and temperature have a significant impact on the occurrence and quantity of insect pests in agricultural plants in diverse agroecosystems, including tea, according to entomologists' research. In the tea ecosystem, Barua et al. (2018) studied the seasonal abundance and predatory capabilities of *Stethorus aptus*, a biocontrol agent of the tea red spider mite, *O. coffeae*. Sarkar et al. (2019, 2021) have underlined the impact of abiotic factors such as temperature, rainfall, and relative humidity in regulating the population trends of two important predators, *Sycanus collaris* and *Eocanthecona furcellata*, in the tea environment. The reproduction rate of insect increases as the temperature rises, followed by the prolonging of their breeding season, resulting in an increase in the number of insect feeding on crops.

The physiology of plants is likewise influenced by high CO_2, with photosynthesis being the most affected. The carbon to nitrogen (C:N) ratio of plants may be influenced by increased carbon, which encourages biting and chewing insects such as caterpillars and beetles to feed more and sucking insects such as plant bugs, thrips, and scale insects to ingest more plant sap and receive more nitrogen (Hare 1992; Musser and Shelton 2005; Caulifield and Bunce 1994; Roth and Lindroth 1995; Coviella and Trumble 1999).

Pest occurrence and activity have changed as a result of climate change in tea plantations in North East India. Some of the pests that have infested the tea plantations and inhabited the same include the black inchworm (*Hyposidra talaca*), litchi looper (*H. infixaria*), hairy caterpillar (*Arctornis submarginata*), sungma caterpillars (*Orgya postica*), green weevil (*Astycus* sp.), white scale insects, and mealybugs (Mukhopadhyay and Sarker 2007; Muraleedharan and Roy 2016; Pathak et al. 2003; Roy et al. 2017, 2020).

Tea growers have had a difficult time managing the larval stages of *Biston suppressaria, H. talaca, H. infixaria*, and a few other species. Looper caterpillars of *B. suppressaria* were once considered a minor nuisance in locations such as Assam's North Bank and Sibsagar districts, with uncommon occurrences in Cachar and the Dooars (Das 1957). Its rise in frequency during the 1950s was mostly owing to its migration from the forest to tea plantations, as well as the deforestation of shade trees such *Dalbergia assamica* (Das 1957), which serve as their host plants. Loopers' voracious leaf-feeding activity and adaptation to a variety of alternate hosts have resulted in heavy defoliation of tea bushes almost throughout the year, as well as the migration of *H. talaca* towards the north-east and Terrai and Dooars, which can be attributed to the current climate change trends such as global warming causing a rise in ambient temperature coupled with extensive deforestation.

By the mid-nineteenth century, the red spider mite, *O. coffeae*, had been observed in all tea districts of the Dooars and Terai, under all climatic circumstances; however, the severity of the attack was mostly dependent on local characteristics (Das 1957). Infestation of mites on tea bushes was lower in the 1950s due to a low multiplication rate (Das 1957), but red spider mite population density increased even during the winter months in north-east tea plantations over the previous 50 years due to decreased rainfall and rising temperature (Roy et al. 2014, 2020)

The increased incidence of the tea mosquito bug (*H. theivora*) in North East India during the late monsoons and autumn months, as compared to its higher incidence during May to July, when the number of rainy days was higher, rainfall was consistent, and cloudy weather conditions prevailed (Das 1957, 1965), is primarily due to a shift in rainfall caused by climate change (Mukhopadhyay and Roy 2009). As a result, it can be concluded that the increase in the tea mosquito bug population is due to its high fecundity and shorter development period, which is induced by increased temperature and a shift in heavy rainfall to the late summer and autumn months, as seen during this decade as a result of climate change trends (Roy et al. 2020).

Pests such as tea thrips (*Scirtothrips dorsalis*) and green leafhoppers (*Empoasca flavescens*), which were once considered insignificant and infrequent, with distribution limited to certain parts of tea plantations, have now become serious pests in North East India's tea plantations (Muraleedharan and Roy 2016; Roy et al. 2009; Saha et al. 2012). In the tea plantations of North Bengal, they are plentiful all year. In the months of March, April, and May, the population of thrips is higher, and the population of green hoppers is higher. Without a doubt, climate change is a major factor in the current pest situation in tea plants (Muraleedharan 2011a). As a result, crucial studies on the influence of climate change on the tea plant's physiology as well as its associated pests are required.

13.7 Probable Adaptive Strategies for Climate Change in Relation to Tea Pest Management

The term 'climate change adaptation' refers to efforts that could aid in the management of climate change's social, environmental, and economic consequences. Adaptation actions lessen climate change susceptibility. Climate change, which includes rising temperatures and changing rainfall patterns, has resulted in shorter crop growing seasons and early pest infestations.

The previous study had shown that agroforestry systems have various advantages, including climate change resilience, biodiversity preservation, and higher-quality botanicals. For effective forecasting of insect/natural enemy population-level responses to climate change, thorough assessments of changes in both mean and variability of temperature on pests and their natural enemies are key components. Understanding and addressing the problem of abiotic stress and crop bug pest interactions under changing climate conditions is a difficult undertaking. As a result, both adaptation and mitigation strategies are viable options for dealing with climate change.

Regular monitoring and ecological evaluations, in addition to predictive models, can be a significant tool for precisely exposing the influence of climate change on the distribution and abundance of pest populations. Rather than changing crop calendars, where crop growers must adjust insect management strategies to account for projected changes in pest incidence and crop losses due to climate change, the best use of IPM basics such as field monitoring, pest forecasting, record keeping, and selecting economically and environmentally sound control measures will be more appropriate. Babu (2021) discussed recent advancements and ongoing issues in India's tea integrated pest management. Pruning at regular intervals is an important agronomic method for restoring the vegetative development of tea bushes and thus enhancing crop output, followed by plucking to reduce insect density and population. Keeping the fields clean might also help to prevent disease spread. Weeds serve as the main foci for the spread of pests on tea because pests do not damage weeds to the same extent as they do tea plants and compete for nutrients; on the other hand, some weeds in the tea ecosystem also provide benefits for tea plantations by conserving natural predators and providing protective ground cover (Hazarika et al. 2009). On small gardens or plantations with a large staff, manual pest removal is possible. Using insect traps can also be an effective method (Hazarika et al. 2009). Traditional tea farms with high and unstable pest populations can conserve natural resources through organic farming, water management through efficient water harvesting, artificial irrigation, and drainage systems to prevent waterlogging, which provides an ideal environment with a suitable microclimate for some pests. Biological control, which depends on predation, parasitism, herbivory, and other natural mechanisms to manage pests, can be an effective technique of pest control. More natural enemies can help to reduce insect populations below economic harm thresholds, which is a vital part of integrated pest management systems. The use of fertilizers, manure, and organic wastes in the tea plant can assist balance nutrient levels, maintain plant health, and boost pest tolerance.

Another option is to plant shade trees at specific spacing in tea-growing areas; however, because the influence of shade trees on pests and illnesses is complex, the research varies widely on how effective shade trees are at controlling pests. In tea plantations, a moderate quantity of shade trees can help to maintain a good balance between insect pests and predators. Pests on intermittent shade trees, shelter belts, and boundary plantings may not only harm the crop in their immediate area, but they may also operate as insect barriers. It has also been documented that thinning shade trees can reduce pests, notably mirids, and that similar research has indicated that increasing the amount of shade has the same impact. Increased heterogeneity in a complex system can restrict the transmission of insects and raise the occurrence of pest predators, but it is the combination of specific species that reduces the pest population.

In the tea environment, when shade trees compete for resources and create low-quality leaf litter, crop vitality, and thus their resistance to pests may deteriorate, interactions of soil with temperature and humidity within the field also play a big impact.

Enormous amounts of pesticides are also used in tea production, as a result of the habit of cultivating tea in large monoculture plantations with minimal natural resistance to pests and illnesses. As a result, various pesticides can be evaluated for new modes of action. Independent of their insecticidal activity, the use of neonicotinoid pesticides to suppress sucking pests causes salicylic acid-related plant defence responses that improve plant vigour and abiotic stress tolerance. This opens up new avenues for research into the role of pesticides in plant stress tolerance. Many more similar chemicals must be discovered in order to be used in crop pest management in the future. Pests are becoming a growing issue as a result of global climate change, since precipitation and temperature variations affect their populations. However, studies on pest prediction models and pest forecasting systems in India and other tea-producing countries are not clearly defined, and the baseline knowledge in this area is still very much a 'black box'.

Very effective actions must be done to safeguard the 'world's best tea', which is grown in the Darjeeling hills, which occupy an area of 18 000 ha. Since they oriented, more than 30 cm of precious rich topsoil has been lost in Darjeeling due to soil erosion by water, according to some estimates. Because the tea industry employs the majority of the Darjeeling hills' inhabitants, any loss of tea will have a direct impact on the region's economy and people's livelihood. More than 76 000 people are employed directly in the Darjeeling tea sector. Every piece of land must be used in accordance with its capabilities and limits in order to have a successful soil conservation programme. Soil conservation principles include not cultivating land with slopes steeper than 50% and improving land-use methods on-farm areas, such as contours and strip-cropping on sloping soils. Forests are being planted on steep slopes, fallow ground, and garden border areas. Plant erosion-controlling grasses such as Lemmon grass, vetiver, and palmarosa in voids and along water channels' edges. Soil conservation in Darjeeling's mountainous terrain necessitates a well-planned and logical land use programme, as well as engineering and cultural approaches.

13.8 Decisive Measures to Combat the Impact of Climate Change on Tea Cultivation in North-Eastern India

a) Given the potential consequences of future climate change, it is critical to educate growers and other supply chain stakeholders on climate change, as well as changes in the occurrence of pests in tea in various regions and the many adaptation measures available to deal with the problem. This can be accomplished through organizing public awareness campaigns, training and capacity-building programmes, and the provision of learning materials and support guides for various pest risk scenarios, among other things.
b) The development of new tea clones that can withstand abiotic stresses such as water-logging, drought tolerance, and susceptibility to pests/diseases, as well as validation of their sustainability under varying geographical conditions with changing climate scenarios, will go a long way toward combating the effects of climate change on productivity.
c) The use of jute geotextile, an environmentally favourable input, for soil conservation and soil erosion mitigation is an effective and alternative method for preventing soil degradation. Jute geotextile can make fields more resistant to erosion, preserving soil health and quality. Jute geotextile can also provide some additional benefits that are difficult to quantify in absolute terms and may have to be viewed as a way of erosion control on slopes.
d) The Vetiver system could be a useful tool for reducing erosion (by up to 90%), reducing and conserving rainfall-runoff (by up to 70%), improving groundwater recharge, removing pollutants from water, reducing the risk of flooding, and improving community economic advantages (Figure 13.12)

Figure 13.12 (a–d) Vetiver System in tea plantation: Vetiver grass planting to avoid top soil erosion in tea gardens in the hill and also for reinforcing contour stone wall.

e) Drainage (Figure 13.13) is now widely regarded as one of the most important inputs for boosting tea output in flatland areas. The goal of surface drainage is to remove excess water from the surface of the ground by grading it, building ditches, and enhancing natural channels, as well as to prevent important topsoil from being lost to erosion. Subsurface drainage is also necessary for removing excess gravitational water from the root zone in order to achieve the soil–air–water equilibrium required for plant growth; Beneficial soil microbes can be introduced to the soil through the use of organic matter such as decomposed cattle manure vermicompost, which can improve soil fertility and nutrient content.
f) Afforestation with *Cassia siamea*, *Acacia auriculiformis*, and other trees in marginal soil, fallow land, and around water bodies can improve microclimate and provide a wind barrier (Figure 13.14).
g) Irrigation should be used in tea-growing areas that are exposed to abiotic stress such as moisture and drought. In drought conditions, skiffed or unpruned irrigation has been found to be advantageous. Flooding, sprinkling (Figure 13.16), furrowing, and large-scale irrigation can all be accomplished with drip irrigation (Figure 13.15).
h) It is vital to plant shade trees in areas where they are lacking and to maintain proper shade status in tea plantations (Figure 13.17).
i) Rainwater collection is a potential option for maintaining humidity in tea plantations during periods of high moisture and drought. This lowers the vapour pressure deficit

Figure 13.13 Proper drainage to remove excess water from the surface of the land.

Figure 13.14 Wind barrier creation in tea plantation.

Figure 13.15 Subsurface drip irrigation to mitigate prolong drought effect.

Figure 13.16 Sprinkler irrigation in tea plantations to mitigate prolong drought effect.

Figure 13.17 Maintenance of proper shade status in tea plantations.

gradient between the leaf surface and the atmosphere, lowering evapotranspirational water loss. Rainwater harvesting options such as the construction of huge water bodies by filling depressions (Figure 13.18) and rainwater collection by blocking collector and main drains can all be used as a water supply depending on the topography.

13.9 Summary

Tea, *C. sinensis* (L.) O. Kuntze, is one of the perennial plantation crops which is cultivated and grown as a monoculture on a large contiguous area and is considered as an important cash crops in major tea growing countries such as China, India, Sri Lanka, and Kenya. Abiotic factors such as extreme temperatures, change in rainfall pattern, increased relative humidity, waterlogging, and extended drought driven by climate change are considered as the limiting factors in agro-ecosystems affecting the crop production and tea is no exception. As a consequence of climate change, biotic factors like increased pest incidence due to an increase in reproductive rate, high feeding efficiency, migration, etc. also affect tea cultivation. The rapidly changing climate poses a serious threat to the sustainability of tea production and is emerging as a real challenge to tea growers. Understanding

Figure 13.18 Maintenance of suitable micro-climate in tea garden by artificially created water bodies (rain water harvesting).

the physiology of tea plant, dynamics of pest incidence, and abundance due to changing climate may bring in some long-term eco-friendly management programs, making tea more exportable and acceptable to health-conscious domestic consumers. This chapter illustrates climate change trends in major tea growing countries affecting tea production, changes in pest scenario, and adaptive or mitigation strategies to face future challenges with special emphasis on Northeast India.

Acknowledgements

The authors wish to thank all the scientists and researchers of the Entomology Department, Climate and GIS Lab (Soils department), Agronomy Department and Advisory Department of Tocklai Tea Research Institute. Authors also extend their thanks to Mr. I.B. Ubhadia, General Manager, Rosekandy Tea Estate, Cachar, Assam, Mr. R.K. Rungta, Senior Manager Kurti Tea Estate, Dooars, Mr. Mohammad Fareed, Ex Advisory Officer, Tea Research Association and Mr. P.K. Murari, Executive, McLeod Russel Co. (P) Ltd. for their inputs and providing some of the photographs to improve the quality of this book chapter.

References

Babu, A. (2021). Integrated pest management in tea: recent advances and current challenges. *Indian Journal of Entomology*, Online Published Ref. No. e201803 DOI: https://doi.org/10.5958/0974-8172.2021.00007.9.

Barua, D.N. (1989). *Science and Practice in Tea Culture*, First Publish. Calcutta-Jorhat: Tea Research Association.

Barua, D.N. (2008). *Science and Practices in Tea Culture*, 119–120. Kolkata, India: Publ. Tea Research Association.

Barua, D.N. and Barua, S.C. (1969). Seasonal dormancy in tea. *Two and a Bud* 16: 41–45.

Barua, A., Babu, A., and Rajkhova, R.C. (2018). Seasonal abundance and predatory potential of *Stethorus aptus* Kapur (Coleoptera: Coccinellidae): a biocontrol agent of tea red spider mite *Oligonychus coffeae* Nietner (Acarina: Tetranychidae). *Proceedings of the Zoological Society (Calcutta)* 71 (3): 224–228.

Bore, J.K., Cheserek, B.C., and Ngeno, P. (2013). Long term impact of climate change on tea yields. *Tea* 34 (2): 59–69.

Caulifield, F. and Bunce, J.A. (1994). Elevated atmospheric carbon dioxide concentration affects interactions between *Spodoptera exigua* (Lepidoptera: Noctuidae) larvae and two host plant species outdoors. *Environmental Entomology* 23: 999–1005.

Cheserek, B., Aziz, E., and Bore, J. (2015). Analysis of links between climate variables and tea production in the recent past in Kenya. *Donnish Journal of Research in Environmental Studies* 2 (1): 12–24.

Coviella, C.E. and Trumble, J.T. (1999). Effects of elevated atmospheric carbon dioxide on insect-plant interactions. *Conservation Biology* 13: 700–712.

Das, G.M. (1957). Pests in relation to environment. *Two and a Bud* 4: 14–16.

Das, G.M. (1965). *Pest of Tea in North-East India and there control*, 169–173. Assam, India: Tea Research Association.

Das, S. (2015). A study on variability of isozymes in population of major Lepidoptera and mite pests of tea from Darjeeling foothills and adjoining Terai regions. PhD Thesis, Department of Zoology, University of North Bengal, India.

Goswami, B.K. (ed.) (2011). *Tea Field Management*, 204. Tea Research Association, Assam.

Gutierrez, A.P., Pointi, L., d'Oultremont, T., and Ellis, C.K. (2008). Climate change effects on poikilotherm tritrophic interactions. *Climatic Change* 87: 67–92.

Hare, J.D. (1992). Effects of plant variation on herbivore-natural enemy interactions. In: *Plant Resistance to Herbivores and Pathogens: Ecology, Evolution and Genetics* (ed. R.S. Fritz and E.L. Simms), 278–298. Chicago, IL, USA: University of Chicago Press.

Hazarika, L.K., Bhuyan, M., and Hazarika, B.N. (2009). Insect pests of tea and their management. *Annual Review of Entomology* 54: 267–284.

Jain, N.K. (1999). *Tea industry in India*. In: *Global Advances in Tea Science*, 43–56. Pub. Aravali Books International (P) Ltd.

Lal, R. (2005). Climate change, soil carbon dynamics, and global food security. In: *Climate Change and Global Food Security* (ed. R. Lal, B. Stewart and N. Uphoff), 113–143. Boca Raton, FL: CRC Press.

Li, Y.J., Wang, C.Y., Zhao, B., and Liu, W.J. (2010). Effects of climate change on agricultural meteorological disaster and crop insects diseases. *Transactions of the CSAE* 26 (Supp.1): 263–271.

Li, X., Ahammed, G.J., Li, Z.X. et al. (2016). Increased biosynthesis of jasmonic acid via lipoxygenase pathway mediates caffeine-induced resistance to *Colletotrichum gloeosporioides* under elevated CO in tea seedlings. *Phytopathology* 106 (11): 1270–1277.

Mitra, H. (1974). Factors affecting early flush crop on North Bank. *Two and a Bud* 21 (1): 22–23.

Mukhopadhyay, A. and Roy, S. (2009). Changing dimensions of IPM in the tea plantations of the north eastern sub-Himalayan region. In: *IPM Strategies to Combat Emerging Pests in the Current Scenario of Climate Change* (ed. V.V. Ramamurthy, G.P. Gupta and S.N. Puri), 290–302. IARI, New Delhi, India: Entomological Society of India.

Mukhopadhyay, A. and Sarker, M. (2007). *Natural enemies of some tea pests with special reference to Darjeeling, Terai and the Dooars*. Kolkata, India: National Tea Research Foundation, Tea Board.

Muraleedharan, N. (2011a). Climate change and plantation crops. *Planters' Chronicle* 107 (1): 5–6.

Muraleedharan, N. and Roy, S. (2016). *Arthropods pests and natural enemies communities in tea ecosystems of India*. In: *Economic and Ecological Significance of Arthropods in Diversified Ecosystems: Sustaining Regulatory Mechanisms* (ed. A.K. Chakravarthy and S. Sridhara), 361–392. Springer DOI:https://doi.org/10.1007/978-981-10-1524-3, ISBN 978-981-10-1524-3.

Musser, F.P. and Shelton, A.M. (2005). The influence of post-exposure temperature on the toxicity of insecticides to *Ostrinia nubilalis* (Lepidoptera: Crambidae). *Pest Management Science* 61: 508–510.

Pathak, S.K., Srilal, S., Dukpa, N. et al. (2003). Biological studies of a defoliator of Darjeeling tea *Orygia* sp. (Lymantriidae: Lepidoptera) under laboratory condition. *Two and a Bud* 50: 31–34.

Prasad, A.K. and Mukhopadhyay, A. (2013). Changing life-cycle pattern of a minor looper pest of tea, *Ectropis* sp. (Lepidoptera: Geometridae) in summer and winter seasons of Darjeeling Terai. *Journal of Animal Science* **7**: 31–34.

Roth, S.K. and Lindroth, R.L. (1995). Elevated atmospheric CO_2 effects on phytochemistry, insect performance and insect-parasitoid interactions. *Global Change Biology* 1: 173–182.

Roy, S., Mukhopadhyay, A., and Gurusubramanian, G. (2009). Seasonal incidence and management of *Scirtothrips dorsalis* hood and *Empoasca flavescens* Fabr.- upcoming pests of tea in North Bengal. In: *Proceedings of the National Symposium on IPM Strategies to Combat Emerging Pests in the Current Scenario of Climate Change* (ed. V.V. Ramamurthy, G.P. Gupta and S.N. Puri), 43–47. New Delhi, India: Entomological Society of India, IARI.

Roy, S., Muraleedharan, N., and Mukhopadhyay, A. (2014). The red spider mite, *Oligonychus coffeae* (Acari: Tetranychidae): its status, biology, ecology and management in tea plantations. *Experimental and Applied Acarology* 63 (4): 431–463. DOI: https://doi.org/10.1007/s10493-014-9800-4.

Roy, S., Das, S., Handique, G. et al. (2017). Ecology and management of the black inch worm, *Hyposidra talaca* Walker (Geometridae: Lepidoptera) infesting *Camellia sinensis* (Theaceae), a review. *Journal of Integrative Agriculture* 16 (10): 2115–2127.

Roy, S., Barooah, A.K., Ahmed, K.Z. et al. (2020). Impact of climate change on tea pest status in northeast India and effective plans for mitigation. *Acta Ecologica Sinica* 40 (6): 432–442.

Saha, D., Mukhopadhyay, A., and Bahadur, M. (2012). Variation in the detoxifying enzymes activity in tea greenfly, *Emposca flavescens* Fabricius (Homoptera: Jassidae) from sub

Himalayan organic and conventional tea plantations. In: *Biodiversity Status and Conservation Strategies with Special Reference to North-East India* (ed. R. Varatharajan), 257–266. Manipur, India: CAS in Life Sciences, Manipur University.

Saren, J., Das, S., and Mukhopadhyay, A. (2015). A temperature-dependent development of red spider mite, *Oligonychus coffeae* (Acari: Tetranychidae) using degree-day model. *Journal of Applied Bioscience* 41 (2): 163–167.

Sarkar, S., Babu, A., Chakraborty, K., and Deka, B. (2019). Study on the biology, feeding behaviour and predatory potential of *Sycanus collaris* (Fabricius) (Heteroptera: Reduviidae), a new Predator of *Hyposidra talaca* (Walk.) (Lepidoptera: Geometridae), a major Tea pest and mass rearing on *Corcyra cephalonica* (Stainton) in laboratory. https://doi.org/10.24327/ijcar.2019.19262.3705.

Sarkar, S., Babu, A., Kaushik, C. et al. (2021). *Eocanthecona furcellata* (Wolff) (Hemiptera: Pentatomidae), a potential biocontrol agent of the black inch worm, *Hyposidra talaca* Walker (Lepidoptera: Geometridae) infesting tea. *Phytoparasitica*. DOI: https://doi.org/10.1007/s12600-021-00888-x.

Thushara, S.C. (2015). Sri Lankan tea industry: prospects and challenges. *Proceedings of the Second Middle East Conference on Global Business, Economics, Finance and Banking*, August.

Wang, S.B. and Jin, Z.F. (2010). *Climate and Tea Cultivation with High Yield and Better Quality*. Beijing: *China Meteorological Press*.

Wang, C.Y., Zhao, J.B., and Li, X.Y. (2006). Study on agricultural adaptation to warming and drying climate in North China. *Arid Land Geography* 29 (5): 646–652.

Wijeratne, M.A. and Chandrapala, L. (2013). Analysis of rainfall and temperature in tea growing agro-ecological regions and assessment of vulnerability of tea plantations in Sri Lanka to climate change. *Sri Lanka Journal of Tea Science* 78 (1/2): 42–59.

Wijeratne, M.A. and Fordham, R. (1996). Effect of environmental factors on growth and yield of tea (*Camellia sinensis* (L) Kuntze) in the low-country wet zone of Sri Lanka. *S. L. J. Tea. Sci.* 64 (1/2): 21–34.

Willson, K.C. and Clifford, M.N. (1992). *Tea: Cultivation to Consumption*. London, United Kingdom: Chapman & Hall.

Yang, Y. (2005). *China Tea Cultivation*. Shanghai: Shanghai Scientific & Technical Publishers.

14

Impact of Climate Change on Integrated Pest Management Strategies

Sonja Gvozdenac, Boško Dedić, Sanja Mikić, Jelena Ovuka, and Dragana Miladinović

Institute of Field and Vegetable Crops, Novi Sad, Serbia

CONTENTS

Climate Change and Agriculture: Perspectives, Sustainability and Resilience,
First Edition. Edited by Noureddine Benkeblia.

14.1 Introduction

Climate change substantially influences biology, ecology, occurrence, and distribution of plant pests (insects, pathogens, and weeds), pest–host plant interactions, and activity of natural enemies. Indirectly, altered climate conditions have an influence on the efficacy of control measures applied within Integrated Pest Management (IPM) programs. Some of the IPM techniques, primarily plant resistance/tolerance and biocontrol measures are highly susceptible to environmental changes. Therefore, there is a need to evaluate the efficacy of IPM techniques under diverse environmental conditions and adapt pest management strategies to changing climate. To address this challenge, a new approach, climate-smart pest management (CSPM) has been developed, which provides more focus on the management of various plant pests in the context of climate change.

The CSPM is a new interdisciplinary approach that requires the adaptation of cultivation and pest management practices in primary crop production, by relying on extension services, research institutions, and with involvement of policy makers. This approach strengthens the role of research, extension, and the policy makers in pest surveillance, detection, forecasting, and control enabling the increase of pest management resilience to climate change.

This chapter aims to give an insight into the major impacts of climate change on agricultural production, plant pests (insects, pathogens, and weeds), with special emphasis on how IPM strategies and techniques may be affected by altered climate conditions. Additionally, within this chapter, we will provide information about the adaptation options, and present new approaches to managing the uncertainty associated with climate change such as the development of climate-smart crops and CSPM using novel technologies and appropriate tools.

14.2 Climate Change Impact on Agro-Ecosystems: Present Knowledge, Implications, and Uncertainties

Climate change has been a constant process on Earth. However, the pace of this change today is one of the most defining concerns and the greatest challenge facing human society. The term 'climate change' encompasses several variables: changes in temperature globally, precipitation amount and patterns, increase in the frequency of extreme weather incidences (droughts, floods, heat waves, etc.), soil erosion, and desertification (Heeb et al. 2019). These changes cause alterations in environmental conditions, as well as more frequent biotic and abiotic stresses (Bebber et al. 2013; Trenberth et al. 2014), that are challenging to both undisturbed and managed ecosystems, including agriculture.

The fact that the climate and agriculture are strongly connected implies that fast changes in the climate conditions pose a threat to food security on a global scale. According to Qafoku (2015), the changes in climate variables (elevated concentrations of atmospheric carbon dioxide – CO_2 ($\geq$400 ppm); increasing average air temperatures (2–4 °C or higher); wet and dry weather cycles; frequent and heavy rainfall and/or heavy storms; prolonged drought periods; extreme frost and heat waves) have a significant influence on water resources, soil properties and fertility, food quantity and quality, and the environmental quality globally. As a consequence of altered climatic conditions, new pests have emerged, crop cultivation practices have changed (War et al. 2016; ESA 2019), and these trends will continue in the future. However, it is still difficult to predict to what extent the changes in climate conditions will affect crop production in a certain region. Therefore, multi-sectoral actions are needed to respond to this problem, create climate-resilient strategies for each agricultural sector, and mitigate future impacts of climate change on agroecosystems.

14.2.1 Direct Climate Change Indicators Affecting Agricultural Production

14.2.1.1 Temperature

The pace of climate change has increased multiple times in the last decades. The figures imply that the Earth's atmosphere has been warming since the beginning of the industrial revolution. The mean temperatures have increased by more than 1 °C since the 1850s, globally. The hottest years on record have all occurred since 2005, namely 2015, 2016, 2017, and 2018. According to the NOAA (2019), the combined land and ocean temperature have been increasing at an average rate of 0.07 °C per decade since 1880. However, the increase since 1981 (0.18 °C) is more than double. Karmalkar and Bradly (2017) suggest that the increase of 1.5–2.0 °C in the mean temperature for the last century is presumed to be the result of both natural events and anthropogenic activities. It is projected that the global mean temperature will further rise by 2–3 °C in the twenty-first century, mainly due to higher use of fossil fuels and intensified conventional agriculture (Walthall et al. 2012), and if continued at the present pace, the global warming will affect the agriculture production in the future.

14.2.1.2 Water Cycle and Precipitation Patterns

Climate change can affect the intensity and frequency of precipitation as reported by EPA (2016). In the northern regions, the precipitation is projected to increase, while in other areas a higher incidence of drought is expected, as well as the changes in rain/snow occurrence. Due to the mentioned events, providing water for the crops timely will become a challenge (EPA 2016).

The precipitation has a direct influence on agriculture. Heavy rains (precipitation) can cause crop damage, soil erosion, and increase flood risks, which can be as damaging as low precipitation (Hatfield and Prueger 2015). In addition, as reported by EPA (2016), run-off caused by precipitation could spoil water quality due to leaching and washing-out of pollutants from land into water bodies. Evapotranspiration increase can cause lower water availability even when the precipitation increases, particularly in the soils with limited soil water holding capacity. Hatfield and Prueger (2015) reported a reduction in maize growth or even plant death if the excess in water occurs during the early growth stages. On the other hand, water deficit in the soil may lead to reduced maize growth and yields if the drought stress occurs during the grain-filling period.

14.2.1.3 Atmospheric Carbon Dioxide (CO_2)

Accelerated climate change has been projected by the end of this century because the concentration of some atmospheric 'greenhouse' gases (GHG), primarily CO_2, is increasing (Bacastow et al. 1985; IPCC 2001). Delmas et al. (1980) speculated that the CO_2 concentration during the last ice age ranged from 160 to 200 μmol mol^{-1}. After the industrial revolution, it increased significantly from 315 μmol mol^{-1} in 1958, to 345 μmol mol^{-1} in 1985 (Bacastow et al. 1985). Sombroek and Gommes (1996) reported an increase of 0.4–0.5% per year, while according to Mahlman (1997), the current 360 μmol mol^{-1} of CO_2 is expected to double by the end of this century if the burning of fossil fuels continues and the air temperature rises for 1.5–5 °C, as predicted by 2100 (Baes et al. 1977; Mahlman 1997).

The evident increase in the atmospheric CO_2 levels and associated changes in the climate variables initiated a number of studies aiming to predict the effects of these events on crop growth, development, and yield. According to Fiscus et al. (2005), higher CO_2 levels could modify the crop responses to environmental stress. For example, elevated CO_2 causes partial stomatal closure that reduces ozone penetration into leaves, which causes ozone overload resulting in yield losses.

14.2.1.4 Ozone and Solar UV-B Radiation

Ozone: The importance of scientific and public awareness of ozone effects on plants was emphasized by Dentener et al. (2006). Recently, a number of studies enhanced the understanding and knowledge of crop responses to ozone under the conditions of changing climate. The ozone and its precursors are characterized by high mobility and are transported hundreds of miles into agricultural areas negatively affecting native and managed forests, crop production, pastures, and farms. Several other studies showed that the current ambient ozone levels caused yield reduction of alfalfa, bean, clover, cotton, peanut, potato, rice, soybean, sugar cane, and wheat in many regions worldwide (Booker et al. 2009; Grantz and Vu 2009). Muntifering et al. (2006) reported changes in leaf chemistry due to the elevated ozone exposure in common grassland which reduced its nutritional quality for grazing animals.

Solar radiation: During the past 50 years, Stanhill and Cohen (2001) reported a 2.7% reduction in solar radiation per decade. The changes in the solar radiation may directly affect crop water balance and evapotranspiration, but the effect on productivity would be lower because of other factors, such as water and temperature (Hatfield et al. 2011). The decrease in the solar radiation in agricultural areas in the last 60 years was modelled by Pan et al. (2004) and was predicted to continue due to the increasing concentrations of the atmospheric GHGs.

Finally, the changes in the temperature, precipitation, CO_2 level, and radiation in the future will be accompanied by other changes that will have certain implications on agriculture. The number and the complexity of these interactions pose a need for systemic analyses of the potential impact of altered climate conditions on agriculture.

14.2.2 Indirect Effects of Climate Change on Agricultural Production

14.2.2.1 Effect on Soil

The climate is one of the most important factors affecting the formation of soil, soil structure, soil stability, top-soil water holding capacity, nutrient availability, and erosion. The soil response to climate change is complex and accompanied by many uncertainties. According to Qafoku (2015), the changes in the level of atmospheric CO_2 (concentrations above 400 ppm), increasing and fluctuating daily, seasonal, and interannual air temperatures, alterations in the wet and/or dry cycles, heavy rains and/or storms, prolonged drought periods, extreme frost and heat waves did and would continue to significantly affect the soil properties and fertility. Karmakar et al. (2016) suggested that a significant change in organic matter turnover and CO_2 dynamics would occur if the constant inputs of carbon to soils from the vegetation continued and if the predicted changes in the temperature, precipitation, and evaporation continued.

14.2.2.2 Effect on Agricultural Production in General

Agriculture is exceptionally vulnerable to climate change (Deshar and Koirala 2019). It has been estimated that the changes in the climate conditions have both positive and negative impacts on the agricultural production systems globally, but unfortunately, the negative impacts are outweighing the positive ones (Müller 2013). The temperature increase, altered precipitation patterns, and increased CO_2 concentrations significantly affect ecosystems as well as agroecosystems, from the species- to ecosystem level (Stevens et al. 2004).

It is evident that climate change has and will have a significant impact on global crop yields in the future (Lobell and Field 2007; Challinor et al. 2014). Moreover, it is, directly and indirectly, affecting pest occurrence, distribution, and severity worldwide (Juroszek and von Tiedemann 2011; FAO 2013b; Lamichhane et al. 2015; Macfadyen et al. 2018), which will continue to additionally affect crop yields.

World Food Programme report from 2018 revealed that the increase in the crop yield per hectare was significantly slower compared to the increase in the human population. According to the FAO data published in 2016, if climate change continued at the present pace by 2100, a reduction in the production of major cereal crops could be expected (20–45% in maize, 5–50% in wheat, and 20–30% in rice yields). Hence, the crop losses would increase, agricultural production would reduce at the global level, food prices would rise, and it would be difficult to cope up with the rising needs of the growing population (Arora 2019).

Altogether, the changes in agricultural productivity can be the result of a direct effect of climate changes at the global level, or an indirect effect at the ecosystem level. Climate change will evidently directly have a negative effect on yields and quality of the final products and may necessitate changes in cultivation and pest management practices. These indirect effects may also include higher production costs (additional inputs to control invasive species), and major changes in crop protection and have an implication on food security (IPCC 2014), particularly in the developing countries where the need to increase the food production and obtain stable and high yields is of great importance.

14.3 Crop Responses to Changing Climate

The joint impacts of climate change variables on crop production will primarily depend on a complex of adaptive actions taken and climate conditions in general. These adaptive actions may include: (i) adjusting cultivation practices and soil management to cope with variable weather conditions; (ii) investing in the development of varieties tolerant to abiotic stress, primarily drought; (iii) growing demand for national risk management programs and (iv) adjustments in international trade (Walthall et al. 2012).

Climate change is expected to put a strong pressure on wild and cultivated plant communities. It will continue to create a changing environment and challenge crops for stability in producing high sustainable yields. Direct influence can be summarized as 'abiotic stress' that can indirectly make the plant more prone to other limiting factors. In their adjustment to the environment, the crops can tolerate a certain degree of fluctuation in environmental factors. The stress is commonly associated with the formation of reactive oxygen species and their presence in a plant cell is related to the stress tolerance up to a certain threshold level above which the cell can suffer severe damage (Shah et al. 2019). Other imbalances in plant morphological, biochemical, and physiological traits have also been observed. Sources of stress, previously described and studied separately, are forming a complex relationship in each field at the plant and canopy level. Because of the significance of this topic and the predicted scale of change in weather patterns, a lot of research effort was put into clarifying of underlying mechanisms of plant adaptation to the environment and modelling the variations commonly associated with climate changes (increasing levels of atmospheric CO_2, temperature variability, etc.).

Today, worldwide crop production and species distribution are a result of continuous long-term selection of plants adapted to regional weather patterns. The area of crop distribution was partly formed on the environmental requirements of each plant species. The main trait, namely expected crop yield, was thus dependent on mean weather patterns in a particular region. Challenges in a prediction of climate change impact on the crop production arouse from potential constraints such as the type of used model, quality of available climate data and data from farms, and potential for cross-study research. The prediction of changes in the environment on crops, therefore, represents a necessity for sustainable plant production (Konduri et al. 2020).

Models indicate that, for example in China, a decline in the yield of major crops (rise, maize, etc.) is predicted, while only in particular regions, the yield of wheat is expected to increase (Ju et al. 2013). The models used to predict the changes in the average temperature and precipitation on maize, peanut, and cotton in the southeast United States are implying a yield decrease in rainfed fields with a possibility of limiting the impact using adjusted planting dates (Shin et al. 2020). On the other hand, some data suggest that the crops in some regions may benefit from the modelled changes in weather patterns (Ahmed 2021). As a consequence of shortening the developmental cycle and elevated CO_2, wheat yields are expected to increase in the monitored region of Morocco (Bouras et al. 2019).

14.3.1 How Changes in Temperature Affect Crops

Plant response to climate change is dictated by a complex of single variables (temperature, precipitation, CO_2 concentration, and solar radiation) and their interactions. The temperature is considered a major factor affecting the distribution of both native plants and also crops, due to the physiological limits of each species (Parmesan and Yohe 2003; Rehman et al. 2015). Each plant species has upper and lower temperature thresholds that define the geographical ranges suitable for its growth and reproduction but also has the optimal temperatures for each developmental phase. The plants are currently grown in the areas in which the temperatures suit their development and fit into the thresholds. If the temperatures continue to increase and exceed the thresholds, the changes in the distribution of certain crops may occur (Walthall et al. 2012). Gregory et al. (2009) calculated that every degree of increase in the temperature above 32 °C would reduce the crop production of wheat and rice by 5%. Additionally, temperature extremes (low and high temperatures) can negatively affect crop development, growth, and yield, especially during critical stages (Luo 2011). According to predictions presented by Tubiello et al. (2002), the higher temperatures will most probably occur during the spring and summer season and will be beneficial for the crop production in northern regions where the duration of the growing season is a major limiting factor at the moment.

14.3.2 How Changes in Precipitation Affect Crops

It is projected that climate change will alter precipitation patterns, rainfall events, soil moisture depo, evaporation, groundwater level, and runoff. According to estimations, more than 80% of the global crop production is supplied by rainfall. Hence, the changes in the total seasonal precipitation or rain patterns may be more influential to agriculture production than the changes in other climatic variables, especially in regions where the occurrence and the duration of dry seasons is a limiting factor for crop production (Olesen and Bindi 2002). The models presented by Lickley and Solomon (2018) predict a significant drying trend in Africa, parts of Latin America, Australia, southern parts of Europe, and North America by the middle of this century, with an increase in drought risks by >10% and a water deficit higher than 200 mm per year.

Strong regional differences on the global level are also predicted under various climate change scenarios for the water-use efficiency of terrestrial ecosystems with an overall decreasing trend (Pan et al. 2018). The crops in different environments have different water use efficiency. It is highly dependent on the weather and agronomic practices. The crops are grown in drier areas generally are more efficient in water use compared to wetter areas. However, under the conditions of changing climate and if combined with unsustainable agricultural practices this adaptability could be severely impaired leading to susceptibility to drought stress (Zhang et al. 2015a). Introducing the cover crop was shown to be an effective strategy for mitigation of changing climate on water losses in the maize-soybean cropping system (Li et al. 2021). Wheat grown in CO_2-enriched atmosphere expressed lower requirements in water although this effect was diminished when a factor of increased canopy temperature was introduced (Wang et al. 2018), as the efficiency increased under optimal temperatures for leaf growth and with further increase in temperature, the efficiency will decline (Hatfield and Dold 2019).

The increase in drought frequency and severity will affect the production of major crops, under so far tested climate change scenarios, resulting in inevitable yield reduction (Li et al. 2009). The risk of maize yield reduction is predicted to increase under dry and hot conditions in all major producing countries (Feng et al. 2021). Extreme heat stress during anthesis will further increase losses in maize yield (Deryng et al. 2014). The direct impact of drought is a reduction in water uptake from the soil. This leads to lower water content in the whole plant and a higher concentration of cellular components. During the drought periods, mitosis is also inhibited resulting in the reduction of plant growth (Camaille et al. 2021). Plants can tolerate drought conditions to a certain degree by osmotic adjustment (Blum 2016). This type of stress is regulated by a number of regulating and functional genes (Joshi et al. 2016), and plants can adjust to drought conditions if drought stress induces enhanced root growth in young plants thus priming them for later drought events (Fu et al. 2021). Some stages in plant development are more susceptible to drought events, particularly the grain filling stage in various crops as a result of lower photosynthesis via stomatal closure (Agurla et al. 2018) or influence on photosystem II and antioxidant enzyme system (Li et al. 2017). The changes in aboveground plant parts are synchronized with roots in response to drought which includes the response to reduced water availability, salinity stress, and reduced availability of nutrients. Climate change will further enhance the impact of drought and salinity stress on plants (Kaushal and Wani 2016). In arid regions and the regions that are predicted to suffer from the lack of precipitation, and/or limited rainfall, the affords should be invested in developing irrigation techniques that enable the efficient use of available water resources, as well as good agronomic practices that emphasize conservation of soil moisture (Kirda et al. 2007).

The extreme precipitation can be damaging for the crops due to a combination of heavy rain that physically injures plants, the excessive water introduced in the root zone, and/or as a result in physical damage caused by storms. Additionally, the risk of the occurrence of some fungal and bacterial diseases is higher under wet conditions and on already damaged plants. On the other hand, the lack of water in the soil impairs plant biological functions making them more susceptible to diseases and pests (Zayan 2019).

14.3.3 How Changes in CO_2 Affect Crops

CO_2 in the atmosphere is the essential chemical compound in the processes of photosynthesis. The main effects of elevated CO_2 on plants are related to a reduction in transpiration and stomatal conductance, improved water and light-use efficiency, and consequentially increase in photosynthetic rate. The direct effect of increased atmospheric concentration of CO_2 could be beneficial because it stimulates plant development and growth, due to rapid carbon assimilation, since carbon is the key element in the structure of the plant (Woodward 1990). The type and quantity of this effect is, however, a subject of a debate. It is evident that higher CO_2 concentrations could increase crop yields. However, the extent of this effect is uncertain. The elevated CO_2 levels will have different effects on C3 and C4 plants as the C4 plants are less sensitive to an increase in atmospheric CO_2 levels compared to the C3 plants (Tubiello et al. 2002). Fleisher et al. (2011) suggested that the C3 crops were more stimulated compared to C4 crops where the elevated CO_2 will have a minor effect. The effect of CO_2 also depends on whether the experiments were conducted in a controlled environment or based on open atmosphere trials, in addition, are the other factors like temperature, water, and plant nutrition a part of the estimation process (Degener 2015). Studies that have the basic assumption of not limiting nutrients, tend to overestimate the effect of CO_2 "fertilization effect" (McGrath and Lobell 2013).

14.3.4 How Changes in Ozone Affect Crops

Climate change is expected to stimulate ozone formation and together with increased emissions of ozone, can pose the risk for vegetation (Fuhrer 2009). The detrimental effect would depend on the frequency and length of ozone exposures and crop damage could be substantial if the emissions are not significantly decreased (Chuwah et al. 2015). Wheat, maize, and soybean were found to have stronger tolerance to ozone in the case of higher consumptive water use (Tai and Martin 2017). Although differences were determined among annual and perennial crops and regions, the effect of ambient ozone on plants is negative and could lead to a substantial crop loss (Fumagalli et al. 2001).

Overall, the exact form of the future climate changes is still uncertain, but the current projections indicate that they will probably have serious negative impacts on the crops, agricultural production, and food security in total.

14.4 Effects of Climate Change on Plant Pests

Pest organisms (insects, pathogens, and weeds) significantly influence agricultural production by reducing the quality and quantity of food and feed produced. Oerke (2006) reported that globally, the highest crop losses were caused by weeds (34%), followed by herbivorous insects (18%) and plant pathogens (16%). Climate change can affect plant pests in both direct and indirect manner. The ***direct*** effects result in the alteration of pest biology, distribution and outbreak potential, the development and survival of herbivores and pathogens, as well as physiological changes in plant defences (Ayres and Lombardero 2000). The changes in the temperature and precipitation patterns, associated with increasing CO_2 levels create altered environmental conditions that induce the changes in the intensity of weed infestation, the incidence of plant pathogens, and the geographic distribution of a number of pests (Skendžić et al. 2021). Additionally, climate change is facilitating the emergence of new invasive species in many regions of the world. Without the effective monitoring and management practices, these species have the potential to become significant pests in the new regions (Hellmann et al. 2008). Predicting the effects of climate change on pests is not easy due to the complicated interactions between altered climate variables and the frequency of extreme weather events (Gregory et al. 2009; Bebber et al. 2013).

Climate change can also have an ***indirect*** effect on pests, for example through the varying responses of host plants, abundance of natural enemies (parasitoids and predators), emergence of vectors of plant pathogens and competitors (Ayres and Lombardero 2000), and the loss of synchronization between the pest and the enemy (Hance et al. 2007; Furlong and Zalucki 2017). The indirect pest responses also result from the changes in the efficacy of pest control measures (biological control, synthetic pesticides, etc.), as well as from the changes in crop management practices (Hoffmann et al. 2008; Cock et al. 2013; Barzman et al. 2015; Lamichhane et al. 2015).

The efficacy of IPM measures, as mentioned, changes under altered environmental conditions. Therefore, IPM measures must be adapted to changing climate, offering new approaches and solutions. One of them is a CSPM. As an interdisciplinary strategy, CSPM aims to reduce the crop losses, enhance ecosystem services, reduce the GHGs emissions, and strengthen the resilience of agricultural systems to climate change. Through the implementation of CSPM, the crop production, extension services, researchers, and policy makers act in coordination with a unique goal of creating efficient and resilient food production systems that are able to cope with the conditions of changing climate.

14.4.1 Insects and the Environment

Weather conditions and the climate in general significantly affect insects' bionomy. Changes in the climate may cause alterations in geographical distribution, population dynamics, population growth rates, increases in voltinism (number of generations), prolonged development, insect-host plant interactions, increased overwintering, loss of synchrony of crop-pest phenology, increased risk of invasion by emerging species, activity and abundance of natural enemies (Figure 14.1), the frequency and intensity of pest outbreaks as well as efficacy of crop protection measures (Porter et al. 1991; Reilly 1995; Ayres and Lombardero 2000; Klok and Chown 2001; Parmesan 2007; Sharma 2014).

14.4.1.1 How Rising Temperatures Affect Arthropods

Insects are cold-blooded organisms (poikilotherms), which mean that the temperature of their bodies is about the same as that of the surrounding environment. Hence, the effect of the temperature on insects' bionomy exceeds the effects of other environmental factors (Bale et al. 2002; DeLucia et al. (2008).

Expansion of geographical ranges: Climate change, primarily increasing temperatures, positively influences the expansion of geographical ranges of insect pests, as documented by Walther et al. (2009), Parmesan and Yohe (2003), and Parmesan (2007). According to Walthall et al. (2012), the geographic ranges of pests are limited by the presence of host plants, and the ability to survive winter temperatures. Species presently limited to tropical and subtropical regions will move to temperate regions along with the expansion of production areas of the host plants. On the other hand, the distribution and population abundance of species vulnerable to high temperatures in the temperate regions may decrease as a result of increasing mean annual temperatures due to global warming. Bebber et al. (2013) predicted a poleward shift of pests as a response to the global warming. According to Porter et al. (1991), the geographical ranges of insect pests were expected to move to higher altitudes in central Europe by 2055, as was the example of the European corn borer (*Ostrinia nubilalis*) that shifted more than 1000 km northward in Europe. Lopez-Vaamonde et al. (2010) listed 97 non-native Lepidoptera species from 20 families that were

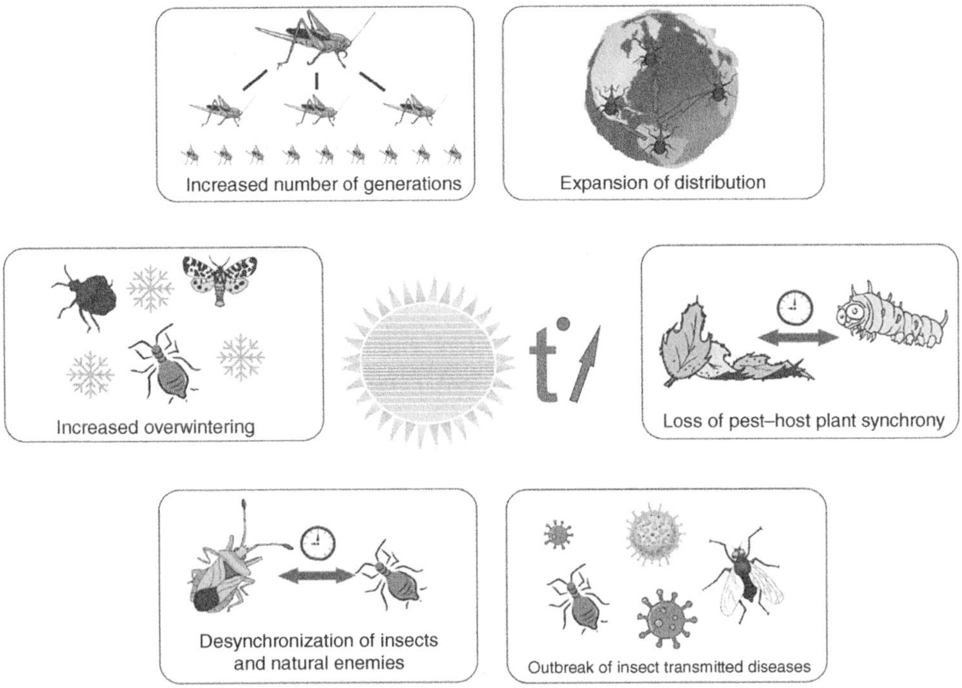

Figure 14.1 How do increasing temperatures affect insect pests?

established in Europe, and 88 European Lepidoptera species in 25 families that expanded their ranges in Europe, with 74% of new species established in the last century. The Diamondback moth (*Plutella xylostella*) shifted its former geographical range of 800 km, from western Russia and got established north in the Arctic Ocean on the Norwegian islands of Svalbard, due to increased fluctuations of warm air towards higher latitudes (Coulson et al. 2002).

Increase of insect populations: It has been documented that rising temperatures speed up the insect life cycle, as long as the upper critical limits are not exceeded. The '**ambient energy hypothesis**' suggests that insect growth and reproduction are faster at higher temperatures, leading to higher population abundance and eventually a higher number of species in dynamic equilibrium (Menéndez et al. 2007; Menéndez 2007). Depending on the development 'strategy' of an insect species, rising temperatures have different implications (Bale et al. 2002). Insects with a life cycle that lasts for several years (cicadas, moths, coleopteran species, etc.) tend to moderate temperature variability during the life cycle. On the other hand, the increased temperatures speed up the development of species with a shorter life cycle (Bale et al. 2002) as well as species that have multiple generations per year (Tobin et al. 2008; Altermatt 2010), which leads to higher population abundance. Also, warmer winter temperatures cause lower winter mortality of insects resulting in increased populations (Harrington et al. 2001).

From another point of view, as outlined by Deutsch et al. (2018), insect populations in tropical zones that already live close to their developmental optimum will decrease as a result of climate changes, primarily the global warming. The opposite is expected for the insects in temperate zones, namely the increase in the growth rates and finally population abundance. The authors confirmed this theory by estimating the changes in the growth of major grain (wheat, rice, and maize) pest populations under different climate change scenarios. For wheat, which is cultivated in temperate climates, warming is projected to speed up the growth of pest populations. For rice cultivated in tropical climates, a decrease in pest populations is predicted. For maize grown in both climates (temperate and tropical), various responses of pest populations are expected, in terms of its growth (Deutsch et al. 2018).

Although increased temperatures during summer favour the growth of pest populations, the prolonged growing season has a proportionately greater effect on the damage insects cause to host plants (Bradshaw and Holzapfel 2010), which should not be overlooked.

Decrease of insect populations: Increased temperatures can cause a decrease in insect populations in several ways. First, the lack of host plants insects feed upon, for mono- and oligophagous species causes a decrease in the pest populations. Secondly, since the same environmental factors impact both insect pest and their natural enemies (predators, parasitoids, and entomopathogens), it results in an increased attack on pest populations. Awmack et al. (1997) reported that at higher temperatures, the aphids were less responsive to the aphid alarm pheromone naturally released when under the attack by natural enemies, resulting in greater predation, and reduction of the aphid's population.

Increased number of generations (voltinism): The above-mentioned 'ambient energy hypothesis' also implies that the high temperatures speed up insect reproduction, leading to an increase in the number of generations and consecutively to higher crop damages (Yamamura and Kiritani 1998). It is expected that future temperature increase affects differently univoltine (one generation per year) and multivoltine (several generations per year) species from the temperate regions. For multivoltine insects, such as aphids and some lepidopteran species, the increase only in the temperature accelerates the development that would result in additional generations in one year (Pollard and Yates 1993; Bale et al. 2002; Menéndez 2007). Univoltine species generally develop more rapidly than those with a longer life cycle (Bale et al. 2002). Yamamura and Kiritani (1998) reported that a 2 °C increase in the temperature could result in four to five additional aphid generations per year due to the short generation time. Lepidopteran species are another good example of phenological changes under altered climatic conditions. In the United Kingdom, 26 of 35 observed butterfly species advanced their first emergence (Roy and Sparks 2000; Menéndez 2007). A similar phenomenon was recorded in Spain. The first emergence of 17 butterfly species was one to seven weeks earlier in only 15 years (Stefanescu et al. 2003; Menéndez et al. 2007), as was also recorded for the European grapevine moth, *Lobesia botrana* (Martín-Vertedor et al. 2010). Altermatt (2010) pointed out that since the 1980s, for many central European lepidoptera species, the voltinism has increased, with some univoltine becoming bivoltine and bivoltine becoming multivoltine.

Increased overwintering: Insects have developed a number of strategies to survive under stressful environmental conditions (González-Tokman et al. 2020). For many species, winter is the most critical season because low temperatures can significantly increase mortality and reduce the population's abundance (Hill 1987). In terms of overwintering, strategy insects are classified into two groups: freeze-avoidant (migratory species) and freeze-tolerant (diapausing species), as presented by Bale and Hayward (2010). Diapause is a physiologically adaptive strategy that is influenced by environmental factors such as temperature, photoperiod, and humidity (Gill et al. 2017). Some insect species enter diapause during the inactive stages (egg or pupa), while others do it as larvae, nymphs, or adults (Bale and Hayward 2010). For many temperate insect species, delayed diapause poses a risk of experiencing cold stress before the cold tolerance mechanisms are initiated (Bradshaw and Holzapfel 2010). This is believed to determine the current northern range of the Green stink bug (*Nezara viridula*) in Japan (Musolin 2007). Non-diapausing species and those that can overwinter in active stages, during warmer winters have higher survival rates, hence their population is increasing and geographic ranges are expected to expand to higher altitudes as the increase of the average temperature (Bale et al. 2002), which can lead to outbreaks and earlier spring migrations (Harrington et al. 1995; Cannon 1998). Climate change, primarily rising temperatures, can disrupt the metabolic balance during the diapause, which significantly influences the timing of the pest's emergence resulting in a loss of synchrony with the host plant (Parmesan 2007; Bale and Hayward 2010). For example, Harrington et al. (2001) determined that 2 °C increase in the temperature would alter the timing of a bud burst of the Pedunculate oak (*Quercus robur*), while the hatching of its major pest – winter moth (*Operophtera brumata*) larvae would be significantly advanced. Early larval hatching before the bud burst, leaves the larvae without the food, causing high mortality and eventually, solving the problem with this specific pest.

Impact on pest–natural enemy interactions: Phytophagous insects are naturally controlled by top-down (natural enemies) and bottom-up (host plant availability and quality) mechanisms (Jamieson et al. 2012). The temperature changes can differently affect the biology of each species in the system, thus destabilizing their population dynamics (Hance et al. 2007) and causing temporal desynchronization. According to Furlong and Zalucki (2017), climate change is expected to significantly affect natural enemies. Parasitism may be reduced if host populations emerge and pass-through vulnerable stages before the parasitoids emerge. The most obvious example is the interaction between cereal crop aphid *Sitobion avenae* and its main natural enemy, the parasitoid *Aphidius rhopalosiphi*. The interactions between the aphid and its parasitoid were altered by only a 5 °C variation in the temperature, because it caused higher metabolic rates of aphids and the defence against parasitoids. Eventually, it led to a reduction in the parasitoid efficiency of aphid control (Le Lann et al. 2014). In a case presented by Vučić-Peštić et al. (2011) the attacks of predatory ground beetles on mobile Drosophila species increased as the temperature rose from 5 to 30 °C, while the attacks on a resident prey, the larvae of Alphitobius spp. remained consistent along the temperature gradient. Reddy (2015) presented how the changes in the temperature affect tri-tropic interactions (host–pathogen–biocontrol agent). In wheat, a rise in the temperature from 17 to 22 °C resulted in increased reproduction of *S. avenae* aphid by 10%, and at the same time, the activity of a predatory ladybug (*Coccinella septempunctata*) adults increased by 250%, while the damage caused by the aphids was reduced because the crop reached maturity earlier. Evans et al. (2013) showed that a rise in the temperature had more influence on the Cereal leaf beetle (*Oulema melanopus*) than on its natural enemy, resulting in phenological changes between the prey and the predator, which resulted in low efficacy of the biological control. According to Hance et al. (2007), if the natural enemy starts to develop at a lower temperature than the prey and develops faster, the early emergence will eventually result in high mortality of the natural enemy, due to the absence of the prey, leading to the extinction of the natural enemy in several years. Gilman et al. (2010)

suggested that the specialist natural enemy species were more affected by climate change than the generalist species since they were less adaptable to spatial desynchronization with the host. In such a case, biological control that is relying on generalists' species might be more resilient to climate change (Crowder and Harwood 2014).

Invasive alien insect risks (IASs): IASs are by the definition the species introduced unintentionally by human activities (international travel, global trade, etc.) or intentionally (with plant material, food, pets, livestock, etc.) outside their natural habitat (Shine et al. 2000; Ricciardi 2013). IAS are usually agricultural, stored-product, forestry, household, or structural pests and often vectors of pathogens or parasites (Ward and Masters 2007). It is hypothesized that only a small part of the introduced IAS (the 'rule of 10') manage to establish and only a small part of these species spread and become of economic importance (Vander Zanden 2005). A number of authors predict expansion in distribution, increase in population abundance, and voltinism for IAS under predicted climate changes scenarios (Bale et al. 2002; Harrington et al. 2007; Walther et al. 2009; Hill et al. 2016), which could have severe consequences on crop production (Ziska et al. 2011). However, it is important to emphasize that climate change is not the main driver force of biological invasions. In combination with landscape characteristics, climate change defines the limits for the geographical distribution of IAS and seasonal conditions for their development and survival in a new habitat (Masters and Norgrove 2010). Parmesan and Yohe (2003) found that insects are spreading at an average rate of 6.1 km per decade due to temperature increases. Invasive species usually have a wider tolerance to climatic conditions than the native species, enabling them to invade a wider range of suitable habitats (Walther et al. 2009).

14.4.1.2 How Changes in Precipitation Affect Arthropods

The effect of moisture (air and soil) on insects' bionomy can be a potentially important aspect of the global climate change scenario (Hunter 2001; Hamilton et al. 2005), although, in comparison to temperature studies, fewer scientific studies refer to it. Yihdego et al. (2019) describe three mechanisms by which the drought affects herbivorous insect species: (i) dry climates provide suitable conditions for the development and growth of some species; (ii) drought-stressed plants attract specific pest species; (iii) drought-stressed plants produce less defensive compounds of secondary metabolism and thus are more susceptible to insect attack. Excess in precipitation affects small-bodied pests (aphids, mites, whiteflies, trips, etc.) which can be removed from crops and killed during heavy rainfall (Reiners and Petzoldt 2005; Pathak et al. 2012). Other insects, like the pea aphid, are not tolerant to drought (Mcvean and Dixon 2001). The insect species that spend winter diapause in the soil as well as soil-dwelling species are under the direct influence of rainfall, as it causes flooding and stagnation of water that results in higher insect mortality (Figure 14.2). The flooding method was used effectively to control some species that overwinter in the soil (e.g. cranberry fruit worm and other cranberry pests) (Vincent et al. 2003). Staley et al. (2007) studied the effects of increased summer rainfall and drought on *Agriotes lineatus*, a soil-dwelling wireworm species in grassland, and detected a rapid population growth in the upper part of the soil as a result of increased summer rainfall events, in comparison to drought conditions. Wireworms are very damaging pests and are predicted to become a greater problem under climate change (Johnson et al. 2008).

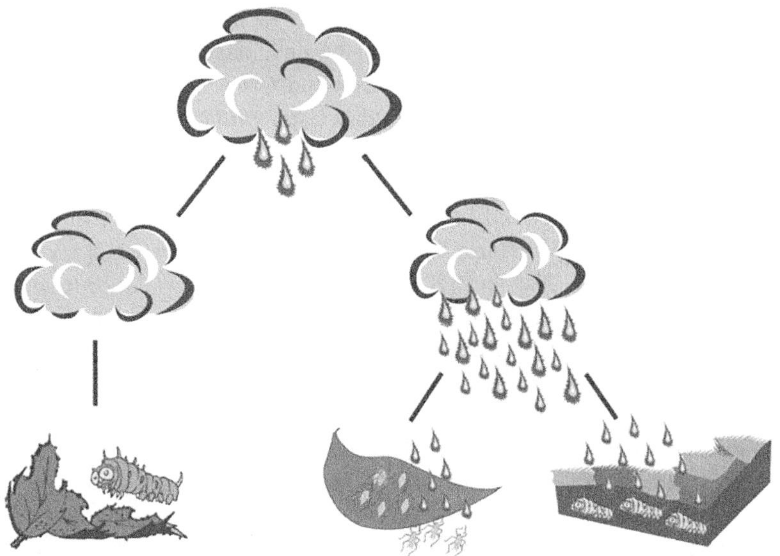

Figure 14.2 How do changes in precipitation affect insect pests?

14.4.1.3 How Rising CO_2 Levels Affect Arthropods

Rising levels of atmospheric CO_2 were proven to affect the **distribution**, **abundance**, **growth rates, fecundity,** and **consumption rates** of crop pests (Fuhrer 2003). Coviella and Trumble (1999) suggested that the effect of higher CO_2 concentration on herbivory was species-specific but also dependent on the particular pest–host plant system. As mentioned in previous sections, the increased CO_2 levels had a greater positive impact on C3 crops (wheat, rice, cotton, etc.) than on C4 crops (corn, sorghum, etc.). Accordingly, the response of insects feeding on C3 and C4 plants may differ. C4 crops often accumulate sugars and starches in leaves, which reduce palatability by altering the carbon/nitrogen ratio (Cotrufo et al. 1998). Additionally, the insect consumption rate increases as the nitrogen content in plants increase, due to elevated CO_2, because nitrogen is the major element in insect bodies involved in body development (Bezemer et al. 1998; Hamilton et al. 2005; DeLucia et al. 2008). Stiling and Cornelissen (2007) conducted a meta-analytic study on the insect pests' responses to increased CO_2 levels and revealed: (i) 17% increase in consumption rates, (ii) 22% decrease in pest abundance, (iii) 4% increase in developmental time, and (iv) about 9% decrease in pests relative growth rate. Hamilton et al. (2005) carried out a research in which soybean was grown under the conditions of elevated CO_2 levels in the atmosphere, and experienced 57% higher damage from insects (primarily Japanese beetle, potato leafhopper, western corn rootworm, and Mexican bean beetle) than when grown under the conditions of the normal atmosphere, most probably as a result of increased levels of monosaccharides in the leaves that stimulated insect feeding. Bezemer et al. (1998) showed that the green peach aphid (*Myzus persicae*) populations increased under the conditions of elevated temperature and CO_2 levels. The elevated temperature and CO_2 reduced the nutritional value of alfalfa plants which resulted in a shorter developmental time of the Beet armyworm (*Spodoptera exigua*) larvae, while the larvae of its natural enemy, the parasitic wasp (*Cotesia marginiventris*) were unable to enter the next stage which resulted in high mortality and eventually the extinction of local wasp population (Dyer et al. 2013).

Effect on pollinators: Honeybees, bumble bees, wasps, butterflies, flies, etc. are the most common pollinators. Approximately 73% of cultivated crops worldwide are pollinated by different bee species, 19% by flies, 6.55 by bats, 5% by wasps, 5% by beetles, 4% by birds, and 4% by butterflies and moths (Abrol 2009). Pollination with insects contributes for about one-third of global food production. However, the changing climate has already initiated the changes in pollinators' distribution ranges, decrease in population abundance, and pollination activities (Parmesan et al. 1999). The loss in pollinator biodiversity occurs at an unusual rate globally, with especially severe declines documented in the North Temperate Zone. Some climate change variables, such as fluctuations in winter weather conditions, changes in the duration of the vegetation season, and increased prevalence of extreme weather events, that rarely draw attention in the studies, are especially detrimental to pollinators (Vasiliev and Greenwood 2021). Reddy (2018) emphasize that the temperature and water availability significantly affect the flowering, pollination, and fruiting of plants. It should not be overlooked that the quality and quantity of pollination have multiple implications for food security, ecosystem stability, and resilience to climate changes in general (Reddy 2015).

14.4.2 Plant Pathogens and the Environment

The development of plant disease requires several factors to be met, the presence of virulent pathogen, susceptible host, and the third factor, often described as 'conductive environmental conditions' (Figure 14.3). Changes in each part of this system will favour or suppress pathogen growth and development which will affect the appearance of disease symptoms. Pathogens are adapted to a fluctuation of environmental factors in each stage of the pathogen life cycle whether it is expressed in a range of temperatures and humidity for

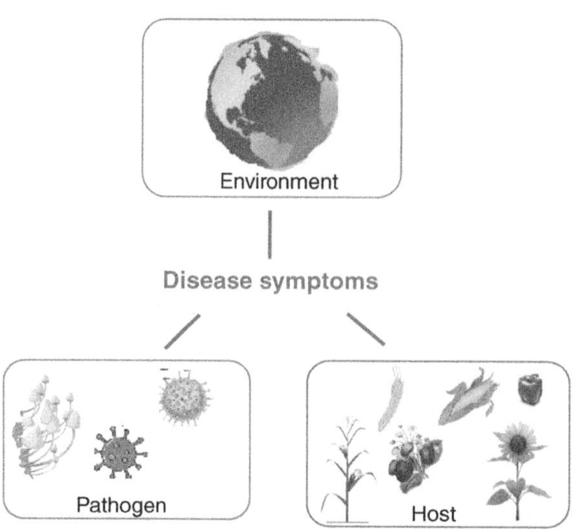

Figure 14.3 Common presentation of factors necessary for plant disease development

spore germination and entering host tissues or formation of structures to a wider range of environmental conditions usually in periods without the presence of host plant. Given enough time, strains of existing pathogens are able to adapt to changing environments.

14.4.2.1 Climate Change and Plant Disease Epidemiology

Climate change as a consequence of rising GHG concentrations in the atmosphere is commonly presented as a general upward trend on an increase in the mean global temperature. In plant disease epidemiology, for the disease prediction, variations on a local scale of this global trend are more important. Even the regional level is often not sufficient enough for a punctuality of monitoring disease occurrence and prediction of possible changes in pathogen presence and its aggressiveness and virulence. Not every pathogen will react in the same way or with the same level of change in the climate presented as a different duration and type of weather patterns. Therefore, the potential changes are also a reflection of each particular pathosystem. Changes in the climate when combined with the environmental requirements of plant pathogens are a potential trigger of changes in the disease distribution, the appearance of new pathogens, pathogen survival, population size, inoculum build-up, decrease in host resistance, increase in the availability of susceptible host, introduction of the alternative host as green bridge, etc. (Sharma et al. 2020). In general, research data indicate that a number of increased risks from the diseases outweigh a number of the diseases with predicted decreasing risk (Juroszek et al. 2020). However, this conclusion is mostly based on fungal pathogens and oomycetes, while groups like viruses and bacteria are still behind in a number of published researches.

The plant pathogens are under the influence of a different environment, in terms of periodic and seasonal fluctuations, during their life cycle. A part of it is bound for a suitable host and the other is in a form of various structures outside the host. Variations in the lifestyles of pathogens reflect the current collection of adaptation strategies.

14.4.2.2 How Rising Temperatures Affect Plant Pathogens

The temperature is the most commonly investigated parameter in climate change and plant disease risk modelling. In general, considering major consequences of climate change, such as an increased temperature, will lead to a higher incidence of some diseases, better conditions for pathogen overwintering, potential emergence of new pathogen strains, and rise of populations of pathogen vectors potentially lowering the quality of final crop product (Velasquez et al. 2018). There is evidence, although a number of investigated pathosystems needs to be increased, that the pathogens can quickly adapt to temperature change (Wu et al. 2019). The changes in temperature and precipitation patterns resulting in the increase in disease frequency and severity were already recorded in some regions (Torres Castillo et al. 2020). In addition to this, the distribution of diseases is expected to change, with an example of the predicted northward direction of pathogens' areal dispersion in the northern hemisphere, as the environment in new areas becomes conductive for pathogens (Harvell et al. 2002). The research based on modelling disease frequency and severity and comparison between the present climate conditions and predicted future scenarios show that the effect of the temperature on plant pathogens can be direct with changes in pathogen life-history traits or indirect through genotype–temperature interaction (Vamourin and Laine 2018). Some models suggest an increase in days with favourable conditions for the occurrence of disease (Ferreira and Miranda 2020). In some plant–pathogen systems, the reduction of latency period as a result of climate change will lead to earlier and faster development of the disease through more inoculum build-up, and consequently an increase in disease severity (Wójtowicz et al. 2020). Projected mild winter conditions that are expected for some regions will represent favourable conditions for the survival of overwintering pathogen structures in the absence of susceptible hosts (Juroszek and von Tiedemann 2015) as well as the earlier occurrence of plant diseases. In case of plant viruses, climate change will have both direct and indirect effects, affecting the distribution and survival of viruses and their vectors with a tendency to increase (Trebicki 2020).

The plant pathogens require a conductive environment on the canopy level. Different approaches have been in use to model this microenvironment and plant architecture (Moraes et al. 2019; Kochi et al. 2021; Teichman and Muhr 2015), pathogen dispersal (Schmale and Ross 2015; Bicout and Sache 2003) and disease development (Fenu and Malloci 2021; Newlands 2018). These models are available for better understanding and prediction of the effects of plant diseases under the changing climate. The plant architecture and its influence on pathogens can be analysed on the level of separate plant organs with specificities in shape and surface characteristics, individual plants with specificities in form and branching, and on the level of a group of plants with special distribution (Costes et al. 2013). Combined together these traits are building specific environmental patterns or microclimates. Vice versa, plant pathogens, depending on the disease severity, could have a significant impact on the plant canopy thus changing the microclimatic conditions.

14.4.2.3 How Changes in Precipitation and Moisture Affect Plant Pathogens

A large group of pathogens requires free water available in a disease cycle. Oomycete motile zoospores and fungal spores need free water for a certain minimum period of time at a particular temperature for germination and entering the susceptible plant tissues. Described as leaf wetness, a presence of water after a rainfall, dew formation, or guttation, is directly connected to other environmental conditions like the temperature and air currents and thus is susceptible to every change. Disturbance in the leaf wetness could lead to a significant decrease in the expected incidence of some diseases as some models suggested (Juroszek and von Tiedemann 2015). On the other hand, the potentially favourable influence of the increased CO_2 and increase in plant growth could end with the opposite effect.

14.4.2.4 How Rising CO_2 Levels Affect Plant Pathogens

One of the drivers of a plant canopy change is the direct impact of elevated CO_2 levels on the plant canopy. This process can have a positive effect on plant growth and is usually referred as a 'fertilization effect'. In a case study with soybean, the growth of this plant in an atmosphere with elevated CO_2 led to an increase in leaf area index and canopy photosynthesis with a significant increase in gross primary productivity (Song et al. 2020). A similar effect was notified in other crops (Degener 2015; Jennings et al. 2020), and the increase is expected to be higher in temperate regions compared to the tropics (Chaloner et al. 2021). Some experiments indicate that increased CO_2 will have a positive effect on the leaf mass per area, up to one-third (Kovenock and Swann 2018). Such an increase would be accompanied by a higher demand for available nutrients and water, making the interaction complex and unpredictable in field conditions. However, data also point to a yield reduction with climate change as a result of changes in a leaf structure, stomatal processes, nitrogen availability, and efficiency of photosynthesis with increasing CO_2 concentrations (Zheng et al. 2019). The plant canopy architecture and microclimate changes will without a doubt affect pathogen dispersal and disease development. The increase of CO_2 concentrations and temperature in controlled conditions resulted in an increase of powdery and downy mildew and several leaf spot diseases on various crops (Pugliese et al. 2018). The effect of the higher levels of CO_2 and temperature could potentially lead to an increase in the disease severity as a result of changes in the microclimate and pathogen life cycle (Debela and Tola 2018). On the other hand, modelling of potato late blight (*Phytophthora infestans*) considering perspectives of climate change in particular regions revealed a significant decrease in the risk of this disease (Gaucher et al. 2018). More examples of predictions on climate change impact are presented in Table 14.1.

Given the spatial and temporal magnitude of the accelerated climate change, its impact on crop production in terms of plant diseases is inevitable. Plant pathogens evolve a large number of adaptive strategies suitable for parasitic lifestyles and are expected to adapt to current changes. The importance of the change is how it will be reflected in relation to a particular crop. The current research on the impact of the changing climate models indicates both decrease and increase in the occurrence and severity of particular plant diseases. With more research and higher prediction power hope is that the changes, regarding plant diseases, can be anticipated with enough certainty to enable implementing control strategies for stable crop production.

Table 14.1 Examples of model predictions of climate change effect on selected crop pathogens.

Crop	Pathogen	Estimated effect of climate change on diseases	References
Rice	*Magnaporthe grisea* *Rhizoctonia solani*	Decrease epidemics incidence in South Korea	Kwang-Hyung et al. (2015)
Maize	*Aspergillus flavus*	Increase the appearance of and higher risk of aflatoxin contamination	Moretti et al. (2019)
Wheat	Plant residue-borne pathogens	Increase in pathogen load	Wahdan et al. (2020)
	Puccinia graminis f. sp. *tritici*	Expanding overwintering areas of pathogens	Prank et al. (2019)
Soybean	Soybean cyst nematode	Increase in number of generations and in distribution areal in Canada	Gendron St-Marseille et al. (2019)
Potatoes	*Phytophthora infestans*	Decrease risk of infection for regions in France	Launay et al. (2014)
Cucumber	Melon yellow spot virus (MYSV)	Higher temperature facilitates viral spread in host	Sugiyama et al. (2009)

14.4.2.5 How Climate Change Impacts Host Resistance and Pathogen Virulence Factors

Plants are sensing the temperature fluctuations via phytochrome-B (PHYB), through a reversion of their active and inactive state (Legris et al. 2016). The changes in the temperature are influencing the activity of a number of temperature-responsive genes affecting plant growth and responses to the environment, referred as thermomorphogenesis (Casal and Balasubramanian 2019). The role of PHYB is related to phytochrome-interacting transcription factors (PIF) and is essential for the termomorphogenesis process PIF4 (Koini et al. 2009; Qiu et al. 2021). Host–pathogen interactions possess a certain range of flexibility within specific temperature and humidity variation. The interdependence is in the range of minimum and maximum value for a particular pathogen. The contact with pathogens triggers immune response in plants and a number of resistant genes are temperature responsive. Recent research indicates that PIF4 negatively regulates plant immunity while positively regulates plant growth at higher temperatures (Gangappa et al. 2017). In general, signalling in effector-triggered immunity requires lower temperatures while signalling in pattern-triggered immunity requires higher temperatures (Cheng et al. 2013). Under the higher temperature regime, ETI-related *RPS4/RRS1* resistance signalling was inhibited in Arabidopsis plants after the inoculation with *Ralstonia solanacearum* (Auon et al. 2017). There are indications that NB-LRR *R* genes are temperature-responsive and that in many cases of plant defence temperature sensitivity is through this type of coding gene (Zhu et al. 2010). Data from a number of studies show that the increase in temperature alters the resistance response of a host (Desaint et al. 2021; Cunchun et al. 2021). The changes in the climate will be reflected in the response of host to a pathogen through the ability to induce a resistance mechanism based on the pathogen's general or specific recognition and in the case of pathogen production of virulence factors (Velasquez et al. 2018). The host resistance can be severely impaired via temperature-induced lowering of resistance gene expression (Peng et al. 2019). However, gene *Xa7* confers rice resistance to *Xanthomonas oryzae* pv. *oryzae* was more effective at high temperatures, which is an example of the opposite effect (Webb et al. 2009). Similar results were found in the case of gene $TaRPM_1$ that confers the resistance of wheat to *Puccinia striiformis* f. sp. *tritici* (Jiahui et al. 2020). The resistance stability under various temperature regimes is genetically determined. The stability of the resistance to the yellow rust in wheat under changing temperatures is determined by genetic background (Bryant et al. 2014). The loss of resistance in *Arabidopsis thaliana* plant grown at higher temperatures and inoculated with *Pseudomonas syringae* pv. *tomato* was connected to an enhanced translocation of bacterial effector proteins to host and loss of salicylic acid biosynthesis (Huot et al. 2017). A different example is a case of increased susceptibility of *Brassica rapa* to the necrotrophic fungal pathogen as a result of trade-off between the adaptation to drought conditions and disease resistance induced by climate change (O'Hara et al. 2021).

In the future plants will, under changes derived from the current trend in climate change, face potentially increased risk of highly aggressive pathogens and unpredictable epidemic events. There is an indication of heat-resistant plant defence mechanisms and resistance that is expressed better under higher temperature. However, a significant volume of available data indicates the increase in plant susceptibility as a result of the disruptive effect of higher temperature on plant immunity and the presence of trade-offs in favour of plant adaptation to abiotic stress. The direction for developing heat resilient and disease-resistant crops might lie in further elucidation of the underlying mechanism for temperature sensing and widening knowledge of temperature non-responsive types of plant resistance. Newly available sequencing technologies, along with studies of crop epialleles and their potential role in adaptation to changes in the environment are facilitating the study of these mechanisms and, combined with new (epi)genome editing tools, could lead to targeted manipulation for the creation of crops with improved and/or combined heat resilience and disease resistance (Miladinović et al. 2021; Varotto et al. 2020).

14.4.3 Weeds and the Environment

Weeds are one of the main factors that compromise crop production and quality by competing for essential resources, such as light, nutrients, and water. Moreover, they can hamper crop harvest, host crop pests, and pest vectors, and produce allelochemicals, indirectly affecting crop yields and quality. Yield losses due to weeds were estimated at 34% on a global scale (Oerke 2006).

Climate change affects weed populations by altering species abundance, richness, geographic range, and phenology (Anonymous 2013), as presented in Figure 14.4. Climate change directly affects weed establishment, growth, reproduction, and survival. In addition, the indirect influence of climate change on weeds is reflected in species interactions with other members of the ecological communities. Any environmental changes can impact not only the growth and interactions among weeds but of the whole ecosystem.

The effects of environmental factors, such as temperature, water, CO_2, and their interactions, on crops, weeds, and weed-crop competition under changing climate also depend on

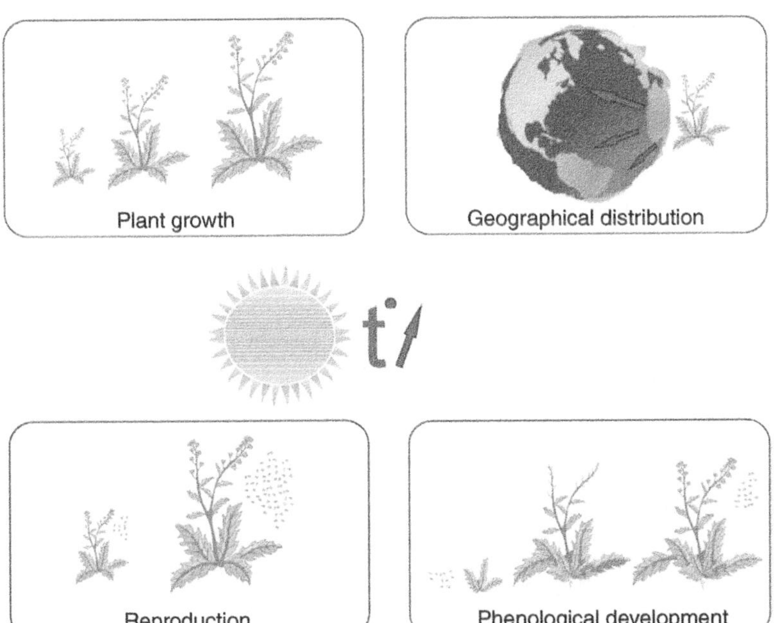

Figure 14.4 How do increasing temperatures affect weeds?

their photosynthesis processes. Based on their photosynthetic pathway, crops and weeds can be classified as C3 and C4 plants. Unlike C3 plants, plants with the C4 pathway have the enzyme phosphoenolpyruvate that concentrates CO_2 around ribulose-1,5-bisphosphate carboxylase /oxygenase (Rubisco) and minimizes CO_2 lost through photosynthetic carbon oxidation. For this reason, the photosynthetic rate under current CO_2 levels in C4 plants is higher than in C3 plants. Although the C3 cycle for carbon fixation is the dominant mechanism in both crops and weeds, C4 weeds are considered the most notorious and are predominantly from the tropics (Holm et al. 1977). Nevertheless, not all crops are based on C3 pathways, and not all weeds are C4 based (Ziska 2010). There is an equal number of important C3 weeds.

Modified environmental conditions in soil, as a natural seeds depository for a great number of weeds, may impact weed seeds' diversity and longevity. Changes in soils due to human activities, such as structure agitation via tillage, and natural processes related to climate change, can cause exposure of weed seeds to altered light, water, and temperature conditions in soil. This can subsequently break seed dormancy and initiate excessive weed germination (Ziska and Dukes 2011).

As a reaction to the effect of climate change, invasive weeds are expected to shift their geographic distributions (Hellmann et al. 2008; Walther et al. 2009), some to expand and some to reduce their range (Bradley et al. 2009). Broad-spectrum invasion mechanisms and circumstances can favour the weed dispersal, such as superior morphological and physiological traits, a satisfactory competing ability, an efficient photosynthesis mechanism, tolerance to numerous abiotic stresses, shorter life cycle, rapid dispersal, shrubiness, adaptation to a wide range of environmental conditions, ability to produce and release allelochemicals, improved growth and reproductive capacity and the absence of its natural enemies in the climate altered environments (Dukes and Mooney 1999; Bajwa et al. 2016). For example, parthenium weed (*Parthenium hysterophorus*) was predicted to be a problem in the future due to its improved photosynthetic rate, water use efficiency, increased biomass, branching and leaf area under a high CO_2 concentration, high temperatures, water-deficit conditions (Khan et al. 2015; Shabbir et al. 2014), and ability to suppress the growth and development of neighbouring seedlings (Belgeri and Adkins 2015). The prediction of combined effects of environmental parameters due to global changes on invasive and non-invasive weeds is rather challenging and would depend on the species and the agroecosystem.

Many environmental factors are important for weed control. Temperature, precipitation, wind, soil moisture, and atmospheric humidity determine the time of herbicides application, as well as their uptake and metabolism by crops and target weeds. Besides, climate change may alter the synchronous development and reproduction of weeds and their natural enemies, affecting the effectiveness of biological weed control.

Different predictions were modelled for invasive and for established species based on ecologically meaningful environmental data (WordClim, GBIF). Environments under climate change would be more suitable for highly invasive species than for established species. Although political boundaries do not match the ecological barriers, they define countries with different conservation resources and policies. Thus, the estimation of invasion risks at the national level is essential for monitoring and outlining the priority management (Kuebbing and Simberloff 2015). Targeted strategic control in climate change threatened areas could be useful for reducing future impacts, while surveillance for new infestations should be given a high priority (Wilson et al. 2009).

Australia can be used as a good example of a country that has taken measures to address the problem of invasive weeds under climate change, by defining a list of priority weed species that encompass the Weeds of National Significance from in The National Biodiversity and Climate Change Action Plan and from the National Weed Alert list for Environmental Weeds. Information on both thc habitats most at risk from these weed species and the native biodiversity likely to be threatened will be used to define areas where management can be prioritized. Species with wide environmental tolerance, the capacity for long-distance dispersal, and good colonizing ability are expected to be the most responsive to changing climate and to take advantage of new environmental conditions (Gallagher et al. 2006).

14.4.3.1 How Rising Temperatures Affect Weeds

Temperature is one of the key determinants of weed species establishment, and as such its predicted increase will undoubtedly affect weeds growth, phenological development, reproduction, and geographical distribution. The most competitive C4 weeds are likely to thrive under increased temperature conditions considerably reducing crop yields (Ramesh et al. 2017). A considerable number of weeds that originate from tropical or subtropical areas (Holm et al. 1977) have higher requirements for accumulated degree days (Patterson et al. 1999), and thus cannot tolerate the lower temperature extremes at higher latitudes. Nevertheless, many warm-season weeds react well to even a small rise in temperature (Flint et al. 1984). It is, therefore, expected that warming associated with climate change would cause the spread of these tropical or subtropical weeds that cohabit with warm-season crops northward (Rahman and Wardle 1990). Some notorious C4 weeds, such as redroot pigweed (*Amaranthus retroflexus*), foxtail (*Setaria* sp.), crabgrass (*Digitaria* sp.), and Johnson grass (*Sorghum halepense*), would expand in northern regions that used to be cooler and less favourable for these species (Clements and DiTommaso 2011). Temperature increases are also expected to expand the geographical distribution of itchgrass (*Rottboellia cochinchinensis*), cogongrass (*Imperata cylindrica*), and witchweed (*Striga asiatica*), and increase their management costs (Korres et al. 2016). In addition, weed species susceptible to high temperatures are expected to shift to higher altitudes (Korres et al. 2016).

Although there are some views that cohabiting weeds and crops would similarly respond to drought (Patterson 1995), it generally observed that drought differently affected various crops and weeds, both increasing and decreasing the competitive ability of weeds, depending on the species (McGiffen et al. 1997; Leguizamón et al. 2011). Rising temperatures may favour C4 invasive weeds in C3 grasses and impede C3 weeds competing with C4 grasses (Bijoor et al. 2008; Williams et al. 2007). Increased temperatures and precipitations during winters may advance the survival of winter annual weeds and their emergence and flowering time later on (Lee 2011), while increased summer temperatures may stimulate the spread of thermophile summer annual weeds (Walck et al. 2011). In addition, the changes in temperature may either accelerate or delay the time of weed emergence so as to be synchronized with the stage of crop emergence when the crop is a weak competitor.

Increased temperatures can reduce lipid viscosity and facilitate flow through the plant cuticle, and thus speeding up herbicide uptake, translocation, and effectiveness (Sharma and Singh 2001). On the other hand, high temperatures may stimulate metabolic activities in weed plants, reducing herbicide activity and effectiveness through pesticide biotransformation and degradation (Johnson and Young 2002). Moreover, high temperatures can cause a considerable evaporation of spray droplets before they reach the weed surface or can cause volatilization losses from soil, reducing the herbicide uptake by weeds (Atienza et al. 2001).

Another important aspect of the effect of rising temperatures and water shortage on weeds is the threat of fire outbreaks. Fires may clear agricultural and other landscapes leaving them exposed to invasive weeds, which may speed at the expense of native species, suppressing them almost completely (Ziska et al. 2005; Bradley 2009).

14.4.3.2 How Changes in Moisture/Precipitation Affect Weeds

Water stress can limit photosynthesis in both C3 and C4 pathways, by reducing the stomatal aperture and CO_2 availability (Lawlor 2002). In general, C4 weeds have greater photosynthetic efficiency and thus higher water use efficiency, meaning they transpire less water for carbon assimilation during photosynthesis than C3 weeds. In water-limited conditions, when stomatal closure is an effective mechanism to prevent water loss through transpiration, the higher water use efficiency of C4 weeds could prove advantageous.

More frequent extreme weather events, such as droughts, floods, and storms, could alter weed invasion (Diez et al. 2012). The effects of changes in the precipitation on invasive weeds are determined by the season and the ecosystem. Increased snowfall during winter and rainfall during spring and summer in different grassland ecosystems intensified early growth of invasive weeds, such as white knapweed (*Centaurea diffusa*), common gypsophila (*Gypsophila paniculate*), broadleaf toadflax (*Linaria dalmatica*), and cheatgrass (*Bromus tectorum*), at the expense of later-growing natives (Blumenthal et al. 2008; Thomsen and D'Antonio 2007; Prevéy and Seastedt 2014). A global meta-analysis, showed that invasive weeds use more water than native species at leaf, plant, and ecosystem levels (Cavaleri and Sack 2010), suggesting that in the light of the changing climate invasive weed management would be required in areas with increased precipitation. An increase in rainfall or in irrigation conditions will also favour hydromorphic weeds. For example, common cocklebur (*Xanthium strumarium*) would reduce yields more in well-watered soybeans than in water-stressed soybeans (Mortensen and Coble 1989), whereas wheat growth and yield will not be compromised by Canada thistle (*Cirsium arvense*) in increased rainfall condition (Donald and Khan 1992).

On the other hand, reduced water availability caused by droughts would intensify the crop-weed competition between crops and weeds (Ramesh et al. 2017), promoting species with better abilities to assess resources and to tolerate water deficits (Patterson 1995). It is observed that some invasive species had better recovery potential than natives after the end of severe and extended drought periods (Scott et al. 2010).

Since rising temperatures cause greater evaporation, the competitive balance would change in favour of deep-rooted plants, if exposed to the conditions of frequent and severe drought periods induced by climate change (Stratonovitch et al. 2012). In addition, dry soil conditions extend the vitality of weed seeds, as the conditions are not suitable for seed predators and for germination (Korres et al. 2016). Summer droughts will affect weed management in spring crops (Peters and Gerowitt 2014) and favour C4 weeds with increased water use efficiency over C3 weeds. Nguyen et al. (2017) reported that under the conditions of low soil moisture predicted within climate change trend, the life span of parthenium (*P. hysterophorus*) was reduced by up to 43% while the plant growth was stimulated by 20% due to alterations in the plants vegetative and reproductive biology. However, Bajwa et al. (2016) showed a potential tolerance of parthenium weed to drought stress.

14.4.3.3 How Rising CO_2 Levels Affect Weeds

The increased level of CO_2 increases photosynthetic rate and reduces stomatal conductance and transpiration through leaves. This is known as CO_2 fertilization effect, which consequently improves water use efficiency leading to faster plant growth (Ziska and Runion 2006; Hatfield and Dold 2019). Improved water-use efficiency could favour invasive weeds in water-limited conditions (Dukes 2000). In a study quantifying the responses in biomass of six most notorious C3 weeds to current and projected elevated CO_2 concentrations, three weed species, namely *C. arvense*, *Centaurea maculosa*, and *Sonchus arvensis*, showed the largest average relative increase in biomass, suggesting a potential threat from these invasive weeds in the conditions of elevated CO_2 (Ziska 2003a). The risk of significant reduction of soybean seed yield and biomass from *C. arvense* under increased CO_2 was also determined for no-tillage practices, indicating possible difficulties in its chemical control (Ziska 2010).

Although both crops and weeds could prosper from the CO_2 fertilization effect, the outcome of the weed-crop interaction depends on the competing ability of each species under enhanced CO_2 (Reilly et al. 1996). Generally, plants with C3 photosynthetic pathways perform better C4 plants than under elevated atmospheric CO_2 (Ziska 2000). This is because C4 plants have already greater photosynthetic efficiency than plants with C3 pathway, and the increased CO_2 would not simulate photosynthesis in C4 weeds much further. No effect was determined of elevated CO_2 on seed yield and biomass in weed-free C4 dwarf sorghum (*Sorghum bicolor*) crop (Ziska 2003b). The elevated CO_2 caused a significant increase in aboveground biomass of C3 weed velvetleaf (*Abutilon theophrasti*) but nonsignificant increase in the aboveground biomass of C4 redroot pigweed (*Amarantus retroflexus*). The competition of dwarf sorghum with each of the two weeds leads to a considerable reduction of dwarf sorghum yield and biomass under increased CO_2 level (Ziska 2003b).

Vegetative growth and yield of important C4 crops could be hampered by C3 weeds with raised level of carbon dioxide (Ziska 2001). Under elevated CO_2 concentration, C3 soybean yield reduction from C3 weed *Chenopodium album* was higher than under ambiental CO_2, while the yield losses from the C4 weed *A. retroflexus* were much lower under increased CO_2 comparing to the ambiental conditions (Ziska 2000). Nevertheless, it is difficult to predict which species will prosper in the conditions of elevated CO_2 merely based on their photosynthetic pathway. The interactions with other species should be also considered, since in some ecological communities CO_2 enrichment favours C3 plants, while in other ecosystems C4 species are favoured. A comparison of studies evaluating different combinations of C4 and C3 crops-weeds interactions revealed that in most cases weeds were favoured under increased CO_2 (Walthall et al. 2012).

The C3 invasive weed species, such as *Avena fatua*, *C. album*, *C. arvense*, and *Convolvulus arvensis*, are supposed to increase their photosynthetic rates, and thus biomass and seed production under an elevated CO_2 and temperature regime (Sheppard and Stanley 2014). The plants with a C4 photosynthetic pathway manifest little response to elevated CO_2 as they have an internal mechanism to concentrate CO_2 at the site of CO_2 carboxylation (Ramesh et al. 2017). Weed-related yield losses of C4 crops (maize, sorghum, sugarcane) will tend to increase under elevated CO_2, but this will not occur with C3 crops (three-quarters of global agriculture is represented by C3 crops), as elevated CO_2 will be a crucial factor in realizing the potential benefits of CO_2 fertilization. In addition to the interaction between weeds and other species, the different responses to elevated CO_2 depend also on the interactions with resources such as temperature, light, water, and nutrients. For example, elevated CO_2 could aggravate nitrogen limitation (Luo et al. 2004).

In general, plants increase their water-use efficiency under CO_2-enriched conditions (Korres et al. 2016). If increased, CO_2 slows down the transpiration and the soils beneath those plants dry out more slowly. In ecosystems where plant growth is limited by water availability, species that can take advantage of the extra moisture might eventually become more abundant (Dukes and Mooney 1999). The ability to capture and efficiently use limited or altered resources is related to increased biomass and reproductive capacity and subsequently to increased weed abundance (Korres and Norsworthy 2015). High CO_2 also increased root, rhizome, and tuber biomass (Chandrasena 2009) suggesting that weeds with strong, deep roots, such as in perennial weeds, might be more difficult to control (Ziska 2004).

14.5 General IPM Principles

The Food and Agriculture Organization of the United Nations (FAO) under IPM considers 'an ecosystem approach to crop production and protection that combines different management strategies and practices to grow healthy crops and minimize the use of pesticides. IPM is an approach-based method for analysis of the agro-ecosystem and the management of its different elements to control pest and keep them at an acceptable level (action threshold) with respect to the economic, health and environmental.' A comprehensive categorization of IPM strategy can be summarized in eight general principles (Barzman et al. 2015), which also present a successive way for integrated management implementation (Table 14.2). The concept of integrated management was created in the United States in the late 1950s, and accepted by FAO in 1968 (Peshin et al. 2009). It stemmed mainly from a need to find a solution for the pest problem in some crops beyond an ineffective application of insecticides, and today, it is involving pathogens and weed control. The novel approach was aiming at the integration of a diverse set of methods to achieve a stable crop production. It integrates ways of pest control in a holistic manner which includes the impact on the farm economy and the social level of perceived ways of food production. Driven by the need for stable crop production and demand for the rational use of pesticides as a control measure this concept was incorporated into the legislation as mandatory by the national and supranational institutions such as the European Union. The concept is in a constant process of rethinking and addition of new aspects particularly emphasizing management, business, and sustainability (Dara 2019).

Table 14.2 General IPM principles.

1. Prevention and suppression
2. Monitoring
3. Decision making
4. Non-chemical methods
5. Pesticide selection
6. Reduced pesticide use
7. Anti-resistant strategies
8. Evaluation

Source: Adapted from Barzman et al. (2015).

14.5.1 Prevention and Suppression

The first level of crop production in the IPM system is a set of measures that aim to prevent and disrupt processes that build up the pest populations above damage thresholds, commonly expressed as yield quality and quantity decrease. Cropping systems designed with narrow objectives to gain high yields of highly demanded cash crops, usually lack crop diversity. They often heavily rely on pesticide application as a mean for control weeds, arthropod (insect and mite) pests, and pathogens which this production system greatly favours. To reduce the pressure of harmful organisms on crops, basic measures are consisted of preventing the problem or applying measures that keep the population of pests under threshold levels that do not cause production losses. To achieve this goal, a diverse set of measures is at disposal to ensure plant growth and development under minimal stressful conditions, from crop rotation to adjust practices.

Crop rotation is a basic tool for keeping a wide range of pests, weeds, and pathogens under control. It can be highly effective in maintaining a low level of pests and pathogens that have a narrow host range. Implementing it, the life cycle of these organisms is disrupted enough to lower the possibility of the negative effect on a crop. However, crop rotation is not universally effective, e.g., in the control of multi-host pathogens such as the fungus *Sclerotinia sclerotiorum* that can thrive on dozens of plant hosts (Wahmhoff et al. 1999). Indirectly, crop rotation has a positive influence on the soil and soil nutrient content by preventing the depletion of nutrients required by a particular crop. It is important for this method not to cause major disruption in the population of beneficial organisms, and there are examples of some diverse rotations to foster these populations (Busch et al. 2020).

Conventional agriculture often implies the use of sometimes excessive amounts of fertilizers. It affects plant growth and architecture and can lead to an increase in susceptibility to some plant diseases. Although the general trend of a fertilizer effect on plants is known for a long time, a challenge is to tune time, amount, and type of fertilizer to the needs of a particular crop, thus making plants vigorous and resilient to stress. By changing the application time at potato emergence, for instance, one can achieve higher starch content and lower susceptibility to early blight affecting at the same time the quality and disease presence (Abuley et al. 2019). The negative effect of an excessive fertilization particularly the application of high concentrations of nitrogen besides the increase in the plant biomass can have an indirect effect. For example, in fruit trees, in orchards, nitrogen over-fertilization has an increasing effect on red spider mite egg production making the application of acaricides unavoidable. At the same time, by chemical control of pest mites is harming predatory mites as their natural enemies (Papp et al. 2001). On the other hand, changes in the fertilizer application in the pathosystem of carrot and *Alternaria dauci* led to minor changes in the disease severity that had no effect on fungicide application (Vintal et al. 1999). Most research is focused on one pest or a disease while in reality crops are challenged by several pests. In a study on the effect of fertilizer rates and insecticides application on the major constraints in onion production, it was concluded that the type of the studies that include more constraints is more suitable for the development of good IPM strategies (Leach et al. 2020).

The implementation of different soil conservation methods is recommendable in IPM, as it decreases the possibility of pest outbreaks via an increase in natural enemy populations and improvement of physical characteristics of the agroecosystem (Alyokhin et al. 2019). In addition, the measures such as reduced tillage or no-tillage system provide the environment means for the preservation of the soil community, which can have a significant effect on the control of pests and pathogens (Pearsons and Tooker 2017). There is an approach useful in IPM, a habitat management, that includes practices focused on the conservation of natural enemies (enemy hypothesis) and practices that reduce the herbivore impact on the crop by the host selection behaviour, as explained in a 'resource concentration' hypothesis (Gonzalez-Chang et al. 2019). In a combination with other IPM practices, the monitoring and sustaining of natural enemy populations provide good biological control (DuPont and Strohm 2020). Manipulation with a various companion plant in some cases yielded to a significant success in pest control. The application of a 'push–pull' system, that includes the cultivation of trap crop Napier grass as a border and intercrop from genus *Desmodium* as a repellent for pests and at the same time that inhibits attachment of *Striga* plant parasite, is an example of an excellent functional control strategy (Smart et al. 2014). The introduction of a cover crop has also the beneficial effect on the natural enemies of pests (Chen et al. 2019; Kahl et al. 2019). In general, more diversity in a farm plant community will have a stabilizing effect on the herbivore pests lowering the chance for the increase in level that has a negative impact on the crop yield and/or quality. The research and observations on the reaction of plants to various forms of biotic stress have a long history.

The differences in the plant host resistance expressed as the absence of symptoms of the slow progress of the disease are still widely used as a control method for a number of plant pathogens (Nelson et al. 2017). It is based on the phenomenon that genotypes of particular crops differ in the expression of disease symptoms, and therefore by using data on susceptibility yield loss can be predicted (Dedić et al. 2012; Miladinović et al. 2012). As a powerful tool and with a low negative impact on the environment, it has an important place in the IPM. As for the resistance to herbivore insects, the resistance can be described as an antibiosis (host plant has a negative effect on herbivore biology), antixenosis (herbivore behaviour is affected resulting in lower damage to host plant), or tolerance (host plant sustain herbivore feeding without major damage) (Stout and Davis 2009). The resistance of varieties and the duration of this trait are crucial for IPM. During the process of breeding, varieties were selected and bred with an emphasis on quantitative traits like the yield. Consumer preferences also played a major role in breeding. In the process, as a result of trade-offs and narrow selection, genetic variability has declined (Dimitrijević and Horn 2018). This reflected among others on the resistance to abiotic and biotic stress. Some diseases of major crops were put under control by this method via the discovery and introduction of resistance genes (Cvejić et al. 2020). Breeding efforts using approaches like 'genetic rewilding' that involves the use of wild relatives in breeding for the resistance, or evolutionary approach in breeding to resistance can help in creating a sustainable IPM with further decreasing of pesticide use as a last resort (Anđelković et al. 2020; Karlsson Green et al. 2020; Terzić et al. 2007).

The use of certified seed and plant material to control pathogens, disable the introduction and spread of new insect and mite pests, and combined with quarantine measures are also an example of preventive measures. A well-organized system of this control hampers the spread of potentially destructive pests and pathogens in new areas. This is of high importance for harmful organisms entering regions with no natural enemies and with an abundance of susceptible host. Such entries in a course of time could mean the destruction of the host plant in a particular area and stop cropping.

Climate change in agricultural systems presents a great challenge. However, there are various ways to influence a negative outcome by implementing some or all of the described prevention measures, through breeding strategies, planting time, pesticide use, and manipulation of the environmental parameters (Burdon and Zhan 2020).

14.5.2 Monitoring and Decision Making

The applicability and efficiency of the control methods in IPM heavily depend on field data and observations. The observations are made based on a general look of a crop, survey of insects on plants, and examination of various types of traps. The monitoring of plant diseases and predicting the potential appearance is usually based on weather data and history or data from various types of spore trapping devices. Disease-warning systems are an essential part of IPM which can be based on leaf wetness duration (Rowlandson et al. 2015). In all cases, the aim of scouting and forecasting potential biotic stress is to apply the control measures at the exact moment to have maximum efficiency. One other potential benefit of this approach is also lowering the dose and the number of pesticide treatments. Automatization of data collecting is increasingly available for the detection of pests via infrared sensors, audio sensors, and photo-based identification (Lima et al. 2020). Marković et al. (2017) assessed the potential use and the precision of a Raspberry PI sensor device with camera, in monitoring the Western corn rootworm (*Diabrotica virgifera* sp. *virgifera*) flight on pheromone traps. The proposed method for monitoring the variations in the number of caught specimens on sticky surfaces of pheromone traps, based on the variations of the dark surface on the images, can be a reliable tool for further work. It is also possible and more available to use remote sensing and airborne technologies in the detection of, for example invasive weed species (Lam et al. 2021), as well as to map and identify the weeds in arable crops (Fernandez-Quintanilla et al. 2018). Collecting and analysing data enables accurate and timely decisions for the application of direct control measures. Making decisions based on the type, level, and frequency of the control measures can be viewed in broad and narrow sense. Putting in effect each step of the IPM program and its adjustment for a particular crop is a decision-making process. One of the key elements of IPM is to provide enough data to select the appropriate measure and estimate a moment of its application. Using the paradigm of aiming not to eradicate pest species but establish the level of its presence above which probability of the crop damage will significantly increase, resulting in the reduced and effective use of pesticides. The application of this principle was satisfied in the greenhouse (Brun et al. 2012), and field crop production. Based on this, the decision support systems were developed to provide necessary and timely information on pest status (Mahaman et al. 2003; Buckmann et al. 2018; Rossi et al. 2019; Martin et al. 2020). They can be supported by manually collected data about the pest status or more present automatic and remote gathering of information (Petric et al. 2019). It enables producers to target a specific pest and to have a timely recommendation on when to apply control measures (Kinjo and Ebina 2018).

14.5.3 Direct Control Measures in IPM

Non-chemical methods include practices for targeted pest control. The approaches and methods are various and differ in efficacy and include mating disruption, pest mass trapping, steaming, flooding, biofumigation, soil solarization, use of natural enemies, interrow precision weeding, tools for intrarow weeding, etc. Some of the methods like pest mating disruption were found to be highly effective resulting in significant pesticide use reduction (Harrari et al. 2007; Miller et al. 2019; Jallow et al. 2020). The application of biocontrol agents as living organisms and their derivatives that act as parasites, through antibiosis or competition for nutrients or space is one non-pesticide option. The use of inert dusts and plant-based preparations (botanicals) represents bio-rational options that can be alternative to chemically synthetisized pesticides (Gvozdenac et al. 2012, 2018a,b, 2021; Vukajlović et al. 2018). IPM includes the use of chemical pesticides when necessary. In the case of pesticide selection of choice of chemicals, should be made based on their efficacy on target organism and their impact on the environment and human health. Broad-spectrum pesticides are excluded from this approach due to the negative effect on non-target organisms. The recommendation is to use biorational pesticides that are pest specific (Kaur Gill and Garg 2014) and with the least side effect. These products include a wide range of microbial pesticides, plant extracts, and biochemical pesticides. However, use of organic certified biopesticides does not necessarily involve lower negative impact on beneficial organisms (Biondi et al. 2012). On the other hand, natural enemies in IPM can be used together with some selective insecticides for efficient pest management (Naranjo 2001).

The implementation of IPM strategies will preferably result in the reduction of pesticide use. Together with the selection of a pesticide with the lower impact on human health and the environment, a decrease in pesticide application is going to enhance this effect. The introduction of IPM approaches, such as pest monitoring and decision support systems based on the pest thresholds, should reduce the pesticide application frequency. This can be achieved with introduction of IPM approaches such as pest monitoring and decision support systems based on the pest thresholds. Another possibility to achieve the pesticide reduction is a dose reduction strategy that also needs to be based on data obtained by monitoring. However, the pesticide dose reduction can potentially increase the risk of resistance developing in targeted pests, to a particular compound. The combination of a biocontrol agent with fungicide could lower the dose of a single fungicide (below the maximum residue limit) or lower the number of fungicide applications while not affecting control efficiency. The reduction is also achievable by the application of compounds that stimulated the activation of *in planta* defence mechanisms.

A widely documented and frequently observed effect of pesticide use is a pest adaptation to particular active ingredient. The pesticide resistance is a significant challenge in pest control. Frequent, extensive, and exclusive use of a single management solution for the control of a particular pest, plant disease, or weed resulted in a number of cases of resistance development (Garcia-Rojas et al. 2017). Decreased susceptibility to pesticides is documented in a number of cases making the products previously effective, and obsolete for pest control. The adaptation appears to mostly be a result of point mutations in target-site genes and selection of standing variation (Hawkins et al. 2019). The resistance is a result of pest's natural response to environmental stress and it can be slowed down (Ma 1998). Mitigation of the resistance is possible in the application of the control measures combined and in the avoidance of a single-tactic strategy. The effective implementation of the previously described prevention practices and decision-based control known as the IPM are the key elements of the anti-resistance strategy.

14.5.4 Evaluation

The practices chosen and implemented as a part of the IPM system on farm are advisable to be assessed comprehensively and over multiple years. The evaluation based only on a yield or short-term targets such as a field cleared of pests can provide only partial and insufficient information. The focusing only on a few indicators can hide the effects of the applied IPM. Therefore, standards and evaluation on several levels such as social and economic impact as well as an impact on human health and the environment is recommended for implementation. In this way, a long-term effect can reveal a number of improvements and positive trends as the number of IPM practices represent are the foundation for the long-term stability of the cropping system.

14.6 Evaluation and Adaptation of IPM Practices Under the Conditions of Changing Climate and 'Climate-Smart Pest Management'

Within the context of sustainable agriculture, the IPM puts emphasis on **preventive** (indirect) plant protection measures, which must be fully utilized before the **control** (direct) measures are applied. The concept of IPM implies that the direct pest control measures should be the last option when economically high losses cannot be prevented by the indirect once (Boller et al. 2004).

The IPM strategy is already widely implemented. However, it is mainly designed for the specific crop, by farmers and researchers, enabling minimal negative impacts on the environment and at the same time maximizing yields and profit (FAO 2021). Under the conditions of changing climate, especially the global warming, the relative efficacy of such context-specific pest control practices (chemical pesticides, biopesticides, natural enemies, and host–plant resistance) is likely to change. Therefore, there is an urgent need to: (i) assess the efficacy of each IPM practice under different environmental conditions; (ii) adapt the existing practices to be efficient in changing climate; (iii) develop new IPM practices and assess the risks of their implementation, and (iv) define the impact of newly developed practices on the environment, productivity, and profitability.

Huge scientific efforts have been undertaken to address the issue of pest management under changed environmental conditions, and the need to revise the existing IPM practices and programs, aiming to develop agroecosystems resilient enough to cope with climate change, at local level (Chakraborty and Newton 2011; Lamichhane et al. 2015; Barzman et al. 2015; Macfadyen et al. 2018). This often refers to a specific pest species or geographic region. However, there is not a single strategy that brings all of the existing IPM practices and recommendations together with an emphasis on their adaptation to climate changes. In order for adapted pest management practices to be successful and applicable on a large scale, it will require both political and public support (Heeb et al. 2019).

14.6.1 Impact of Climate Change on Ecological Processes Important for Successful IPM

Pest population dynamics: The knowledge of insects' thermal biology and the effect of rising temperatures on the pest population growth and fluctuations represent additional and valuable information facilitating the adaptation of IPM practices and programs to cope with climate changes. This research area will continue to provide forecasts of pest outbreaks under global warming conditions in the future (Denlinger and Lee 2010; Bale and Hayward 2010; Colinet et al. 2015). As explained in Section 14.3.1, the exposure of insects to both rising, high and extreme temperatures will continue to influence pests' biology, survival, feeding, growth rates, reproduction, and population dynamics (Ma et al. 2004; Zani et al. 2005; Kingsolver et al. 2011; Zhang et al. 2015b).

Pest–host plant interactions: The nutritional quality of agricultural products will most probably change as a result of higher temperatures, elevated atmospheric CO_2 concentrations, and variable moisture and precipitation patterns, resulting in the changed palatability of host plant for the insect pest. For example, the effects of elevated CO_2 levels and reduced content of soil water, modified the major hormone signalling pathways of jasmine acid (JA), salicylic acid (SA), and ethylene before and afterwards the damage caused by the Japanese beetle (*Papilla japonica*), as reported by Casteel et al. (2012).

Pest–pathogen interactions: The impact of climate change on plant pathogens transferred by the insect pests has been relatively understudied, especially their role in reducing global food security (Chakraborty and Newton 2011; Juroszek and von Tiedemann 2013). It is expected that, as the geographical ranges of pests move, the pathogens associated with these pests will also move into new areas, causing new diseases to different plant species and varieties (Coakley et al. 1999). Since 1960, hundreds of pests and pathogens spread their distribution ranges on average $2.7 \pm 0.8\,km\,year^{-1}$, which is similar to the spreading of wild insect species, regardless of the changes in land use, crop varieties, or cultivation practices applied (Bebber et al. 2013). Global harvest losses as a result of plant diseases, including those transferred by insects, are estimated to amount from 10 to 16% (Strange and

Scott 2005; Oerke 2006). However, with the changes in climate conditions, economic losses are likely to increase due to increased insect activity and spreading.

Intraguild predation: Polis et al. (1989) explained that intraguild predation occurred when a parasitoid or a predator attacked the host pest species and continued to feed on other parasitoid/predatory species within the biocontrol system. Introduced biological control agents can have serious consequences on intraguild predation and negative impacts on the native species found in agro-ecosystems (Rosenheim et al. 1995). For example, the introduction and the establishment of a primary aphidfeeder, the Asian ladybug (*Harmonia axyridis*) in North and South America and Europe reduced the populations of native coccinellids, and based on laboratory assays, this species could feed on other predatory species, affecting biocenotic balance (Pell et al. 2008). The presented results imply that climate changes and the expansion of geographical ranges of introduced biological control agents into new habitats, may intensify competition and enhance intraguild predation of native species.

Tritrophic interactions: Parasitoids are influenced by both host plant and insect herbivore in complex interactions (Harvey et al. 2003), and climate change has and will continue to have a significant impact on these interactions (Facey et al. 2014) since it affects each component of the complex. Stireman III et al. (2005) assessed caterpillar–parasitoid interactions across 15 databases, from central Brazil to Canada, and revealed that with the increase in the variability of climate factors, the ability of parasitoids to track the host population decreased. The authors predicted more frequent herbivore outbreaks in managed systems, as natural enemy–herbivore interaction was disrupted due to changes in temperature and precipitation. There is an evident shortage of studies assessing the changes in trophic interactions with the effects of interrelated climate factors. As suggested by Jamieson et al. (2012) and Facey et al. (2014), although a species' response to altered environmental conditions can be highly distinctive, the response of tritrophic interactions is still highly unpredictable.

Pest–symbiont interactions: The symbiont assemblage in an insect is crucial for its digestion and the studies imply that the interaction of an insect and a symbiont is also subjected to climate changes. Shan et al. (2014) assessed the responses of three symbionts in the whitefly *Bemisia tabaci* complex, to high and low temperatures: two *Candidatus* species and a body cavity resident *Rickettsia* spp. The continuous exposure of whiteflies to 40 °C caused the reduction in infection rates of *Candidatus hamiltonella defensa*, while the infection rate of *Candidatus portiera aleyrodidarum* was significantly reduced after 3–5 days of exposure. On the other hand, *Rickettsia* infection rates were not significantly impacted by the increased temperature (Shan et al. 2014). Prado et al. (2010) showed that two stink bug species *Acrosternum hilare* and *Murgantia histionica* lost their gut-associated symbionts when reared constantly at 30 °C (in comparison to those at 25 °C), which resulted in lower survival and reproduction.

Based on all the above mentioned, there is an increasing need to understand and assess more thoroughly fundamental crop species, biology ecology, physiology, and behaviour of insect pests and their natural enemies, potential modifications of interactions between species and among species under the conditions of the changing climate. This will allow the **adaptation of IPM** programs and more focused IPM practices efficient enough to manage plant pests in different climate change scenarios.

14.6.2 Adaptation of Cultivation and IPM Practices

The issue of the pest management efficacy in the changing environment and the need to reassess the existing agricultural and IPM practices in order to enhance agroecosystems' resilience to climate change was comprehensively elaborated by Barzman et al. (2015). The authors emphasize that the adaptation of cultivation and IPM practices is needed to mitigate the impacts and losses caused by plant pests in the changing climate.

Adaptation of cultivation practices includes: (i) growing climate-resilient and pest-tolerant varieties; (ii) altering planting/sowing dates to minimize the exposure to potential pest outbreaks and (iii) implementing some of the agroecological techniques which increase biodiversity and the diversity of natural enemies, such as field margins or flower strips (Andrew and Hill 2017).

Adaptation of IPM practices includes several options such as (i) development of new pesticidal formulations, as biopesticides and chemical pesticides are susceptible to the environmental changes, and the increase in temperature and decrease in relative humidity may lower their efficacy (Vadez et al. 2012; Wenda-Piesik et al. 2015; Nihal 2020); (ii) the use of biotechnical methods (insect pheromones) for biological control, mass-trapping, mating disruption, 'push–pull' strategies, and monitoring (Heuskin et al. 2011); (iii) the introduction of pest-tolerant varieties (Gomez-Zavaglia et al. 2020); (iv) changing the intervention thresholds for a specific pest; and (v) conduct continuous monitoring and surveillance of present and emerging pest occurrence. The IPM programs mostly form the decisions based on the established thresholds for economic yield losses, but they are not always relevant, sufficient, or possible to define (Barzman et al. 2015). Trumble and Butler (2009) reported that due to the faster insect development at higher temperatures, the population developed faster and crop damage occurred earlier than expected. Therefore, the treatment thresholds based on the number of insects per plant must be adapted and lowered in order to prevent yield losses.

The adaptation of IPM practices and programs to climate change should be based on the comprehension of how the environmental conditions influence the development of crops and associated pests, and their interactions with the surrounding environment. Lamichhane et al. (2015) presented a good example of how the climate conditions such as drought-affected recommendations for crop protection in general. A drought-stressed crop is more susceptible to additional stress caused by herbivorous insects, which requires a lower economic threshold for the specific pest.

Monitoring pest distribution and abundance is a key component of the IPM. In this regard, one of the crucial preconditions for determining whether climate change is altering pests' population dynamics is to establish a long-term monitoring system (Yamamura et al. 2006; Andrew and Hill 2017). The long-term monitoring, especially in the climate change-sensitive regions, provides an insight into the biological responses of pests to climate change (Heeb et al. 2019). Invasive alien species are being spread constantly worldwide aided by the changes in weather (climate) conditions. Therefore, to prevent the IAS from establishing in new geographic regions and becoming pests of economic importance, the efficient monitoring and pest management systems are required (Hellmann et al. 2008; Heeb et al. 2019). Global information sharing system on IAS (insects, pathogens, and weeds) and current weather conditions and expected changes, need to be established among regions as support to adapting IPM to the conditions of changing climate. Hence, it is necessary to enhance networking and improve collaboration between scientific institutions, policy makers, and farmer's organizations across countries and geographical regions (Perrings et al. 2010).

Climate forecasting and model development: It is impossible to design a reliable adapted IPM program for the specific climate change scenarios on both national and/or global levels due to diverse alterations in all climate variables. Pest management programs must be adapted to cope with regional climate change and its uncertainties. Some of the available options include sensitivity analyses and results obtained using the projected climate change scenarios that include the sensitivity analyses for a given area. This option could provide practical aid for pest management experts when adapting IPM to new environmental conditions (Sutherst et al. 2011). Climate prediction models combined with the information on the environmental demands of a specific pest species can be useful in projecting the potential extent of pest risks globally (Raza et al. 2014). The potential distribution of insect pest species is mainly defined using ecological niche models: correlative and mechanistic. Correlative models use the values of the climate variables and information on pest occurrence to predict potentially suitable habitats for the particular species in the future (Pacifici et al. 2015). Correlative models (MaxEnt, Bioclim, Random Forest) are mainly used for predicting the effect of climate change on pests' biodiversity, projecting changes in the geographic distribution, estimating elimination rates, and creating a roadmap for biodiversity conservation (Gillson et al. 2013; Porfirio et al. 2014; Kumar et al. 2014). The final output of these models is usually a map presenting future suitable regions for a given species, in terms of climatic conditions, the total area that can be compared with the current geographic ranges enabling the estimation of the future risk of pests introduction and establishment (Evans et al. 2015). Mechanistic models use environmental variables of a given area combined with information on a specific pest's tolerance to environmental factors (Kumar et al. 2014). CLIMEX is an example of semi-mechanistic modelling software that uses physiological and behavioural parameters of the pest and environmental (climate) variables to predict suitable habitats or regions for the specific species (Kriticos et al. 2015). Mentioned models could be key points in developing new programs for pest prevention and control under the conditions of changing climate (Yonow et al. 2018).

14.6.3 'Climate-Smart Agriculture' (CSA) and 'Climate-Smart Pest Management' (CSPM)

'Climate-smart agriculture' (CSA), as promoted by FAO (2010), is a well-documented approach aiming to enable the transformation of the entire agricultural systems in order to ensure food security in the changing climate. However, the issue of pest management within this approach receives relatively little attention and is mainly based on the recommendations from the IPM programs. In response, a new approach of CSPM has been designed, to address this need and provide more detailed and focused pest management guidance within the CSA. The CSPM is expected to become a key component of the CSA.

The CSPM encompasses short-term or long-term adaptive measures to climate change (Juroszek and von Tiedemann 2011; Olesen and Bindi 2002). Short-term adaptations include the optimization of plant production with minor system changes (climate-resilient varieties, or the use of biological or chemical pesticides to control certain pests), while long-term adaptations include major changes such as introducing crop rotations, intercropping, or changing to a new crop (Olesen and Bindi 2002).

The most important actions of the CSPM are:

- Prediction and recognition of the future pest outbreaks
- Suppression of pest outbreaks and pathogen transmissions
- Enhancement of health and pest resilience of farms and landscapes
- Strengthening a climate-responsive extension system

The CSPM is a new interdisciplinary approach that involves practices that need to be adapted in primary crop production (i), by relying on extension services (ii), research institutions (iii), and policy makers (iv). It strengthens the role of research, extension, and the policy makers (public and private sectors) in pest surveillance, detection, forecasting, and control (Figure 14.5). By implementing CSPM, the crop producers, research institutions, extension services, and policy makers act in coordination enabling more effective management of the threat posed by new and existing crop pests. In this way, the resilience of the agricultural production to climate change is enhance and overall, food security is provided (Heeb et al. 2019).

Adaption of CSPM requires the actions of each of the above-mentioned services (Figure 14.5).

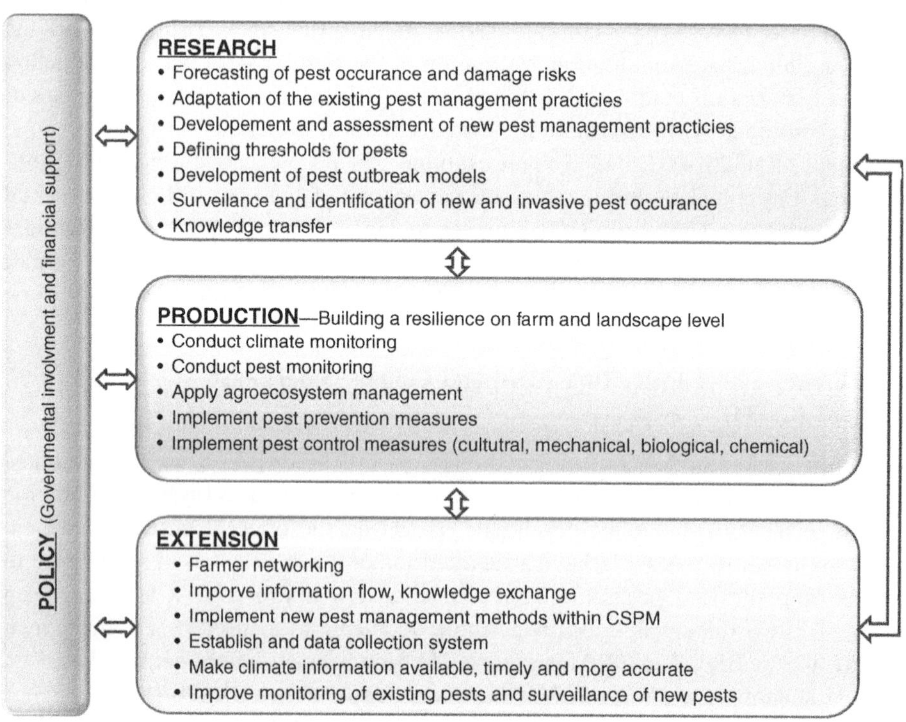

Figure 14.5 Climate smart pest management structure.

14.6.3.1 Crop Production (Farm and Landscape Level)

One of the most significant barriers in the adoption of CSPM practices, on both local (farm) and global levels, are insufficient knowledge about climate-smart practices, the lack of resources, and potentially high costs (Macfadyen et al. 2018). Adoption of CSPM practices is also depending on the farmers will and capability to finance new technologies. Also, it is strongly dependant on implementation costs, risks associated with the implementation and efficacy of new practices, as well as on a farm size, farmers' recognition and knowledge about climate changes, and farmers' incomes. Market accessibility and availability of information about climate conditions, as well as accessibility of extension services to farmers also play a significant role in adaptation of CSPM (FAO 2013a; Belay et al. 2017; Boansi et al. 2019). The crucial point in overcoming some of these barriers and adaptation of the CSPM practices is the increasing farmers' consciousness and the knowledge about climate change effects and benefits of the CSPM. Promotion of the public/private financing mechanisms (credits) should also be a priority as this can motivate the farmers to adopt new practices.

The following actions should be performed 'on farm' enabling the CSPM adoption:

a) conduct continuous monitoring of the climate variables and pest occurrence, aiming to predict and respond rapidly to new, emerging, and/or existing pests;
b) implement pest prevention measures to disable the establishment and development of pest populations;
c) apply agroecosystem management; and implement mechanical, cultural, biological, and chemical (last option) measures to control pests.

14.6.3.2 Extension Service

The role of the extension service is one of the most important in supporting and aiding farmers to adopt the CSPM practices and programs under the conditions of changing climate. The extension service and its staff should be well-informed, effective, responsive, and accessible to farmers. A competent extension service should be digitally developed, and equipped with early warning systems based on climatic data analysis and modelling. The service should issue monthly bulletins on expected climate variabilities and crop management advices and share this information also through different communication channels (phone hotlines, radios, TV shows, etc.). However, in many countries, the extension systems lack these characteristics which limit the potential of farmers to adapt their agricultural production to the changing climate. Some of the main reasons are the lack of climate literacy of the extension staff, limited operational funds, weak linkages with the research institutions and policy makers, the absence of access to advisory services due to farmers' restricted mobility, etc. (Heeb et al. 2019). To facilitate the CSPM adoption for farmers, According to Heeb et al. (2019), the extension services should be responsible for:

a) Farmer networking. This action should improve farmers' and farmer organizations' networks and linkage to technology providers, enhance knowledge sharing and enable access to the information about CSPM practices, facilitating their implementation.
b) Information sharing and advising about the pest management options. This activity should ensure a constant and frequent two-way information flow (the transfer of

pest-related information and forecasts to farmers and collection of farmers' feedback for validating and adaptation of recommended pest control practices); to establish a connection with research institutions and in coordination to develop CSPM technologies based on the current needs and local contexts, and to establish interdisciplinary knowledge exchange and knowledge management systems.

c) Pest diagnostics and surveillance. One of the responsibilities of the extension service is to develop tools and improve the monitoring of the existing plant pests and establish and/or continuously perform the surveillance of new and invasive pest species; to enable the quick detection of emerging pests; to establish efficient data collection systems for pest surveillance, connected with national/international diagnostic support services; to assess feasibility and efficacy of pest management programs; to enhance the accessibility and reliability of the information on climate conditions (past, monitored and predicted) and, to help farmers to create information-based decisions, successfully manage pest risks, exploit favourable climate conditions and adapt to climate change.

14.6.3.3 Research

In an effective and well-functioning plant health system, the involvement of research institutes is crucial for the development and implementation of the CSPM practices and programs. However, in many countries, this vital service is not functioning adequately, due to reduced funding and lack of appropriate specialist expertise (Lamichhane et al. 2015; Masters and Norgrove 2010), inadequate education and training in the field of pest management to generate skilled specialists (Lamichhane et al. 2015), lack of research devoted to IPM and its adaptation to climate change, and weak regional research collaboration and coordination (Machenko et al. 2017). Within the CSPM strategy, the research institutions should be responsible for:

a) Forecasting pest risks. Research institutions should perform basic research related to determining the impacts of climate change on pests' distribution, occurrence, spreading, development, phenology, behaviour, interactions with host and natural enemies, etc.
b) Developing models of pest outbreak potential
c) Development of early warning systems that will enable timely prevention and rapid response to new and existing pests, and limit regional and cross-border spreading
d) Development/improvement of monitoring and surveillance systems for the existing pest species
e) Identification and surveillance of emerging pest species
f) Implementation of effective data collection systems for pest surveillance connected with national/international diagnostic support services (Heeb et al. 2019).

14.6.3.4 Policy Makers

The implementation of CSPM programs depends on government policies and regulatory instruments. They must create clear guidance to all services (research and extension) and enable the adequate planning of measures and steps to be undertaken in adaptation of the primary production to climate change, as well as efficient targeting of different funds. However, in many countries, the existing institutions are not worn and/or informed about the constraints in the adaptation process at the local level (Ampaire et al. 2017). In addition, there is a shortage of financial resources and crop protection expertise and experts, which are required to develop efficient adaptation plans or policies (Chakraborty and Newton 2011; Liu et al. 2016). Rising awareness of the impacts of climate changes on agricultural production and pest management, as well as the efficient knowledge transfer at the government level, will bolster the capacity for the development of regulatory instruments, well-informed and harmonized policies, addressing both adaptation and mitigation actions and creating the conditions for the implementation of the CSPM. Creating nationally assigned funds that will enable the development of adaptation and mitigation plans is supportive of these policies.

By acting in coordination, the research, extension, and policy services will enable the CSPM practices to be implemented in the primary production thus increasing its resilience to climate changes. The CSPM will issue the information and guidelines on the application of ***prevention practices*** (crop diversification, establishment of natural habitats, etc.) that will help farmers to enhance the health and sustainability of their productions, reduce reliance on pesticides, and susceptibility to pests. Also, this approach will enable farmers to make fast decisions regarding the most appropriate **pest control practices,** in situations when the pest populations reach intervention thresholds (Heeb et al. 2019).

All sustainable agricultural approaches, including CSA, IPM, and finally CSPM are highly context-dependent (Beuchelt and Badstue 2013; FAO 2013a; Neufeldt et al. 2013) and the recommendations will depend on the climatic, agricultural, ecological, socio-economic factors as well as on the political environment, on farm, regional and national levels.

According to Heeb et al. (2019), CSPM has the following benefits:

Food security: CSPM leads to cost-effective management of the existing crop pests, resulting in reduced losses and increased food security and farmers' profitability.

Adaptation: At national and global levels, the CSPM bolsters the important role of the extension services, research institutions, and the public and private sectors in pest forecasting, surveillance, detection, and control, as these are vital services enabling climate resilience.

Mitigation: The CSPM contributes to increased efficacy of agricultural production and promotes rational use of agricultural inputs, reduction of yield losses, and GHG emissions.

Win/win situations: The CSPM involves adaptation and mitigation practices at the same time whenever possible. For example, the prevention practices, such as mulching, minimum tillage, and establishing field margins, increase organic carbon input in the soil, but also enhance the resilience to certain pests. Along with the lower pest-induced plant losses, a triple positive effect is achieved, by adapting to emerging pest threats, mitigating GHG emissions, and enhancing food security.

Delivery of sustainable development goals (SDGs): CSPM directly contributes to a number of SDGs such as SDG 1 – No Poverty, SDG 2-Zero Hunger (through its direct impact on crop production and income), SDG 12 – Responsible Consumption and Production; SDG 13 – Climate Action (through, among others, reducing GHG emission intensity per unit of food produced), and SDG 15 – Life on Land (through conserving biodiversity and maintaining ecosystem services).

14.6.3.5 Costs of Implementing CSPM Programs

Costs of implementing CSPM are estimated based on the actions needed in a specific crop production, involvement of extension service, research, and public/private actors, and should be considered in relation to short- and long-term financial benefits. According to Macfadyen et al. (2018), for example, if the introduction of a pest-tolerant variety is involved, only minimal investment and knowledge are required. However, the introduction of a new crop species, due to high risks of new and invasive pests, may require higher investments (new plant material, harvesting equipment, implementing of new rotation practices and finding new buyers and markets for the product, etc.)

At the farm level, the adoption of the CSPM can influence the input costs and yields. No cost-benefit analyses yet exist for the CSPM, but according to cost-benefit analyses of the IPM, which partially overlaps with CSPM at the plant production level, show that the IPM increases yields up to 41%, and reduces the pesticide use by 31% on average (Norton and Mullen 1994; Pretty and Bharucha 2015). However, assessing the financial benefits of IPM is complicated, as many IPM programs involve actions beyond the pest management. CSPM and IPM benefit by reducing the external costs (Pimentel 2005). Funding for the CSPM requires joint efforts and coordination between public/private and national/international actors because many pest problems need to be addressed using a cross-border approach. As mentioned, in a number of developing countries, there is a trend of reducing public financing that would facilitate adoption of the CSPM. Therefore, alternative funding models are needed, such as private sector funding. In Europe, the IPM is well established and functional, but it requires consumers who are willing to pay a higher price and a public sector that is willing and able to compensate for positive or negative externalities (Lefebvre et al. 2015).

Relation to other climate-smart agriculture practices: The CSPM is closely related to the existing CSA practices and programs at both farm and national levels. Interactions and connections of CSPM and currently undertaken CSA practices are multiple because on the farm level, the CSPM answers to the risk caused by biotic stressors – pests, but also abiotic factors such as soil management and crop nutrition. Certain CSA practices such as nutrient management, integrated soil fertility management, agriculture conservation, breeding for climate-resilient varieties, or crop diversification are crucial for the success of CSPM (World Bank, FAO, IFAD 2009; FAO 2011).

14.7 Conclusions

Agriculture is heavily dependent on the climate hence the changes in climate conditions pose a threat to primary production and food security. As a consequence of altered climatic conditions, aside from the yield and quality reduction, new pests have emerged and will continue in the future. It is expected: that the geographical ranges of pests move northward, the number of pest outbreaks increases, survival rate rises and more generations develop, new pest-related pathogens occur and weeds expand their distribution limits into regions with suitable conditions. Reducing the impact of pests is of high importance for global food security under the conditions of changing climate. Also, climate change was proven to influence the efficacy of pest control measures (biological control, synthetic pesticides application, etc.), indicating a necessity to develop adaptation and mitigation strategies in the form of modified IPM tactics, climate and pest monitoring, and the use of different modelling tools, that will enable climate-resilient pest management.

One of the IPM adaptation strategies is a CSPM concept. It is a multi-actor concept that requires crop production, well-informed extension services and researchers relying on information and communication technology, and policy makers to act in coordination with a unique goal of creating the efficient and resilient food production and plant health systems that are able to cope with the conditions of changing climate. The main pillars of CSPM are climate-informed forecasting and monitoring system, climate-informed advisory services and tools, and implementation of a wider range of new IPM practices (host plant resistance/tolerance, biological control, habitat manipulation, and climate-smart use of semiochemicals and biopesticides) and early-action against future new or pre-established pest outbreaks, especially the species that have high outbreak potential and the tendency to spread and cause higher damages.

Within CSPM innovative tools must be implemented. For example, tools of precision agriculture, such as non-invasive digital technologies can provide essential information on how genetics, environmental and soil conditions, and crop management can guide selection towards productive plants suitable for their environment under climate change. An important application of innovative techniques in phenotyping is the proximal monitoring of plant variability, which includes the use of a wide range of sensing technologies. Remote sensing techniques could rapidly provide a description of complex plant traits such as growth, development, tolerance, plant architecture, plant physiology, ecology, and yield. In the future, it is expected that the use of remote sensing techniques will be an irreplaceable tool in the modern evaluation of plant traits.

14.8 Summary

Climate change, manifested in an increasingly warmer climate as a direct consequence of higher concentrations of GHG, has been identified as a changing force in crop production. This chapter provides a review of the expected impact of climate change on crop production with respect to limitations of currently used models and an up-to-date summary of the type of its influence on plant growth and development. The authors have focused also on the indirect effect on crop production through the predicted impact of climate change on diverse populations of insect pests, weeds, and pathogens. Based on current knowledge we examined the possibility to anticipate the importance of pests under the perceived magnitude of climate change. In addition, the review provides insight into the application of IPM, an approach aimed to tackle the constraints of crop production as a consequence of stress conditions. The major focus was attributed to modified IPM that takes into consideration climate change and adaptation of IPM methods and is referred as Climate-smart Pest Management (CSPM).

Acknowledgements

The work was supported by the Ministry of Education, Science and Technological Development of the Republic of Serbia, grant number 451-03-9/2021-14/ 200032, the COST action CA18111 'Genome editing in plants – a technology with transformative potential', the COST action CA1912 'EPIgenetic mechanisms of Crop Adaptation To Climate cHange', IAEA project RER5024 'Enhancing Productivity and Resilience to Climate Change of Major Food Crops in Europe and Central Asia' and project No. 50148 'Development of new biopesticides for seed protection – ECOPEST' funded by Innovation Fund of Republic of Serbia and project 'Redesigning the exploitation of small grains genetic resources towards increased sustainability of grain-value chain and improved farmers' livelihoods in Serbia and Bulgaria—GRAINEFIT' (2020–2023), funded by The Benefit-Sharing Fund of the International Treaty on Plant Genetic Resources for Food and Agriculture (project number PR-166-Serbia). The chapter is supported by the Science Fund of the Republic of Serbia, through IDEAS project "Creating climate smart sunflower for future challenges" (SMARTSUN) grant number 7732457, as well as activities of Center of Excellence for Innovations in Breeding of Climate-Resilient Crops - Climate Crops, Institute of Field and Vegetable Crops, Novi Sad, Serbia.

References

Abrol, D.P. (2009). Plant-pollinator interactions in the context of climate change - an endangered mutualism. *Journal of Palynology* 45: 1–25.

Abuley, I.K., Nielsen, B.J., and Hansen, H.H. (2019). The influence of timing the application of nitrogen fertilizer on early blight (*Alternaria solani*). *Pest Management Science* 75: 1150–1158.

Agurla, S., Gahir, S., Munemasa, S. et al. (2018). Mechanism of stomatal closure in plants exposed to droughtand cold stress. In: *Survival Strategies in Extreme Cold and Desiccation*'(eds. M. Iwaya-Inoue, M. Sakurai, M. Uemura), Part of *Advances in Experimental Medicine and Biology* 1081:215–232. Springer, Singapore.

Ahmed, S.M. (2021). Modeling crop yields amidst climate change in the Nile basin (2040–2079). *Modeling Earth Systems and Environment* 159: 385–405.

Altermatt, F. (2010). Climatic warming increases voltinism in European butterflies and moths. *Proceeding. Biological Sciences* 277 (1685): 1281–1287.

Alyokhin, A., Naul, B., and Brown, B. (2019). Soil conservation practices for insect pest management in highly disturbed agroecosystems – a review. *Entomologia Experimentalis et Applicata* 168: 7–27.

Ampaire, E.L., Jassogne, L., Providence, H. et al. (2017). Institutional challenges to climate change adaptation: a case study on policy action gaps in Uganda. *Environmental Science & Policy* 75: 81–90.

Anđelković, V., Cvejić, S., Jocić, S. et al. (2020). Use of plant genetic resources in crop improvement – example of Serbia. *Genetic Resources and Crop Evolution* 67: 1935–1948.

Andrew, N.R. and Hill, S.J. (2017). Effect of climate change on insect pest management. In: *Environmental Pest Management: Challenges for Agronomists, Ecologists, Economists and Policymakers* (ed. M. Coll and E. Wajnberg). Hoboken, NJ: Wiley. Strane.

Anonymous (2013). Climate change consortium for specialty crops: impacts and strategies for resilience. California Department of Food and Agriculture, pp. 1–76.

Arora, N.K. (2019). Impact of climate change on agriculture production and its sustainable solutions. *Environmental Sustainability* 2: 95–96.

Atienza, J., Tabernero, M.T., Alvarez-Benedi, J., and Sanz, M. (2001). Volatilisation of triallate as affected by soil texture and air velocity. *Chemosphere* 42: 257–261.

Auon, N., Tauleigne, L., Lonjon, F. et al. (2017). Quantitative disease resistance under elevated temperature: genetic basis of new resistance mechanisms to *Ralstonia solanacearum*. *Frontiers in Plant Science* 8: 1387.

Awmack, C.S., Woodcock, C.M., and Harrington, R. (1997). Climate change may increase vulnerability of aphids to natural enemies. *Ecological Entomology* 22: 366–368.

Ayres, M.P. and Lombardero, M.J. (2000). Assessing the consequences of global change for forest disturbance from herbivores and pathogens. *Science of the Total Environment* 262: 263–286.

Bacastow, R.B., Keeling, C.T., and Whorf, T.P. (1985). Seasonal amplitude increase in atmospheric CO_2 concentration at Mauna Loa, Hawaii, 1959–1982. *Journal of Geophysical Research* 90: 529–540.

Baes, C.F., Goeller, H.E., Olson, J.S., and Rotty, R.M. (1977). Carbon dioxide and climate: the uncontrolled experiment. *American Scientist* 65: 310–320.

Bajwa, A.A., Chauhan, B.S., Farooq, M. et al. (2016). What do we really know about alien plant invasion? A review of the invasion mechanism of one of the world's worst weeds. *Planta* 244: 39–57.

Bale, J.S. and Hayward, S.A.L. (2010). Insect overwintering in a changing climate. *Journal of Experimental Biology* 213: 980–994.

Bale, J.S., Masters, G.J., Hodkinson, I.D. et al. (2002). Herbivory in global climate change research: direct effects of rising temperature on insect herbivores. *Global Change Biology* 8 (1): 1–16.

Barzman, M., Barberi, P., Nicholas, A. et al. (2015). Eight principles of integrated pest management. *Agronomy for Sustainable Development* 35: 1199–1215.

Bebber, D.P., Ramotowski, M.A.T., and Gurr, S.J. (2013). Crop pests and pathogens move polewards in a warming world. *Nature Climate Change* 3: 985–988.

Belay, A., Recha, J.W., Woldeamanuel, T., and Morton, J.F. (2017). Smallholder farmers' adaptation to climate change and determinants of their adaptation decisions in the Central Rift Valley of Ethiopia. *Agriculture & Food Security* 6 (24): 1–13.

Belgeri, A. and Adkins, S.W. (2015). Allelopathic potential of invasive parthenium weed (*Parthenium hysterophorus* L.) seedlings on grassland species in Australia. *Allelopathy Journal* 36: 1–14.

Beuchelt, T.D. and Badstue, L. (2013). Gender, nutrition-and climate-smart food production: opportunities and trade-offs. *Food Security* 5: 709–721.

Bezemer, T.M., Jones, T.H., and Knight, K.J. (1998). Long-term effects of elevated CO_2 and temperature on populations of the peach potato aphid *Myzus persicae* and its parasitoid *Aphidius matricariae*. *Oecologia* 116: 128–135.

Bicout, D. and Sache, I. (2003). Dispersal of spores following a persistent random walk. *Physical Review E* 67: 031913.

Bijoor, N.S., Czimczik, C.I., Pataki, D.E., and Billings, S.A. (2008). Effects of temperature and fertilization on nitrogen cycling and community composition of an urban lawn. *Global Change Biology* 14: 2119–2131.

Biondi, A., Desneux, N., Siscaro, G., and Zappala, L. (2012). Using organic-certified rather than synthetic pesticides may not be safer for biological control agents: selectivity and side effects of 14 pesticides on the predator *Orius laevigatus*. *Chemosphere* 87: 803–812.

Blum (2016). Osmotic adjustment is a prime drought stress adaptive engine in support of plant production. *Plant, Cell & Environment* 40 (1): 4–10.

Blumenthal, D., Chimner, R.A., Welker, J.M., and Morgan, J.A. (2008). Increased snow facilitates plant invasion in mixedgrass prairie. *New Phytologist* 179: 440–448.

Boansi, D., Tambo, J.A., and Müller, M. (2019). Intra-seasonal risk of agriculturally-relevant weather extremes in West African Sudan Savanna. *Theoretical and Applied Climatology* 135: 355–373.

Boller, E.F., Avilla, J., Joerg, E. et al. (2004). Integrated production: principles and technical guidelines. *IOBC WPRS Bulletin* 27 (2): 1–54.

Booker, F.L., Muntifering, R., McGrath, M.T. et al. (2009). The ozone component of global change: potential effects on agricultural and horticultural plant yield, product quality and interactions with invasive species. *Journal of Integrative Plant Biology* 51: 337–351.

Bouras, E., Jarlan, L., Khabba, S. et al. (2019). Assessing the impact of global climate changes on irrigated wheat yields and water requirements in semi-arid environment of Morocco. *Scientific Reports* 9 (1): 19142.

Bradley, B.A. (2009). Regional analysis of the impacts of climate change on cheatgrass invasion shows potential risk and opportunity. *Global Change Biology* 15: 196–208.

Bradley, B.A., Oppenheimer, M., and Wilcove, D.S. (2009). Climate change and plant invasions: restoration opportunities ahead? *Global Change Biology* 15: 1511–1521.

Bradshaw, W.E. and Holzapfel, C.M. (2010). Light, time, and the physiology of biotic response to rapid climate change in animals. *Annual Review of Plant Biology* 72: 147–166.

Brun, R., Metay, C., Wdziekonski, C. et al. (2012). Use of decision rules to manage IPM on rose and tomato crops: an example with Whitefly. Proceedings of XXVIII International Horticultural Congres on Science and Horticulture for People (IHC2010): International Symposium on Greenhouse 2010 and Soilless Cultivation, Lisbon, pp. 195–202.

Bryant, R., McGrann, G., Mitchell, A. et al. (2014). A change in temperature modulates defence to yellow (stripe) rust in wheat line UC1041 independently of resistance gene Yr36. *BMC Plant Biology* 14: https://doi.org/10.1186/1471-2229-14-10.

Buckmann, H., Bojer, O., Montull, J. et al. (2018). DSS-IWM: an improved European Decision Support System for Integrated Weed Management. Proceedings of the 28th German Conference on Weed Biology and Weed Control, Braunschweig, pp. 205–208.

Burdon, J. and Zhan, J. (2020). Climate change and disease in plant communities. *PLoS Biology* 18: e3000949.

Busch, A.K., Douglas, M.R., Malcolm, G.M. et al. (2020). A high-diversity/IPM cropping system fosters beneficial arthropod populations, limits invertebrate pests, and produces competitive maize yields. *Agriculture, Ecosystems & Environment* 292: 106812.

Camaille, M., Fabre, N., Clement, C., and Barka, E.A. (2021). Advances in wheat physiology in response to drought and the role of plant growth promoting rhizobacteria to trigger drought tolerance. *Microorganisms* 9: 687.

Cannon, R.J. (1998). The implications of predicted climate change for insect pests in the UK, with emphasis on non-indigenous species. *Global Change Biology* 4: 785–796.

Casal, J.J. and Balasubramanian, S. (2019). Thermomorphogenesis. *Annual Review of Plant Biology* 70: 321–346.

Casteel, C.L., Niziolek, O.K., Leakey, A.D.B. et al. (2012). Effects of elevated CO_2 and soil water content on phytohormone transcript induction in *Glycine max* after *Popillia japonica* feeding. *Arthropod-Plant Interactions* 6: 439–447.

Cavaleri, M.A. and Sack, L. (2010). Comparative water use of native and invasive plants at multiple scales: a global meta-analysis. *Ecology* 91: 2705–2715.

Chakraborty, S. and Newton, A.C. (2011). Climate change, plant diseases and challenges in agriculture. *Crop Protection* 89: 139–151.

Challinor, A.J., Watson, J., Lobell, D.B. et al. (2014). A meta-analysis of crop yield under climate change and adaptation. *Nature Climate Change* 4: 287–291.

Chaloner, T.M., Gurr, S.J., and Bebber, D.P. (2021). Plant pathogen infection risk tracks global crop yields under climate change. *Nature Climate Change* 11: 710–715.

Chandrasena, N. (2009). How will weed management change under climate change? Some perspectives. *Journal of Crop and Weed* 5: 95–105.

Chen, L., Yuan, P., You, M. et al. (2019). Cover crops enhance natural enemies while help suppressing pests in a tea plantation. *Annals of the Entomological Society of America* 112: 348–355.

Cheng, C., Gao, X., Feng, B. et al. (2013). Plant immune response to pathogens differs with changing temperatures. *Nature Communications* 4: 2530.

Chuwah, C., van Noije, T., van Vuuren, D.P. et al. (2015). Global impacts of surface ozone changes on crop yields and land use. *Atmospheric Environment* 106: 11–23.

Clements, D.R. and DiTommaso, A. (2011). Climate change and weed adaptation: can evolution of invasive plants lead to greater range expansion than forecasted? *Weed Research* 51: 227–240.

Coakley, S.M., Scherm, H., and Chakraborty, S. (1999). Climate change and plant disease management. *Annual Review of Phytopathology* 37: 399–426.

Cock, M.J.W., Biesmeijer, J.C., Cannon, R.J.C. et al. (2013). The implications of climate change for positive contributions of invertebrates to world agriculture. *CAB International* 8: 1–48.

Colinet, H., Sinclair, B.J., Vernon, P., and Renault, D. (2015). Insects in fluctuating thermal environments. *Annual Review of Entomology* 60: 123–140.

Costes, E., Lauri, P., Simon, S., and Andrieu, B. (2013). Plant architecture, its diversity and manipulation in agronomic conditions, in relation with pest and pathogen attacks. *European Journal of Plant Pathology* 135: 455–470.

Cotrufo, M.F., Ineson, P., and Scott, A. (1998). Elevated CO_2 reduces the nitrogen concentration of plant tissues. *Global Change Biology* 4: 43–54.

Coulson, S.J., Hodkinson, I.D., Webb, N.R. et al. (2002). Aerial colonization of high Arctic islands by invertebrates: the diamondback moth *Plutella xylostella* (Lepidoptera: Yponomeutidae) as a potential indicator species. *Diversity and Distributions* 8: 327–334.

Coviella, C. and Trumble, J. (1999). Effects of elevated atmospheric carbon dioxide on insect-plant interactions. *Conservation Biology* 13: 700–712.

Crowder, D.W. and Harwood, J.D. (2014). Promoting biological control in a rapidly changing world. *Biological Control* 75: 1–7.

Cunchun, Y., Zhongwei, Z., and Wannakuwattewaduge, G.F. (2021). The effect of temperature on the hypersensitive response (HR) in the *Brassica napus–Leptosphaeria maculans* pathosystem. *Plants* 10: 843.

Cvejić, S., Radanović, A., Dedić, B. et al. (2020). Genetic and genomic tools in sunflower breeding for broomrape resistance. *Genes* 11: 152.

Dara, S.K. (2019). The new integrated pest management paradigm for the modern age. *Journal of Integrated Pest Management* 10: 12.

Debela, C. and Tola, M. (2018). Effect of elevated CO_2 and temperature on crop disease interactions under rapid climate change. *International Journal of Environmental Sciences & Natural Resources* 13: IJESNR.MS.ID.555851.

Dedić, B., Maširević, S., Miladinović, D. et al. (2012). Screening of sunflower inbred lines tolerance to Phoma black stem. *Rastenievdni Nauki* 6: 41–44.

Degener, J. (2015). Atmospheric CO_2 fertilization effects on biomass yields of 10 crops in northern Germany. *Frontiers in Environmental Science* 3: 48.

Delmas, R.J., Ascencio, J.M., and Legrand, M. (1980). Polar evidence that atmospheric CO_2 20,000 yr BP was 50% of present. *Nature* 284: 155–157.

DeLucia, E.H., Casteel, C.L., Nabity, P.D., and O'Neill, B.F. (2008). Insects take a bigger bite out of plants in a warmer, higher carbon dioxide world. *Proceedings of the National Academy of Sciences of the United States of America* 105: 1781–1782.

Denlinger, D.L. and Lee, R.E. (2010). *Low Temperature Biology of Insects*. Cambridge, UK: Cambridge University Press.

Dentener, F., Stevenson, D., Ellingsen, K. et al. (2006). The global atmospheric environment for the next generation. *Environmental Science and Technology* 40: 3586–3594.

Deryng, D., Conway, D., Ramankutty, N. et al. (2014). Global crop yield response to extreme heat stress under multiple climate change futures. *Environmental Research Letters* 9: 034011.

Desaint, H., Aoun, N., Deslandes, L. et al. (2021). Fight hard or die trying: when plants face pathogens under heat stress. *New Phytologist* 229: 712–734.

Deshar, R. and Koirala, M. (2019). Climate change and gender policy. In: *Global Climate Change and Environmental Policy: Agriculture Perspectives* (ed. V. Venkatramanan, S. Shah and R. Prasad), 411–422. Singapore: Springer Nature Pte Ltd.

Deutsch, C.A., Tewksbury, J.J., Tigchelaar, M. et al. (2018). Increase in crop losses to insect pests in a warming climate. *Science* 361: 916–919.

Diez, J.M., D'Antonio, C.M., Dukes, J.S. et al. (2012). Will extreme climatic events facilitate biological invasions? *Frontiers in Ecology and the Environment* 10: 249–257.

Dimitrijević, A. and Horn, R. (2018). Sunflower hybrid breeding: from markers to genomic selection. *Frontiers in Plant Science* 8: 2238.

Donald, W.W. and Khan, M. (1992). Yield loss assessment for spring wheat (*Triticum aestivum*) infested with Canada thistle (*Cirsium arvense*). *Weed Science* 40: 590–598.

Dukes, J.S. (2000). Will the increasing atmospheric CO_2 concentration affect the success of invasive species? In: *Invasive Species in a Changing World* (ed. H.A. Mooney and R.J. Hobbs), 95–113. Washington: Island Press.

Dukes, J.S. and Mooney, H.A. (1999). Does global change increase the success of biological invaders? *Trends in Ecology & Evolution* 14: 135–139.

DuPont, S. and Strohm, C. (2020). Integrated pest management programmes increase natural enemies of pear psylla in Central Washington pear orchards. *Journal of Applied Entomology* 144: 109–122.

Dyer, L.A., Richards, L.A., Short, S.A., and Dodson, C.D. (2013). Effects of CO_2 and temperature on tritrophic interactions. *PLoS One* 8: e62528.

EPA (2016). Climate change indicators: heavy precipitation. https://www.epa.gov/climate-indicators/climate-change-indicators-heavy-precipitation (accessed 23 July 2021).

ESA (2019). Position Statement on Climate Change. *Annals of the Entomological Society of America* 112: 288–291.

Evans, E.W., Carlile, N.R., Innes, M.B., and Pitigala, N. (2013). Warm springs reduce parasitism of the cereal leaf beetle through phenological mismatch. *Journal of Applied Entomology* 137: 383–391.

Evans, T.G., Diamond, S.E., and Kelly, M.W.X. (2015). Mechanistic species distribution modelling as a link between physiology and conservation. *Conservation Physiology* 3: cov056.

Facey, S.L., Ellsworth, D.S., Staley, J.T. et al. (2014). Upsetting the order: how climate and atmospheric change affects herbivore-enemy interactions. *Current Opinion in Insect Science* 5: 66–74.

FAO (2010). *"Climate-Smart" Agriculture: Policies, Practice and Financing for Food Security, Adaptation and Mitigation*. Rome, Italy: FAO.

FAO (2011). Knowledge on climate smart agriculture. https://www.fao.org/3/i4226e/i4226e.pdf (accessed 7 May 2021).

FAO (2013a). *Climate-Smart Agriculture: Sourcebook*. Rome, Italy: FAO http://www.fao.org/docrep/018/i3325e/i3325e.pdf.

FAO (2013b). International Standard for Phytosanitary Measure. ISSN 2521-7232

FAO (2021). How to practice integrated pest management. Available online: http://www.fao.org/agriculture/crops/thematic-sitemap/theme/compendium/tools-guidelines/how-to-ipm/en/ (accessed 7 May 2021).

Feng, S.F., Hao, Z.C., Zhang, X., and Hao, F.H. (2021). Changes in climate-crop yield relationships affect risks of crop yield reduction. *Agricultural and Forest Meteorology* 304: 108401.

Fenu, G. and Malloci, F. (2021). Forecasting plant and crop disease: an explorative study on current algorithms. *Big Data and Cognitive Computing* 5: 2.

Fernandez-Quintanilla, C., Pena, J., Andujar, D. et al. (2018). Is the current state of the art of weed monitoring suitable for site-specific weed management in arable crops? *Weed Research* 58: 259–272.

Ferreira, N. and Miranda, J. (2020). Potential occurrence of *Puccinia sorghi* in corn crops in Paraná, under scenarios of climate change. *International Journal of Biometeorology* 64: 1051–1062.

Fiscus, E.L., Booker, F.I., and Burkey, K.O. (2005). Crop responses to ozone: uptake, modes of action, carbon assimilation and partitioning. *Plant, Cell and Environment* 28 (8): 997–1011.

Fleisher, D., Timlin, D., Reddy, K. et al. (2011). Effects of CO_2 and temperature on crops: lessons from SPAR growth chambers. In: *Handbook of Climate Change and Agroecosystems* (ed. D. Hillel and C. Rosenzweig), 55–86. London: Imperial College Press.

Flint, E.P., Patterson, D.T., Mortensen, D.A. et al. (1984). Temperature effects on growth and leaf production in three weed species. *Weed Science* 32: 655–663.

Fu, P., Jaiswal, D., McGrath, J.M. et al. (2021). Drought imprints on crops can reduce yield loss: nature's insights for food security. *Food and Energy Security* 11: e332. https://doi.org/10.1002/fes3.332.

Fuhrer, J. (2003). Agroecosystem responses to combinations of elevated CO_2, ozone, and global climate change. *Agriculture, Ecosystems & Environment* 97: 1–20.

Fuhrer, J. (2009). Ozone risk for crops and pastures in present and future climates. *Naturwissenschaften* 96: 173–194.

Fumagalli, I., Gimeno, B.S., Velissariou, D. et al. (2001). Evidence of ozone-induced adverse effects on crops in the Mediterranean region. *Atmospheric Environment* 35: 2583–2587.

Furlong, M.J. and Zalucki, M.P. (2017). Climate change and biological control: the consequences of increasing temperatures on host–parasitoid interactions. *Current Opinion in Insect Science* 20: 39–44.

Gallagher, R., Beaumont, L., Downey, P.O. et al. (2006). Assessing the potential impacts of climate change on weeds in New South Wales: establishing priorities. Proceedings of the15th Australian Weeds Conference, Adelaide, pp. 35–9.

Gangappa, S.N., Berriri, S., and Vinod Kumar, S. (2017). PIF4 coordinates thermosensory growth and immunity in Arabidopsis. *Current Biology* 27: 243–249.

Garcia-Rojas, J.C., Robles-Bermudez, A., Cambero-Campos, O.J. et al. (2017). Metabolic resistance to pesticides. *Revista Bio Ciencias* 4: UNSP 04.06.01.

Gaucher, D., Deudon, O., Huard, F., and Launay, M. (2018). Effects of climate change on potato late blight risk and adaptation strategies. Proceedings of the 12e Conférence Internationale sur les Maladies des Plantes, Tours, pp. 19–28.

Gendron St-Marseille, A., Burgeois, G., Brodeur, J., and Mimee, B. (2019). Simulating the impacts of climate change on soybean cyst nematode and the distribution of soybean. *Agricultural and Forest Meteorology* 264: 178–187.

Gill, H.K., Goyal, G., and Chahil, G. (2017). Insect diapause: a review. *Journal of Agricultural Science and Technology* 7: 454–473.

Gillson, L., Dawson, T.P., Jack, S., and McGeoch, M.A. (2013). Accommodating climate change contingencies in conservation strategy. *Trends in Ecology & Evolution* 28: 135–142.

Gilman, S.E., Urban, M.C., Tewksbury, J. et al. (2010). A framework for community interactions under climate change. *Trends in Ecology & Evolution* 25: 325–331.

Gomez-Zavaglia, A., Mejuto, J.C., and Simal-Gandara, J. (2020). Mitigation of emerging implications of climate change on food production systems. *Food Research International* 134: 109256.

Gonzalez-Chang, M., Tiwari, S., Sharma, S., and Wratten, S.D. (2019). Habitat management for pest management: limitations and prospects. *Annals of the Entomological Society of America* 112: 302–317.

González-Tokman, D., Córdoba-Aguilar, A., Dáttilo, W. et al. (2020). Insect responses to heat: physiological mechanisms, evolution and ecological implications in a warming world. *Biological Reviews* 95: 802–821.

Grantz, D.A. and Vu, H.B. (2009). O_3 sensitivity in a potential C4 bioenergy crop: sugarcane in California. *Crop Science* 49 (2): 643–650.

Gregory, P.J., Johnson, S.N., Newton, A.C., and Ingram, J.S.I. (2009). Integrating pests and pathogens into the climate change/food security debate. *Journal of Experimental Botany* 60: 2827–2838.

Gvozdenac, S., Inđić, D., Vuković, S. et al. (2012). Antifeeding activity of several plant extracts against *Lymantria dispar* L. (Lepidoptera: Lymantriidae) larvae. *Pesticidi i fitomedicina* 27 (4): 305–311.

Gvozdenac, S., Šućur, J., Manojlović, A. et al. (2018a). Influence of winter savory (*Satureja montana*) aqueous extract on mortality of lesser grain borer (*Rhyzopertha dominica*). *Botanica Serbica* 42 (2): 277–281.

Gvozdenac, S., Tanasković, S., Krnjajić, S. et al. (2018b). Effects of different inert dusts on *Sitophilus oryzae* and *Plodia interpunctella* during contact exposure. Proceedings of the 12th International Working Conference on Stored Product Protection (IWCSPP), Berlin, Germany, vol. 2, pp. 829–834.

Gvozdenac, S., Prvulović, D., Vučinić Vasić, M. et al. (2021). Efficacy of diatomaceous earth in controlling major store product pests: *Plodia interpunctella*, *Tribolium confusum* and *Acanthosclides obtectus*. Proceedings of 7th International Conference Sustainable Postharvest and Food Technologies – INOPTEP 2021, Vršac, Serbia, pp. 7–12.

Hamilton, J.G., Dermody, O., Aldea, M. et al. (2005). Anthropogenic changes in tropospheric composition increase susceptibility of soybean to insect herbivory. *Environmental Entomology* 34: 479–485.

Hance, T., van Baaren, J., Vernon, P., and Boivin, G. (2007). Impact of extreme temperatures on parasitoids in a climate change perspective. *Annual Review of Entomology* 52: 107–126.

Harrari, A., Zahavi, T., Gordon, D. et al. (2007). Pest management programmes in vineyards using male mating disruption. *Pest Management Science* 63: 769–775.

Harrington, R., Bale, J.S., and Tatchell, G.M. (1995). Aphids in a changing climate. In: *Insects in a Changing Environment* (ed. R. Harrington and N.E. Stork), 125–155. London, UK: Academic Press.

Harrington, R., Fleming, R.A., and Woiwod, I.P. (2001). Climate change impacts on insect management and conservation in temperate regions: can they be predicted? *Agricultural and Forest Entomology* 3: 233–240.

Harrington, R., Clark, S.J., Welham, S.J. et al. (2007). Environmental change and the phenology of European aphids. *Global Change Biology* 13: 1550–1564.

Harvell, H., Mitchell, C.E., Ward, J. et al. (2002). Climate warming and disease risks for terrestrial and marine biota. *Science* 296: 2158–2162.

Harvey, J.A., van Dam, N.M., and Gols, R. (2003). Interactions over four trophic levels: foodplant quality affects development of a hyperparasitoid as mediated through a herbivore and its primary parasitoid. *Journal of Animal Ecology* 72: 520–531.

Hatfield, J.L. and Dold, C. (2019). Water-use efficiency: advances and challenges in a changing climate. *Frontiers in Plant Science* 10: 103.

Hatfield, J.L. and Prueger, J.H. (2015). Temperature extremes: effect on plant growth and development. *Weather and Climate Extremes* 10 (Part A): 4–10.

Hatfield, J., Boote, J., Kenneth, J. et al. (2011). Climate impacts on agriculture: implications for crop production. *Agronomy Journal* 103: 351–370.

Hawkins, N.J., Bass, C., Dixon, A., and Neve, P. (2019). The evolutionary origins of pesticide resistance. *Biological Reviews* 94: 135–155.

Heeb, L., Jenner, E., and Cock, M.J. (2019). Climate-smart pest management: building resilience of farms and landscapes to changing pest threats. *Journal of Pest Science* 92: 951–969.

Hellmann, J.J., Byers, J.E., Bierwagen, B.G., and Dukes, J.S. (2008). Five potential consequences of climate change for invasive species. *Conservation Biology* 22: 534–543.

Heuskin, S., Verheggen, F.J., Haubruge, E. et al. (2011). The use of semiochemical slow-release devices in integrated pest management strategies. *Biotechnology, Agronomy, Society and Environment* 15: 459–470.

Hill, D.S. (1987). *Agricultural insect Pests of Temperate Regions and Their Control*. New York, NY, USA: Cambridge University Press.

Hill, M.P., Bertelsmeier, C., Clusella-Trullas, S. et al. (2016). Predicted decrease in global climate suitability masks regional complexity of invasive fruit fly species response to climate change. *Biological Invasions* 18: 1105–1119.

Hoffmann, A.A., Weeks, A.R., Nash, M.A. et al. (2008). The changing status of invertebrate pests and the future of pest management in the Australian grains industry. *Australian Journal of Experimental Agriculture* 48 (12): 1481–1493.

Holm, L.G., Plucknett, D.L., Pancho, J.V., and Herberger, J.P. (1977). *The World's Worst Weeds: Distribution and Biology*. Hawaii: East-West Center/University Press of Hawaii.

Hunter, M.D. (2001). Effects of elevated atmospheric carbon dioxide on insect-plant interactions. *Agricultural and Forest Entomology* 3: 153–159.

Huot, B., Castroverde, C.D.M., Velasquez, A.C. et al. (2017). Dual impact of elevated temperature on plant defence and bacterial virulence in Arabidopsis. *Nature Communications* 8: 1808.

IPCC (2001). *Intergovernmental Panel on Climate Change. Climate Change – The IPCC Scientific Assessment*. Cambridge: Cambridge University, 288 pp.

IPCC (2014). *Climate Change 2014: Synthesis Report. Contribution of Working Groups I, II and III to the Fifth Assessment Report of the Intergovernmental Panel on Climate Change* (ed. R.K. Pachauri and L.A. Meyer), 151. Geneva, Switzerland: IPCC.

Jallow, M., Dahab, A., Albaho, M. et al. (2020). Efficacy of mating disruption compared with chemical insecticides for controlling *Tuta absoluta* (Lepidoptera: Gelechiidae) in Kuwait. *Applied Entomology and Zoology* 55: 213–221.

Jamieson, M.A., Trowbridge, A.M., Raffa, K.F., and Lindroth, R.L. (2012). Consequences of climate warming and altered precipitation patterns for plant-insect and multitrophic interactions. *Plant Physiology* 160: 1719–1727.

Jennings, S.A., Koehler, A.K., Nicklin, K.J. et al. (2020). Global potato yields increase under climate change with adaptation and CO_2 fertilisation. *Frontiers in Sustainable Food Systems* 4: 519324.

Jiahui, W., Wei, T., Fei, T. et al. (2020). TaRPM1 positively regulates wheat high-temperature seedling-plant resistance to *Puccinia striiformis* f. sp. *tritici*. *Frontiers in Plant Science* 10: 1679.

Johnson, B.C. and Young, B.G. (2002). Influence of temperature and relative humidity on the foliar activity of mesotrione. *Weed Science* 50: 157–161.

Johnson, S.N., Anderson, E.A., Dawson, G., and Griffiths, D.W. (2008). Varietal susceptibility of potatoes to wireworm herbivory. *Agricultural and Forest Entomology* 10: 167–174.

Joshi, R., Wani, S.H., Singh, B. et al. (2016). Transcriptional factors and plants response to drought stress: current understanding and future directions. *Frontiers in Plant Science* 7: 1029.

Ju, H., Lin, E., Wheeler, T. et al. (2013). Climate change modelling and its roles to Chinese crops yield. *Journal of Integrative Agriculture* 12: 892–902.

Juroszek, P. and von Tiedemann, A. (2011). Potential strategies and future requirements for plant disease management under a changing climate. *Plant Pathology* 60: 100–112.

Juroszek, P. and von Tiedemann, A. (2013). Plant pathogens, insect pests and weeds in a changing global climate: a review of approaches, challenges, research gaps, key studies and concepts. *Journal of Agricultural Science* 151: 163–188.

Juroszek, P. and von Tiedemann, A. (2015). Linking plant disease models to climate change scenarios to project future risks of crop diseases: a review. *Journal of Plant Diseases and Protection* 122: 3–15.

Juroszek, P., Racca, P., Link, S. et al. (2020). Overview on the review articles published during the past 30 years relating to the potential climate change effects onplant pathogens and crop disease risks. *Plant Pathology* 69: 179–193.

Kahl, H.M., Leslie, A.W., and Hooks, C.R. (2019). Effects of red clover living mulch on arthropod herbivores and natural enemies, and cucumber yield. *Annals of the Entomological Society of America* 112: 356–364.

Karlsson Green, K., Stenberg, J.A., and Lankinen, A. (2020). Making sense of Integrated Pest Management (IPM) in the light of evolution. *Evolutionary Applications* 13: 1791–1805.

Karmakar, R., Das, I., Dutta, D., and Rakshit, A. (2016). Potential effects of climate change on soil properties: a review. *Science International* 4: 51–73.

Karmalkar, A.V. and Bradly, R.S. (2017). Consequences of global warming of 1.5 °C and 2 °C for regional temperature and precipitation changes in the contiguous United States. *PLoS One* 12 (1): e0168697.

Kaur Gill, H. and Garg, H. (2014). Pesticides: environmental impacts and management strategies. In: *Pesticides - Toxic Aspects* (ed. M.L. Larramendy and S. Soloneski), 46083. UK: IntechOpen.

Kaushal, M. and Wani, S.P. (2016). Rhizobacterial-plant interactions: strategies ensuring plant growth promotion under drought and salinity stress. *Agriculture, Ecosystems and Environment* 231: 68–78.

Khan, N., George, D., Shabbir, A. et al. (2015). Rising CO_2 can alter fodder-weed interactions and suppression of *Parthenium hysterophorus*. *Weed Research* 55: 113–117.

Kingsolver, J.G., Arthur Woods, H., Buckley, L.B. et al. (2011). Complex life cycles and the responses of insects to climate change. *Integrative and Comparative Biology* 51: 719–732.

Kinjo, K. and Ebina, T. (2018). Case-based decision model matches ideal point model: application to marketing decision support system. *Journal of Intelligent Information Systems* 50: 341–362.

Kirda, C., Topcu, S., Cetin, M. et al. (2007). Prospects of partial root zone irrigation for increasing irrigation water use efficiency of major crops in the Mediterranean region. *Annals of Applied Biology* 150: 281–291.

Klok, C.J. and Chown, S.L. (2001). Critical thermal limits, temperature tolerance and water balance of a sub-Antarctic kelp fly, *Paractora dreuxi* (Diptera: Helcomyzidae). *Journal of Insect Physiology* 47 (1): 95–109.

Kochi, N., Isobe, S., Hayashi, A. et al. (2021). Review: introduction of all-around 3D modeling methods for investigation of plants. *International Journal of Automotive Technology* 15: 301–312.

Koini, M.A., Alvey, L., Allen, T. et al. (2009). High temperature-mediated adaptations in plant architecture require the bHLH transcription factor PIF4. *Current Biology* 19: 408–413.

Konduri, V.S., Vandal, T.J., Ganguly, S., and Ganguly, A.R. (2020). Data science for weather impacts on crop yield. *Frontiers in Sustainable Food Systems* 4: 52.

Korres, N.E. and Norsworthy, J.K. (2015). Influence of Palmer amaranth interrow distance and emergence date on seed production in wide-row and drill-seeded soybean. Proceedings of the Weed Science Society of America, Annual Meeting, Lexington, Kentucky, USA.

Korres, N.E., Norsworthy, J.K., Tehranchian, P. et al. (2016). Cultivars to face climate change effects on crops and weeds: a review. *Agronomy for Sustainable Development* 36: 12.

Kovenock, M. and Swann, A. (2018). Leaf trait acclimation amplifies simulated climate warming in response to elevated carbon dioxide. *Global Biogeochemical Cycles* 32: 1437–1448.

Kriticos, D.J., Maywald, G.F., Yonow, T. et al. (2015). Exploring the effects of climate on plants, animals and diseases. *CLIMEX* 4: 184.

Kuebbing, S.E. and Simberloff, D. (2015). Missing the bandwagon: nonnative species impacts still concern managers. *NeoBiota* 25: 73–86.

Kumar, S., Neven, L.G., and Yee, W.L. (2014). Evaluating correlative and mechanistic niche models for assessing the risk of pest establishment. *Ecosphere* 5: 1–23.

Kwang-Hyung, K., Jaepil, C., Yong, L., and Woo-Seop, L. (2015). Predicting potential epidemics of rice leaf blast and sheath blight in South Korea under the RCP 4.5 and RCP 8.5 climate change scenarios using a rice disease epidemiology model, EPIRICE. *Agricultural and Forest Meteorology* 203: 191–207.

Lam, O.Y.H., Dogotari, M., Prum, M. et al. (2021). An open source workflow for weed mapping in native grassland using unmanned aerial vehicle: using *Rumex obtusifoliusas* a case study. *European Journal of Remote Sensing* 54: 71–88.

Lamichhane, J.R., Barzman, M., Booij, K. et al. (2015). Robust cropping systems to tackle pests under climate change: a review. *Agronomy for Sustainable Development* 35: 443–459.

Launay, M., Caubel, J., Bourgeois, G. et al. (2014). Climatic indicators for crop infection risk: application to climate change impacts on five major foliar fungal diseases in Northern France. *Agriculture, Ecosystems & Environment* 197: 147–158.

Lawlor, D.W. (2002). Limitation to photosynthesis in water stressed leaves: stomata vs. metabolism and the role of ATP. *Annals of Botany* 89: 871–885.

Le Lann, C., Lodi, M., and Ellers, J. (2014). Thermal change alters the outcome of behavioural interactions between antagonistic partners: thermal change alters species interactions. *Ecological Entomology* 39: 578–588.

Leach, A., Reiners, S., and Nault, B. (2020). Challenges in integrated pest management: a case study of onion thrips and bacterial bulb rot in onion. *Crop Protection* 133: 105123.

Lee, J. (2011). Combined effect of elevated CO_2 and temperature on the growth and phenology of two annual C_3 and C_4 weedy species. *Agriculture, Ecosystems & Environment* 140: 484–491.

Lefebvre, M., Langrell, S.R.H., and Gomez-y-Paloma, S. (2015). Incentives and policies for integrated pest management in Europe: a review. *Agronomy for Sustainable Development* 35: 27–45.

Legris, M., Klose, C., Sethe Burgie, E. et al. (2016). Phytochrome B integrates light and temperture signals in Arabidopsis. *Plant Science* 354: 897–900.

Leguizamón, E., Yanniccari, M., Guiamet, J., and Acciaresi, H. (2011). Growth, gas exchange and competitive ability of *Sorghum halepense* populations under different soil water availability. *Canadian Journal of Plant Science* 91: 1011–1025.

Li, Y.P., Ye, W., Wang, M., and Yan, X.D. (2009). Climate change and drought: a risk assessment of crop-yield impacts. *Climate Research* 39: 31–46.

Li, J., Cang, Z., Jiao, F. et al. (2017). Influence of drought stress on photosynthetic characteristics and protective enzymes of potato at seedling stage. *Journal of the Saudi Society of Agricultural Sciences* 16 (1): 82–88.

Li, Y.Z., Tian, D., Feng, G. et al. (2021). Climate change and cover crop effects on water use efficiency of a corn-soybean rotation system. *Agricultural Water Management* 255: 107042.

Lickley, M. and Solomon, S. (2018). Drivers, timing and some impacts of global aridity change. *Environmental Research Letters* 13: 104010.

Lima, M., Leanardo, M., Valero, C. et al. (2020). Automatic detection and monitoring of insect pests: a review. *Agriculture Basel* 10: 161.

Liu, T., Ma, Z., Huffman, T. et al. (2016). Gaps in provincial decisiio-makers perception and knowledge of climate change adaptation in China. *Environmental Science & Policy* 58: 41–51.

Lobell, D.B. and Field, C.B. (2007). Global scale climate-crop yield relationships and the impacts of recent warming. *Environmental Resources Letter* 2, Article ID: 01400.

Lopez-Vaamonde, C., Agassiz, D., Augustin, S. et al. (2010). Lepidoptera. Chapter 11. *BioRisk* 4: 603–668.

Luo, Q. (2011). Temperature thresholds and crop production: a review. *Climate Change* 109: 583–598.

Luo, Y., Su, B.O., Currie, W.S. et al. (2004). Progressive nitrogen limitation of ecosystem responses to rising atmospheric carbon dioxide. *Bioscience* 54: 731–739.

Ma, H. (1998). Myths, models and mitigation of resistance to pesticides. *Philosophical Transactions of the Royal Society of London Series B-Biological Sciences* 353: 1787–1795.

Ma, C.S., Hau, B., and Poehling, H.M. (2004). The effect of heat stress on the survival of the rose grain aphid, *Metopolophium dirhodum* (Hemiptera: Aphididae). *European Journal of Entomology* 101: 327–331.

Macfadyen, S., McDonald, G., and Hill, M.P. (2018). From species distributions to climate change adaptation: knowledge gaps in managing invertebrate pests in broad-acre grain crops. *Agriculture, Ecosystems & Environment* 253: 208–219.

Machenko, H., Mvumi, B.M., and Nyamukondiwa, C. (2017). Diamondback moth, *Plutella xylostella* (L.) in Southern Africa: research trends, challenges and insights on sustainable management options. *Sustainability* 9: https://doi.org/10.3390/su9020091.

Mahaman, B., Passam, H., Sideridis, A., and Yialouris, C. (2003). DIARES-IPM: a diagnostic advisory rule-based expert system for integrated pest management in Solanaceous crop systems. *Agricultural Systems* 76: 1119–1135.

Mahlman, J.D. (1997). Uncertainties in projections of human-caused climate warming. *Science* 278: 1416–1417.

Marković, D., Ranđić, S., Tanasković, S., and Gvozdenac, S. (2017). Possibility of monitoring *D.v. virgifera* flight by processing image of phero-traps using Raspberry Pi based devices. *Acta Agriculturae Serbica* 22 (44): 207–217.

Martin, P., Silvie, P., Marnotte, P., and Goebel, F. (2020). A decision support system for determining sugarcane pest reservoir. *Sugar Tech* 22: 655–661.

Martín-Vertedor, D., Ferrero-García, J.J., and Torres-Vila, L.M. (2010). Global warming affects phenology and voltinism of Lobesia botrana in Spain. *Agricultural and Forest Entomology* 12: 169–176.

Masters, G. and Norgrove, L. (2010). Climate change and invasive alien species. CABI Working Paper 1; CABI, Wallingford, UK, pp. 30.

McGiffen, M.E., Forcella, F., Lindstrom, M.J., and Reicosky, D.C. (1997). Covariance of cropping systems and foxtail density as predictors of weed interference. *Weed Science* 45: 388–396.

McGrath, J.M. and Lobell, D.B. (2013). Regional disparities in the CO_2 fertilization effect and implications for crop yields. *Environmental Research Letters* 8: 014054.

Mcvean, R. and Dixon, A.F.G. (2001). The effect of plant drought-stress on populations of the pea aphid *Acyrthosiphon pisum*. *Ecological Entomology* 26: 440–443.

Menéndez, R. (2007). How are insects responding to global warming? *Tijdschrift voor Entomologie* 150: 355.

Menéndez, R., González-Megías, A., Collingham, Y. et al. (2007). Direct and indirect effects of climate and habitat factors on butterfly diversity. *Ecology* 88: 605–611.

Miladinović, D., Dedić, B., Quiroz, F. et al. (2012). Orobanche cumana Wallr. resistance of commercial sunflower cultivars grown in Argentina. *Journal of Basic and Applied Genetics* 23: 38–41.

Miladinović, D., Antunes, D., Yildirim, K. et al. (2021). Targeted plant improvement through genome editing: from laboratory to field. *Plant Cell Reports* 40 (6): 935–951.

Miller, B., Dalton, D., Xue, L. et al. (2019). Use of filbertworm (*Cydia latiferreana*) mating disruption within a hazelnut IPM program. *Crop Protection* 122: 118–124.

Moraes, T., Dornelas, M., and Martinelli, A. (2019). FT/TFL1: calibrating plant architecture. *Frontiers in Plant Science* 10: 97.

Moretti, A., Pascale, M., and Logrieco, A. (2019). Mycotoxin risks under a climate change scenario in Europe. *Trends in Food Science & Technology* 84: 38–40.

Mortensen, D.A. and Coble, H.D. (1989). The influence of soil water content on common cocklebur (*Xanthium strumarium*) interference in soybeans (*Glycine max*). *Weed Science* 37: 76–83.

Müller, C. (2013). African lessons on climate change risks for agriculture. *Annual Review of Nutrition* 33: 395–411.

Muntifering, R.B., Chappelka, A.H., Lin, J.C. et al. (2006). Chemical composition and digestibility of trifolium exposed to elevated ozone and carbon dioxide in a free air (FACE) fumigation system. *Functional Ecology* 20: 269–275.

Musolin, D.L. (2007). Insects in a warmer world: ecological, physiological and life-history responses of true bugs (Heteroptera) to climate change. *Global Change Biology* 13: 1565–1585.

Naranjo, S. (2001). Conservation and evaluation of natural enemies in IPM systems for *Bemisia tabaci*. Proceedings of the 21st International Congress of Entomology, Iguassu Falls, pp. 835–852.

Nelson, R., Wiesner-Hanks, T., Wisser, R., and Balint-Kurti, P. (2017). Navigation complexity to breed disease-resistant crops. *Nature Reviews Genetics* 19: 21–33.

Neufeldt, H., Jahn, M., Campbell, B.M. et al. (2013). Beyond climate-smart agriculture: toward safe operating spaces for global food systems. *Agriculture & Food Security* 2: 12.

Newlands, N. (2018). Model-based forecasting of agricultural crop disease risk at the regional scale, integrating airborne inoculum, environmental, and satellite-based monitoring data. *Frontiers in Environmental Science* 6: 63.

Nguyen, T., Bajwa, A.A., Navie, S. et al. (2017). Parthenium weed (*Parthenium hysterophorus* L.) and climate change: the effect of CO_2 concentration, temperature, and water deficit on growth and reproduction of two biotypes. *Environmental Science and Pollution Research* 24: 10727–10739.

Nihal, R. (2020). Global Climate change and its impact on integrated pest management. *International Journal of Agricultural Economics* 7: 133–137.

NOAA (2019). Global Climate Report – Annual 2019.

Norton, G.W. and Mullen, J. (1994). *Economic Evaluation of Integrated Pest Management Programmes: A Literature Review*. Blacksburg, VA, USA: Virginia Cooperative Extension Publication.

Oerke, E.C. (2006). Crop losses to pests. *The Journal of Agricultural Science* 144 (1): 31–43.

O'Hara, N.B., Franks, S.J., Kane, N.C. et al. (2021). Evolution of pathogen response genes associated with increased disease susceptibility during adaptation to an extreme drought in a *Brassica rapa* plant population. *BMC Evolutionary Biology* 21: 61.

Olesen, J.E. and Bindi, M. (2002). Consequences of climate change for European agricultural productivity, land use and policy. *European Journal of Agronomy* 16: 239–262.

Pacifici, M., Foden, W.B., Visconti, P. et al. (2015). Assessing species vulnerability to climate change. *Nature Climate Change* 5: 215–225.

Pan, Z., Segal, M., Arritt, R.W., and Takle, E.S. (2004). On the potential change in solar radiation over the U.S. due to increases of atmospheric greenhouse gases. *Renewable Energy* 29 (11): 1923–1928.

Pan, S.F., Chen, G.S., Ren, W. et al. (2018). Response of global terrestrial water use efficiency to climate change and rising atmospheric CO_2 concentration in the twenty-first century. *International Journal of Digital Earth* 11: 558–582.

Papp, J., Jenser, G., and Haltrich, A. (2001). Effect of nitrogen supply on the population of European red spider mite and green apple aphid in an IPM apple orchard. Proceedings of the 4th International Symposium on Mineral Nutrition of Deciduous Fruit Crops, Penticton, pp. 407–412.

Parmesan, C. (2007). Influences of species, latitudes and methodologies on estimates of phenological response to global warming. *Global Change Biology* 13: 1860–1872.

Parmesan, C. and Yohe, G. (2003). A globally coherent fingerprint of climate change impacts across natural systems. *Nature Cell Biology* 421: 37–42.

Parmesan, C., Ryrholm, N., Stefanescu, C. et al. (1999). Poleward shifts in geographical ranges of butterfly species associated with regional warming. *Nature* 399 (6736): 579–583.

Pathak, H., Aggarwal, P.K., and Singh, S.D. (2012). *Climate Change Impact, Adaptation and Mitigation in Agriculture: Methodology for Assessment and Applications*. New Delhi, India: Indian Agricultural Research Institute.

Patterson, D.T. (1995). Effects of environmental stress on weed/crop interactions. *Weed Science* 43: 483–490.

Patterson, D.T., Westbrook, J.K., Joyce, R.J.C. et al. (1999). Weeds, insects and diseases. *Climatic Change* 43: 711–727.

Pearsons, K.A. and Tooker, J.F. (2017). In-field habitat management to optimize pest control of novel soil communities in agroecosystems. *Insects* 8: 82.

Pell, J., Baverstock, J., Roy, H. et al. (2008). Intraguild predation involving *Harmonia axyridis*: a review of current knowledge and future perspectives. *BioControl* 53: 147–168.

Peng, Z.R., Liu, Y., He, Q. et al. (2019). Population diversity and variation of the rice blast fungus under different climate conditions. *European Journal of Plant Pathology* 155: 881–889.

Perrings, C., Burgiel, S., Lonsdale, M. et al. (2010). International cooperation in the solution to trade-related invasive species risks. *Annals of the New York Academy of Sciences* 1195: 198–212.

Peshin, R., Bandral, R., Zhang, W. et al. (2009). Integrated Pest Management: a global overview of history, programs and adoption. In: *Integrated Pest Management: Innovation-Development Process* (ed. R. Peshin and A. Dhawan), 1–49. Netherland: Springer.

Peters, K. and Gerowitt, B. (2014). Important maize weeds profit in growth and reproduction from climate change conditions represented by higher temperatures and reduced humidity. *Journal of Applied Botany and Food Quality* 87: 234–242.

Petric, M., Vandendriessche, J., Marsboom, C. et al. (2019). Autonomous wireless sensor networks in an IPM Spatial Decision Support System. *Computers* 8: 43.

Pimentel, D. (2005). Environmental and economic costs of the application of pesticides primarily in the United States. *Environment, Development and Sustainability* 7: 229–252.

Polis, G.A., Myers, C.A., and Holt, R.D. (1989). The ecology and evolution of intraguild predation: potential competitors that eat each other. *Annual Review of Ecology and Systematics* 20 (1): 297–330.

Pollard, E. and Yates, T.J. (1993). *Monitoring Butterflies for Ecology and Conservation: The British Butterfly Monitoring Scheme*, 274. Netherlands: Springer.

Porfirio, L.L., Harris, R.M., Lefroy, E.C. et al. (2014). Improving the use of species distribution models in conservation planning and management under climate change. *PLoS One* 9: e113749.

Porter, J., Parry, M., and Carter, T. (1991). The potential effects of climatic change on agricultural insect pests. *Agricultural and Forest Meteorology* 57: 221–240.

Prado, S.S., Hung, K.Y., Daugherty, M.P., and Almeida, R.P.P. (2010). Indirect effects of temperature on stink bug fitness, via maintenance of gut-associated symbionts. *Applied and Environmental Microbiology* 76: 1261–1266.

Prank, M., Kenaley, S., Bergstrom, G. et al. (2019). Climate change impacts the spread potential of wheat stem rust, a significant crop disease. *Environmental Research Letters* 14: 124053.

Pretty, J. and Bharucha, Z.P. (2015). Integrated pest management for sustainable intensification of agriculture in Asia and Africa. *Insects* 6: 152–182.

Prevéy, J.S. and Seastedt, T.R. (2014). Seasonality of precipitation interacts with exotic species to alter composition and phenology of a semi-arid grassland. *Journal of Ecology* 102: 1549–1561.

Pugliese, M., Gullino, M., and Garibaldi, A. (2018). Water management and climate change: effects on plant diseases. *Protezione delle Colture* 3: 27–32.

Qafoku, N.P. (2015). Chapter Two – Climate-change effects on soils: accelerated weathering, soil carbon, and elemental cycling. *Advances in Agronomy* 131: 111–172.

Qiu, Y., Pasoreck, E.K., Yoo, Y.C. et al. (2021). RCB initiates *Arabidopsis thermomorphogenesis* by stabilizing the thermoregulator PIF4 in the daytime. *Nature Communications* 12: 2042.

Rahman, A. and Wardle, D. (1990). Effects of climate change on cropping weeds in New Zealand. In: *The Impact of Climate Change on Pests, Diseases, Weeds and Beneficial Organisms Present in New Zealand Agricultural and Horticultural Systems* (ed. R.A. Prestidge and R.P. Pottinger), 107–112. Hamilton: Ruakura Agricultural Centre.

Ramesh, K., Matloob, A., Aslam, F. et al. (2017). Weeds in a changing climate: vulnerabilities, consequences, and implications for future weed management. *Frontiers in Plant Science* 8: 95.

Raza, M.M., Khan, M.A., Arshad, M. et al. (2014). Impact of global warming on insects. *Archives of Phytopathology and Plant Protection* 48: 84–94.

Reddy, P.P. (2015). *Climate Resilient Agriculture for Ensuring Food Security*. New Delhi, India: Springer.

Reddy, P.P. (2018). *Crop Protection Strategies Under Climate Change Scenarios*. India: Scientific Publishers.

Rehman, M.U., Rather, G.H., Gull, Y. et al. (2015). Effect of climate change on horticultural crops. In: *Crop Production and Global Environmental Issues* (ed. K.R. Hakeem), 211–239. Cham, Switzerland: Springer International Publishing.

Reilly, J. (1995). Climate change and global agriculture: recent findings and issues. *American Journal of Agricultural Economics* 77: 243–250.

Reilly, J., Baethgen, W., Chege, F.E. et al. (1996). Agriculture in a changing climate: impacts and adaptation. In: *Climate Change 1995; Impacts, Adaptations and Mitigation of Climate Change: Scientific-Technical Analyses* (ed. R.T. Watson, M.C. Zinyowera and R.H. Moss), 427–467. Cambridge, UK: Cambridge University Press.

Reiners, S. and Petzoldt, C. (2005). Integrated Crop and Pest Management Guidelines for seasonal timing that matters. *Molecular Ecology* 17: 157–166.

Ricciardi, A. (2013). Invasive species. In: *Ecological Systems* (ed. R. Leemans), 161–178. New York, NY, USA: Springer.

Rosenheim, J.A., Kaya, H.K., Ehler, L.E. et al. (1995). Intraguild predation among biological-control agents: theory and evidence. *Biological Control* 5: 303–335.

Rossi, V., Sprendio, G., Caffi, T. et al. (2019). Critical success factors for the adoption of decision tools in IPM. *Agromony-Basel* 9: 710.

Rowlandson, T., Gleason, M., Sentelhas, P. et al. (2015). Reconsidering leaf wetness duration determination for plant disease management. *Plant Disease 99*: 310–318.

Roy, D.B. and Sparks, T.H. (2000). Phenology of British butterflies and climate change. *Global Change Biology* 6: 407–416.

Schmale, D. and Ross, S. (2015). Highways in the Sky: scales of atmospheric transport of plant pathogens. *Annual Review of Phytopathology* 53: 591–611.

Scott, R.L., Hamerlynck, E.P., Jenerette, G.D. et al. (2010). Carbon dioxide exchange in a semidesert grassland through drought-induced vegetation change. *Journal of Geophysical Research: Biogeosciences* 115: G03026.

Shabbir, A., Dhileepan, K., Khan, N., and Adkins, S.W. (2014). Weed-pathogen interactions and elevated CO_2: growth changes in favour of the biological control agent. *Weed Research* 54: 217–222.

Shah, K., Chaturvedi, V., and Gupta, S. (2019). Advances in rice research for abiotic stress tolerance. In: *Advances in Rice Research for Abiotic Tolerance* (ed. M. Hasanuzzaman, M. Fujita, K. Nahar and J. Krishna Biswas), 505–535. Sawston, UK: Woodhead Publishing.

Shan, H.W., Lu, Y.H., Bing, X.L. et al. (2014). Differential responses of the whitefly *Bemisia tabaci* symbionts to unfavorable low and high temperatures. *Microbial Ecology* 68 (3): 472–482.

Sharma, H.C. (2014). Climate change effects on insects: implications for crop protection and food security. *Journal of Crop Improvement* 28: 229–259.

Sharma, S.D. and Singh, M. (2001). Environmental factors affecting absorption and bio-efficacy of glyphosate in Florida beggarweed (*Desmodium tortuosum*). *Crop Protection* 20: 511–516.

Sharma, M., Sharma, A., and Gupta, S. (2020). Climate change and plant diseases. *Agricultural Research Journal* 57: 1–6.

Sheppard, C.S. and Stanley, M.C. (2014). Does elevated temperature and doubled CO_2 increase growth of three potentially invasive plants? *Invasive Plant Science and Management* 7: 237–246.

Shin, D.W., Cocke, S., Baigorria, G.A. et al. (2020). Future crop yield projections using a multi-model set of regional climate models and a plausible adaptation practice in the southeast United States. *Atmosphere* 11: 1300.

Shine, C., Williams, N., and Gündling, L. (2000). A guide to designing legal and institutional frameworks on alien invasive species, environmental policy and law. Paper No. 40, IUCN, Switzerland; Cambridge, UK; Bonn, Germany.

Skendžić, S., Zovko, M., Živković, I.P. et al. (2021). The impact of climate change on agricultural insect pests. *Insects* 12: 440.

Smart, L., Aradottir, G., and Bruce, T. (2014). Role of semiochemicals in integrated pest management. In: *Integrated Pest Management Current Concepts and Ecological Perspective* (ed. D.P. Abrol), 93–109. London: Academic Press.

Sombroek, W.G. and Gommes, R. (1996). The climate change-agriculture conundrum. In: *Global Climate Change and Agricultural Production* (ed. F. Bazzaz and W. Sombroek), 1–16. New York: FAO and John Wiley & Sons.

Song, Q., Srinivasan, V., Long, S., and Zhu, X. (2020). Decomposition analysis on soybean productivity increase under elevated CO2 using 3-D canopy model reveals synergistic effects of CO_2 and light in photosynthesis. *Annals of Botany* 126: 601–614.

Staley, J.T., Hodgson, C.J., Mortimer, S.R. et al. (2007). Effects of summer rainfall manipulations on the abundance and vertical distribution of herbivorous soil macro-invertebrates. *European Journal of Soil Biology* 43: 189–198.

Stanhill, G. and Cohen, S. (2001). Global dimming: a review of the evidence for a widespread and significant reduction in global radiation with discussion of its probable causes and possible agricultural consequences. *Agricultural and Forest Meteorology* 107 (4): 255–278.

Stefanescu, C., Penuelas, J., and Filella, I. (2003). Effects of climatic change on the phenology of butterflies in the northwest Mediterranean Basin. *Global Change Biology* 9: 1494–1506.

Stevens, C.J., Dise, N.B., Mountford, J.O., and Gowing, D.J. (2004). Impact of nitrogen deposition on the species richness of grasslands. *Science* 303: 1876–1879.

Stiling, P. and Cornelissen, T. (2007). How does elevated carbon dioxide (CO_2) affect plant–herbivore interactions? A field experiment and meta-analysis of CO_2-mediated changes on plant chemistry and herbivore performance. *Global Change Biology* 13: 1823–1842.

Stireman, J.O. III, Dyer, L.A., Janzen, D.H. et al. (2005). Climatic unpredictability and parasitism of caterpillars: implications of global warming. *Proceedings of the National Academy of Sciences* 102: 17384–17387.

Stout, M. and Davis, J. (2009). Keys to the increased use of host plant resistance in integrated pest management. In: *Integrated Pest Management: Innovation-Development Process* (ed. R. Peshin and A.K. Dhavan), 163–181. Dordrecht, Netherlands: Springer.

Strange, R.N. and Scott, P.R. (2005). Plant disease: a threat to global food security. *Annual Review of Phytopathology* 43: 83–116.

Stratonovitch, P., Storkey, J., and Semenov, A.A. (2012). A process-based approach to modelling impacts of climate change on the damage niche of an agricultural weed. *Global Change Biology* 18: 2071–2080.

Sugiyama, M., Yoshioka, Y., and Sakata, Y. (2009). Effect of temperature on symptom expression and viral spread of Melon yellow spot virus in resistant cucumber accessions. *Journal of General Plant Pathology* 75: 381–387.

Sutherst, R.W., Constable, F., Finlay, K.J. et al. (2011). Adapting to crop pest and pathogen risks under a changing climate. *Wiley Interdisciplinary Reviews: Climate Change* 2: 220–237.

Tai, A.P.K. and Martin, M.V. (2017). Impacts of ozone air pollution and temperature extremes on crop yields: spatial variability, adaptation and implications for future food security. *Atmospheric Environment* 169: 11–21.

Teichman, T. and Muhr, M. (2015). Shaping plant architecture. *Frontiers in Plant Science* 6: 223.

Terzić, S., Dedić, B., Atlagić, J., and Maširević, S. (2007). Transfering *Plasmopara halstedii* resistance from annual wild to cultivated sunflower. *Helia* 30: 199–204.

Thomsen, M.A. and D'Antonio, C.M. (2007). Mechanisms of resistance to invasion in a California grassland: the roles of competitor identity, resource availability, and environmental gradients. *Oikos* 116: 17–30.

Tobin, P.C., Nagarkatti, S., Loeb, G., and Saunders, M.C. (2008). Historical and projected interactions between climate change and insect voltinism in a multivoltine species. *Global Change Biology* 14: 951–957.

Torres Castillo, N., Melchor-Martínez, E., Ochoa Sierra, J. et al. (2020). Impact of climate change and early development of coffee rust – an overview of control strategies to preserve organic cultivars in Mexico. *Science of the Total Environment* 738: 140225. https://doi.org/10.1016/j.scitotenv.2020.140225.

Trebicki, P. (2020). Climate change and plant virus epidemiology. *Virus Research* 286: 198059.

Trenberth, K.E., Dai, A., van der Schrier, G. et al. (2014). Global warming and changes in drought. *Nature Climate Change* 4: 17–22.

Trumble, J. and Butler, C. (2009). Climate change will exacerbate California's insect pest problems. *California Agriculture* 63: 73–78.

Tubiello, F., Rosenzweig, C., Goldberg, R. et al. (2002). Effects of climate change on US crop production: simulation results using two different GCM scenarios, Part I: wheat, potato, maize, and citrus. *Climate Research* 20: 259–270.

Vadez, V., Berger, J.D., Warkentin, T. et al. (2012). Adaptation of grain legumes to climate change: a review. *Agronomy for Sustainable Development* 32: 31–44.

Vamourin, E. and Laine, A.L. (2018). Role of temperature and coinfection in mediating pathogen life-history traits. *Frontiers in Plant Science* 9: 1670.

Vander Zanden, M.J. (2005). The success of animal invaders. *Proceedings of the National Academy of Sciences of the United States of America* 102: 7055–7056.

Varotto, S., Tani, E., Abraham, E. et al. (2020). Epigenetics: possible applications in climate-smart crop breeding. *Journal of Experimental Botany* 71: 5223–5236.

Vasiliev, D. and Greenwood, S. (2021). The role of climate change in pollinator decline across the Northern Hemisphere is underestimated. *Science of the Total Environment* 775: 145788.

Velasquez, A.C., Castroverde, C.D., and He, S.Y. (2018). Plant–pathogen warfare under changing climate conditions. *Current Biology* 28: 619–634.

Vincent, C., Hallman, G., Panneton, B., and Fleurat-Lessardú, F. (2003). Management of agricultural insects with physical control methods. *Annual Review of Entomology* 48: 261–281.

Vintal, H., Ben-Noon, E., Shelvin, E. et al. (1999). Influence of rate of soil fertilization on Alternaria leaf blight (*Alternaria dauci*) in carrots. *Phytoparasitica* 27: 193–200.

Vučić-Peštić, O., Ehnes, R.B., Rall, B.C., and Brose, U. (2011). Warming up the system: higher predator feeding tares but lower energetic efficiencies. *Global Change Biology* 17 (3): 1301–1310.

Vukajlović, F., Pešić, S., Tanasković, S. et al. (2018). Efficacy of *Echium* spp. water extracts as post-harvest grain protectants against *Plodia interpunctella*. *Romanian Biotechnological Letters* 24 (5): 761–769.

Wahdan, S., Hossen, S., Tanunchai, B. et al. (2020). Future climate significantly alters fungal plant pathogen dynamics during the early phase of wheat litter decomposition. *Microorganisms* 8: 908.

Wahmhoff, W., Hedke, K., von Tiedemann, A. et al. (1999). Impact of crop rotation and soil cultivation on the development of pests and diseases of rapeseed. *Journal of Plant Diseases and Protection* 106: 57–73.

Walck, J.L., Hidayati, S.N., Dixon, K.W. et al. (2011). Climate change and plant regeneration from seed. *Global Change Biology* 17: 2145–2161.

Walthall, C.L., Hatfield, J., Backlund, P. et al. (2012). *Climate Change and Agriculture in the United States: Effects and Adaptation*. Washington, DC: USDA Technical Bulletin 1935.

Walther, G.R., Roques, A., Hulme, P.E. et al. (2009). Alien species in a warmer world: risks and opportunities. *Trends in Ecology & Evolution* 24: 686–693.

Wang, J.Q., Liu, X.Y., Cheng, K. et al. (2018). Winter wheat water requirement and utilisation efficiency under simulated climate change conditions: a Penman-Monteith model evaluation. *Agricultural Water Management* 197: 100–109.

War, A.R., Taggar, G.K., War, M.Y., and Hussain, B. (2016). Impact of climate change on insect pests, plant chemical ecology, tritrophic interactions and food production. *International Journal of Clinical and Biological Sciences* 1 (2): 16–29.

Ward, N.L. and Masters, G.J. (2007). Linking climate change and species invasion: an illustration using insect herbivores. *Global Change Biology* 13: 1605–1615.

Webb, K., Oña, I., Bai, J. et al. (2009). A benefit of high temperature: increased effectiveness of a rice bacterial blight disease resistance gene. *New Phytologist* 185: 568–576.

Wenda-Piesik, A., Piesik, D., Nowak, A., and Wawrzyniak, M. (2015). *Tribolium confusum* responses to blends of cereal kernels and plant volatiles. *Journal of Applied Entomology* 140: 558–563.

Williams, A.L., Wills, K.E., Janes, J.K. et al. (2007). Warming and free-air CO_2 enrichment alter demographics in four co-occuring grassland species. *New Phytologist* 176: 365–374.

Wilson, P.D., Downey, P.O., Leishman, M. et al. (2009). Weeds in a warmer world: predicting the impact of climate change on Australia's alien plant species using MaxEnt. *Plant Protection Quarterly* 24: 84–87.

Wójtowicz, A., Wójtowicz, M., Sigvald, R. et al. (2020). Assessment of the impact of climate change on the latency period of leaf rust on triticale in Poland. *Acta Agriculturae Scandinavica Section B Soil and Plant Science 70*: 195–207.

Woodward, F.I. (1990). Global change: translating plant ecophysiological responses to ecosystems. *Trends in Ecology & Evolution* 5: 308–311.

World Bank, FAO, IFAD (2009). *Gender in Agriculture Sourcebook*. Washington, DC, USA: The International Bank for Reconstruction and Development/The World Bank.

Wu, E.J., Wang, Y.P., Yahuza, L. et al. (2019). Rapid adaptation of the Irish potato famine pathogen *Phytophthora infestans* to changing temperature. *Evolutionary Applications* 13: 768–780.

Yamamura, K. and Kiritani, K. (1998). A simple method to estimate the potential increase in the number of generations under global warming in temperate zones. *Applied Entomology and Zoology* 33: 289–298.

Yamamura, K., Yokozawa, M., Nishimori, M. et al. (2006). How to analyze long-term insect population dynamics under climate change: 50-year data of three insect pests in paddy fields. *Population Ecology* 48: 31–48.

Yihdego, Y., Salem, H.S., and Muhammed, H.H. (2019). Agricultural pest management policies during drought: case studies in Australia and the state of Palestine. *Natural Hazards Review* 20: 0501802–05018010.

Yonow, T., Kriticos, D.J., Kirichenko, N., and Ota, N. (2018). Considering biology when inferring range-limiting stress mechanisms for agricultural pests: a case study of the beet armyworm. *Journal of Pest Science* 91: 523–538.

Zani, P.A., Cohnstaedt, L.W., Corbin, D. et al. (2005). Reproductive value in a complex life cycle: heat tolerance of the pitcher-plant mosquito, *Wyeomyia smithii*. *Journal of Evolutionary Biology* 18: 101–105.

Zayan, S.A. (2019). Impact of climate change on plant diseases and IPM strategies. In: *Plant Diseases - Current Threats and Management Trends* (ed. S. Topolovec-Pintarić). London, UK: IntechOpen https://doi.org/10.5772/intechopen.87055.

Zhang, J., Ren, W., An, P. et al. (2015a). Responses of crop water use efficiency to climate change and agronomic measures in the semiarid area of northern China. *PLoS One* 10: e0137409.

Zhang, W., Chang, X.Q., Hoffmann, A.A. et al. (2015b). Impact of hot events at different developmental stages of a moth: the closer to adult stage, the less reproductive output. *Scientific Reports* 5: 10436.

Zheng, Y., Li, F., Hao, L. et al. (2019). Elevated CO_2 concentration induces photosynthetic down-regulation with changes in leaf structure, non-structural carbohydrates and nitrogen content of soybean. *BMC Plant Biology* 19: 255.

Zhu, Y., Qian, W., and Hua, J. (2010). Temperature modulates plant defense responses through NB-LRR proteins. *PLoS Pathogens* 6: e1000844.

Ziska, L.H. (2000). The impact of elevated CO_2 on yield loss from a C_3 and C_4 weed in field-grown soybean. *Global Change Biology* 6: 899–905.

Ziska, L.H. (2001). Changes in competitive ability between a C_4 crop and a C_3 weed with elevated carbon dioxide. *Weed Science* 49: 622–627.

Ziska, L.H. (2003a). Evaluation of the growth response of six invasive species to past, present and future atmospheric carbon dioxide. *Journal of Experimental Botany* 54: 395–404.

Ziska, L.H. (2003b). Evaluation of yield loss in field sorghum from a C_3 and C_4 weed with increasing CO_2. *Weed Science* 51: 914–918.

Ziska, L.H. (2004). Rising carbon dioxide and weed ecology. In: *Weed Biology and Management* (ed. Inderjit), 159–176. Dordrecht: Kluwer Academic.

Ziska, L.H. (2010). Elevated carbon dioxide alters chemical management of Canada thistle in no-till soybean. *Field Crops Research* 119: 299–303.

Ziska, L.H. and Dukes, J.S. (2011). *Weed Biology and Climate Change.* Ames, IA: Wiley-Blackwell.

Ziska, L.H. and Runion, G.B. (2006). Future weed, pest and disease problems for plants. In: *Agroecosystems in a Changing Climate* (ed. P. Newton, A. Carman, G. Edwards and P. Niklaus), 262–287. New York: CRC.

Ziska, L.H., Reeves, J.B., and Blank, B. (2005). The impact of recent increases in atmospheric CO_2 on biomass production and vegetative retention of cheatgrass (*Bromus tectorum*): Implications for fire disturbance. *Global Change Biology* 11: 1325–1332.

Ziska, L.H., Blumenthal, D.M., Runion, G.B. et al. (2011). Invasive species and climate change: an agronomic perspective. *Climatic Change* 105: 13–42.

15

Climate Change and Its Effects on Plant Viruses

Aarshi Srivastava, Vineeta Pandey, and Rajarshi K. Gaur

Department of Biotechnology, Deen Dayal Upadhyaya Gorakhpur University, Gorakhpur, Uttar Pradesh, India

CONTENTS

15.1 Introduction

Modernization and human population growth has led to the increase in energy production ensuring anthropogenic causes such as deforestation and burning of fossil fuels are progressively intensified, which causes the difference in global climate and agricultural patterns. The climate change has impact on phenology, geographical distribution, migration potential, and the densities of insect vector population, viruses, and their associated host plants, and from Figure 15.1 we interpret how this triad interacts with each other. The resulted frequent changes in weather events such as atmospheric temperature, increased carbon

Climate Change and Agriculture: Perspectives, Sustainability and Resilience,
First Edition. Edited by Noureddine Benkeblia.

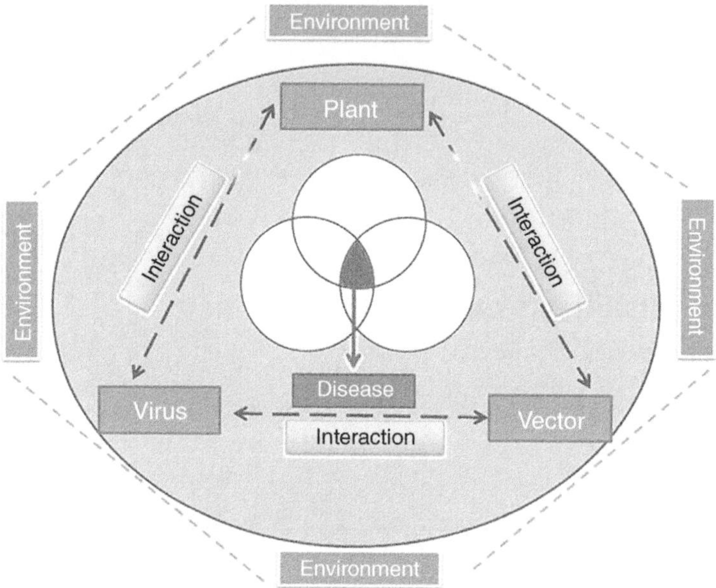

Figure 15.1 Triad interactions (host, virus, and vector) within environment.

dioxide concentration, and change in water availability might have revelatory outcome on food security, distribution, and production in the near future, and similarly may indirectly or directly affect plant viruses. The altered natural structure may lead to altered epidemiology of plant viruses initially with their distribution and finally with their survival. This may project a change in the pathogenicity and virulence of plant viruses, as epidemiology theory depends upon the number of primary infection, transmission efficiency into new host, and then the spread for secondary infection (Hull 2002), thereupon results in increase in prevalence and scale of disease revolution. In same context, it might affect the physiological behaviour of plant that may influence the ability of individual host to control or accelerate viral infections. However, in order to assess the impact of climate change that influences the incidence, emergence, and evolution of plant viruses and establishment of interactions with variable host varieties via accelerating disease in natural vegetation and cultivated plants, the knowledge of ecological changes and their other operating processes predominantly towards universal food security and natural environment that could mitigate or aggravate the ecosystem is necessary (Ahanger et al. 2013; Canto et al. 2009; Jones 2014). Thus, greater understanding of direct and indirect ecological parameters are analysed here to explore the identifying changes in virus–vector epidemiology and interaction are needed to surpass the challenges ahead and to evolve effective and robust control framework that will aid in asserting plant biodiversity and global food production for tomorrow.

15.2 Aspects of Altered Agriculture Pattern

Climate shift is the most challenging environmental issue for ecosystem. Numerous papers have been published on climate change and its adverse impact on disease incidence along with its severity, epidemic frequency, and its spatial dissemination in agriculture

(Islam and Kieu 2020; Newbery et al. 2016). The consequences of climate change mainly focus on agricultural systems and viral population dynamics that are understandable. In the past it has been proved that enormous losses in yield can occur when the climatic events are favourable for boosting disease development. These outbreaks are assumed to occur more often in some plant pathosystem due to the invasion of novel pathogens and because of the increase in the severity of existing pathogens by mutation or recombination. Although major epidemics may occur in agricultural system, in this system manipulation of host genetic factors, environmental parameters, change in planting time, and by the application of pesticide, humans will be able to ameliorate disease development. In case of crops and viruses, less attention has been given to the eco-evolutionary result of climate shift. Moreover, there is evidence aiding the probability of genetically based temperature adaptation (Wu et al. 2020) that may cause the prospect of proliferation in the geographic range of viruses more concerning. Not surprisingly, and somewhat similar to agricultural systems, most of the agricultural studies target the effect of change in climatic conditions on the population dynamics of the compatible pathogen along with the health and survival of the host tree (Sturrock et al. 2011).

15.3 Climate Change

15.3.1 Direct Effects

Global climate shift is occurring, which is a major concern worldwide for agro-ecosystem. Impact of the climate difference has been seen on all the three interactions of agricultural process parts, i.e. environmental conditions–vectors–viruses, which play important role in disease epidemiology. The changes drastically affect crop health and disease susceptibility and also vector–virus longevity, which in turn accelerates the decline in agricultural productivity (Coakley et al. 1999). Intergovernmental Panel on Climate Change 5th Assessment Report, 2014, forecasted a progressive shift in global average temperatures by 2100 by up to 4.6 °C, and beside the completeness of sixth assessment report in 2022, they had released the special report in 2018 on global warming of 1.5 °C, which made sixth assessment cycle most hard-driving. The major uncertainties of climate change involve rate and magnitude of climate shifting, the response of agriculture-based output, and the tendency to respond towards expected impact, which leads to limitations in research and understanding of future climate change (Anita et al. 2010). Direct climate effects such as temperature, greenhouse gases, rainfall, and relative humidity change with wind direction are responsible for change in environmental parameters likely describes the changes associated with vector, viruses, and host and control possibilities for viruses (Figure 15.2). Consequently, understanding of effect of different climatic parameters is listed with reference to virus and vector (Jones 2014).

15.3.1.1 Temperature

Climate change influences temperature and precipitation regimes, and it may also alter the growth stage, rate of development, pathogenicity of infectious agents, with alteration in the physiology and resistance of the host plants as well (Chakraborty and Datta 2003). With

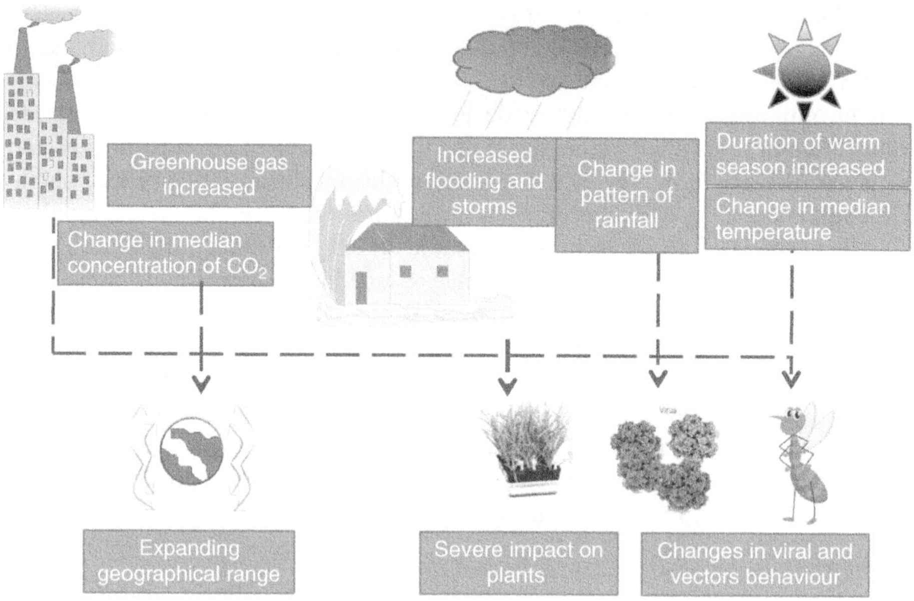

Figure 15.2 Climate parameters affecting environmental components.

increase in temperature, duration of winter and growth rate along with reproduction rate of pathogens may modify (Ladanyi and Horvath 2010). Conversely, geographical distributions of viruses within mountainous regions of tropical and subtropical area are found, and lower temperature for mortification within their host plants are likely to contract to higher latitude areas. Repeatedly, expansion of vector-borne viruses tends to be confined if the ranges of vectors remained stable, but in case of aphids the opposite scenario is predicted for vectors from key temperate region. Elevated temperature affects plants that tend to vary greatly throughout the year; evidence suggests that sunlight influences different pathogens due to the aggregation of phytoalexins or protective pigments in host tissue. During colder time of the year, warming might be relieving plant stress; however, during the hotter parts of the year, it may increase stress. However, there are many aspects whereby rise in temperature is expected to uplift the scale of plant virus epidemics, although it can diminish due to some circumstances as well. However, such predictions exclude those crops that are cultivated under temperature-controlled condition, i.e. protected cropping situations by providing heating under cold climates with cooling in warm climates (Jones and Barbetti 2012). A change in temperature may be favourable for development of different inactive pathogens, which could bring out an epidemic.

15.3.1.1.1 Effect on Vector

Temperature fluctuation is the primary climatic change that affects insect herbivores by altering their development, fecundity, survival, and dispersal, which affects their abundance as well. Simultaneously, rise in mean temperature alters physiology of plant by affecting the secondary metabolite pathways, due to which the nutritious value of leaves concurrently alters the insect vectors and that will cause rise in stomatal conductance and affect photosynthetic efficiency. Moreover, alteration in temperature results in virus amplification within cells, thereby influencing systemic movement of virus and acquisition by vectors. Moreover, in some plants rise in temperature leads to change in route of gene expression for defence signalling against insect vector, and this can expand the efficiency of viral transmission to healthy plants. In addition, this enhancement in transmission efficiencies favours the virus, i.e. they could expand their range to the region of too cold for them to be transmitted efficiently with same effects. Although in some other viruses the transmission is diminished with increase in temperature due to their less transmission efficiencies in high temperature cases. Also, prolonged exposure to heat waves can contract the vector numbers, thus decreasing virus epidemics (Canto et al. 2009; Jones and Barbetti 2012). Hence, we can say that climate can substantially affect the widespread distribution and development of vectors. Many insect vectors like aphid promote virus diseases are more ubiquitously in temperate zone under extreme winter and summer temperature (Canto et al. 2009; Jones 2009; Jones and Barbetti 2012). Likewise, whitefly vectors also respond to climatic changes because of their high reproductive capacity, short generation times, and ability to make quick life history with behavioural changes. Whiteflies *Bemisia tabaci* and *Trialeurodes vaporariorum* are main virus-transmitted vectors in which *T. vaporariorum* increases its dispersal in cooler regions. Thus, this will relocate vector distribution that influences the distribution of viruses transmitted by whitefly worldwide, and *B. tabaci*-transmitted begomoviruses' versatility with its damaging epidemics are more widespread now (Jones and Barbetti 2012) (Table 15.1).

In addition, there are many vectors such as thrips vector whose population has also increased due to the rise in mean temperature, which favours their rate of development leading to an increase in generation per year. Viruses transmitted by mealy-bugs leafhoppers and eriophyid mites are also under similar considerations (Canto et al. 2009; Jones 2009; Jones and Barbetti 2012).

15.3.1.1.2 Effect on Virus

Viruses have distinct temperature optima favourable for their division in host plants; among them some are adapted to cooler region and other to warmer ones. In lowland tropical environments such optima is maximum for the virus to invade the host growing in such regions and lowest in viruses adapted to cold climates. Simultaneously, increasing temperature also affects the host–plant physiology, nutritional status, their metabolic pathways, phenology, and morphology (Canto et al. 2009; Jones and Barbetti 2012). Plant susceptibility to virus infection is generally affected by both rise in mean temperature and heat stress with decrease in effectiveness of temperature sensitivity towards single-gene resistances. In addition to this, as mean temperature increases, the rate of viral division, systemic movement, and transmission in seed are altered, thus affecting the multiplication and systemic movement of individual viruses involved in mixed infection. Furthermore, rate of virus evolution and selection pressure changes, leading to the emergence of a new viral strain with broadened range of natural host, increase in multiplication rates of virus in host reservoir, and increase in transmission efficiencies of virus up to next level. In some instances, prolonged heat waves subjected to plants might be responsible for elimination of virus infections from natural plants where systemic invasion is formerly incomplete or rather the virus is associated with unstable particles. Sometimes prolonged heat waves might be in adequate level, and its prolonged exposure will prevent the transmission of seed with seed coat contaminants containing viruses (Jones and Barbetti 2012).

Table 15.1 Risk analysis studies on the effects of change in climate parameters on vectors.

Climate parameters	Affected vectors	Experimental outcomes	References
Temperature	*Acyrthosiphongossypii (a cotton aphid)*	21–27 °C is optimal temperature range for growth, while 30 °C was beyond its upper limit for reproduction	Gao et al. (2012)
	M. persicae	Under heat waves population growth of *M. persicae* was lower	Gillespie et al. (2012)
	A. pisum	Successful plant colonization unaffected by a rise of up to 30 °C, but root nodulation severely reduced	Ryalls et al. (2013)
	Sitobionavenae (grain aphid) and *R. padi*	Stays at one feeding site for a long time but to avoid heat stress based on the combination of temperature and exposure time, they escape before they are damaged by high temperatures	Ma and Ma (2012)
	Sitobion avenae	Regulating aphid population dynamics	Ma et al. (2015)
	Aphis craccivora (cowpea aphid)	Negative effects on life history traits and demographic parameters	Chen et al. (2013)
	S. avenae	Reduced fecundity and increased physiological development period causing reduced population growth rate	Jeffs and Leather (2014)
	M. persicae	Clonal mixing limited local population adaptation to areas	Alford et al. (2012)
	Aphidiusmatricariae and *Aphelinusabdominalis*	Impacts on aphid populations were greater under heat wave conditions than when there were periodic hot days	Gillespie et al. (2012)
	B. tabaci (Q biotype)	Spread might be diminished by reduced vector fitness during heat waves caused by the global warming process	Pusag et al. (2012)
	B. tabaci (biotype B)	At 35 °C, egg hatching, immature survival rate, and fecundity all declined	Guo et al. (2013a)
	B. tabaci (biotype B)	Its oviposition, nymphal survival, and reproduction diminished	Cumutte et al. (2014)
	B. tabaci (Q biotype)	Negative effects on survival, longevity, and reproduction	Lu et al. (2014)
	Thrips palmi	Population trend index was also highest at 25 °C and lowest at 16 °C, and its optimal developmental temperature was 25 °C	Yadav and Chang (2014)
	F. occidentalis	Survival, longevity, and reproduction diminished	Li et al. (2014)

	F. occidentalis	The longevity of parental adults and larvae number decreased as the period of exposure to heat increased	Wang et al. (2014b)
	Aceria tosichella	Length of survival of host decreased from 106 h at 10 °C to only 17 h at 30 °C	Wosula et al. (2015)
	Xiphenema and Trichodorus	Increase in the importance and distribution of nematode-transmitted viruses	Stvenbak et al. (2012)
Greenhouse gases	*R. padi (oat aphid)*	Relative growth rate of *R. padi* increased in wheat although *M. persicae* decreased in canola	Oehme et al. (2013)
	Brevicorynebrassicae (cabbage aphid)	eCO_2 increased glucosinolate levels; this reduced the performance of the vector species	Klaiber et al. (2013a,b)
	R. padi and Festucaarundinacea, syn. Schedonorusarundinaceus	Aphid abundance diminished at a slightly increased eCO_2 level but increased at a higher level, dependent on host genotype	Ryan et al. (2014a)
	M. persicae	Under eCO_2 conditions host species affected feeding behavior more than changes in plant chemistry	Sudderth and Sudderth (2014)
	M. persicae	Prereproductive period of *M. persicae* increased and decreased its fecundity and salivation into sieve tubes	Dá der et al. (2016)
	Amphorophoraidaei	Lady bird larvae (*Harmonia axyridis*) were severely impaired under eCO_2	Hentley et al. (2014)
	F. occidentalis	Increasing the activities of two types of detoxifying enzymes and decreasing the activity of the protective enzyme	Liu et al. (2014)
Rainfall	*Aphidiuservi*	Altered aphid development rates on drought-stressed plants	Aslam et al. (2013)
	A. pisum and *Therioaphis maculate (spotted alfalfa aphid)*	Effects of drought on mixed populations of virus vectors in the field	Barton and Ives (2014)
Wind speed and direction	*A. pisum*	Aphids adopted a crouching body posture favoring them from being dislodged and carried off by a continuous gust of wind	Ben-Ari et al. (2014)
	Aphis glycines and predator H. axyridis	Effective prediction on the magnitude of arthropod-borne virus disease epidemics	Barton (2014)
Relative humidity and microclimates of leaf	*B. tabaci biotype B*	Humidity influenced the survivability	Chu et al. (2012)

On the other hand, acceleration in temperature is also known to influence antiviral resistance mechanisms that are of two types: RNA interference (RNAi) based, which is acknowledged as gene silencing; and some dominant resistances, which are based on protein–protein identification. RNAi-based defence mechanism of plant against different viruses and pathogens is initiated by dsRNA and mediated by a variety of small RNAs with a cascade of different enzymatic processes, responsible for plant growth and response production in stress as well (Vazquez 2006). Moreover, by targeting and destroying siRNA of virus; RNAi pathway defends the plants in sequence-specific manner although temperature affects these pathways by affecting dsRNA, which is involved in this defence mechanism. Chellappan et al. (2005) and Szittya et al. (2003) prove this point in their studies on virus Cymbidium ring spot virus and African cassava mosaic virus. Furthermore, some dominant resistances also have temperature dependency: over a certain threshold, the resistance is shut down or may weaken like in the case of the N gene (Canto and Palukaitis 2002; Dinesh-Kumar et al. 1995) and theTm-1gene of tobacco mosaic virus and theTm-2gene of tomato mosaic virus (Fraser and Loughlin 1982; Pfitzner 2006). As a result, we can conclude that some proteins elicited are temperature sensitive and their dominant resistances may be negatively affected by ambient temperature rises, whereas gene silencing-based resistance may have a much broader spectrum (Table 15.2).

15.3.1.2 Greenhouse Gases

Attention has rightly focused on most significant greenhouses gases, i.e. carbon dioxide, emitted by the activities of human beings. Around 40% of non-CO_2 greenhouse gases, i.e. ozone, nitrous oxide, and methane, act as long-lasting fluorinated gases and are responsible for direct emissions of greenhouse gases and essentially represent permanent acceleration in global radioactive forcing (Shine and Sturges 2007). Considering, the preindustrial atmospheric CO_2 concentration, found lower when compared to the current level, although it is congruently increasing, by 2050 rise up in global population might be challenging to adjust the basic demands of lives, simultaneously boost in CO_2 again will be a factor of concern (Alexandratos and Bruinsma 2012; Dlugokencky et al. 2015). Thus, enormous pressure on global parameters, adding in harsh weather patterns and events, inducing another level of complexity with vectors, viruses, and diseases, and impacting both ecological and agricultural conditions.

15.3.1.2.1 Effects on Virus

The vigorous and continuous plant virus prevalence and the losses they cause are greatly influenced by indirect activity of transmission and its abundance and directly by consequence of climate change (Jones 2009). Viruses show significant impact on food crops loss, and the severity and consequences of infection are greatly influenced due to climate shift affecting quality, quantity, total crop failure, and yield worldwide (Trębicki and Finlay 2019). In relation to climate change, major focus was shown on economically important viruses, which have significant impact on food crops (Trębicki and Finlay 2019). Despite having poor literature data to predict climate scenarios on plant viruses, in context with elevated CO_2 concentration either as a single factor or as a combination with other events describes its considerable effect in prevalence and epidemiology of viruses (Jones and Naidu 2019). Viruses may get affected by their dispersal, acquisition efficiency by vectors and their

Table 15.2 Risk analysis studies on the effects of change in climate parameters on virus(es).

Climate parameters	Affected viruses	Experimental outcomes	References
Temperature	*Barley yellow dwarf virus*	Developed virus symptoms earlier	Nancarrow et al. (2014)
	Turnip mosaic virus (TuMV)	Rate of systemic infection increased	Chung et al. (2015)
	Turnip mosaic virus (TuMV)	Favourable for virus multiplication	Nyalugwe et al. (2014)
	WSMV-infected wheat plants	Gene wsm1 is temperature sensitive where WSMV resistance gene overcame this resistance gene	Zinati et al. (2012)
	Potato virus X and *plum pox virus*	Titers and virulence of viruses decreased markedly	Aguilar et al. (2015)
	Barley yellow dwarf virus (BYDV)	BYDV epidemics are likely to increase in grass pastures	Rua et al. (2013b)
	Bean yellow mosaic virus (BYMV)	Single infections either with BYMV or with *Kabatiella caulivora* may decline or increase, respectively, and mixed infection becomes less damaging	Guerret et al. (2016)
	Turnip crinkle virus	Loss of function mutations and RNA methyl transferase genes but not in wild type	Zhang et al. (2012)
	Tomato ring spot virus	Induction of RNA silencing	Ghosal and Sanfacon (2014)
	Cucumber mosaic virus, potato virus Y, or potato virus X	Few or no symptoms found while elevating the temperature up to 30 °C	Del Toro et al. (2015)
	Peanut stunt virus	Proteins involved in metabolic processes were all more abundant	Obrepalska-Steploweska et al. (2015)
Greenhouse gases	*Barley yellow dwarf virus (BYDV)*	Increased biomass and water use efficiency	Malmstrom and Field (1997)
	Tobacco mosaic virus(TMV)	Increased virus resistance and lowered disease incidence	Zhang et al. (2015)
	Tomato yellow leaf curl virus (TYLCV)	Lowered disease severity along with increased above-ground plant height and biomass	Huang et al. (2012)

(*Continued*)

Table 15.2 (Continued)

Climate parameters	Affected viruses	Experimental outcomes	References
	CMV, PVY, or a potato virus X (PVX)	Grow larger and develop higher virus titers without affecting expression of symptoms	Del Toro et al. (2015)
	Barley yellow dwarf virus (BYDV)	Virus titers, host resistance, and biomass increase, with virus incidence to decrease, in virus-infected plants	Trębicki et al. 2015
	Potato virus X	CO_2 lowered PVX titer and attenuated its virulence toward the host	Aguilar et al. (2015)
	BYDV infection and *mycorrhizal root colonization*	Plant mutualists and their pathogens interact to alter each other's success under eCO_2	Rua et al. (2013a)
Rainfall	*Barley yellow dwarf virus (BYDV)*	Endophyte infection influences the extent of BYDV spread in grass pastures	Rua et al. (2013b)
	Turnip mosaic virus *(TuMV)*	Drought stress and heat stress deactivate the defense responses and also increase plant susceptibility, neither of which occurred with single-stress treatments	Prasch and Sonnewaald (2013)
	Myzus persicae	Drought conditions reduced the size, fecundity, and intrinsic rates of increase of the aphid population. The aphid numbers were positively correlated with moisture of soil	Simson et al. (2012)
	M. persicae and *Brevicoryne brassicae*	Influence aphid growth due to change in metabolite concentrations in phloem sap	Mewis et al. (2012)
	M. persicae and *B. brassicae*	Fecundity of both species increased with increasing nitrogen content and glucosinolate content with increase in development times	Tariq et al. (2013)
	Medicago truncatula and *Acyrthosiphon pisum*	Significant effect on virus transmission by aphids	Guo et al. (2016)
	Potato spindle tuber viroid (PSTVd)	Detections increased and the viroid found for the first time in some cropping areas and also in an uncropped native bushland area	Mackie et al. (2016)

	White clover mosaic virus (WClMV)	Combined stresses of virus infection and alterations in rainfall patterns are likely to enhance the herbage yield losses	Gerrard (2013)
Wind speed and direction	*Pea seed-borne mosaic virus (PSbMV)*	Increasing leaf due to PSbMV spread increased in blown plants in the presence of an abrasive	Congdon et al. (2016)
Relative humidity and microclimates of leaf	*Acyrthosiphon gossypii*	Influencing its numbers	Patel (2013)
	F. occidentalis	Effects on the spatiotemporal distribution	Fatnassi et al. (2015)

biology, their transmission, their distribution, and their persistence. The synergism association between viruses, increases and decreases in virulence capacity, were not observed (Aguilar et al. 2015). Some conditions may increase the viral titer but symptomology is not affected (Del Toro et al. 2015). Malmstrom and Field (1997) concluded from their study that the infections of viruses may increase in future due to elevated CO_2 concentration, and again the incidence and epidemiology have increased by up to 36% due to increased CO_2 level when compared with ambient concentration (Trębicki et al. 2015) and cause early and more severe expression of symptoms on host (Vassiliadis et al. 2016). This explains the increases in incidence of natural virus transmission, change in vector preferences, acquisition efficiency, and increased virus titer subjected to increased CO_2 level.

15.3.1.1.2 Effects on Vector

Vector housings in order Hemipetra accounting around 64% are critical and effective for majority of plant viruses' dispersal and transmission. Shift in climatic conditions alters the distribution of these vectors that significantly show impact on virus transmission and incidence. Potentially, vectors are a major controlling factor and a major driver for the expansion and emergence of viruses (Elena et al. 2011; Fereres 2015). In context with elevated CO_2 concentration, the population of vectors such as whitefly, aphids, and leafhopper may decrease or increase or remain unchanged (Robinson et al. 2012), and the relationship developed between host and vectors may get disrupted. Alternation in behavioural change, metabolism, development time, fecundity, and longevity can be seen at individual level, whereas in case of population the seen impact can be found associated with genotypic variations and population dynamics (Ryalls and Harrington 2017). It might also increase the number of insect generation, extend the length of flying season of vector, accelerate the overwintering rate, and also might affect the behaviour and fitness of vectors (Canto et al. 2009). Largely, the feeding capacity of vector increases with increase in CO_2. The vectors such as *Myzuspersicae* and *Rhopalosiphum padi* are most studied in context with CO_2 levels, despite limited case studies being found with whitefly, leafhopper, and aphids. Thus, understanding the change in behavioural pattern with respect to prevalence, population size, and effectiveness in context of climate change is accurately needed to work at scientific advancements.

15.3.1.3 Change in Precipitation (Rainfall)

Viruses' infections are often regarded as most important driver for around 47% of emerging infectious diseases. Their spread and survival were entirely dependent on appropriate vector and host, thus emerging as a threat for agricultural systems. For an infection to befall, disease triad involves a suitable environment, a pathogen (viruses), and a susceptible host, and this happens due to favourable climatic conditions. Therefore, change in climate in any way can be driver or controller for this triad relationship (Zayan 2019). The direct consequence, i.e. rainfall may influence the vector–virus–environmental relationship. This predicts how climate shifting might pull the infection rate of viruses and their biological behaviour to infect host (Jones and Barbetti 2012).

15.3.1.3.1 Effect on Viruses

There can be two events related to rainfall depending on the range of precipitation between an extreme rainfall and mean rainfall pattern. This alternation in pattern and amount due to climate shift may likely support the incidence and spread of plant viruses (Schumann and D'Arcy 2006). Altered rainfall may project the altered way of spreading of viruses in soil through nematodes or fungi (Jones 2009). The incidence and transmission of virus get affected in mid-latitude area due to less activity and movement of their suitable vector; this happens due to decline in rainfall and moisture content of soil (Delfosse et al. 1997). Seasonal change in rainfall pattern has always been found to be critical in the distribution of pathogens. The events related with rainfall create conditions like flood or drought influencing viral dispersal by diminishing or enhancing their epidemics. It has been found that combinational events such as flood or drought with viral infection can be more damaging rather than by either of the individual parameters, thus additive effect would greatly be a curse (Clover et al. 1999; Olson et al. 1990). Consequences like flood and drought can pull the ability of unprotected viruses to act as inoculum in soil and their survival towards their susceptible food crops (Johnson and Ogden 1929).

In case of water-borne viruses, combinational events can promote their distribution via rivers, streams, and drainage channels (Koenig 1986; Sevik 2011; Tomlinson and Faithfull 1984). Similarly, the prevalence of contact-transmitted viruses shows greater intensity and frequency with combinational events such as heavy rain associated with summer storms, which would provide moisture and wounding of host and also expect to favour their spread, thus enhancing severe epidemics (Boland et al. 2004). Heavy rainfall due to climate change can anticipate the epidemiology of virus–vector in many ways. It would splash the vectors off, thereby lagging virus transmission by decreasing vector population size (Layland and Thripspalmi 1991).

15.3.1.3.2 Effects on Vector

Vectors are important carriers of plant viruses; consequently changes in climatic condition such as rainfall pattern may alter the transmission rate of plant viruses. In some cases, decrease in rainfall may favour the increase in the population rate of some vectors (Morales and Jones 2004). The increased movement and activity of vectors and their zoospores as a result of soil moisture content will increase the spread of economically important viruses and have an impact on food crops in various parts of the world (Henry and Adams 2004; Jones 1988, 2009; Kuhn 2009; Schlosser 1988; Wissett 1991). In those areas where there is decline in mean rainfall, the mid-latitude crops get less affected by nematode-transmitted viruses as their incidence is diminished due to less moisture content (Jones 2009). In a same way, there would be less prevalence of virus with fungal vector as the movement and activity of vector seems decreased in soil (Delfosse et al. 1997; Thouvenel et al. 1988) Decline in rainfall can also limit vector population size, thus preventing economic losses due to epidemiology of virus (Matthews 1970). Consequently, both increase or decrease in rainfall pattern either inhibits the movement of vector or lowers the moisture content of soil, thus hindering the circulation of viruses (Johnson and Ogden 1929).

15.3.1.4 Effects of Direction and Speed of Wind

Regular spike in velocity of wind linked with climate alteration is likely associated with expansion and dispersal of mite and insect vectors via wind flow, especially species of those vectors that already disperse readily. Virus spread through arthropod can disperse over long distances through wind. In addition viruliferous aphids also can be distributed over long distances and are responsible for the initiation of new viral infection from the initial virus inoculum source (Jones 2016). Depending on the speed and direction of wind, the spread of arthropod-borne viruses is inhibited when the speed of wind is high and down-wind, which is responsible for distribution of disparate insects and mites vectors as well. Simultaneously these winds also favour the spread of viruses with tiny, wingless mite vectors. Fluctuation in speed and direction of wind due to global climate change is likely to have intense implications for virus-borne diseases through arthropod vectors that spread locally as well as over distance (Jones and Barbetti 2012).

15.3.1.5 Relative Humidity and Microclimates of Leaf

The relative humidity of leaves regulate factors such as span of leaf wetness, which is coupled with temperature, when compared with other climate variables, it is more predictable. High relative humidity is linked with typified local microclimates of moist leaf surfaces, and when compared with low humidity it is found associated with dry microclimate of leaf surface. Relative humidity and microclimates of leaf surface can have noticeable effects on the rate of population survival, growth, behaviour, and drift of arthropod virus vectors. Entomopathogenic fungi that kill aphid vectors are also favoured by high humid environment. Growth of lush plant and its soft tender leaves that grow under high relative humidity condition favour the spread of contact-transmitted viruses in crops and pastures. These plants are more susceptible as compared with hard leaf plants; in case of infection, typical humidity conditions as wounds develop more frequently when grown on soft leaf as viruses feed on damaged cuticle layer before they attack the infected cells (Jones and Barbetti 2012). Relative humidity at evening, the number of rainy days, and minimum temperature played considerable roles for influencing virus numbers. Moreover, under different climate change scenarios the consequences on relative humidity and microclimate of leaf need to be focused especially (i) spread of the viruses and their vectors via which they are transmitted and (ii) contact-transmitted viruses and their spread

15.3.2 Indirect Effects

Shaping the structure and nature of whole community and individual species, pathogens are ruled out (Burdon and Laine 2019). These are found to impact the genetic structure and population size of individual vector and host (Croll and Lain 2016). Thus, greater diversity accelerates the prevalence and spread of infectious diseases and the evolution of pathogens (Ostfeld and Keesing 2012) in agriculture systems contributed by climatic conditions (Yang et al. 2019). Meta population is a community where host, vector, and its associated pathogens live, and it is important in regulating local plant–pathogen dynamics in response to climatic shift (Burdon and Thrall 2014; Smith et al. 2011). The eventuality of climate change is recognized to cause circuitous impact on virus–vector interaction in natural plant community (Parmesan 2006), altering the relative frequency of viruses and their nature of interactions with vector in associated ecology (Canto et al. 2009). Furthermore, higher latitudinal and longitudinal expansion in their geographical areas are observed, as are temporal shifts in seasonal peak, which can reshape the geographical limits of species, alternation in their geographical distribution, which can alter the relative frequency and nature of interaction with other individuals, and can generate a new reservoir combination.

Climate instability and global shift are always a challenge for humankind to achieve disease management of natural vegetation and cultivated crops. This instability is due to rapid population growth that consequently increases the global food insecurity leading to the replacement of natural biodiversity. Input of diverse climatic parameters anticipating its effects on plant vectors and plant viruses is likely less available; however, some reviews provide the critical impact of climatic conditions on viruses and vectors and total effects on plant pathosystem changing agricultural pattern (Garrett et al. 2006; Southerst et al. 2011). The indirect and direct shift in climatic parameters influence vector and virus transmission behaviour, as well as altering their origin, incidence, emergence, and evolution, thus enhancing epidemics (Jones and Barbetti 2012). Epidemic of nonvector-borne and vector-borne plant viruses are influenced by many aspects and geographically diverse consequences of climate change worldwide, which would provide knowledge of how climatically influenced virus disease changes the pattern of natural vegetation and cultivated plants. The changing climatic pattern often results in critical spreading of viruses, but some time reverse effects may be seen depending on conditions and pathosystem. The pathosystem, such as vector-borne viruses, may add significant variations in the emergence of previously unknown viruses, further increasing the adversity for domineering epidemics of infectious plant viruses predicted to emerge as a result of climate change. This would ensure significant research effort to take preventative measures for global biodiversity and food security.

15.3.2.1 Alteration in Cultivation Systems

Modernization and rapid growth of human activity has led to the overexploitation of natural resources to fulfil their basic demands that complicate agricultural practices. This also initiates wide use of monoculture. Changes in cultivation systems involve inappropriate large-scale plant cultivation in an area due to climate shifting. Climatic consequences include use of protected cropping increase, agricultural diversification, extensification, intensification, use of irrigation in arid regions to furnish all-year-round cropping, and use of chemicals to provide control measures. These changes in cultivation pattern often cause constant epidemics of damaging viruses (Jones 2009). In mid-latitudinal, semi-arid, and arid regions, climate becomes hotter and drier that limits the choice of cultivating rained crops. Land degradation acts as a critical concern, and dry land cultivation in many such areas gives rise to difficulties. Such cultivation patterns involve use of improved rainfall efficiency, thus increasing maximum availability of rainfall proportion to crops by decreasing the loss of moisture content from soil. Although such practices support crop development on one hand, but on the other hand there is an increase in the incidence of vectors around the crops, accelerating epidemics of damaging and infectious viruses (Jones and Barbetti 2012).

Though protected cropping is beneficial, at the same time they are highly facilitative for the incidence of vectors and viruses. The epidemics are supported by hot temperatures, high relative humidity, cultural practices, and irrigation mode, which favour spreading of vectors and viruses. However, potential climatic shift are abrasive.

15.3.2.2 Cultivated Plant or Weed Acting as Reservoir Host

During transportation of plants and plant products, seeds of weed species get introduced in them as contaminants, which promotes the introduction of seed-borne viruses. Weed species requires a high proportion of rainfall and specific temperature to establish themselves, otherwise they will not be stable. This requirement may be ensured by change in climatic condition that subsequently made them establish themselves. Seeds as contaminants can introduce viruses that start adapting in surrounding environment by invading indigenous vegetation of particular region, and the native vegetation promotes epidemiology of viruses by providing an attractive vector to the host plants. Subsequently, newly introduced weeds get infected by particular viruses and start acting as a reservoir for particular viruses and vector. This accelerates the emergence of viruses to newly introduced food crops from native vegetation (Jones 2009). However, vectors and viruses that are already present in an area might get a favourable host as introduced weeds and concerned viruses might damage locally grown important crops. If an area lacks weed management, this would accelerate the epidemic of damaging virus on vulnerable vegetation grown in a particular area. Moreover, presence of virus reservoir would accelerate the prevalence and epidemiology of virus to and within, which might replace the traditionally grown crop into newly introduced crops. Similarly, a condition would also arise when newly introduced crops turn out to be a fit host for native arthropod vector, thus serving as a reservoir and increasing population size of vector, and it also drives the virus incidence from introduced cultivated plant to vulnerable traditional crops (Jones and Barbetti 2012).

15.3.2.3 Changing Range of Regional Areas Cultivation and Cultivated Plants Grown

Previously, plants of regional areas are grown in higher latitude regions that were cooler, but due to shift in climatic conditions these regions have gradually become hotter, and during annual cropping period the availability of rainfall increases that supports growth diverse crops. This increase in the level of virus inoculum and extent of the range of vector-borne and nonvector-borne plant viruses thus favour virus epidemics. Conversely, in subtropical, semi-arid, and arid regions, due to climate change, there is an increase in drought and a reduction in rainfall condition; in such areas irrigation is very difficult to run out, therefore, area is crucial for crop diversity. Due to small growing period and field size, it is less favourable for incidence, distribution, activity, and diversity of vectors and viruses. However, such situation might not apply always because some vectors like *B. tabaci* are actively found transmitting viruses (Jones 2016).

Talking about some areas such as North Ostrobothnia, a region in Europe, that are mainly potato seeds producing area. Its growing season starts from June and lasts till early September, a favourable environment limits the incidence of virus (Kirchner et al. 2011, 2013, 2014). Future climate shift can increase the area of cultivation, annual cropping period, and crop diversity, which might favour increasing incidence of vectors such as aphids, and their abundance and diversity eventually result in transmission of aphids-borne potato viruses. Further, Finland has a cool annual cultivating season where forest area is 75% and farming area is 10%. By 2050, the annual cropping period is likely to expand by five weeks, and an increment in both the range of crops grown and cropping area is expected (Forsius et al. 2013). Nevertheless, the wheat belt region of southwest Australia is a mid-latitude and semi-arid region, affected greatly by climatic change. Due to the lack of irrigation facilities there, in winter/spring crops are grown between June and November. The rainfall pattern results in low, medium, and high rainfall zones of wheat belts. However, under such circumstances, low rainfall zones are expanded and high and medium regions go down towards coast. This causes uncultivated land and reduction in crop diversity. Thus, wool production and diversified cropping were replaced by wheat cultivation and also affected the diversity, incidence, and distribution of virus and vector due to diminishing of green bridge (Hawkes and Jones 2005; Jones et al. 1990; McKirdy and Jones 1996).

As a range of cultivated plants were restricted to hotter regions, expansion is seen due to acceleration in rainfall and temperature simultaneously with the distribution of vectors and viruses via infected vegetative propagules, seed, dust, or soil and wind. This is also supported by expanding transportation of plant and its products. Therefore, it results in exposure of native plants contaminants to new vegetation and vice versa. Such new encounters might project to new openings for rapid evolution of adaptive viruses. Prejudicial epidemic is likely to occur between generalist and specialist viruses. Specialist viruses are less adapted with new encounters, because of vector and geographical range of their host plant rather than generalist virus. However, as a result of climate change, epidemic-favoured and native vegetation are exposed to new vectors and viruses introduced through new encounters, limiting their natural management strategies, which would harm native vegetation through introduced virus (Jones 2016).

In arid, semi-arid, and subtropical regions of drying middle latitude, climate shift reduced the new encounter between native plants and introduced viruses, hosts, and indigenous viruses. Such situations occur due to decrease in rainfall and increase in drought condition, which reduce the use of land for cropping (Jones 2009; Jones and Barbetti 2012). Gilbertson et al. (2015) explain that two economically important vectors *Frankliniella occidentalis* and *B. tabaci* have driven the emergence of viruses in past 20 years disproportionately. It is also remarked that genome mutation of virus increases their host range (Bedhomme et al. 2012; Kehoe et al. 2014; Vassilakos et al. 2016). Necessarily, progressive climatic change might introduce vector to drive virus emergence from native to introduced plants leading to substantial loss of food crops and its production (Jones 2014; Krishnareddy 2013). As, host adaptation is high in case of generalist viruses, transportation of plant and plant products may cause serious consequences due to introduction of viruses through trading (Jones and Coutts 2015; Vincent et al. 2014). The most recent review highlights the interrelated insect vector for driving virus emergence from native cultivation (Fereres 2015). They are potentially involved in host-to-host transmission of virus and their emergence.

15.4 Research Impediments and Precedence

To address and understand the influence of climate shift on ecology and evolution of natural plant communities and pathogens, researchers have to exponent an essential resource exploring climatic fluctuations from short term to long term year-to-year studies. This will acknowledge the ecological and evolutionary fluctuations in progressing manner. Circumstances addressing such studies need punctuality and longevity and funding in a very significant manner, which is again a difficulty. To achieve the viability of such studies, encouragement must be given to short-term inputs from scholars, postdoctorals, specialists of evolutionary and epidemiology, and researchers to achieve long-term information (Burdon and Thrall 2014). Climatic parameters get influenced by multifactorial phenomenon and simultaneously influence multiple biological and ecological behaviours. Changing climatic condition attracts interest of plants, pathogens, and their interactions, which supports species adaption and evolution.

Unfortunately, all these factors are at micro level, which creates complexity for monitoring them. For this, development of rapid environmental sensors together with effective reporter is advantageous. Knowledge related to plant–pathogen interaction and mechanisms are important to understand their changing behaviour towards changing environment. Lack of sufficient understanding related to them becomes a barrier to predict response of both wild plant and agricultural communities (Zhan et al. 2014). Besides, there is an increasing interest in the area to oppose the risk posed by infectious pathogens in agriculture by applying eco-evolutionary strategies. The approaches are organized in an effective way to evaluate and minimize the evolutionary changes, epidemics frequency, and enhance stability of yield (Zhang et al. 2015).

Since effective adaptations of species towards fluctuating environment are annoying, it leads to a change in genetic mechanisms causing novel mutational change. Considering those changes favouring disease evolution would need empirical study along with experimental approaches to give a significant framework for future challenges (Hoffmann et al. 2011, Yang et al. 2016). Human attention is needed in those areas that can mitigate disease emergence and evolution together with using better adapted crops, crop management, and use of agricultural land for cultural practices. These are needed for integrated strategies of agriculture, transgenic avenue for better monitoring, and management of natural ecology and agriculture biodiversity.

15.5 Conclusion

Potential effects of climate change on environmental components, i.e. virus, vector, and plants, can be understood by their severe impact on the geographical distribution of plant viruses with incidence of diseases caused by them, which is also an important issue. Adverse climate condition will affect the quality and quantity of crops. Although our awareness regarding multifactor climate changes that may affect the crop health is limited, from the aforementioned review we can predict that climate change influences development of both the virus and vector and variation in host resistance, consequently leading to change in plant–vector–virus triad interaction and increasing the severity of plant diseases. Moreover, on account of specific nature of their interaction, alteration in climate has positive, negative, and some time neutral impact on this triad interaction. The subsequent outcome of triad interactions that may be triggered by climate modification remains far undiscovered. From the standpoint of disease management, a basic disease field scale is required to evaluate the mechanisms and dynamics that drive epidemics. Discovering these patterns will play a meaningful role in generating response for biodiversity conservation of natural communities that are very effective, especially in those areas that are currently sitting on crucial environmental boundaries or particularly in an area where pronounced impact of climate change is expected.

15.6 Summary

Complexity of agricultural act such as globalization, intensification, and diversification in an interrelated way support the disease emergence. Additionally, accelerating climate shift acts as a potential inducer of agricultural complexity, thus enhancing viruses' infection and projecting threat to the diversity of indigenous vegetation and global food security. This review summarizes information of how climate changes are subjected to influence native vegetation and cultivated plants by virus epidemics. The changing parameters such as greenhouse gases, temperature, rainfall, humidity, and direction of wind and also alternation in cultivation system, cultivated plant or weed acting as reservoir host, and changing range of regional areas and cultivated plants grown are emphasized here, subjected to cause abrasive effect and providing knowledge about influence of environmental events upon plant pathosystem, vector–virus interactions, and environment–virus–vector relationships. This also explains the appraisal of direct and indirect impact of climatic parameters on a number of vectors representing diversity of viruses. Considerably, overall analysis suggests critical viral epidemics in different ways having reverse and forward effect depending upon circumstances and pathosystem type. Future research is required to fill multiple void with information of identifying and improving unborn virus epidemics due to climate instability, though this review summarizes knowledge of possible effects of climatic change scenarios on plant viruses and global loss associated with it. It also provides up-to-date assessed data from previous reviews adding for gradual betterment of agricultural practice towards significantly important plant viruses as climatic shifting is in progress.

References

Aguilar, E., Allende, L., del Torro, F.J. et al. (2015). Effects of elevated CO_2 and temperature on pathogenicity determinants and virulence of Potatovirus X/potyvirus-associated synergism. *Molecular Plant-Microbe Interactions* 28: 1364–1373.

Ahanger, R.A., Bhat, A.H., Bhat, T.A. et al. (2013). Impact of climate change on plant diseases. *International Journal of Modern Plant & Animal Sciences* 1 (3): 105–115.

Alexandratos, N. and Bruinsma, J. (2012). *World Agriculture Towards 2030/2050: The 2012 Revision*. FAO.

Alford, L., Blackburn, T.M., and Bale, J.S. (2012). Effect of latitude and acclimation on the lethal temperatures of the peach potato aphid Myzuspersicae. *Agricultural and Forest Entomology* 14: 69–769.

Anita, W., Dominic, M., and Neil, A. (2010). *Climate Change and Agriculture: Impacts, Adaptation and Mitigation*. Paris: OECD Publishing. doi: https://doi.org/10.178 7/9789264086876-en.

Aslam, T.J., Johnson, S.N., and Karley, A.J. (2013). Plant-mediated effects of drought on aphid population structure and parasitoid attack. *Journal of Applied Entomology* 137: 136–145.

Barton, B.T. (2014). Reduced wind strengthens top-down control of an insect herbivore. *Ecology* 95: 2375–2381.

Barton, B.T. and Ives, A.R. (2014). Direct and indirect effects of warming on aphids, their predators, and mutualists. *Ecology* 95: 1479–1484.

Bedhomme, S., Lafforgue, G., and Elena, S.F. (2012). Multihost experimental evolution of a plant RNA virus reveals local adaptation and host specific mutations. *Molecular Biology* 29: 1481–1492.

Ben-Ari, M., Talal, S., and Inbar, M. (2014). Anticipatory and reactive crouching of pea aphids in response to environmental perturbations. *Environmental Entomology* 43: 1319–1326.

Boland, G.J., Melzer, M.S., Hopkin, A. et al. (2004). Climate change and plant diseases in Ontario. *Canadian Journal of Plant Pathology* 26: 335–350.

Burdon, J.J. and Laine, A.L. (2019). *Evolutionary Dynamics of Plant–Pathogen Interactions.* Cambridge: Cambridge University Press.

Burdon, J.J. and Thrall, P.H. (2014). What have we learned from studies of wild plant-pathogenic associations?—the dynamic interplay of time, space and life-history. *European Journal of Plant Pathology* 138: 417–429.

Canto, T. and Palukaitis, P. (2002). Novel N gene-associated, temperature-independent resistance to the movement of Tobacco mosaic virus vectors neutralized by a Cucumber mosaic virus RNA 1transgene. *Journal of Virology* 76: 12908–12916.

Canto, T., Aranda, M.A., and Fereres, A. (2009). Climate change effects on physiology and population processes of hosts and vectors that influence the spread of hemipteran-borne plant viruses. *Global Change Biology* 15: 1884–1894.

Chakraborty, S. and Datta, S. (2003). How will plant pathogens adapt to host plant resistance at elevated CO_2 under a changing climate. *The New Phytologist* 159: 733–742.

Chellappan, C., Vanitharani, R., Ogbe, F., and Fauquet, C.M. (2005). Effect of temperature on Geminivirus-induced RNA silencing in plants. *Plant Physiology* 138: 1828–1841.

Chen, C.Y., Chiu, M.C., and Kuo, M.H. (2013). Effect of warming with temperature oscillations on a low-latitude aphid, *Aphis craccivora*. *Bulletin of Entomological Research* 103: 406–413.

Chu, D., Tao, Y.L., Zhang, Y.J. et al. (2012). Effects of host, temperature and relative humidity on competitive displacement of two invasive *Bemisia tabaci* biotypes [Q and B]. *Insect Science* 19: 595–603.

Chung, B.N., Choi, K.S., Ahn, J.J. et al. (2015). Effects of temperature on systemic infection and symptom expression of Turnip mosaic virus in Chinese cabbage (*Brassica campestris). Plant Pathology Journal* 31: 363–370.

Clover, G.R.G., Smith, H.G., Azam-Ali, S.N., and Jaggard, K.W. (1999). The effects of drought on sugar beet growth in isolation and in combination with beet yellows virus infection. *Journal of Agricultural Science (Cambridge)* 133: 251–261.

Coakley, S., Scherm, H., and Chakraborty, S. (1999). Climate change and plant disease Management. *Annual Review of Phytopathology* 37 (1): 399–426.

Congdon, B.S., Coutts, B.A., Renton, M., and Jones, R.A.C. (2016). Pea seed-borne mosaic virus: stability and wind-mediated contact transmission in field pea. *Plant Disease* 100: 953–958.

Croll, D. and Lain, A.L. (2016). What the population genetic structures of host and pathogen tell us about disease Evolution. *New Phytologist* 212: 537–539.

Cumutte, L.V., Simmons, A.M., and Abd-Rabbou, S. (2014). Climate change and *Bemisia tabaci* (Hemiptera: Aleyrodidae): impacts of temperature and carbon dioxide on life history. *Annals of the Entomological Society of America* 107: 933–943.

Del Toro, F.J., Aguilar, E., Hernández-Walias, F.J. et al. (2015). High temperature, high ambient CO_2 affect the interactions between three positive-sense RNA viruses and a compatible host differentially, but not their silencing suppression efficiencies. *PLoS ONE* 10 (8): e0136062.

Delfosse, P., Reddy, A.S., Legrève, A. et al. (1997). Indian peanut clump virus (IPCV) infection on wheat and barley: symptoms, yield loss and transmission through seed. *Plant Pathology* 48: 273–282.

Dinesh-Kumar, S.P., Whitman, S., Choi, D. et al. (1995). Transposon tagging of Tobacco mosaic virus resistance gene N: its possible role in the TMV-N-mediated signal transduction pathway. *Proceedings of the National Academy of Sciences United States of America* 92: 4175–4180.

Dlugokencky, E., Lang, P., Masarie, K. et al. (2015). *Atmospheric Carbon Dioxide Dry Air Mole Fractions from the NOAA ESRL Carbon Cycle Cooperative Global Air Sampling Network 1968–2014*. Boulder, CO, USA: NOAA ESRL Global Monitoring Division.

Elena, S.F., Bedhomme, S., Carrasco, P. et al. (2011). The evolutionary genetics of emerging plant RNA viruses. *Molecular Plant-Microbe Interactions* 24 (3): 287–293.

Fatnassi, H., Pizzol, J., Senoussi, R. et al. (2015). Within-crop air temperature and humidity outcomes on spatio-temporal distribution of the key rose pest *Frankliniella occidentalis*. *PLoS ONE* 10: e0126655.

Fereres, A. (2015). Insect vectors as drivers of plant virus emergence. *Current Opinion in Virology* 10: 42–46.

Forsius, M., Anttila, S., Arvola, L. et al. (2013). Impacts and adaptation options of climate change on ecosystem services in Finland: a model based study. *Current Opinion in Environmental Sustainability* 5: 26–40.

Fraser, R.S.S. and Loughlin, S.A.R. (1982). Effects of temperature on the Tm-1gene resistance to tobacco mosaic virus in tomato *Lycopersicum esculentum* cultivar cragiella. *Physiological and Molecular Plant Pathology* 20: 109–118.

Gao, G.Z., Perkins, L.E., Zalucki, M.P. et al. (2012). Effect of temperature on the biology of *Acyrthosiphon gossypii* (Mordvilko) (Homoptera: Aphididae) on cotton. *Journal of Pesticide Science* 86: 167–172.

Garrett, K.A., Dendy, S.P., Frank, E.E. et al. (2006). 2010 Climate change effects on plant disease: genomes to ecosystems. *Annual Review of Phytopathology* 44: 489–509.

Gerrard, M.G. (2013). Effects of WClMV and root flooding on white clover. MSc Thesis, University of Otago, New Zealand.

Ghosal, B. and Sanfacon, H. (2014). Temperature-dependent symptom recovery in *Nicotiana benthamiana* plants infected with tomato ringspot virus is associated with reduced translation of RNA2 and requires ARGONAUTE 1. *Virology* 456–457: 188–197.

Gilbertson, R.L., Batuman, O., Webster, C.G., and Adkins, S. (2015). Role of insect supervectors *Bemisia tabaci* and *Frankliniella occidentalisin* in the emergence and global spread of plant viruses. *Annual Review of Phytopathology* 2: 67–93.

Gillespie, D.R., Nasreen, A., Moffat, C.E. et al. (2012). Effects of simulated heat waves on an experimental community of pepper plants, green peach aphids, and two parasitoid species. *Oikos* 1321: 149–159.

Guerret, G.L., Barbetti, M.J., You, M.J., and Jones, R.A.C. (2016). Effects of temperature on disease severity in plants of subterranean clover infected singly or in mixed infection with Bean yellow mosaic virus and *Kabatiell acaulivora*. *Journal of Phytopathology* 164: 608–619.

Guo, J.Y., Cong, L., and Wan, F.H. (2013a). Multiple generation effects of high temperature on the development and fecundity of *Bemisia tabaci (*Gennadius) (Hemiptera: Aleyrodidae) biotype B. *Insect Science* 20: 541–549.

Guo, H., Sun, Y., Peng, X. et al. (2016). Up-regulation of abscisic acid pathway facilitates aphid xylem absorption and osmoregulation under drought stress. *Journal of Experimental Botany* 67: 681–693.

Hawkes, J.R. and Jones, R.A.C. (2005). Incidence and distribution of Barley yellow dwarf virus and Cereal yellow dwarf virus in over-summering grasses in a Mediterranean type environment. *Australian Journal Agriculture Research* 56: 257–270.

Henry, M. and Adams, M.J. (2004). Other cereals. In: *Virus and Virus-Like Diseases of Major Crops in Developing Countries* (ed. G. Loebenstein and G. Thottappilly), 337–354. Dordrecht, The Netherlands: Kluwer.

Hentley, W.T., Vanbergen, A.J., Hails, R.S. et al. (2014). Elevated atmospheric CO_2 impairs aphid escape responses to predators, and conspecific alarm signals. *Journal of Chemical Ecology* 40: 1110–1114.

Hoffmann, A.A., Carla, M., and Sgro, C.M. (2011). Climate change and evolutionary adaptation. *Nature* 470: 479–485.

Huang, L., Ren, Q., Sun, Y. et al. (2012). Lower incidence and severity of tomato virus in elevated CO_2 is accompanied by modulated plant induced defence in tomato. *Plant Biology* 14: 905–913.

Hull, R. (2002). *Matthews Plant Virology*. San Diego, CA, USA: Academic Press.

Islam, M.S. and Kieu, E. (2020). Tackling regional climate change impacts and food security issues: A critical analysis across ASEAN, PIF, and SAARC. *Sustainability* 12: 883.

Jeffs, C.T. and Leather, S.M. (2014). Effects of extreme, fluctuating temperature events on life history traits of the grain aphid, Sitobionavenae. *Entomologia Experimentalis et Applicata* 150: 240–249.

Johnson, J. and Ogden, W.B. (1929). *The Overwintering of Tobacco Mosaic Virus*, vol. 95, 25. University of Wisconsin Agricultural Experiment Station Bulletin.

Jones, R.A.C. (1988). Epidemiology and control of potato mop top virus. In: *Viruses with Fungal Vectors* (ed. J.I. Cooper and M.J.C. Asher), 255–270. Wellesbourne, UK: *Association of Applied Biologist.*

Jones, R.A.C. (2009). Plant virus emergence and evolution: origins, new encounter scenarios, factors driving emergence, effects of changing world conditions, and prospects for control. *Virus Research* 141: 113–130.

Jones, R.A.C. (2014). Plant virus ecology and epidemiology: historical perspectives, recent progress and future prospects. *Annals of Applied Biology* 164: 320–347.

Jones, R.A.C. (2016). Future scenarios for plant virus pathogens as climate change progresses. *Advances in Virus Research* 95: 87–147.

Jones, R.A.C. and Barbetti, M.J. (2012). Influence of climate change on plant disease infections and epidemics caused by viruses and bacteria. *CAB Reviews* 7 (22): 1–32.

Jones, R.A.C. and Coutts, B.A. (2015). Spread of introduced viruses to new plants in natural ecosystems and the threat this poses to plant biodiversity. *Molecular Plant Pathology* 16: 541–545.

Jones, R.A.C. and Naidu, R.A. (2019). Global dimensions of plant virus diseases: current status and future perspectives *Annual Review. Virology* 6 (1): 387–409.

Jones, R.A.C., McKirdy, S.J., and Shivas, R.G. (1990). Occurrence of barley yellow dwarf virus in over-summering grasses and cereal crops in Western Australia, Australas. *Plant Pathology* 19: 90–96.

Kehoe, M.A., Coutts, B.A., Buirchell, B., and Jones, R.A.C. (2014). Hardenbergia mosaic virus: crossing the barrier between native and introduced plant species. *Virus Research* 184: 87–94.

Kirchner, S.M., Doring, T.F., Hiltunen, L.H. et al. (2011). Information-theory-based model selection for determining the main vector and period of transmission of Potato virus Y. *Annals of Applied Biology* 159: 414–427.

Kirchner, S.M., Hiltunen, L.H., Doring, T.F. et al. (2013). Seasonal phenology and species composition of the aphid fauna in a northern crop production area. *PLoS ONE* 8 (8): e71030.

Kirchner, S.M., Hiltunen, L.H., Santala, J. et al. (2014). Comparison of straw mulch, insecticides, mineral oil, and birch extract for control of transmission of Potato virus Y in seed potato crops. *Potato Research* 57: 59–75.

Klaiber, J., Dorn, S., and Najar-Rodriguez, A.J. (2013a). Acclimation to elevated CO_2 increases constitutive glucosinolate levels of Brassica plants and affects the performance of specialised herbivores from contrasting feeding guilds. *Journal of Chemical Ecology* 39: 653–665.

Klaiber, J., Najar-Rodriguez, A.J., Piskorski, R., and Dorn, S. (2013b). Plant acclimation to elevated CO_2 affects important plant functional traits, and concomitantly reduces plant colonization rates by an herbivorous insect. *Planta* 237: 29–42.

Koenig, R. (1986). Plant viruses in rivers and lakes. *Advances in Virus Research* 31: 321–333.

Krishnareddy, M. (2013). Impact of climate change on insect vectors and vector-borne plant viruses and vectors. In: *Climate-Resilient Horticulture: Adaptation and Mitigation Strategies* (ed. H.P. Singh, H.H.K. Rao and K.S. Shivashankar), 255–277. India: Springer.

Kuhn, T. (2009). Soil-borne viruses affecting cereals: known for Long but still a threat. *Virus Research* 141: 174–183.

Ladanyi, M. and Horvath, L. (2010). A review of the potential climate change impact on insect populations –general and agricultural aspects. *Applied Ecology and Environmental Research* 8: 143–152.

Layland, J. and Thripspalmi (1991). *AgNote*. Northern Territory, Australia: Department of Primary Industry and Fisheries.

Li, H.B., Zheng, Y.T., Sun, D.D. et al. (2014). Combined effects of temperature and avermectins on life history and stress response of the western flower thrips, *Frankliniella occidentalis*. *Pesticide Biochemistry and Physiology* 108: 42–48.

Liu, J.Y., Quian, L., Jiang, X.C. et al. (2014). Effects of elevated CO_2 concentration on the activities of detoxifying enzymes and protective enzymes in adults of *Frankliniella occidentalis* and *F. intonsa* (Thysanoptera: Thripidae). *Acta Entomologica Sinica* 57: 745–761.

Lu, Z.H., Wang, Y.M., Zhu, S.G. et al. (2014). Trade-offs between survival, longevity, and reproduction, and variation of survival tolerance in Mediterranean *Bemisia tabaci* after temperature stress. *Journal of Insect Science* 14: 1–11.

Ma, G. and Ma, C.S. (2012). Effect of acclimation on heat-escape temperatures of two aphid species: implications for estimating behavioral response of insects to climate warming. *Journal of Insect Physiology* 58: 303–309.

Ma, G., Hoffman, A.A., and Ma, C.S. (2015). Daily temperature extremes play an important role in predicting thermal effects. *The Journal of Experimental Biology* 218: 2289–2296.

Mackie, AE, Rodoni, BC, Barbetti, MA, Mc Kirdy, SJ, Jones, RAC 2016, Potato spindle tuber viroid: alternative host reservoirs and strain found in an isolated sub-tropical irrigation area, *European Journal of Plant Pathology* DOI:https://doi.org/10.1007/s10658-016-0857-2.

Malmstrom, C.M. and Field, C.B. (1997). Virus-induced differences in the response of oat plants to elevated carbon dioxide. *Plant, Cell & Environment* 20: 178–188.

Matthews, R.E.F. (1970). *Plant Virology*, 778. New York: *Academic Press*.

McKirdy, S.J. and Jones, R.A.C. (1996). Use of imidacloprid and newer generation pyrethroids to control the spread of barley yellow dwarf luteovirus in cereals. *Plant Diseases* 80: 895–901.

Mewis, I., Khan, M.A.M., Glawischnig, E. et al. (2012). Water stress and aphid feeding differentially influence metabolite composition in Arabidopsis thaliana (L.). *PLoS ONE* 7: e48661.

Morales, F.J. and Jones, P.G. (2004). The ecology and epidemiology of whitefly-transmitted viruses in Latin America. *Virus Research* 100: 57–65.

Nancarrow, N., Constable, F.E., Finlay, K.J. et al. (2014). The effect of elevated temperature on Barley yellow dwarf virus-PAV in wheat. *Virus Research* 186: 97–103.

Newbery, F., Qi, A., and Fitt, B.D.L. (2016). Modelling impacts of climate change on arable crop diseases: progress, challenges and applications. *Current Opinion in Plant Biology* 32: 101–109.

Nyalugwe, E.P., Barbetti, M.J., and RAC, J. (2014). Preliminary studies on resistance phenotypes to Turnip mosaic virus in *Brassica napus* and *B. carinata* from different continents and effects of temperature on their expression. *European Journal of Plant Pathology* 139: 687–706.

Obrepalska-Steploweska, A., Renault, J., Planchon, S. et al. (2015). Effect of temperature on the pathogenesis, accumulation, of viral, and satellite RNAs and on plant virus proteome in peanut stunt virus and satellite RNA-infected plants. *Frontiers in Plant Science* 6: 903.

Oehme, V., Hogy, P., Zebitz, C.P.W., and Fangmeier, A. (2013). Effects of elevated CO_2 concentrations on phloem concentrations of spring crops and aphid performance. *Journal of Plant Interactions* 8: 74–84.

Olson, A.J., Pataky, J.K., D'Arcy, C.J., and Ford, R.E. (1990). Effects of drought stress and infection by maize dwarf mosaic virus on sweet corn. *Plant Disease* 74: 147–151.

Ostfeld, R.S. and Keesing, F. (2012). Effects of host diversity on infectious disease. *Annual Review of Ecology, Evolution, and Systematics* 43: 157–182.

Parmesan, C. (2006). Ecological and evolutionary responses to recent climate change. *Annual Review of Ecology, Evolution, and Systematics* 37: 637–669.

Patel, Y. (2013). Effect of environmental factors on population dynamics of *Aphisgossypii* (Glover) (Aphididae; Hemiptera), in cotton under agro-climatic conditions of Madhya Pradesh. *International Journal of Plant Production* 6: 394–397.

Pfitzner, A.J.P. (2006). Resistance to Tobacco mosaic virus and Tomato mosaic virus in tomato. In: *Natural Resistance Mechanisms to Plant Viruses*, 299–413. Dordrecht, The Netherlands: *Springer*.

Prasch, C.M. and Sonnewaald, U. (2013). Simultaneous application of heat, drought and virus to Arabidopsis plants reveals significant shifts in signaling networks. *Plant Physiology* 162: 1849–1866.

Pusag, J.C.A., Jahan, S.M.H., Lee, K.S., and Li, K.Y. (2012). Up regulation of temperature susceptibility in *Bemisia tabaci* upon acquisition of Tomato yellow leaf curl virus. *Journal of Insect Physiology* 58: 1343–1348.

Robinson, E.A., Ryan, G.D., and Newman, J.A. (2012). A meta-analytical review of the effects of elevated CO_2 on plant–arthropod interactions highlights the importance of interacting environmental and biological variables. *New Phytologist* 194: 321–336.

Rua, M.A., Umbanhowar, J., Hu, S. et al. (2013a). Elevated CO_2 spurs reciprocal positive effects between a plant virus and an arbuscular mycorrhizal fungus. *The New Phytologist* 199: 441–549.

Rua, M.A., Mc Culley, R.L., and Mitchell, C.E. (2013b). Climate drivers, host identity, and fungal endophyte infection determine virus prevalence in a grassland ecosystem. *Journal of Ecology* 102: 690–699.

Ryalls, J.M.W. and Harrington, R. (2017). Climate and atmospheric change impacts on aphids as vectors of plant diseases. In: *Global Climate Change and Terrestrial Invertebrates* (ed. S.N. Johnson and T.H. Jones), 148–175. *Wiley*.

Ryalls, J.M.W., Riegler, M., Moore, B.D. et al. (2013). Effects of elevated CO_2 on above ground-below ground systems: a case study with plants, their mutualistic bacteria and root/shoot herbivores. *Frontiers in Plant Science* 4: 445.

Ryan, G.D., Emiljanowicz, L., Harri, S.A., and Newman, J.A. (2014a). Aphid and host-plant genotype interactions under elevated CO_2. *Ecological Entomology* 39: 309–315.

Schlosser, E. (1988). Epidemiology and management of Polymyxabetae and beet necrotic yellow vein virus. In: *Viruses with Fungal Vectors* (ed. J.I. Cooper and M.J.C. Asher), 281–292. Wellesbourne, UK: *Association of Applied Biologists*.

Schumann, G.L. and D'Arcy, C.J. (2006). *Essential Plant Pathology*, 338. St. Paul, MN: *American Phytopathological Society*.

Sevik, M.A. (2011). Water pollution: water-borne plant viruses. *Erciyes Universitesi Fen Billericay Enstitusii Dergisi* 27: 40–47.

Shine, K.P. and Sturges, W.T. (2007). CO_2 is not the only gas. *Science* 315: 1804–1805.

Simson, K.L.S., Jackson, G.E., and Grace, J. (2012). The response of aphids to plant water stress the case of *Myzuspersicae* and *Brassica oleraceavarcapitata*. *Entomologia Experimentalis et Applicata* 142: 191–202.

Smith, D.L., Ericson, L., and Burdon, J.J. (2011). Co-evolutionary hot and cold spots of selective pressure move in space and time. *Journal of Ecology* 99: 634–641.

Southerst, R.W., Constable, F., Finlay, K.J. et al. (2011). Adapting to crop pest and pathogen risks under changing climate. *Wiley Interdisciplinary Reviews: Climate Change* 2: 220–237.

Sturrock, R.N., Frankel, S.J., Brown, A.V. et al. (2011). Climate change and forest diseases. *Plant Pathology* 60: 133–149.

Stvenbak, K., Maraldo, K., Georgieva, S. et al. (2012). Suppression of soil decomposers and promotion of long-lived root herbivorous nematodes by climate change. *European Journal of Soil Biology* 52: 1–7.

Sudderth, E.A. and Sudderth, E.B. (2014). Quantifying aphid behavioral responses to environmental change. *Entomologia Experimentalis et Applicata* 150: 7–18.

Szittya, G., Silhavy, D., and Molnar, A. (2003). Low temperature inhibits RNA silencing-mediated defense by the control of siRNA generation. *EMBO Journal* 22: 6233–6640.

Tariq, M., Rossiter, J.T., Wright, D.J., and Staley, J.T. (2013). Drought alters interactions between root and foliar herbivores. *Oecologia* 172: 1095–1104.

Thouvenel, J.C., Fauquet, C., Fargette, D., and Fishpool, L.D.C. (1988). Peanut clump virus in West Africa. In: *Viruses with Fungal Vectors* (ed. J.I. Cooper and M.J.C. Asher), 247–254. Wellesbourne, UK: Association of Applied Biologist.

Tomlinson, J.A. and Faithfull, E.M. (1984). Studies on the occurrence of tomato bushy stunt virus in English rivers. *Annals of Applied Biology* 104: 485–495.

Trębicki, P. and Finlay, K. (2019). Pests and diseases under climate change; its threat to food Security, *Food Security. Climate Change* 229–250.

Trębicki, P., Nancarrow, N., Cole, E. et al. (2015). Virus disease in wheat predicted to increase with a changing climate. *Global Climate Change* 21: 3511–3519.

Vassilakos, N., Simon, V., Tzima, A. et al. (2016). Genetic determinism and evolutionary reconstruction of a host species jump in a plant virus. *Molecular Biology and Evolution* 33: 541–553.

Vassiliadis, S., Plummer, K.M., Powell, K.S. et al. (2016). The effect of elevated CO_2 and virus infection on the primary metabolism of wheat. *Functional Plant Biology* 43 (9): 892–902.

Vazquez, F. (2006). Arabidopsis endogenous small RNAs: high-ways and by ways. *Trends in Plant Science* 11: 460–468.

Vincent, S.J., Coutts, B.A., and Jones, R.A.C. (2014). Effects of introduced and indigenous viruses on native plants: exploring their disease causing potential at the agro-ecological interface. *PLoS ONE* 9 (6): e91224.

Wang, J.C., Zhang, B., Li, H.G. et al. (2014b). Effects of exposure to high temperature on *Frankliniella occidentalis* (Thysanoptera: Thripidae), under Arrhenotoky and sexual reproduction conditions. *Florida Entomologist* 97: 504–510.

Wissett, M.V. (ed.) (1991). *Compendium of Wheat Diseases*, 2e. Minnesota: The American Phytopathological Society.

Wosula, E.N., Mc Mechan, A.J., and Hein, G.L. (2015). The effect of temperature, relative humidity and virus infection status on off-host survival of the wheat curl mite (Acari: Eriophyidae). *Journal of Economic Entomology* 108: 1545–1552.

Wu, E.J., Wang, Y.P., Yahuza, L. et al. (2020). Rapid adaptation of the Irish potato famine pathogen *Phytophthora infestans* to changing temperature. *Evolutionary Applications* 13: 768–780.

Yadav, R. and Chang, N.T. (2014). Effects of temperature on the development and population growth of melon thrips, *Thrips palmi*, on eggplant, *Solanum melongena*. *Journal of Insect Science* 14.

Yang, L.N., Zhu, W., Wu, E.J. et al. (2016). Trade-offs and evolution of thermal adaptation in the Irish potato famine pathogen. *Molecular Ecology* 25: 4047–4058.

Yang, L.N., Pan, Z.C., Zhu, W. et al. (2019). Enhanced agricultural sustainability through within-species diversification. *Nature Sustainability* 2: 46–52.

Zayan, S.A. (2019). Impact of climate change on plant diseases and I'm strategies. doi: https://doi.org/10.5772/intechopen.87055.

Zhan, J., Thrall, P.H., and Burdon, J. (2014). Achieving sustainable plant disease management through evolutionary principles. *Trends in Plant Science* 19 (9): 570–575.

Zhang, X., Zhang, X., Singh, J. et al. (2012). Temperature-dependent survival of Turnip crinkle virus-infected Arabidopsis plants relies on an RNA silencing-based defense that requires DCL2, AGO2, and HEN1. *Journal of Virology* 86: 6847–6854.

Zhang, S., Li, X., Sun, Z. et al. (2015). Antagonism between phytohormone signaling underlies the variation in disease susceptibility of tomato plants under elevated CO_2. *Journal of Experimental Botany* 66: 1951–1963.

Zinati, Z., Assad, M.T., Masumi, M. et al. (2012). The effect of high temperature treatment on Wheat streak mosaic virus *resistance* and certain resistance-related chemicals in bread wheat. *Iranian Journal of Virology* 6: 27–35.

16

Utilization of Biowastes in Green Chemistry

Pratibha Deka[1], Mayuree Gohain[1], Nilutpal Bhuyan[2], Nirmali Gogoi[1], and Rupam Kataki[2]

[1] *Department of Environmental Science, Tezpur University, Tezpur, Assam, India*
[2] *Department of Energy, Tezpur University, Tezpur, Assam, India*

CONTENTS

16.1 Introduction

Green chemistry is an environmentally sustainable branch of chemistry which emphasizes on designing chemical products and processes by the use/generation of hazardous substances (United States Environmental Protection Agency (USEPA) 2017). This branch of chemistry follows 12 principles which encourages environment friendly and sustainable chemistry practices to reduce or eliminate the hazardous waste generation (Anastas and Warner 1998).

Among different environmentally conscious terms in chemistry such as clean chemistry, environmental chemistry, benign chemistry, and sustainable chemistry, a linear fashion of growing trend exists for the term green chemistry in articles since 1998 (Linthorst 2010).

Climate Change and Agriculture: Perspectives, Sustainability and Resilience,
First Edition. Edited by Noureddine Benkeblia.

Green chemistry utilizes engineering concepts in pollution prevention and focuses on zero waste generation both at laboratory and industrial scales. Thus, by using economical and eco-compatible techniques, green chemistry improves the yield of useful products and decreases the cost of waste disposal simultaneously at the end of a chemical process (Vert et al. 2012).

Waste can be defined as any unwanted or unusable material or a substance that is discarded after primary use. Wastes of biological nature are called biowaste. Biowaste is also known as green waste. Biowaste can be decomposed aerobically or anaerobically and they are biodegradable in nature. When not properly treated, they produce hazardous substances as well as greenhouse gases (GHGs). The generation of these substances from biowaste can be reduced by adopting optimal biowaste management techniques that transform this waste into new resources like green energy or green products. The conversion of waste to wealth is very essential for the economic prosperity for which scientific intervention is important. A huge volume of biowaste is generated during the household, agricultural, and commercial activities. Selection of proper biological, physico-chemical, and thermo-chemical treatment in recovering of different products from biowaste (Lohri et al. 2017) is important. Therefore, identification of the important compound present in biowaste, and the technique for recovering it is very crucial for its sustainable utilization.

16.2 Classification and Composition of Biowaste

16.2.1 Classification

Based on the sources, biowaste can be classified into three categories:

1) **Household biowaste**: Different day-to-day activities in our households generate a huge amount of biowaste. Grass, plants, leaves, wood, kitchen scraps, fruits and vegetables, other foodstuffs, paper, cardboard, natural textiles, garden waste, soil, etc. constitute the household biowaste. Kitchen waste comprises a high percentage of household waste (Qu et al. 2009). Researchers found seasonal variation in the composition of biowaste (Hanc et al. 2011). Researchers demonstrated that up to 68% of household waste is biodegradable (Parfitt 2002).
2) **Industrial/commercial biowaste:** This includes biowaste generated from different industrial and commercial activities. Mainly food-related industries generate a considerable amount of biowaste. This sector includes the wholesale sector, hotel, and catering industry, retail sectors, municipal authorities and services, and food industries.
3) **Biowaste generated from agricultural fields/farms**: Agricultural field and farm activities also generate biowaste. The generation of this waste depends on the types of agricultural activities. For example, cultivation practices, aquaculture, and livestock production generate different biowastes (Obi et al. 2016). This may include parts of vegetation, agricultural residues, animal excreta, urine, etc. Agro-residues have attracted a large amount of research interest due to their high volume.

16.2.2 Biomolecules in Biowaste

Although composition varies, biowaste contains the same types of major molecules namely protein, carbohydrate, and lipids with lesser amounts of speciality molecules (Dugmore et al. 2017). The main constituents of agro-waste materials are lignin and cellulose and minor constituents such as lipids, proteins, simple sugars, starches, hemicellulose, extractives, water, hydrocarbons, ash, and compounds containing different functional groups (Ali and Hassaan 2017).

16.3 Methods Used for Extraction of Usable Components from Biowaste

Valuable materials can be extracted from biowaste and can be used in different fields of green chemistry. Biowaste valorization is an essential step to promote a future sustainable lifestyle, which is based on 'waste to wealth' concept. Waste valorization not only provides its intrinsic benefits to the environment but also helps to develop new technologies, livelihoods, and jobs (Xu et al. 2019). Various methods are being used in the utilization of different biowastes to obtain useful green materials. Different biowaste valorization methods are listed in Table 16.1 and discussed in a later section.

Recently, nanoparticles receive a wide application in areas such as biomedicine, biocatalysis, electronics, chemistry, energy, etc. Synthesis of Ag nanoparticles by the chemical method is toxic, flammable, and has many adverse effects in medical applications, whereas, green synthesis of AgNPs is eco-friendly, non-toxic, cost-effective, and safe for human beings (Vivek et al. 2011). Moreover, greener syntheses of nanoparticles pose other advantages namely, simple and one-step, minimum amount of chemical usage, environment friendly, cost-effective, reproducible, and avoids mandatory use of chemical reducing agents and stabilizers (Mittal et al. 2014). Jasuja et al. (2014) synthesized stable AgNPs by fast reduction and stabilization of silver ions of different molar concentrations of NaOH in aqueous solution of $AgNO_3$ using a pomegranate (*Punica granatum*) peel extract. Antimicrobial activity of the synthesized AgNPs was tested on different gram-positive and gram-negative bacteria. The study documented a significant antimicrobial activity of AgNPs compared to control antibiotic Streptomycin and inhibitory effects of AgNPs were maximum when prepared with higher molar concentration of NaOH. In a study conducted by Nikhila et al. (2018), silver nanoparticles were synthesized using a very common biowaste banana blossom peel (*Musa acuminata* Colla) that was non-toxic, eco-friendly, or green approach and economically feasible being a zero-cost method. They studied the behaviour of these green synthesized nanoparticles towards naked-eye detection of metal ions by exploiting a phenomenon called surface plasmon resonance (SPR). They also evaluated the larvicidal as well as the antibacterial properties of these nanoparticles. The results showed good antibacterial properties against Gram-positive bacteria *Escherichia coli* and mosquito larval toxicity.

Biowaste materials have been recognized as valuable sources to produce renewable, low cost, and sustainable adsorbents using nanotechnology in water purification. These

Table 16.1 Use of biowaste and their extraction methods.

Biowaste	Aim	Methods of extraction	Use	References	Source
Biowaste peel extract of *P. granatum*	Synthesis of AgNPs stabilized with biowaste	Rapid reduction and stabilization of Ag^+ ions with different NaOH molar concentrations in the aqueous solution of silver nitrate by the biowaste peel extract of *P. granatum*	Antimicrobial activities	Jasuja et al. (2014)	Original research paper
Banana blossom peel	Synthesis of silver nanoparticles using a common biowaste banana blossom peel	Preparation of banana blossom peel extract – bioreduction of Ag^+ using the extract	Detection of metal ions, larvicidal potential, and antibacterial property	Nikhila et al. (2018)	Original research paper
Agriculture biomass e.g. banana and orange peels, Bagasse pith, sawdust wastes, maize cob, coconut husk fibres, nutshells	Synthesis of nanoparticles	–	Water purification	Ali and Hassaan (2017)	Review article
Depectinized apple pomace	Production of hard carbon materials	Pyrolysis of the biowaste to produce hard carbon	Battery electrodes	Baldinelli et al. (2018)	Original research paper
Diverse waste streams including plant matter, algae, eggshell, animal bone, clay, ash, and industrial waste/by-products	Biodiesel production	Low-temperature esterification of free fatty acids (FFAs) or transesterification of triacyl glycerides (TAGs)	Production of solid acid and base catalysts	Bennett et al. (2016) and references therein	Review paper
Urban biowaste	Production of chemical auxiliaries of biological origin	Light-induced generation of reactive oxygenated species from biowaste	Photodegradation of organic pollutants	Prevot et al. (2011)	Original research paper

Eggshell biowaste and brewery wastewater	Enhanced production of fumaric acid	Eggshells were used both as a source of CaCO and as immobilizing support, brewery wastewater was used as fermentation medium for fumaric acid production through submerged fermentation		Das et al. (2015)	Original research paper
Natural biomasses e.g., agricultural crop residues, wood wastes, animal wastes, aquatic plants, and algae	Production of renewable carbon materials	Biochar production by burning or pyrolysis	Used in electrochemical energy storage devices, such as supercapacitors and lithium-sulphur batteries	Gao et al. (2017)	Review article
Cashew nutshell	Production of bio-sourced chemicals, materials, and fuels	Extraction of cashew nut shell liquid (CNSL) from cashew shells by mechanical, thermal, chemical (solvent extraction), and pyrolysis method	Applications such as prevention of termite attack, tsetse fly traps (kairomone), and synthesis of polymers	Mgaya et al. (2019)	Review article
Pine needles, a forest biowaste	Production of dialdehyde cellulose ethers	Extracted cellulose from pine needles, first etherified with different alkyl bromides (RBr) and then further modified to dialdehyde via oxidation with $NaIO_4$ to get the desired products, dialdehyde cellulose ethers	Supports for α-chymotrypsin immobilization	Kumari et al. (2016)	Original research paper

waste-based adsorbents can be used for the efficient removal of pollutants and bacteria from water and can be much more economical compared to other physico-chemical processes available. Some examples of agricultural wastes that have been analyzed and tried for their adsorption property to decontaminate waste from water are banana and orange peels (Annadurai et al. 2003; Mandina et al. 2013), sawdust wastes (Zarraa 1995), sugarcane bagasse, maize cob, coconut husk fibres (Babarinde 2002), nutshells (Demirbas 2003), soybeans and cottonseed hulls, *Datura innoxia* (Drake et al. 1995), wheat and rice bran, oat fibre, etc. (Ali and Hassaan 2017 and references therein). Researchers found that these agro-wastes can adsorb both organic as well as inorganic pollutants from water (ions of heavy metals, dyestuff, and toxic pollutants) (Abia and Asuquo 2006).

Different functional groups present in the biowaste help in the biosorption process by facilitating their complexation with heavy metals and finally sequester them. For instance, rice husk produced as a by-product of the rice milling industry is an agricultural waste. However, this is a cheaper biowaste material, but biotechnologically modified rice husks are found to be potentially very effective as adsorbents in removing lead from aqueous systems (Williams and Nugranad 2000). Kumar and Dara (1982) reported sawdust and tree barks as efficient removers of cadmium species from aqueous systems. Synthesized nanoparticles are very helpful due to eco-friendliness, economic, biocompatibility, low cytotoxicity, high antioxidant, and antimicrobial activity (Ali and Hassaan 2017).

The increasing demand and growing scarcity of natural resources led to constant search for alternatives and renewable sources of materials. Researchers are engaged in exploring sustainable alternatives to meet the increasing demand of battery electrode materials. Biowaste-derived hard carbon as a renewable and widespread source of battery electrode material is under investigation. Baldinelli et al. (2018) presented a model of hard carbon production via pyrolysis of biowaste for energy sustainability. In this model, hard carbon derived from depectinized apple pomace generated from the food industry was used to produce electrode materials. They stated that this model is very useful to estimate the energy balance of a process that uses biowaste as primary feedstock.

Catalysts derived from the transformation of biowaste materials can be utilized to obtain energy economy in chemical processes (Bennett et al. 2016). Waste biomass is being used as renewable feedstocks for the production of fuels and many chemicals, but in these processes, catalysts are used. Although, a small amount of catalysts is required, the high cost of the catalysts makes it expensive. So, various researches have been going on for a cost-effective and environmentally friendly way to obtain catalysts using waste materials in addition to the target products. There are many reports on the preparation of heterogeneous catalysts from waste materials.

Prevot et al. (2011) tested the use of waste-derived bioorganic materials. They documented the considerable efficacy of these materials in the remediation of light-induced reactive oxygen species generated from several technological applications in chemical industries and in environmental remediation. They also reported the applicability of urban waste-derived bioorganic substances (UW-BOS) as chemical auxiliaries in the photodegradation of organic pollutants. Here, the photolysis mechanism was promoted by AC8, a photosensitizer molecule (derived from a mixture of food and green residues). It has the ability of chemical change to another molecule through photochemical process and hence can be used in environmental remediation.

Two industrial biowastes (eggshells and brewery wastewater) were valorized to enhance the production of fumaric acid (FA), the strongest organic food acid and fundamental metabolite with a prime role in food acidity regulation (Das et al. 2015). It has wide application in medicinal, food, and feed industries apart from use as unsaturated polyester, cleansing agents, alkyd resins, and printing inks. In the production process, an immobilization strategy was adopted where a filamentous fungus, *Rhizopus oryzae* 1526 was developed into thin biofilms after its immobilization on eggshell surfaces. Here, eggshells served double functions: (i) source of $CaCO_3$ and (ii) support immobilization while brewery wastewater acts as a fermentation medium. Biofilms were used for FA production through submerged fermentation at optimized fermentation parameters. They found that biofilms mediated submerged fermentation significantly increased the production of FA by many times as compared to free-cell fermentation. It was a new approach to the valorization of biowaste which can provide ecological as well as economic benefits.

Supercapacitors and batteries are electrochemical energy storage devices which are the most effective technologies for practical application in energy conversion and storage (Larcher and Tarascon 2015). But the carbon materials used in these devices as an electrode are derived from non-renewable resources and are non-sustainable. Natural biomass is a good alternative of carbon source and researches have been going on to obtain renewable carbon materials using natural biomass for various applications particularly in lithium-sulphur batteries and supercapacitors.

Pyrolysis is an important thermochemical energy recovery process from biomass which produces different kinds of products such as char, oil, and gas (Pütün et al. 2005). During thermal pyrolysis of biomass materials, the moisture and volatile matter present are eliminated, and the solid char remains as a by-product. By adopting different well-designed synthesis routes these can be utilized to produce various carbon-rich materials such as activated carbon, porous carbon, and even nanostructured carbons such as graphene, fullerenes, and carbon nanotubes (CNT)s (Marriott et al. 2014; Primo et al. 2012; Su 2009; Tay et al. 2009). Surface functional groups and texture of the carbon materials were modified to get the final products using appropriate activation agents and methods (Titirici and Antonietti 2010). These products have potential utility in the control of air pollution, water pollution, catalyst support, and storage for hydrogen and energy conversion.

Cashew nutshell (CNS) and associated liquid extract (CNSL) produced from cashew nut processing factories are rich in valuable and functional renewable products (Mgaya et al. 2019). Biorenewable chemicals, materials, and energy can be obtained from these green products. The extraction techniques for CNSL include mainly pyrolysis, mechanical, thermal, and chemical (solvent extraction). Major components of CNSL are anacardic acid, cardanol, cardol, and 2-methylcardol. CNSL is used as a corrosion inhibitor of carbon steel, used in storage and transportation material. On ethenolysis reaction and in presence of suitable catalysts CNSL produces 1-octene and detergents. CNSL is also used for the synthesis of insecticides, larvicide, acaricide, and other biological applications. These are also used for the synthesis of composite nanomaterials and metal–organic precursors for nanomaterials.

Large-scale application of enzymes is limited due to higher cost, selective nature of the narrow operational range, and complicacy in recovery from the reaction medium. In biotechnology and biomedicine, the use of techniques like enzyme immobilization places a crucial role in which enzymes are immobilized onto a solid support to increase stability, reusability, and storability. Enzyme immobilization is very economical as compared to free enzymes as it improves pH and thermal stability, better resistance to environmental changes, easy recovery of the enzymes, and increases reusability and storage stability (Brady and Jordaan 2009; Mateo et al. 2007). Kumari et al. (2016) conducted a study in which pine needles were used to synthesize new cellulose-rich materials (dialdehydecellulose ethers) to support the immobilization of an enzyme α-chymotrypsin (CT). CT activity was later tested for its reusability and storage stability. It was also tested in artificial fluids such as gastric and intestinal fluid to simulate physiological conditions.

16.4 Biowaste-Derived Products

16.4.1 Biofuels

Biofuels are solid, liquid, or gaseous compounds generated primarily from biomass. They are environment friendly, abundantly available in most of the areas, and are a sustainable form of energy (Demirbas et al. 2011). Present-day concern on the emission of hazardous substances from the use of fossil fuels and its impact on global warming and climate change demands on the use of an alternative cleaner fuel. Production of biofuels through the principles of green chemistry, i.e. use of renewable feedstocks, catalysts, safer solvents, etc. can meet this demand. Bioethanol and biodiesel are the most commonly used biofuels (Chakraborty et al. 2016). Biodiesel is made up of methyl esters of fatty acids (FAME) presently generated from animal fat, waste cooking oil, and the vegetable oil extracted from green sources such as *Jatropha curcas, Azardirachta indica, Cascabela thevetia, Madhuca indica, Pongamia pinnata, Calophyllum inophyllum,* silk cotton tree, rubber seed, tall oil, microalgae, etc. (Banković-Ilić et al. 2012; Sut et al. 2016). Biodiesel can be combined with petroleum diesel at different proportions and these biodiesel blends are used in conventional diesel engines which can significantly reduce the emission of unburned hydrocarbons, polycyclic aromatic hydrocarbons (PAHs), nitrated derivatives of PAHs, CO, sulfates, and other particulate matters (Demirbas et al. 2011). A vast number of literatures are available on the production of biodiesel from different biowastes using green chemistry principles. The advantages of green chemistry rely on feedstocks, procedures, and products. Processes that involve multiple steps for conversion and separation affect the energy efficiency, overall atom economy, and waste emissions (Ballini 2009; Donate 2014). Heterogeneous structure of renewables demands the use of energy efficient and clean techniques for separation and purification (Donate 2014; Gallezot 2012). For conversion of biomass, simple process with fewer steps or having no intermediate steps are efficient in terms of biomass usage and waste economy compared to the traditional methods (Donate 2014; Xie and Gathergood 2013). To achieve the greenness of the production process, keen monitoring is required at different phases of production such as selection of feedstock, use of biobased catalyst, use of minimum reaction temperature, and utilization of reaction by-product (Dugmore et al. 2017). Chakraborty and Das (2012) obtained a higher percentage of FAME from waste soybean oil using a lower amount of methanol and hydroxyapatite-supported nickel catalyst. In another study, Outili et al. (2020) produced biodiesel from waste frying soybean oil by means of a transesterification reaction. They used five green metrics such as atom economy, reaction mass efficiency, carbon efficiency, stoichiometric factor, and environmental factor to interpret the greenness of the reaction. This process of biodiesel production provides optimum conditions with reduced temperature (around 45 °C) and reduced quantity of excess alcohol (molar alcohol/oil ratio 4.7), leading to lower consumption of energy and waste generation than the common method of biodiesel production (Outili et al. 2020). Figure 16.1 shows various conversion methods of biowaste and their energy application (Bhuyan et al. 2019). Conversion of different biowaste to biofuels using various conversion technologies is presented in Table 16.2.

Alcohols are organic molecules containing hydroxyl groups and can be used as a fuel. Bioethanol, bio-methanol, propanol, and bio-butanol are though used as motor fuels, only the first two is technically and economically suitable for internal combustion (IC) engines (Demirbas et al. 2011). Ethanol, a significant liquid biofuel had been used as a vehicle fuel since 1896 when Henry Ford designed the first car (Datta et al. 2011). Ethanol is blended with gasoline at different ratios (5–100%) to use in vehicles (Matsakas et al. 2017). A blend of 10% can be used directly in standard vehicles, whereas some modification of the engine is needed when the percentage of ethanol exceeds 25 (Pickett et al. 2008). Besides increasing the octane number in the blends, the use of ethanol has higher oxygen content, flammability range, flame speeds, and heat of vaporization than gasoline (Datta et al. 2011; Demirbas et al. 2011; Niven 2005). These characteristics enable a greater compression ratio and enhanced efficiency of combustion in IC engines with a short burn time (Balat 2007; Balat 2009). Moreover, the use of ethanol also reduces the emission of hazardous air pollutants (Matsakas et al. 2017).

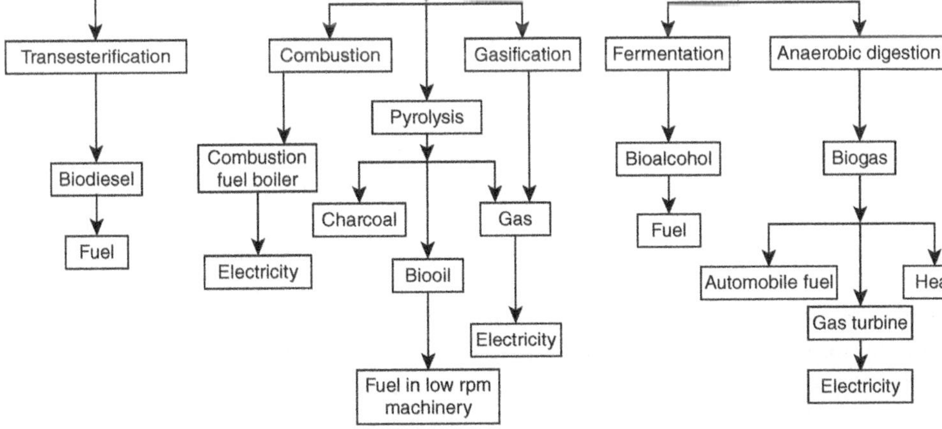

Figure 16.1 Various conversion methods of biowaste and their energy application. *Source:* Adapted from Bhuyan et al. (2019).

Table 16.2 Various biowaste and their conversion methods to biofuel.

Biowaste	Conversion method/ technology	Fuel types	References
Rubber seed shell	Pyrolysis	Pyrolytic oil	Reshad et al. (2018)
Neem seeds and waste nitrile gloves	Pyrolysis semi-batch reactor	Pyrolytic oil	Mishra et al. (2019)
Waste cooking oil	Hydro-processing	Aviation fuel	Chen and Wang (2019)
Cornstalk	Catalytic conversion using autoclave (catalyst: Ru/C, $LiTaMoO_6$)	Phenols (35.7%) and gasoline (82.4%)	Liu et al. (2015)
Waste plastic and used cooking oil	Microwave vacuum pyrolysis	Liquid oil	Lam et al. (2019)
Castor seeds	Low-temperature pyrolysis	Pyrolytic oil	Figueiredo et al. (2009)
Ipomoea carnea	Pyrolysis	Pyrolytic oil	Saikia et al. (2015)
Dried black liquor solids	Pyrolysis	Pyrolytic oil	Chutia et al. (2018)
Tea factory waste	Pyrolysis	Pyrolytic oil	Basumatary et al. (2018b)
Saccharum ravannae	Pyrolysis	Pyrolytic oil	Saikia et al. (2018)
Oil palm empty fruit bunches	Pretreatment, hydrolysis, and fermentation	Bioethanol	Derman et al. (2018)
Seaweed solid wastes Micro-algae *Spirulina*	Pretreatment, hydrolysis, fermentation, distillation, transesterification	Bioethanol	Tan and Lee (2014)
Palm oil mill effluent	Anaerobic batch fermentations	Bio-butanol	Al-Shorgani et al. (2015)

Biowaste consisting of cellulose and hemicellulose are potential raw materials for ethanol production as the polysaccharides can be converted to fermentable sugars in the presence of bacteria or yeast. Microbial conversion process can change the fermentable sugars into liquid fuels such as ethanol or other higher alcohols. Hydrolysis methods like acid/alkali hydrolysis and enzyme hydrolysis are engaged to convert the polysaccharides into simple sugars to overcome the difficulty in microbial metabolism of polysaccharides. Although the acid treatment is economical and fast, the use of acid is not a green process and it may convert sugars into undesirable forms. On the other hand, enzymatic treatments are effective and do not produce unwanted by-products apart from higher-cost involvement and time consumption (Lee et al. 2019). To reduce the hydrolysis time and to enhance the efficiency, cell disruption methods can be used prior to the hydrolysis process (Günerken et al. 2015). The left-over solid waste can be used to transform into valuable products using techniques such as liquefaction, gasification, or microwave-assisted pyrolysis. Some portions of sugars may be transformed to hydrocarbons after aqueous reformation or for the production of biohydrogen required for furfural upgradation (Donate 2014). Due to the higher chemical functionality and reactivity, furfurals can be upgraded catalytically to various value-added products. For instance, furfural and 5-hydroxymethyl-2-furfural (HMF) when undergoing catalytic hydrogenation/hydrogenolysis, it produces 2-methylfuran and 2,5-dimethylfuran, respectively, both having higher octane number and good miscibility with gasoline (Dugmore et al. 2017). Figure 16.2 shows the production of ethanol fuel from cellulosic biomass.

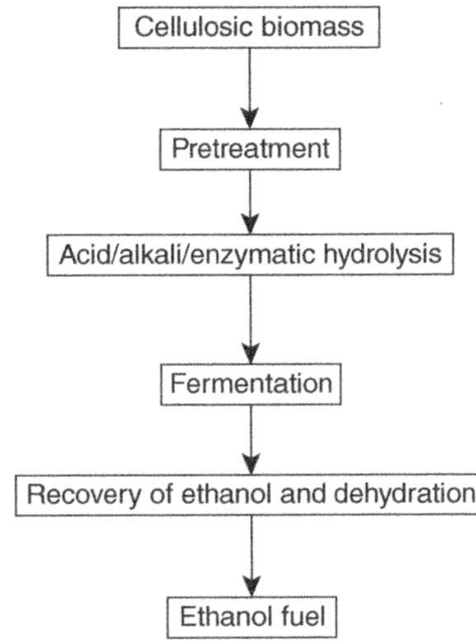

Figure 16.2 Flow chart for the production of ethanol fuel from cellulosic biomass.

Biogas is a renewable form of energy and a gaseous biofuel to replace the conventional fuels (Beenu and Sodhi 2016; Demirbas et al. 2011). Methane and CO_2 gas are generated when bio-organic matter is digested in anaerobic environment during biogas production. Other compounds like alcohols, ammonia, compounds of sulphur (H_2S and alkythiols), carboxylic acids (such as formic acid and acetic acid), carbonyl compounds (ketones and aldehydes), terpenes, and aromatic compounds are also generated in the process (Papurello et al. 2012). Methane production is more efficient than other fuels such as ethanol (Jeihanipour et al. 2013; Zheng et al. 2014). The anaerobic digestion is a complex route involving various phases, such as hydrolysis, acidogenesis, acetogenesis, and methanogenesis. Consortia of different microorganisms (Franchi et al. 2018) catalyze all these steps. After a proper gas clean-up process, biogas can be used as a fuel for microgas turbines, spark ignition gas engines (SIGEs), boilers, solid oxide fuel cells (SOFCs), industrial heaters, and for the manufacturing of chemicals (Demirbas et al. 2011; Kan 2009; Papurello et al. 2014). Modifying the compositions of biogas, it can use as a vehicle fuel and to generate heat and electricity (Papurello et al. 2014). Production of biogas in anaerobic digestion is shown in Figure 16.3. No extensive vehicular modification is needed for using methane as vehicle fuel (Aslam et al. 2006).

Pyrolysis is one of the emerging thermochemical decomposition processes for the sustainable amalgamation of waste biomass. Generally, this process is carried out at a temperature range of 300 to 650 °C in an inert atmosphere (Mohan et al. 2006). Compared to the biochemical treatment processes, thermochemical conversion processes are occurred at elevated temperatures with fast conversion rates. The products of pyrolysis can be divided into three parts: solid (char), liquid (bio-oil or pyrolysis oil), and gases. Usually, liquid products are formed from condensable gases and higher yield of liquid products is achieved at faster heating rates under mild temperatures. However, raw bio-oil is not used as a fuel or chemical without upgradation owing to many constraints such as high oxygen content, acidity and humidity; lower calorific value and oxidative stability. After upgradation of the bio-oil, it is used as transport grade fuel or added to oil refinery feedstock, or as an intermediate for synthesized value-added chemicals and hydrocarbons (Demirbas 2008;

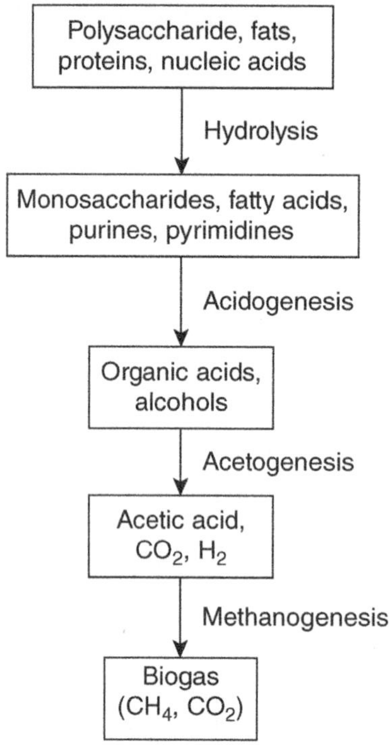

Figure 16.3 Flowchart for the production of biogas in anaerobic digestion.

Qi et al. 2007). Upgraded bio-oils can be used in some diesel engines or low-rpm machinery to run turbines and firing boilers in the agricultural sector (Czernik and Bridgwater 2004; Tang et al. 2009). On the other hand, the solid product, biochar has various other uses including as heating fuel, energy carrier, feedstock for manufacturing several value-added materials such as adsorbents, catalysts or their supports, supercapacitors, and soil amendments for long-term carbon sequestration in agricultural and forest soil (Bhuyan et al. 2019).

In the gasification process, synthesis gas or syngas is produced in presence of a controlled oxidizing agent such as oxygen, vapour, or mixture of gases. Syngas is a mixture of CO, CO_2, and H_2 and the production process is strongly endothermic. In the steam oxidation process, syngas with high heating value are evolved, while in the air gasification process, syngas exhibits lower heating value (Kataki et al. 2018). Large-scale production of syngas can be obtained by some 'non-gasification' processes, such as partial oxidation, autothermal reforming, and steam methane reforming. Feedstock material and pure oxygen gas are mixed and reacted together in a partial oxidation process to produce syngas containing a higher percentage of hydrogen (H_2). Syngas is used as fuel in IC engine in combination with a synchronous alternator that ultimately generates electricity (Kataki et al. 2018). Thermal energy obtained as hot water, steam, thermal oil, etc. on combustion of syngas can be used in various industries such as chemical, cement, and food. Moreover, syngas can be transformed into other biofuel such as bioethanol and bio-synthetic natural gas (SNG) through chemical treatment, which has found application as transportation fuels, or in the synthesis of various chemicals, and plastics (Naik et al. 2010; Sikarwar et al. 2017). Although thermochemical conversion of waste biomass has many advantages in energy and material recovery, still these conversion methods have some drawbacks regarding the emission of trace quantity of inorganic and organic pollutants into the environment.

Due to the above-mentioned drawbacks, the use of some upgraded green methods like microwave synthesis has gained publicity to conduct chemical transformations sustainably (Dugmore et al. 2017). Moreover, microwave energy is considered as an attractive mode because of its high efficiency in energy transfer and selectivity, which in turn significantly reduces the reaction time (Dugmore et al. 2017). Microwave-assisted acetalization of glycerol and carbonyl compounds results high yields of cyclic acetals or ketals, and fuel additive precursors. This valorization process is considered as a clean and green approach which can be achieved without the use of large volume of solvents or precious catalysts (Dugmore et al. 2017). As a source of energy, ultrasound is also used in chemical processes that reduce energy consumption and increase the selectivity of the product. Already various researchers produced biodiesel through ultrasound-assisted transesterification methods (Malani et al. 2019; Veljković et al. 2012).

16.4.2 Catalysts

Catalyst plays a vital role in the advancement of greener industry with their assistance in chemical synthesis by accelerating the reaction rate while maintaining the temperature and thus provide sustainability to the system (Tariq et al. 2012). In addition to assisting green chemical reactions, catalysts can decrease the environmental effects as well as the costs of the processes (Clark 2001). Therefore, the application of waste biomass as catalytic components or as precursors to catalytic products continues to public attention. On combustion, biowaste or any plant part produces ash that consists primarily of inorganic components. Majorly SiO_2, Al_2O_3, Fe_2O_3, Na_2O, K_2O, CaO, MgO, and other metal oxides depending on the type of sources (Rashid et al. 2011). In the production of biodiesel, both homogeneous and heterogeneous catalysts were used which helps in the transesterification reaction (Gohain et al. 2017). Various homogeneous and heterogeneous biobased catalysts used in the production of biodiesel are presented in Table 16.3. Homogeneous catalysts are widely used in large-scale biodiesel production, although they are extremely corrosive, flammable, and poisonous in nature. In addition, the use of these catalysts produces soaps as a by-product along with large quantities of wastewater involving extra cost for proper disposal. Su and Guo (2014) studied a sustainable method to obtain biodiesel by applying viable solid acid catalysts instead of liquid acid catalysts to eliminate the generation of harmful substances and toxic wastes. Hence, heterogeneous catalysts can overcome the limitations of homogeneous catalysts and offer multiple benefits such as non-toxic, reusable, high catalytic activity, stability in both acidic and basic conditions, and high-water tolerance, which varies with various catalysts properties (Bennett et al. 2016; Gohain et al. 2017).

Biowaste including plant matter, algae, shell of molluscs or eggs, animal bone, etc. can be used to produce various heterogeneous catalyst through simple and cost-effective processes (Bennett et al. 2016). Due to its higher efficiency in transesterification as well as recycling properties, the use of these catalysts can considerably reduce the environmental pollution and total cost of the produced biodiesel. These green catalysts assist to provide necessary conditions (acid or basic) in the low-temperature esterification of free fatty acids (FFAs) or transesterification of triacyl glycerides (TAGs). During the production of biodiesel, several reports documented the efficient use of calcium oxide-based solid catalysts produced from waste biomass. For example, Wei et al. (2009) reported the efficacy of eggshell as rich sources of CaO solid catalyst for biodiesel production. Another competent calcium-based solid catalyst for biodiesel production was developed by Viriya-empikul et al. (2012) with the use of egg-shells, golden apple snail, and meretrix venus where the major constituent was $CaCO_3$. Lee et al. (2015) fabricated a heterogeneous solid base (CaO) catalyst from an obtuse horn shell, which showed insignificant catalyst reusability with a FAME output of 70% even after three experiments without additional

Table 16.3 Various homogeneous and heterogeneous biobased catalyst are used in the production of biodiesel.

Catalyst source	Feedstock for biodiesel	Optimum yield (wt%)	References
Coconut chaff	Waste frying oil	91.05	Yusuff and Owolabi (2019)
Chicken manure	Waste cooking oil	90.8	Maneerung et al. (2016)
Chicken eggshell	Waste frying oil	77.56	Yusuff and Popoola (2019)
Waste orange peel	Corn acid oil (by-product of corn refinery)	91.68	Lathiya et al. (2018)
Ostrich eggshells	Waste cooking oil	96	Tan et al. (2015)
Chicken eggshells	Waste cooking oil	94	Tan et al. (2015)
Cement wastes	Karanja oil	76 ± 0.3	Kumar et al. (2018)
Chicken bones	Waste cooking oil	89.33	Farooq et al. (2015)
Peanut husk ash	Soybean oil	98.4	Dai et al. (2014)
Kola nut pod husk	Kariya seed oil	98.67 ± 0.01	Betiku et al. (2019)
Waste Carica papaya stem	Waste cooking oil	95.23	Gohain et al. (2020)
Waste Carica papaya stem	*Scenedesmus obliquus* lipid	93.33	Gohain et al. (2020)
Chicken-eggshells	*Chlorella vulgaris* microalgae	87.8	Pandit and Fulekar (2019)
Coconut coir husk	Waste palm oil	89.8	Thushari et al. (2019)
Eggshell and peat clay	Waste cooking oil	91	Putra et al. (2018)
Snail shell ash	Waste cooking oil	92.5	Kaewdaeng et al. (2017)
Waste tucumã peel ash	Soybean oil	97.3	Mendonça et al. (2019)

activation. The biowaste-based heterogeneous catalyst containing potassium oxide as major constituents shows higher basicity (Kawashima et al. 2008). Several studies confirmed the role of potassium in providing basicity and thus higher catalytic activity in biodiesel production. Hence, several catalysts derived from agro-wastes containing higher amount of potassium are responsible for their efficacy in biodiesel production (Basumatary et al. 2018a; Odude et al. 2019). Researchers reported the higher activity of heterogeneous catalysts in biodiesel conversion if calcined at 800 and 1000 °C. However, a calcination temperature of 1200 °C results in decreased alkalinity, subsequently decreasing the catalytic activity with a biodiesel conversion of 60.5% (Basumatary et al. 2018a). Catalysts obtained from agro-wastes also showed convenient reusability. Aslam et al. (2014) achieved a three-fold higher reusability with a 20% decrease in efficiency in the third-time use of the solid catalyst obtained from *Musa balbisiana* underground stem ash for the synthesis of biodiesel using *Mesua ferrea* seed oil. This reduction in catalytic activity could be due to leaching of potassium from the catalyst. In another study, Gohain et al. (2017) obtained five times reusability of catalysts prepared from *Musa balbisiana* peels. The decreased catalytic efficiency in the repeated reaction cycles was due to the leaching of K and Ca. Ash obtained from biowastes such as different parts of the banana plant (trunk, leaf, underground steam, and peel) can use as an active catalyst due to its basic nature (Kawashima et al. 2008). Similarly, ash from common duckweed, cocoa pod, and coconut husk was used as an effective catalyst (Odude et al. 2019). Biomass wastes can also be transformed into activated carbons through different chemical treatments in the pyrolysis or carbonization process where subsequent sulfonation generates solid acid catalysts to use for esterification. Moreover, water extracts from agro-wastes-derived ashes can be utilized as a catalyst in reactions such as oxidation, hydroxylation, and cross-coupling (Sarmah et al. 2017). In another study, Park et al. (2003) employed rice husk for the synthesis of zeolites, mesoporous silicates, and carbon honeycomb monoliths with different catalytic properties. The Henry reaction (nitroaldol reaction) is among the most important synthetic tools for C—C bond formation and is largely used in chemical industries (Abdellattif and Mokhtar 2018). It is extremely desirable to use biowaste catalysts in this response from the ethos of sustainability and green credentials. As found by researchers, *M. acuminata* peel ash could be a promising green solid catalyst for biodiesel production as well as for the synthesis of nitroalcohols via sustainable Henry reaction (Rajkumari et al. 2019).

Another biofuel-bioethanol can be achieved efficiently from waste biomass using carbon-based solid catalyst containing $-SO_3H$, $-COOH$, and $-OH$ functional groups (Luterbacher et al. 2014). These catalysts documented very good acid density and higher stability in terms of reusability and catalytic activity compared to conventional acid catalysts like corrosive homogeneous H_2SO_4 catalyst. Suganuma et al. (2010) observed the usability of carbon-based sulfonated catalyst for the production of fermentable sugar at lesser activation energy compared to conventional H_2SO_4.

Thus, the generation of catalysts from agro-waste significantly subsidizes the overall reuse and minimization of waste. In this area, more research at the pilot scale is needed to find more efficient and highly active heterogeneous catalysts using waste biomass. The feasibility of heterogenous catalysts from various biowaste as a base to catalyze organic reactions should be given attention.

16.4.3 Nanocomposites

Nanocomposites are the materials having multiphase solids with nano-size particles (<100 nm). The existence of nano-size particles creates a discontinuous phase over a matrix of standard material, which provides mechanical strength, toughness, and the expansion capability in both mechanical and thermal conductivity of nano-composite (Camargo et al. 2009). These features of nanocomposites approve their application in areas such as functional and structural material, as catalysts, sorption and separation material, fuel cell, etc. (Camargo et al. 2009). Biowaste such as lignocellulosic agricultural wastes mostly contains holo-cellulose (cellulose, hemicellulose) and lignin with some amount of sugars, proteins, extractives, and ash (Hamza et al. 2013) which is similar to plant leaf fibres such as oil palm, banana, sisal, pineapple, abaca, etc.; vast fibres such as kenaf, jute, and flax; and seed fibres such as cotton, rice husk, kapok, and coir. Due to the wide availability, cost-effectiveness, eco-friendliness, biodegradable and suitable mechanical properties, natural fibres from agricultural wastes are considered as one of the potential and effective

Table 16.4 Biomass-derived nano-composites and their area of application.

Biomass	Nano-composites	Application	References
Jute stick	Silver loaded activated carbon (Ag-AC) nanocomposite	Antibacterial agent; exhibit photocatalytic activity	Devi et al. (2019)
Oak-based biomass (OBM)	CuO/OBM nanocomposite	Electrochemical hydrogen storage	Seifi and Masoum (2019)
Cellulose	Reduced graphene oxide, rGO/Cu_2O nanocomposite	Strong antibacterial activity	Chen et al. (2020)
Bamboo	Biomass charcoal/TiO_2 composites	Photocatalytic activities	Liu et al. (2018)
Cassia fistula, Aloe vera	A magnetic nanocomposite (NC-ALV-450)	Water and wastewater treatment	Kirti et al. (2018)
Grapefruit peel	SnO_2@ grapefruit peel carbon (GPC) nanocomposites	As an adsorbent to remove 2,4-dichlorophenol from aqueous solution	Batool et al. (2020)
Tannery sludge	ZnO-SBAC(sludge biomass activated carbon) nanocomposites	Environmentally benign photocatalyst; reduction of Cr(VI) metals; and oxidation of organic molecules	Ramya et al. (2018)
Glucose	Carbon@montmorillonite nanocomposites	Adsorbent to remove organic dyes from aqueous solution	Ai and Li (2013)
Murdannia simplex stalks	Carbon/Co_3O_4 nanocomposites	Potentially useful for rechargeable Lithium Ion Batteries	Wu et al. (2016)

contenders for polymer-based nano-composite. Table 16.4 shows some biomass-derived nano-composites with their probable area of applications.

In agricultural wastes, cellulose fibres are linked with other constituents such as hemicellulose, lignin, and pectin (Pickering 2008). Therefore, some intensive methods such as chemical (alkali and acid hydrolysis) (Fahma et al. 2010; Johar et al. 2012), mechanical (homogenization, grinding, ultra-sonication, and electrospinning) (Abdul Khalil et al. 2014; Stelte and Sanadi 2009) and combination of both chemical and mechanical methods (Jonoobi et al. 2011) are used as isolation technique for segregating the cellulose fibres. However, the use of sulphuric and hydrochloric acid is found very effective, but the use of concentrated acid is not under green principle (Brinchi et al. 2013). Hence, the combined chemical and mechanical methods gained popularity (Jonoobi et al. 2011).

In the past few years, special emphasis has been given on the valorization of waste materials (such as lignocellulosic fibres of agricultural waste) to produce polymer reinforcements which can substitute glass and carbon. Nanda et al. (2013) reported the use of soy hull fibres, switchgrass, miscanthus, and their hybrids for the fabrication of renewable green composites in poly (3-hydroxybutyrate-*co*-valerate) and polylactide blends through melt mixing. They observed that the hybrid composites exhibited typical physico-mechanical properties. Biodegradable green composites have the potential of being a better and cost-effective alternative in automotive industries instead of conventional petrol-based composites in terms of cost versus mechanical performance. Some other residues, such as corn stalk, soy stalk, wheat straw, and some perennial grasses had been also investigated as reinforcement into biodegradable polymer matrices (Nagarajan et al. 2013). Cellulose can be transformed to nanocellulose that would be exceptional if mixed with inorganic nanoparticles to achieve nanocomposites with nanocellulose and nanoparticle characteristics. Helmiyati et al. (2017) used nanocellulose from rice husk-derived cellulose, which acts as a support in manufacturing superabsorbent nanocomposite for absorbing urea fertilizer carried by rainwater and irrigation in agricultural field. Moreover, nanofibers obtained from waste biomass (as reinforcing fillers) such as rice straw, rice husk, bamboo, wood, ramie, abaca, jute, oil palm, and flax using electrospinning technique have been also gained attention (Alemdar and Sain 2008). In this way, the utilization of biowaste provides viable and less costly material as well as helps in the waste management system.

16.5 Conclusion

Generation of valuable products such as catalysts, biofuels, and nanocomposites from biowaste using green chemistry subsidizes the overall reuse and minimization of waste substantially. Extraction methods used for obtaining valuable compounds from biowaste need more research attention at the pilot scale. Emphasis should be paid to explore more efficient and highly active catalysts, biofuels, and nanocomposites through waste valorization.

16.6 Summary

Management of biowaste is a great challenge in the agrarian economy. Due to the bulky nature, slow degradation and lack of proper market to sell in developing countries, burning and landfilling are the normal practices to reduce its volume which leads to the generation of greenhouse gases. Effective application of green chemistry can reduce the load of biowaste while strengthen the country's economy by producing valuable products. Compounds present in biowaste can be successfully extracted by using appropriate methods to use as catalysts, nanocomposites, and biofuels. This not only reduces pollution and greenhouse gas emission but also helps to attain sustainability, one of the waste management principles of India. This chapter summarizes about different biowaste, the valuable chemicals present, and the methods used for their extraction. Discussion on important biowaste-derived products such as catalysts, nanocomposites, and biofuels are also presented.

References

Abdellattif, M. and Mokhtar, M. (2018). MgAl-layered double hydroxide solid base catalysts for henry reaction: a green protocol. *Catalysts* 8: 133. https://doi.org/10.3390/catal8040133.

Abdul Khalil, H.P.S., Hossain, M.S., Rosamah, E. et al. (2014). High-pressure enzymatic hydrolysis to reveal physicochemical and thermal properties of bamboo fiber using a supercritical water fermenter. *Bio Resources* 9: 7710–7720.

Abia, A.A. and Asuquo, E.D. (2006). Lead (II) and nickel (II) adsorption kinetics from aqueous metal solutions using chemically modified and unmodified agricultural adsorbents. *African Journal of Biotechnology* 5: 1475–1482.

Ai, L. and Li, L. (2013). Efficient removal of organic dyes from aqueous solution with ecofriendly biomass-derived carbon@ montmorillonite nanocomposites by one-step hydrothermal process. *Chemical Engineering Journal* 223: 688–695. https://doi.org/10.1016/j.cej.2013.03.015.

Alemdar, A. and Sain, M. (2008). Biocomposites from wheat straw nanofibers: morphology, thermal and mechanical properties. *Composites Science and Technology* 68: 557–565. https://doi.org/10.1016/j.compscitech.2007.05.044.

Ali, H.R. and Hassaan, M.A. (2017). Applications of bio-waste materials as green synthesis of nanoparticles and water purification. *Advances in Materials* 6: 85–101. https://doi.org/10.11648/j.am.20170605.16.

Al-Shorgani, N.K.N., Shukor, H., Abdeshahian, P. et al. (2015). Process optimization of butanol production by *Clostridium saccharoperbutylacetonicum* N1-4 (ATCC 13564) using palm oil mill effluent in acetone–butanol–ethanol fermentation. *Biocatalysis and Agricultural Biotechnology* 4: 244–249. https://doi.org/10.1016/j.bcab.2015.02.004.

Anastas, P.T. and Warner, J.C. (1998). *Green Chemistry: Theory and Practice*. New York: Oxford University Press.

Annadurai, G., Juang, R.S., and Lee, D.L. (2003). Adsorption of heavy metals from water using banana and orange peels. *Water Science & Technology* 47: 185–190. https://doi.org/10.2166/wst.2003.0049.

Aslam, M.U., Masjuki, H.H., Kalam, M.A. et al. (2006). An experimental investigation of CNG as an alternative fuel for a retrofitted gasoline vehicle. *Fuel* 85: 717–724. https://doi.org/10.1016/j.fuel.2005.09.004.

Aslam, M., Saxena, P., and Sarma, A.K. (2014). Green technology for biodiesel production from *Mesua ferrea* L. seed oil. *Energy and Environment Research* 4: 11–21. https://doi.org/10.5539/eer.v4n2p11.

Babarinde, N.A.A. (2002). Adsorption of zinc (II) and cadmium (II) by coconut husk and goat hair. *Journal of Pure and Applied Sciences* 5: 81–85.

Balat, M. (2007). Global bio-fuel processing and production trends. *Energy Exploration & Exploitation* 25: 195–218. https://doi.org/10.1260/014459807782009204.

Balat, M. (2009). Bioethanol as a vehicular fuel: a critical review. *Energy Sources, Part A: Recovery, Utilization, and Environmental Effects* 31: 1242–1255. https://doi.org/10.1080/15567030801952334.

Baldinelli, A., Dou, X., Buchholz, D. et al. (2018). Addressing the energy sustainability of biowaste-derived hard carbon materials for battery electrodes. *Green Chemistry* 20: 1527–1537. https://doi.org/10.1039/C8GC00085A.

Ballini, R. (ed.) (2009). *Eco-Friendly Synthesis of Fine Chemicals*. Cambridge: Royal Society of Chemistry Publications.

Banković-Ilić, I.B., Stamenković, O.S., and Veljković, V.B. (2012). Biodiesel production from non-edible plant oils. *Renewable and Sustainable Energy Reviews* 16: 3621–3647. https://doi.org/10.1016/j.rser.2012.03.002.

Basumatary, S., Nath, B., and Kalita, P. (2018a). Application of agro-waste derived materials as heterogeneous base catalysts for biodiesel synthesis. *Journal of Renewable and Sustainable Energy* 10: 043105. https://doi.org/10.1063/1.5043328.

Basumatary, V., Saikia, R., Narzari, R. et al. (2018b). Tea factory waste as a feedstock for thermo-chemical conversion to biofuel and biomaterial. *Materials Today: Proceedings* 5: 23413–23422. https://doi.org/10.1016/j.matpr.2018.11.081.

Batool, S., Idrees, M., Ahmad, M. et al. (2020). Design and characterization of a biomass template/SnO_2 nanocomposite for enhanced adsorption of 2,4-dichlorophenol. *Environmental Research* 181: 108955. https://doi.org/10.1016/j.envres.2019.108955.

Beenu and Sodhi, S. (2016). Generation of biogas energy and its uses. *International Journal for Research in Applied Science & Engineering Technology (IJRASET)* 4: 323–329. https://www.ijraset.com/fileserve.php?FID=5370.

Bennett, J.A., Wilson, K., and Lee, A.F. (2016). Catalytic applications of waste derived materials. *Journal of Materials Chemistry A* 4: 3617–3637. https://doi.org/10.1039/C5TA09613H.

Betiku, E., Okeleye, A.A., Ishola, N.B. et al. (2019). Development of a novel mesoporous biocatalyst derived from kola nut pod husk for conversion of Kariya seed oil to methyl esters: a case of synthesis, modeling and optimization studies. *Catalysis Letters* 149: 1772–1787. https://doi.org/10.1007/s10562-019-02788-6.

Bhuyan, N., Sut, D., Gogoi, L. et al. (2019). Rural biorefinery: a viable solution for production of fuel and chemicals in rural India. In: *Sustainable Bioenergy* (ed. M. Rai and A.P. Ingle), 21–47. Amsterdam, Netherlands: https://doi.org/10.1016/B978-0-12-817654-2.00002-2.

Brady, D. and Jordaan, J. (2009). Advances in enzyme immobilization. *Biotechnology Letters* 31: 1639–1650. https://doi.org/10.1007/s10529-009-0076-4.

Brinchi, L., Cotana, F., Fortunati, E., and Kenny, J.M. (2013). Production of nanocrystalline cellulose from lignocellulosic biomass: technology and applications. *Carbohydrate Polymers* 94: 154–169. https://doi.org/10.1016/j.carbpol.2013.01.033.

Camargo, P.H.C., Satyanarayana, K.G., and Wypych, F. (2009). Nanocomposites: synthesis, structure, properties and new application opportunities. *Materials Research* 12: 1–39. https://doi.org/10.1590/S1516-14392009000100002.

Chakraborty, R. and Das, S.K. (2012). Optimization of biodiesel synthesis from waste frying soybean oil using fish scale-supported Ni catalyst. *Industrial & Engineering Chemistry Research* 51: 8404–8414. https://doi.org/10.1021/ie2030745.

Chakraborty, R., Chatterjee, S., Mukhopadhyay, P., and Barman, S. (2016). Progresses in waste biomass derived catalyst for production of biodiesel and bioethanol: a review. *Procedia Environmental Sciences* 35: 546–554. https://doi.org/10.1016/j.proenv.2016.07.039.

Chen, R.X. and Wang, W.C. (2019). The production of renewable aviation fuel from waste cooking oil. Part I: bio-alkane conversion through hydro-processing of oil. *Renewable Energy* 135: 819–835. https://doi.org/10.1016/j.renene.2018.12.048.

Chen, M., Li, Z., and Chen, L. (2020). Highly antibacterial rGO/Cu_2O nanocomposite from a biomass precursor: synthesis, performance, and mechanism. *Nano Materials Science* 2: 172–179. https://doi.org/10.1016/j.nanoms.2019.09.005.

Chutia, S., Narzari, R., Bordoloi, N. et al. (2018). Pyrolysis of dried black liquor solids and characterization of the bio-char and bio-oil. *Materials Today: Proceedings* 5: 23193–23202. https://doi.org/10.1016/j.matpr.2018.11.050.

Clark, J.H. (2001). Catalysis for green chemistry. *Pure and Applied Chemistry* 73: 103–111. https://doi.org/10.1351/pac200173010103.

Czernik, S. and Bridgwater, A.V. (2004). Overview of applications of biomass fast pyrolysis oil. *Energy & Fuels* 18: 590–598. https://doi.org/10.1021/ef034067u.

Dai, Y.M., Chen, K.T., Wang, Y.J., and Chen, C.C. (2014). Application of peanut husk ash as a low-cost solid catalyst for biodiesel production. *International Journal of Chemical Engineering and Applications* 5: 276. https://doi.org/10.7763/IJCEA.2014.V5.393.

Das, R.K., Brar, S.K., and Verma, M. (2015). Valorization of eggshell biowaste and brewery wastewater for the enhanced production of fumaric acid. *Waste and Biomass Valorization* 6: 535–546. https://doi.org/10.1007/s12649-015-9377-0.

Datta, R., Maher, M.A., Jones, C., and Brinker, R.W. (2011). Ethanol-the primary renewable liquid fuel. *Journal of Chemical Technology & Biotechnology* 86: 473–480. https://doi.org/10.1002/jctb.2580.

Demirbas, E. (2003). Adsorption of cobalt (II) from aqueous solution onto activated carbon prepared from hazelnut shells. *Adsorption Science & Technology* 21: 951–963. https://doi.org/10.1260/02636170360744380.

Demirbas, A. (2008). Present and future transportation fuels. *Energy Sources, Part A: Recovery, Utilization, and Environmental Effects* 30: 1473–1483. https://doi.org/10.1080/15567030701258519.

Demirbas, M.F., Balat, M., and Balat, H. (2011). Biowastes-to-biofuels. *Energy Conversion and Management* 52: 1815–1828. https://doi.org/10.1016/j.enconman.2010.10.041.

Derman, E., Abdulla, R., Marbawi, H., and Sabullah, M.K. (2018). Oil palm empty fruit bunches as a promising feedstock for bioethanol production in Malaysia. *Renewable Energy* 129: 285–298. https://doi.org/10.1016/j.renene.2018.06.003.

Devi, T.B., Mohanta, D., and Ahmaruzzaman, M. (2019). Biomass derived activated carbon loaded silver nanoparticles: an effective nanocomposites for enhanced solar photocatalysis and antimicrobial activities. *Journal of Industrial and Engineering Chemistry* 76: 160–172. https://doi.org/10.1016/j.jiec.2019.03.032.

Donate, P.M. (2014). Green synthesis from biomass. *Chemical and Biological Technologies in Agriculture* 1: 1–8. https://doi.org/10.1186/s40538-014-0004-2.

Drake, L.R., Lin, S., Rayson, G.D., and Jackson, P.J. (1995). Chemical modification and metal binding studies of *Datura innoxia*. *Environmental Science & Technology* 30: 110–114. https://doi.org/10.1021/es950131d.

Dugmore, T.I.J., Clark, J.H., Bustamante, J. et al. (2017). Valorisation of biowastes for the production of green materials using chemical methods. *Topics in Current Chemistry* 375 (46): https://doi.org/10.1007/s41061-017-0133-8.

Fahma, F., Iwamoto, S., Hori, N. et al. (2010). Isolation, preparation, and characterization of nanofibers from oil palm empty-fruit-bunch (OPEFB). *Cellulose* 17: 977–985. https://doi.org/10.1007/s10570-010-9436-4.

Farooq, M., Ramli, A., and Naeem, A. (2015). Biodiesel production from low FFA waste cooking oil using heterogeneous catalyst derived from chicken bones. *Renewable Energy* 76: 362–368. https://doi.org/10.1016/j.renene.2014.11.042.

Figueiredo, M.K., Romeiro, G.A., Damasceno, R.N., and Franco, A.P. (2009). The isolation of pyrolysis oil from castor seeds via a Low Temperature Conversion (LTC) process and its use in a pyrolysis oil–diesel blend. *Fuel* 88: 2193–2198. https://doi.org/10.1016/j.fuel.2009.05.025.

Franchi, O., Rosenkranz, F., and Chamy, R. (2018). Key microbial populations involved in anaerobic degradation of phenol and p-cresol using different inocula. *Electronic Journal of Biotechnology* 35: 33–38. https://doi.org/10.1016/j.ejbt.2018.08.002.

Gallezot, P. (2012). Conversion of biomass to selected chemical products. *Chemical Society Reviews* 41: 1538–1558. https://doi.org/10.1039/C1CS15147A.

Gao, Z., Zhang, Y., Song, N., and Li, X. (2017). Biomass-derived renewable carbon materials for electrochemical energy storage. *Materials Research Letters* 5: 69–88. https://doi.org/10.1080/21663831.2016.1250834.

Gohain, M., Devi, A., and Deka, D. (2017). *Musa balbisiana* Colla peel as highly effective renewable heterogeneous base catalyst for biodiesel production. *Industrial Crops and Products* 109: 8–18. https://doi.org/10.1016/j.indcrop.2017.08.006.

Gohain, M., Laskar, K., Paul, A.K. et al. (2020). Carica papaya stem: a source of versatile heterogeneous catalyst for biodiesel production and C–C bond formation. *Renewable Energy* 147: 541–555. https://doi.org/10.1016/j.renene.2019.09.016.

Günerken, E., D'Hondt, E., Eppink, M.H.M. et al. (2015). Cell disruption for microalgae biorefineries. *Biotechnology Advances* 33: 243–260. https://doi.org/10.1016/j.biotechadv.2015.01.008.

Hamza, S., Saad, H., Charrier, B. et al. (2013). Physico-chemical characterization of Tunisian plant fibers and its utilization as reinforcement for plaster-based composites. *Industrial Crops and Products* 49: 357–365. https://doi.org/10.1016/j.indcrop.2013.04.052.

Hanc, A., Novak, P., Dvorak, M. et al. (2011). Composition and parameters of household bio-waste in four seasons. *Waste Management* 31: 1450–1460. https://doi.org/10.1016/j.wasman.2011.02.016.

Helmiyati, H.H., Abbas, G.H., and Kurniawan, S. (2017). Superabsorbent nanocomposite synthesis of cellulose from rice husk grafted poly (acrylate acid-co-acrylamide)/bentonite. *IOP Conference Series: Materials Science and Engineering* 188: 012024. https://doi.org/10.1088/1757-899X/188/1/012024.

Jasuja, N.D., Gupta, D.K., Reza, M., and Joshi, S.C. (2014). Green synthesis of AgNPs stabilized with biowaste and their antimicrobial activities. *Brazilian Journal of Microbiology* 45: 1325–1332. https://doi.org/10.1590/S1517-83822014000400024.

Jeihanipour, A., Aslanzadeh, S., Rajendran, K. et al. (2013). High-rate biogas production from waste textiles using a two-stage process. *Renewable Energy* 52: 128–135. https://doi.org/10.1016/j.renene.2012.10.042.

Johar, N., Ahmad, I., and Dufresne, A. (2012). Extraction, preparation and characterization of cellulose fibres and nanocrystals from rice husk. *Industrial Crops and Products* 37: 93–99. https://doi.org/10.1016/j.indcrop.2011.12.016.

Jonoobi, M., Khazaeian, A., Tahir, P.M. et al. (2011). Characteristics of cellulose nanofibers isolated from rubberwood and empty fruit bunches of oil palm using chemo-mechanical process. *Cellulose* 18: 1085–1095. https://doi.org/10.1007/s10570-011-9546-7.

Kaewdaeng, S., Sintuya, P., and Nirunsin, R. (2017). Biodiesel production using calcium oxide from river snail shell ash as catalyst. *Energy Procedia* 138: 937–942. https://doi.org/10.1016/j.egypro.2017.10.057.

Kan, A. (2009). General characteristics of waste management: a review. *Energy Education Science and Technology Part A: Energy Science and Research* 23: 55–69.

Kataki, R., Bordoloi, N.J., Saikia, R. et al. (2018). Waste valorization to fuel and chemicals through pyrolysis: technology, feedstock, products, and economic analysis. In: *Waste to Wealth* (ed. R.R. Singhania, R.A. Agarwal, R.P. Kumar and R.K. Sukumaran), 477–514. Singapore: Springer.

Kawashima, A., Matsubara, K., and Honda, K. (2008). Development of heterogeneous base catalysts for biodiesel production. *Bioresource Technology* 99: 3439–3443. https://doi.org/10.1016/j.biortech.2007.08.009.

Kirti, S., Bhandari, V.M., Jena, J., and Bhattacharyya, A.S. (2018). Elucidating efficacy of biomass derived nanocomposites in water and wastewater treatment. *Journal of Environmental Management* 226: 95–105. https://doi.org/10.1016/j.jenvman.2018.08.028.

Kumar, P. and Dara, S.S. (1982). Utilization of agricultural wastes for decontaminating industrial /domestic wastewaters from toxic metals. *Agricultural Wastes* 4: 213–223. https://doi.org/10.1016/0141-4607(82)90013-0.

Kumar, D., Singh, B., Banerjee, A., and Chatterjee, S. (2018). Cement wastes as transesterification catalysts for the production of biodiesel from Karanja oil. *Journal of Cleaner Production* 183: 26–34. https://doi.org/10.1016/j.jclepro.2018.02.122.

Kumari, S., Chauhan, G.S., Ahn, J.H., and Reddy, N.S. (2016). Bio-waste derived dialdehyde cellulose ethers as supports for α-chymotrypsin immobilization. *International Journal of Biological Macromolecules* 85: 227–237. https://doi.org/10.1016/j.ijbiomac.2015.12.063.

Lam, S.S., Mahari, W.A.W., Ok, Y.S. et al. (2019). Microwave vacuum pyrolysis of waste plastic and used cooking oil for simultaneous waste reduction and sustainable energy conversion: recovery of cleaner liquid fuel and techno-economic analysis. *Renewable and Sustainable Energy Reviews* 115: 109359. https://doi.org/10.1016/j.rser.2019.109359.

Larcher, D. and Tarascon, J.M. (2015). Towards greener and more sustainable batteries for electrical energy storage. *Nature Chemistry* 7: 19–29. https://doi.org/10.1038/nchem.2085.

Lathiya, D.R., Bhatt, D.V., and Maheria, K.C. (2018). Synthesis of sulfonated carbon catalyst from waste orange peel for cost effective biodiesel production. *Bioresource Technology Reports* 2: 69–76. https://doi.org/10.1016/j.biteb.2018.04.007.

Lee, H.V., Juan, J.C., Taufiq-Yap, Y.H. et al. (2015). Advancement in heterogeneous base catalysed technology: an efficient production of biodiesel fuels. *Journal of Renewable and Sustainable Energy* 7: 032701. https://doi.org/10.1063/1.4919082.

Lee, S.Y., Sankaran, R., Chew, K.W. et al. (2019). Waste to bioenergy: a review on the recent conversion technologies. *BMC Energy* 1: 1–22. https://doi.org/10.1186/s42500-019-0004-7.

Linthorst, J.A. (2010). An overview: origins and development of Green Chemistry. *Foundations of Chemistry* 12: 55–68. https://doi.org/10.1007/s10698-009-9079-4.

Liu, Y., Chen, L., Wang, T. et al. (2015). One-pot catalytic conversion of raw lignocellulosic biomass into gasoline alkanes and chemicals over $LiTaMoO_6$ and Ru/C in aqueous phosphoric acid. *ACS Sustainable Chemistry & Engineering* 3: 1745–1755. https://doi.org/10.1021/acssuschemeng.5b00256.

Liu, B., Liu, S., Meng, L.Y. et al. (2018). Microwave-hydrothermal synthesis and photocatalytic properties of biomass charcoal/TiO_2 nanocomposites. *Journal of Saudi Chemical Society* 22 (5): 509–518. https://doi.org/10.1016/j.jscs.2017.09.003.

Lohri, C.R., Diener, S., Zabaleta, I. et al. (2017). Treatment technologies for urban solid biowaste to create value products: a review with focus on low- and middle-income settings. *Reviews in Environmental Science and Bio/Technology* 16: 81–130. https://doi.org/10.1007/s11157-017-9422-5.

Luterbacher, J.S., Rand, J.M., Alonso, D.M. et al. (2014). Nonenzymatic sugar production from biomass using biomass-derived γ-valerolactone. *Science* 343: 277–280. https://doi.org/10.1126/science.1246748.

Malani, R.S., Umriwad, S.B., Kumar, K. et al. (2019). Ultrasound–assisted enzymatic biodiesel production using blended feedstock of non-edible oils: Kinetic analysis. *Energy Conversion and Management* 188: 142–150. https://doi.org/10.1016/j.enconman.2019.03.052.

Mandina, S., Chigondo, F., Shumba, M. et al. (2013). Removal of chromium (VI) from aqueous solution using chemically modified orange (*citrus cinensis*) peel. *IOSR Journal of Applied Chemistry (IOSR JAC)* 6: 66–75.

Maneerung, T., Kawi, S., Dai, Y., and Wang, C.H. (2016). Sustainable biodiesel production via transesterification of waste cooking oil by using CaO catalysts prepared from chicken manure. *Energy Conversion and Management* 123: 487–497. https://doi.org/10.1016/j.enconman.2016.06.071.

Marriott, A.S., Hunt, A.J., Bergström, E. et al. (2014). Investigating the structure of biomass-derived non-graphitizing mesoporous carbons by electron energy loss spectroscopy in the transmission electron microscope and X-ray photoelectron spectroscopy. *Carbon* 67: 514–524. https://doi.org/10.1016/j.carbon.2013.10.024.

Mateo, C., Palomo, J.M., Fernandez-Lorente, G. et al. (2007). Improvement of enzyme activity, stability and selectivity via immobilization techniques. *Enzyme and Microbial Technology* 40: 1451–1463. https://doi.org/10.1016/j.enzmictec.2007.01.018.

Matsakas, L., Gao, Q., Jansson, S. et al. (2017). Green conversion of municipal solid wastes into fuels and chemicals. *Electronic Journal of Biotechnology* 26: 69–83. https://doi.org/10.1016/j.ejbt.2017.01.004.

Mendonça, I.M., Paes, O.A.R.L., Maia, P.J.S. et al. (2019). New heterogeneous catalyst for biodiesel production from waste tucumã peels (*Astrocaryum aculeatum* Meyer): parameters optimization study. *Renewable Energy* 130: 103–110. https://doi.org/10.1016/j.renene.2018.06.059.

Mgaya, J., Shombe, G.B., Masikane, S.C. et al. (2019). Cashew nut shell: a potential bio-resource for the production of bio-sourced chemicals, materials and fuels. *Green Chemistry* 21: 1186–1201. https://doi.org/10.1039/C8GC02972E.

Mishra, R.K., Iyer, J.S., and Mohanty, K. (2019). Conversion of waste biomass and waste nitrile gloves into renewable fuel. *Waste Management* 89: 397–407. https://doi.org/10.1016/j.wasman.2019.04.032.

Mittal, J., Batra, A., Singh, A., and Sharma, M.M. (2014). Phytofabrication of nanoparticles through plant as nanofactories. *Advances in Natural Sciences: Nanoscience and Nanotechnology* 5: 043002. https://doi.org/10.1088/2043-6262/5/4/043002.

Mohan, D., Pittman, C.U. Jr., and Steele, P.H. (2006). Pyrolysis of wood/biomass for bio-oil: a critical review. *Energy & Fuels* 20: 848–889. https://doi.org/10.1021/ef0502397.

Nagarajan, V., Mohanty, A.K., and Misra, M. (2013). Sustainable green composites: value addition to agricultural residues and perennial grasses. *ACS Sustainable Chemistry & Engineering* 1: 325–333. https://doi.org/10.1021/sc300084z.

Naik, S.N., Goud, V.V., Rout, P.K., and Dalai, A.K. (2010). Production of first and second generation biofuels: a comprehensive review. *Renewable and Sustainable Energy Reviews* 14: 578–597. https://doi.org/10.1016/j.rser.2009.10.003.

Nanda, M.R., Misra, M., and Mohanty, A.K. (2013). Performance evaluation of biofibers and their hybrids as reinforcements in bioplastic composites. *Macromolecular Materials and Engineering* 298: 779–788. https://doi.org/10.1002/mame.201200112.

Nikhila, P.S., Satheesh, N., Sreejitha, V.S. et al. (2018). Sustainable utilization of bio waste towards the green synthesis of nanoparticles and its utility in the naked eye detection of metals coupled with its larvicidal and antimicrobial properties. *IOP Conference Series: Materials Science and Engineering* 310: 012089. https://doi.org/10.1088/1757-899X/310/1/012089.

Niven, R.K. (2005). Ethanol in gasoline: environmental impacts and sustainability review article. *Renewable and Sustainable Energy Reviews* 9: 535–555. https://doi.org/10.1016/j.rser.2004.06.003.

Obi, F.O., Ugwuishiwu, B.O., and Nwakaire, J.N. (2016). Agricultural waste concept, generation, utilization and management. *Nigerian Journal of Technology* 35: 957–964. https://doi.org/10.4314/njt.v35i4.34.

Odude, V.O., Adesina, A.J., Oyetunde, O.O. et al. (2019). Application of agricultural waste-based catalysts to transesterification of esterified palm kernel oil into biodiesel: a case of banana fruit peel versus cocoa pod husk. *Waste and Biomass Valorization* 10: 877–888. https://doi.org/10.1007/s12649-017-0152-2.

Outili, N., Kerras, H., Nekkab, C. et al. (2020). Biodiesel production optimization from waste cooking oil using green chemistry metrics. *Renewable Energy* 145: 2575–2586. https://doi.org/10.1016/j.renene.2019.07.152.

Pandit, P.R. and Fulekar, M.H. (2019). Biodiesel production from microalgal biomass using CaO catalyst synthesized from natural waste material. *Renewable Energy* 136: 837–845. https://doi.org/10.1016/j.renene.2019.01.047.

Papurello, D., Soukoulis, C., Schuhfried, E. et al. (2012). Monitoring of volatile compound emissions during dry anaerobic digestion of the organic fraction of municipal solid waste by proton transfer reaction time-of-flight mass spectrometry. *Bioresource Technology* 126: 254–265. https://doi.org/10.1016/j.biortech.2012.09.033.

Papurello, D., Borchiellini, R., Bareschino, P. et al. (2014). Performance of a solid oxide fuel cell short-stack with biogas feeding. *Applied Energy* 125: 254–263. https://doi.org/10.1016/j.apenergy.2014.03.040.

Parfitt, J. (2002). *Analysis of Household Waste Composition and Factors Driving Waste Increases*. London: WRAP for the Strategy Unit, Government Cabinet Office.

Park, B.D., Wi, S.G., Lee, K.H. et al. (2003). Characterization of anatomical features and silica distribution in rice husk using microscopic and micro-analytical techniques. *Biomass Bioenergy* 25: 319–327. https://doi.org/10.1016/S0961-9534(03)00014-X.

Pickering, K.L. (ed.) (2008). *Properties and Performance of Natural-Fibre Composites*. New York, Washington DC: CRC Press.

Pickett, J., Anderson, D., Bowles, D. et al. (2008). *Sustainable Biofuels: Prospects and Challenges*. London: The Royal Society.

Prevot, A.B., Avetta, P., Fabbri, D. et al. (2011). Waste-derived bioorganic substances for light-induced generation of reactive oxygenated species. *ChemSusChem* 4: 85–90. https://doi.org/10.1002/cssc.201000237.

Primo, A., Atienzar, P., Sanchez, E. et al. (2012). From biomass wastes to large-area, high-quality, N-doped graphene: catalyst-free carbonization of chitosan coatings on arbitrary substrates. *Chemical Communications* 48: 9254–9256. https://doi.org/10.1039/C2CC34978G.

Putra, M.D., Irawan, C., Udiantoro et al. (2018). A cleaner process for biodiesel production from waste cooking oil using waste materials as a heterogeneous catalyst and its kinetic study. *Journal of Cleaner Production* 195: 1249–1258. https://doi.org/10.1016/j.jclepro.2018.06.010.

Pütün, A.E., Özbay, N., Önal, E.P., and Pütün, E. (2005). Fixed-bed pyrolysis of cotton stalk for liquid and solid products. *Fuel Processing Technology* 86: 1207–1219. https://doi.org/10.1016/j.fuproc.2004.12.006.

Qi, Z., Jie, C., Tiejun, W., and Ying, X. (2007). Review of biomass pyrolysis oil properties and upgrading research. *Energy Conversion and Management* 48: 87–92. https://doi.org/10.1016/j.enconman.2006.05.010.

Qu, X.Y., Li, Z.S., Xie, X.Y. et al. (2009). Survey of composition and generation rate of household wastes in Beijing, China. *Waste Management* 29: 2618–2624. https://doi.org/10.1016/j.wasman.2009.05.014.

Rajkumari, K., Das, D., Pathak, G., and Rokhum, L. (2019). Waste-to-useful: a biowaste-derived heterogeneous catalyst for a green and sustainable Henry reaction. *New Journal of Chemistry* 43: 2134–2140. https://doi.org/10.1039/C8NJ05029E.

Ramya, V., Murugan, D., Lajapathirai, C., and Sivasamy, A. (2018). Activated carbon (prepared from secondary sludge biomass) supported semiconductor zinc oxide nanocomposite photocatalyst for reduction of Cr (VI) under visible light irradiation. *Journal of Environmental Chemical Engineering* 6: 7327–7337. https://doi.org/10.1016/j.jece.2018.08.055.

Rashid, U., Anwar, F., and Knothe, G. (2011). Biodiesel from Milo (*Thespesia populnea* L.) seed oil. *Biomass & Bioenergy* 35: 4034–4039. https://doi.org/10.1016/j.biombioe.2011.06.043.

Reshad, A.S., Tiwari, P., and Goud, V.V. (2018). Thermo-chemical conversion of waste rubber seed shell to produce fuel and value-added chemicals. *Journal of the Energy Institute* 91: 940–950. https://doi.org/10.1016/j.joei.2017.09.002.

Saikia, P., Gupta, U.N., Barman, R.S. et al. (2015). Production and characterization of bio-oil produced from *Ipomoea carnea* bio-weed. *Bioenergy Research* 8: 1212–1223. https://doi.org/10.1007/s12155-014-9561-2.

Saikia, R., Baruah, B., Kalita, D. et al. (2018). Pyrolysis and kinetic analyses of a perennial grass (*Saccharum ravannae* L.) from north-east India: optimization through response surface methodology and product characterization. *Bioresource Technology* 253: 304–314. https://doi.org/10.1016/j.biortech.2018.01.054.

Sarmah, M., Mondal, M., and Bora, U. (2017). Agro-waste extract-based solvents: emergence of novel green solvent for the design of sustainable processes in catalysis and organic chemistry. *ChemistrySelect* 2: 5180–5188. https://doi.org/10.1002/slct.201700580.

Seifi, S. and Masoum, S. (2019). Preparation of copper oxide/oak-based biomass nanocomposite for electrochemical hydrogen storage. *International Journal of Hydrogen Energy* 44: 11979–11988. https://doi.org/10.1016/j.ijhydene.2019.03.083.

Sikarwar, V.S., Zhao, M., Fennell, P.S. et al. (2017). Progress in biofuel production from gasification. *Progress in Energy and Combustion Science* 61: 189–248. https://doi.org/10.1016/j.pecs.2017.04.001.

Stelte, W. and Sanadi, A.R. (2009). Preparation and characterization of cellulose nanofibers from two commercial hardwood and softwood pulps. *Industrial & Engineering Chemistry Research* 48: 11211–11219. https://doi.org/10.1021/ie9011672.

Su, D.S. (2009). The use of natural materials in nanocarbon synthesis. *ChemSusChem* 2: 1009–1020. https://doi.org/10.1002/cssc.200900046.

Su, F. and Guo, Y. (2014). Advancements in solid acid catalysts for biodiesel production. *Green Chemistry* 16: 2934–2957. https://doi.org/10.1039/C3GC42333F.

Suganuma, S., Nakajima, K., Kitano, M. et al. (2010). Synthesis and acid catalysis of cellulose-derived carbon-based solid acid. *Solid State Sciences* 12: 1029–1034. https://doi.org/10.1016/j.solidstatesciences.2010.02.038.

Sut, D., Chutia, R.S., Bordoloi, N. et al. (2016). Complete utilization of non-edible oil seeds of *Cascabela thevetia* through a cascade of approaches for biofuel and by-products. *Bioresource Technology* 213: 111–120. https://doi.org/10.1016/j.biortech.2016.02.066.

Tan, I.S. and Lee, K.T. (2014). Enzymatic hydrolysis and fermentation of seaweed solid wastes for bioethanol production: an optimization study. *Energy* 78: 53–62. https://doi.org/10.1016/j.energy.2014.04.080.

Tan, Y.H., Abdullah, M.O., Nolasco-Hipolito, C., and Taufiq-Yap, Y.H. (2015). Waste ostrich- and chicken-eggshells as heterogeneous base catalyst for biodiesel production from used cooking oil: catalyst characterization and biodiesel yield performance. *Applied Energy* 160: 58–70. https://doi.org/10.1016/j.apenergy.2015.09.023.

Tang, Z., Lu, Q., Zhang, Y. et al. (2009). One step bio-oil upgrading through hydrotreatment, esterification, and cracking. *Industrial & Engineering Chemistry Research* 48: 6923–6929. https://doi.org/10.1021/ie900108d.

Tariq, M., Ali, S., and Khalid, N. (2012). Activity of homogeneous and heterogeneous catalysts, spectroscopic and chromatographic characterization of biodiesel: a review. *Renewable and Sustainable Energy Reviews* 16: 6303–6316. https://doi.org/10.1016/j.rser.2012.07.005.

Tay, T., Ucar, S., and Karagöz, S. (2009). Preparation and characterization of activated carbon from waste biomass. *Journal of Hazardous Materials* 165: 481–485. https://doi.org/10.1016/j.jhazmat.2008.10.011.

Thushari, I., Babel, S., and Samart, C. (2019). Biodiesel production in an autoclave reactor using waste palm oil and coconut coir husk derived catalyst. *Renewable Energy* 134: 125–134. https://doi.org/10.1016/j.renene.2018.11.030.

Titirici, M.M. and Antonietti, M. (2010). Chemistry and materials options of sustainable carbon materials made by hydrothermal carbonization. *Chemical Society Reviews* 39: 103–116. https://doi.org/10.1039/B819318P.

United States Environmental Protection Agency (USEPA) (2017). Basics of Green Chemistry, Green Chemistry, USEPA, https://www.epa.gov/greenchemistry/basics-green-chemistry (viewed 11 May 2019)

Veljković, V.B., Avramović, J.M., and Stamenković, O.S. (2012). Biodiesel production by ultrasound-assisted transesterification: state of the art and the perspectives. *Renewable and Sustainable Energy Reviews* 16: 1193–1209. https://doi.org/10.1016/j.rser.2011.11.022.

Vert, M., Doi, Y., Hellwich, K.H. et al. (2012). Terminology for biorelated polymers and applications (IUPAC Recommendations 2012). *Pure and Applied Chemistry* 84: 377–410. https://doi.org/10.1351/PAC-REC-10-12-04.

Viriya-empikul, N., Krasae, P., Nualpaeng, W. et al. (2012). Biodiesel production over Ca-based solid catalysts derived from industrial wastes. *Fuel* 92: 239–244. https://doi.org/10.1016/j.fuel.2011.07.013.

Vivek, M., Kumar, P.S., Steffi, S., and Sudha, S. (2011). Biogenic silver nanoparticles by *Gelidiella acerosa* extract and their antifungal effects. *Avicenna Journal of Medical Biotechnology (AJMB)* 3: 143–148.

Wei, Z., Xu, C., and Li, B. (2009). Application of waste eggshell as low-cost solid catalyst for biodiesel production. *Bioresource Technology* 100: 2883–2885. https://doi.org/10.1016/j.biortech.2008.12.039.

Williams, P.T. and Nugranad, N. (2000). Comparison of products from the pyrolysis and catalytic pyrolysis of rice husks. *Energy* 25: 493–513. https://doi.org/10.1016/S0360-5442(00)00009-8.

Wu, J., Zuo, L., Song, Y. et al. (2016). Preparation of biomass-derived hierarchically porous carbon/Co_3O_4 nanocomposites as anode materials for lithium-ion batteries. *Journal of Alloys and Compounds* 656: 745–752. https://doi.org/10.1016/j.jallcom.2015.10.063.

Xie, H. and Gathergood, N. (ed.) (2013). *The Role of Green Chemistry in Biomass Processing and Conversion*. Hoboken: Wiley.

Xu, C., Nasrollahzadeh, M., Selva, M. et al. (2019). Waste-to-wealth: biowaste valorization into valuable bio (nano) materials. *Chemical Society Reviews* 48: 4791–4822. https://doi.org/10.1039/C8CS00543E.

Yusuff, A.S. and Owolabi, J.O. (2019). Synthesis and characterization of alumina supported coconut chaff catalyst for biodiesel production from waste frying oil. *South African Journal of Chemical Engineering* 30: 42–49. https://doi.org/10.1016/j.sajce.2019.09.001.

Yusuff, A.S. and Popoola, L.T. (2019). Optimization of biodiesel production from waste frying oil over alumina supported chicken eggshell catalyst using experimental design tool. *Acta Polytechnica* 59: 88–97. https://doi.org/10.14311/AP.2019.59.0088.

Zarraa, M.A. (1995). A study on the removal of chromium (VI) from waste solutions by adsorption on to sawdust in stirred vessels. *Adsorption Science & Technology* 12: 129–138. https://doi.org/10.1177/026361749501200205.

Zheng, Y., Zhao, J., Xu, F., and Li, Y. (2014). Pretreatment of lignocellulosic biomass for enhanced biogas production. *Progress in Energy and Combustion Science* 42: 35–53. https://doi.org/10.1016/j.pecs.2014.01.001.

17

Assessing Satellite-Based Products in Characterizing Agricultural Drought Under Climate Change in Northeast Brazil

Franklin Paredes-Trejo[1,2], Humberto Alves Barbosa[2], Jason Giovannettone[3], Tumuluru Venkata Lakshmi Kumar[4], Catarina de Oliveira Buriti[5], and José Prieto[6]

[1] *Department of Civil Engineering, Institute for Sustainable Agroindustry, University of the Western Plains Ezequiel Zamora, San Carlos, Venezuela*
[2] *Laboratório de Análise e Processamento de Imagens de Satélites (LAPIS), Instituto de Ciências Atmosféricas, Universidade Federal de Alagoas, Alagoas, Brazil*
[3] *Dewberry, Fairfax, VA, USA*
[4] *Atmospheric Science Research Laboratory, SRM University, Chennai, Tamilnadu, India*
[5] *National Semi-Arid Institute (INSA), Ministry of Science, Technology, Innovations and Communications (MCTIC), Campina Grande, Brazil*
[6] *Eumetsat (European Organisation for the Exploitation of Meteorological Satellites), Darmstadt, Germany*

CONTENTS

Climate Change and Agriculture: Perspectives, Sustainability and Resilience, First Edition. Edited by Noureddine Benkeblia.

17.1 Introduction

According to the Intergovernmental Panel on Climate Change (IPCC), the incidence and severity of drought may increase in northeast Brazil (NEB) due to human-induced climate change. Recent droughts in NEB, like the one in 2010–2016, triggered severe socio-economic impacts along with extensive damage to rainfed crops and pastures. In this semiarid region, drought impacts are significantly aggravated by poor policies, conflicts over limited water resources, and its little resilience and preparedness. Most Brazilian agencies and research institutions are working on the development of operational applications to track agricultural drought and provide early warning of deteriorating crop moisture conditions. These applications are based on drought indices and environmental variables such as soil moisture, vegetation greenness, temperature, rainfall, and evapotranspiration deficits derived from observational networks, model output, and satellite estimates. Unlike other approaches, satellites can be used for sensing large regions with a high temporal and spatial resolution. However, there remain knowledge gaps about the reliability of satellite-based products in monitoring agricultural drought under climate change. The purpose of this chapter is to present an overview of the strengths, weaknesses, and limitations of some state-of-the-art satellite-based agricultural drought products in NEB.

17.2 Droughts and Impacts on Agriculture in NEB

Precipitation is of paramount importance for a wide range of economic sectors in drought-prone regions (Sheffield and Wood 2012). Northeast Brazil (NEB) is recognized as one of the most vulnerable regions to droughts in the world and the more important drought hotspot in Brazil. It is a region politically divided into nine states (Alagoas, Bahia, Ceará, Maranhão, Paraíba, Piauí, Rio Grande do Norte, Sergipe, and Pernambuco; Figure 17.2a), localized between 1.3–18.2°S and 34.4–48.4°W. This region covers an area of 1,553,860 km^2 (nearly 19% of Brazilian territory) with more than 53 million inhabitants, representing about 30% of the Brazilian population (Instituto Brasileiro de Geografia e Estatística 2010). The climate type of the NEB belongs to a semiarid climate, with an average annual precipitation ranging from less than 250 to 750 mm (Figure 17.2b), which decreases gradually from the eastern coastline inland (Marengo and Bernasconi 2015). The driest region of the NEB (known locally as sertão, about 980,133 km^2) is the world's most densely populous dryland region. For centuries, the sertão region has experienced recurrent extreme droughts, whose impacts adversely affect their water resources and crop production (Correia et al. 2019; Figure 17.1).

The NEB contains the largest concentration of rural poverty and social inequality (Campos and Carvalho 2008). Several factors are related to the poverty, including the lack of institutional capacity of the public and private sectors to deal with drought events (Buriti et al. 2020; Campos 2015) and different natural barriers such as shallow soils and the prevalence of crystalline rocks that reduce groundwater storage (Herwehe and Scott 2018). These natural barriers, together with the high interannual and seasonal variability of precipitation (Souza et al. 2005; Oliveira et al. 2017) and the fact that the rainy season is concentrated in a few months, result in the dominance of intermittent rivers in all of its municipalities (Campos et al. 2014).

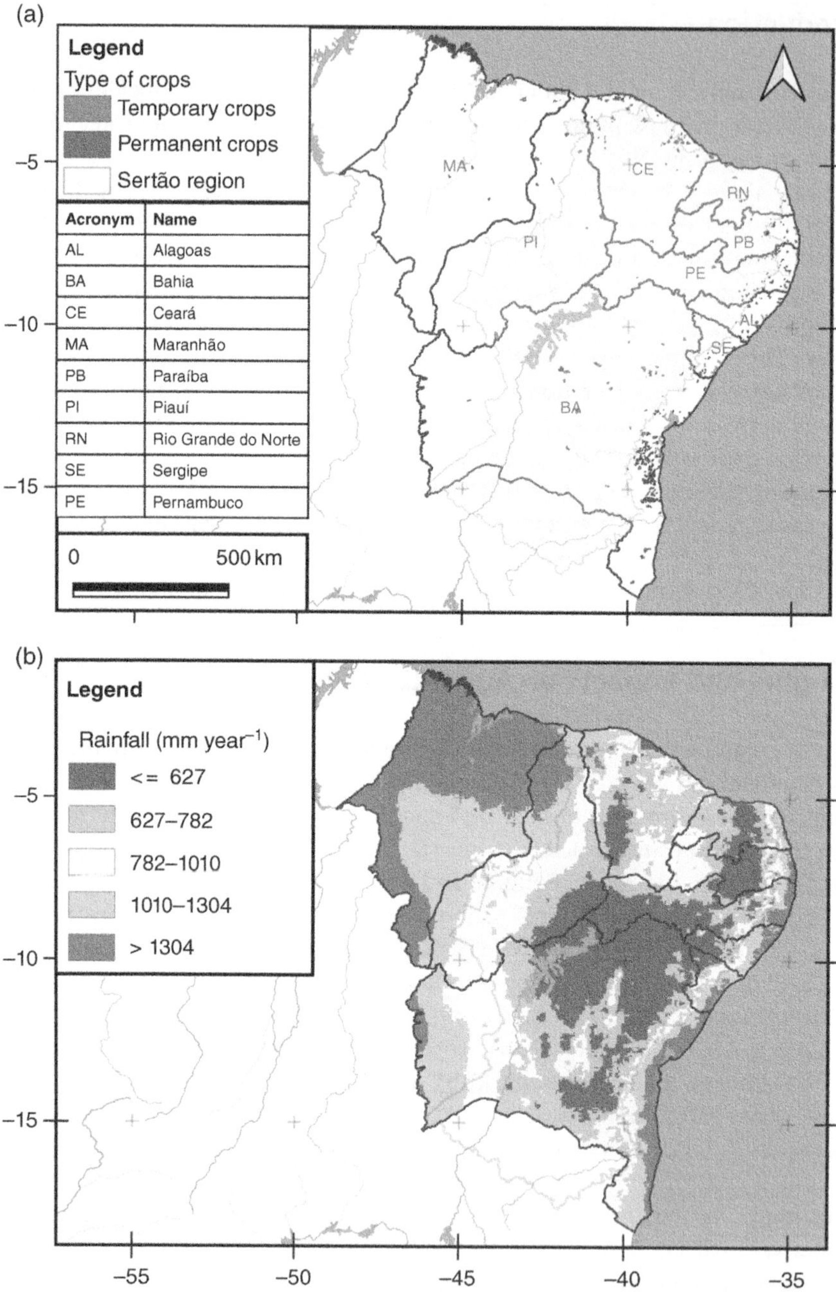

Figure 17.1 The NEB: (a) its nine states, including the Sertão region, along with the spatial distribution of temporary and permanent crops in 2006. *Source:* IBGE – Instituto Brasileiro de Geografia e Estatística (2009). (b) Mean annual precipitation from the Climate Hazards Group InfraRed Precipitation with Stations (CHIRPS) for the period 1981–2020. *Source:* Adapted from Funk et al. (2015).

The temporal variability of precipitation in the NEB is primarily controlled by the Inter-Tropical Convergence Zone (ITCZ), which in turn, is modulated by sea surface temperature (SST) anomalies over the tropical Atlantic region, the El NiñoSouthern Oscillation (ENSO) phenomenon (Andreoli and Kayano 2006; Silva et al. 2018), and the Madden Julian Oscillation (MJO) (Giovannettone et al. 2020; Valadão et al. 2017). Although ENSO modulates the precipitation regime by modifying the Walker and Hadley circulations, the Atlantic SST anomalies affect the precipitation regime by shifting the ITCZ (Andreoli and Kayano 2006). Both ENSO and Atlantic SST anomalies are the major drivers of the interannual variability of precipitation (Giovannettone et al. 2020; Pereira et al. 2014a), while the MJO influences the intraseasonal variability of precipitation (Souza and Ambrizzi 2006).

In recent years, an increase in the number of drought events, such as those that occurred between 2010 and 2015, has been associated with increased climate variability in the NEB, which in turn, is attributed to recent changes in climate (Brito et al. 2018) and exacerbated by anthropogenic factors such as land-use changes associated with agriculture (Marengo et al. 2017). In this context, it has been shown that the region's agriculture sector is highly vulnerable (less resilient), particularly in the sertão region, where subsistence farming practices and rain-fed crops are dominant (Brabo et al. 2009). It is worthwhile emphasizing that both the federal and state governments have built dams and reservoirs, among other hydraulic infrastructures, on perennial and seasonal rivers (e.g. São Francisco and Parnaiba); currently, however, many of these rivers flow dry or almost dry throughout the year due to uncontrolled water extraction (Buriti et al. 2020). In a similar way to surface waters, the overexploitation of groundwater for irrigation and livestock has had terrible impacts such as irreversible depletion of groundwater levels in certain aquifers of the NEB (Cirilo 2008).

Despite severe water scarcity, rain-fed agriculture plays a key role in the socioeconomic development in most cities and towns across the NEB. The warm climate favours the production of maize, beans, cotton, and sugarcane as well as fruit and vegetable crops grown under sprinkler and localized irrigation in drought-prone areas (Burney et al. 2014). Nevertheless, as can be seen from Figure 17.2, the high dependence of rain-fed agriculture

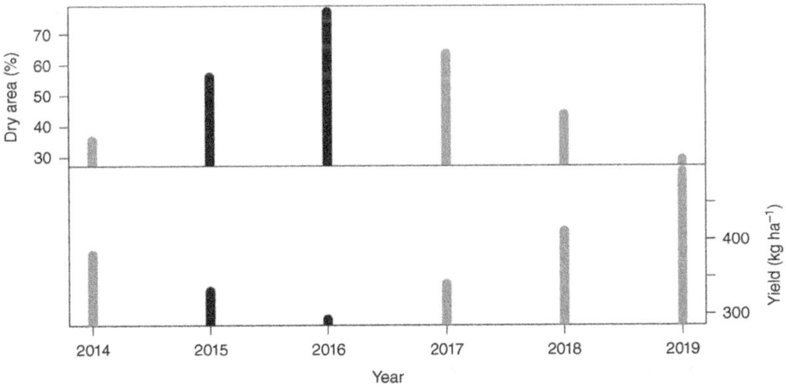

Figure 17.2 Temporal variation of the NEB regionally averaged values of: (a) dry area with severity levels of drought classified as severe or extreme [in %]; (b) yield of bean [in kg ha^{-1}] for the period 2014–2019. The extreme drought events in 2015 and 2016 are shown with black bars. Dry area is based on the maps of drought published by the Brazilian Drought Monitor (https://bit.ly/3b9ZIkN), whereas the bean yield data were obtained by the Brazilian National Institute of Geography and Statistics (IBGE) (https://bit.ly/3rZsTh9).

on weather conditions has raised concern about the risk climate change poses to the entire regional economy (Buriti et al. 2020). These concerns are even greater due to the fact that climate change projections also suggest that future droughts will be more frequent and intense (Marengo and Bernasconi 2015, Marengo et al. 2017, 2020).

Due to the reasons mentioned above, the NEB is advancing in the implementation of appropriate mitigation and adaptation strategies, particularly for smallholder farmers in the sertão region (Buriti et al. 2020). To help achieve this goal, considerable attention has been devoted to implementing a wide variety of operational products based on remote sensing data for monitoring drought conditions (Barbosa et al. 2019; Barbosa et al. 2006).

17.3 Satellite-Based Agricultural Drought Indices

The scientific literature specifies four types of drought according to their persistence over time and as well as their impacts: meteorological, hydrological, agricultural, and socio-economic (Mishra and Singh 2010). Initially, a prolonged deficit of precipitation is categorized as a meteorological drought. This decrease in precipitation together with high temperature conditions may lead to a situation where the water supply is insufficient to satisfy the water demand of crops during their growing season, resulting in lower productivity or even the loss of entire harvests, which is known as agricultural drought (Zargar et al. 2011). Agricultural drought can often be visually assessed based on the 'greenness' of the crop leaves.

The deficit of precipitation and high temperatures are the major drivers of water stress in vegetation, making them useful for assessing the drought conditions through quantitative metrics commonly known as agricultural drought indices (Liu et al. 2016). Its traditional calculation uses field measurements as inputs (Zargar et al. 2011). Nevertheless, these measurements are costly, time consuming, and often are not available for smallholder subsistence farms.

The field measurements, albeit important, are being gradually complemented with remote sensing estimates due to the enormous advances in remote sensing technology. This technological milestone is opening new opportunities for retrieving climatic and biophysical variables from space, especially over areas where measurements are scarce, such as the NEB (Barbosa et al. 2006; Paredes-Trejo et al. 2017; Spatafora et al. 2020). Currently, most of the drought indices can be calculated using remote sensing data. Consequently, a wide variety of satellite-based drought indices have been adopted by the Brazilian federal and state agencies for monitoring agricultural drought (Buriti et al. 2020; Rossato et al. 2017). We will focus attention on a few of these indices in the following sections.

17.4 Precipitation-Based Drought Indices

17.4.1 Standardized Precipitation Index (SPI)

The Standardized Precipitation Index (SPI) was developed by McKee et al. (1993) and has been widely used to assess drought impacts in many agricultural regions, especially in the NEB (Brito et al. 2018). Besides its simplicity due to the fact that it only uses historical precipitation records as input, the SPI can identify any type of drought (i.e. meteorological, agricultural, and hydrological) due to its multi-scalar nature. In general, the lower the time scale, the higher its capacity to identify short-term droughts (Brito et al. 2018; Vicente-Serrano 2006). Previous studies have shown that the short-term SPI (i.e. six months or less) has a strong relationship with soil moisture and vegetation response in the NEB (Barbosa et al. 2019; Barbosa and Kumar 2016). The SPI also has been recommended by the World Meteorological Organization (WMO) as the main meteorological drought index that countries should use to monitor drought conditions (Hayes et al. 2011).

The SPI takes into account accumulated precipitation at different time scales and is fitted with a nonnormal probability distribution (e.g. Gamma) for a baseline period (at least 30 years of data). The fitted values are then standardized and classified into drought categories (McKee et al. 1993) as shown in Table 17.1. Regardless of the time scale of the SPI, a drought event begins when SPI ≤ −1.00 and persists until SPI > −1.00; the value of −1.00 is the threshold value that differentiates dry vs. non-dry events (Brito et al. 2018).

Several satellite-derived products with high spatial and temporal resolution and low latency currently provide reliable precipitation estimates over the entire extent of Brazil (Nogueira et al. 2018; Paredes-Trejo et al. 2019; Rozante et al. 2018). One state-of-the-art satellite-based precipitation product that has shown high accuracy compared to in situ measurements over the NEB has been the Climate Hazards Group InfraRed Precipitation with Station data (CHIRPS) (Paredes-Trejo et al. 2017). For this reason, CHIRPS is being widely exploited by some agricultural institutions to obtain precipitation-based drought indices (including the SPI) for agricultural drought monitoring in large regions with scarce ground-based climate data, such as the sertão region.

CHIRPS precipitation estimates from 1981 to near-present are available to the public at the UCSB-Climate Hazards Group (CHG) webpage (https://www.chc.ucsb.edu/data; version 2 released in February 2015) at a daily time scale and a spatial resolution of 0.05°. SPI can then be easily calculated from the CHIRPS precipitation estimates using different tools, such as the SPI Generator application developed by the National Drought Mitigation Center (NDMC) at the University of Nebraska-Lincoln (available online at https://bit.ly/3buqFj4).

As an example, Figure 17.3 shows the temporal variation of the CHIRPS-derived SPI at a three-month time scale (SPI-3) over an agricultural region in the

Table 17.1 SPI classifications for dry and wet periods.

Class names	SPI
Extreme wet	>2.00
Severe wet	1.50–1.99
Moderate wet	1.00–1.49
Near normal	0.99 to −0.99
Moderate drought	−1.00 to −1.49
Severe drought	−1.50 to −1.99
Extreme drought	<−2.00

Source: Adapted from McKee et al. (1993).

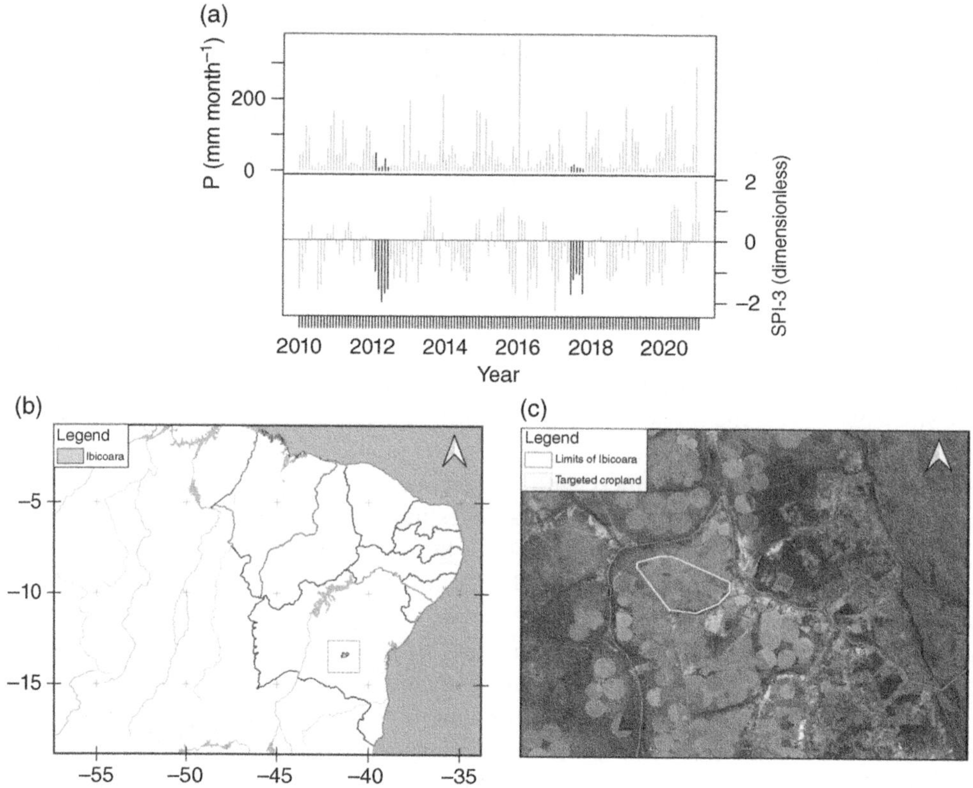

Figure 17.3 (a) Temporal variation of the area-averaged monthly values of precipitation (P) [in mm] and SPI at a three-month time scale (SPI-3) [dimensionless; base period 1981/01–2012/12]; over an agricultural region in the municipality of Ibicoara – Bahia (identified in b, c) for the period 2010–2020. The longest drought spells are shown with black bars. The high-resolution image in panel (c). *Source:* (c) Google LLC.

municipality of Ibicoara – Bahia (13.38°S, 41.32°W). This region (hereinafter referred to as the targeted region) is located in the sertão region and is characterized by the presence of irrigated bean crops. According to CHIRPS precipitation estimates, the mean annual precipitation for the 11-year period (2010–2020) was 653.69 mm, with the highest and the lowest values in 2020 (1029.56 mm) and in 2012 (344.77 mm), respectively.

A visual inspection of Figure 17.3 shows that in the last 10 years this region has had two relevant drought spells, which occurred from February 2012 to June 2012 and from June 2017 to October 2017 (duration = 5 months in both cases). The second drought event was partially concurrent with the second season bean crop, whose growing period in the state of Bahia extends from April to June (Pereira et al. 2014b).

17.4.2 Standardized Precipitation Evapotranspiration Index (SPEI)

The fact that the SPI does not take into account temperature in its formulation is its main weakness since it cannot be compared between sites within different temperature regimes (Wilhite and Pulwarty 2017). To overcome this limitation, the Standardized Precipitation Evapotranspiration Index (SPEI) was developed based on the original SPI calculation procedure while also considering the difference between precipitation (P) and potential evapotranspiration (PET) (Vicente-Serrano et al. 2010). The PET is the sum of the amount of evaporation and transpiration from a reference vegetation of grass. It can be calculated with the Thornthwaite, Hargreaves, or Penman–Monteith equations (Allen et al. 1998). The first equation is the simplest and is ideal when only temperature data are available (Stephens and Stewart 1963). The Thornthwaite equation is computed as follows:

$$\text{PET} = 16\left(\frac{L}{12}\right)\left(\frac{N}{30}\right)\left(\frac{10\text{Td}}{\text{I}}\right)^{\alpha} \tag{17.1}$$

$$\alpha = \left(6.75\times10^{-7}\right)I^3 - \left(7.71\times10^{-5}\right)I^2 + \left(1.792\times10^{-2}\right)I + 0.49239 \tag{17.2}$$

$$I = \sum_{i=1}^{12}\left(\frac{\text{Tmi}}{5}\right)^{1.514} \tag{17.3}$$

where PET is the estimated potential evapotranspiration (mm/month), Td denotes the average monthly temperature (°C), N is the number of days for each month, L indicates the local insolation (hours/month), and I is the annual heat index, which depends on the monthly mean temperatures (Tmi in °C).

With the inclusion of temperature data through the PET, the SPEI is ideal for assessing the impact of climate change in crop models under future scenarios (Wang et al. 2017). Similar to the SPI, the SPEI requires the time series of P and PET with no missing months. P estimates can be derived from the CHIRPS precipitation product using a pixel-to-site extraction, while the PET can be extracted from any satellite-derived PET product, such as the global MODIS MOD16A2 product (He et al. 2019), which has shown an acceptable performance when compared to in situ flux-tower measurements in different parts of the NEB (Miranda et al. 2017). Version 6 (MOD16A2v006) is freely available at https://bit.ly/3aGGJ20 from 2001 to present with an eight-day temporal and 500 m spatial resolution. Full details about the main steps carried out to produce MOD16A2v006 can be found in Mu et al. (2013).

In order to compare values of P and the SPEI-3, monthly values of both in the targeted region are shown in Figure 17.4. In this case, P and PET were derived from the CHIRPS and MOD16A2v006 products, respectively. Unlike the SPI, the SPEI reveals only one relevant drought spell between March 2012 and June 2012 (duration = 4 months). Even though the area-averaged values of SPI and SPEI have a strong relationship over the common time period (Pearson r: 0.725), there is a slight difference in terms of drought duration as seen by the SPI (Figure 17.3a) and SPEI (Figure 17.4). This result can be associated with the effect of the PET on the SPEI and the base period used in its calculation (i.e. 2010/01–2020/12).

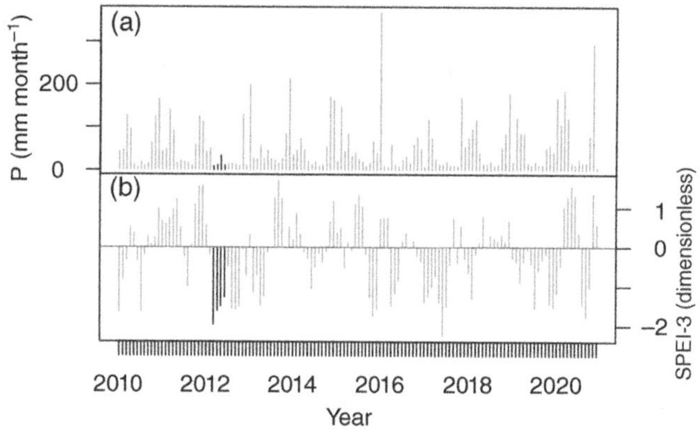

Figure 17.4 As in Figure 17.3, but for SPEI-3 [dimensionless; base period 2010/01–2020/12]. P and PET were derived from the CHIRPS and MOD16A2v006 products, respectively.

17.5 Radiance-Based Drought Indices

17.5.1 Normalized Difference Vegetation Index (NDVI)

The SPI and SPEI do not provide direct information on vegetation status, which limits their operational use for crops and pastures. In these cases, the Normalized Difference Vegetation Index (NDVI) is the preferred option due to its capacity to monitor the health of vegetation in relation to drought events (Barbosa et al. 2006; Kogan 1995). The NDVI is a vegetation-based drought index derived from radiance values measured in both the visible and near-infrared wavelengths over an adequate aggregation period (e.g. eight day) as a way of minimizing cloud contamination (Julien and Sobrino 2010).

Currently, there is a wide variety of satellite-derived NDVI products with varying spatial resolutions and the temporal frequencies. Among them, the Moderate Resolution Imaging Spectroradiometer – Vegetation Indices product (termed as MOD13Q1.006) – provides both NDVI and Enhanced Vegetation Index (EVI; discussed in the following section) data from February 2000 to the present at a temporal frequency of 16 days and a spatial resolution of 250 m. The MOD13Q1.006 is freely available at https://bit.ly/3umtco6

17.5.2 Enhanced Vegetation Index (EVI)

Despite the fact that the NDVI is extensively used for agricultural drought assessment throughout the world, it has a marked saturation effect over densely vegetated areas, especially when the leaf area index (LAI) is higher than 4 (Sun et al. 2018). The EVI was developed by Huete et al. (2002) to overcome this shortcoming of the NDVI. The EVI includes radiance values measured by the sensor instrument aboard the Terra and Aqua satellites in the blue band to correct for atmospheric and soil background influences (Huete et al. 2002). A comparison of the area-averaged maximum values of NDVI and EVI from the MOD13Q1.006 product within the study region is shown in Figure 17.5. In order to provide

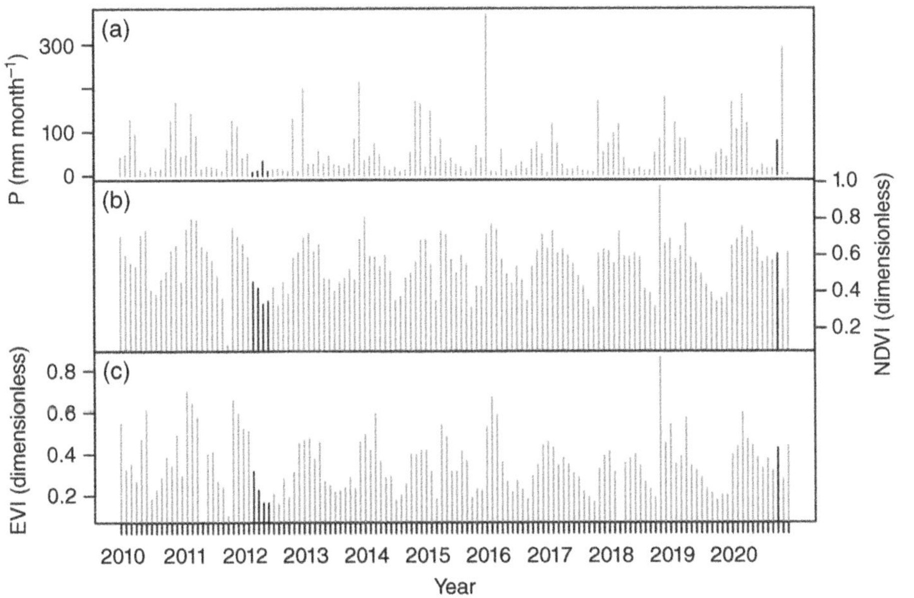

Figure 17.5 As in Figure 17.3, but for NDVI [dimensionless] and EVI [dimensionless]. NDVI and EVI were derived from the MOD13Q1.006 product. The dry spells occurred from March 2012 to June 2012 and October 2020 are shown with black bars.

a clear comparison of vegetation health against the area-averaged values of the NDVI and EVI, Figure 17.6 shows the spatial distribution of the NDVI at a 30-m spatial resolution based on spectral data captured by the multi-spectral instrument (MSI) aboard the Sentinel-2A satellite on 20 October 2020 (available at the Sentinel Data Hub of the European Space Agency via https://bit.ly/2NjcRjL). When looking at this image, the vegetation suffering from soil water stress exhibits NDVI values lower than 0.20 (brown tones).

Note that both EVI and NDVI show a relatively low coupling with P over time (Pearson $r = 0.321$ and 0.332 for EVI and NDVI, respectively). This can be mainly attributed to a delay in the response of vegetative health to rain events (including dry spells) or to the water supplied by the centre pivot irrigation systems evident in the circles observed in Figure 17.6 (Barbosa et al. 2015). On the other hand, Figure 17.5 also reveals that the EVI results have a lower range than those for the NDVI (from 0.110 to 0.970 and from 0.102 to 0.863 for the NDVI and EVI, respectively). This is an expected effect, since the EVI calculation algorithm minimizes the noise effect caused by the sensor saturation aboard the Terra and Aqua satellites (Solano et al. 2010).

The time lag in the response of vegetative health to the precipitation deficit may become an important operational limitation in detecting the early emergence of agricultural drought in crops highly sensitive to moisture deficiencies. An alternative approach involves the use of soil moisture-based drought indices, such as surface soil moisture (SSM) estimates from the NASA Soil Moisture Active Passive (SMAP) satellite mission (O'Neill et al. 2016).

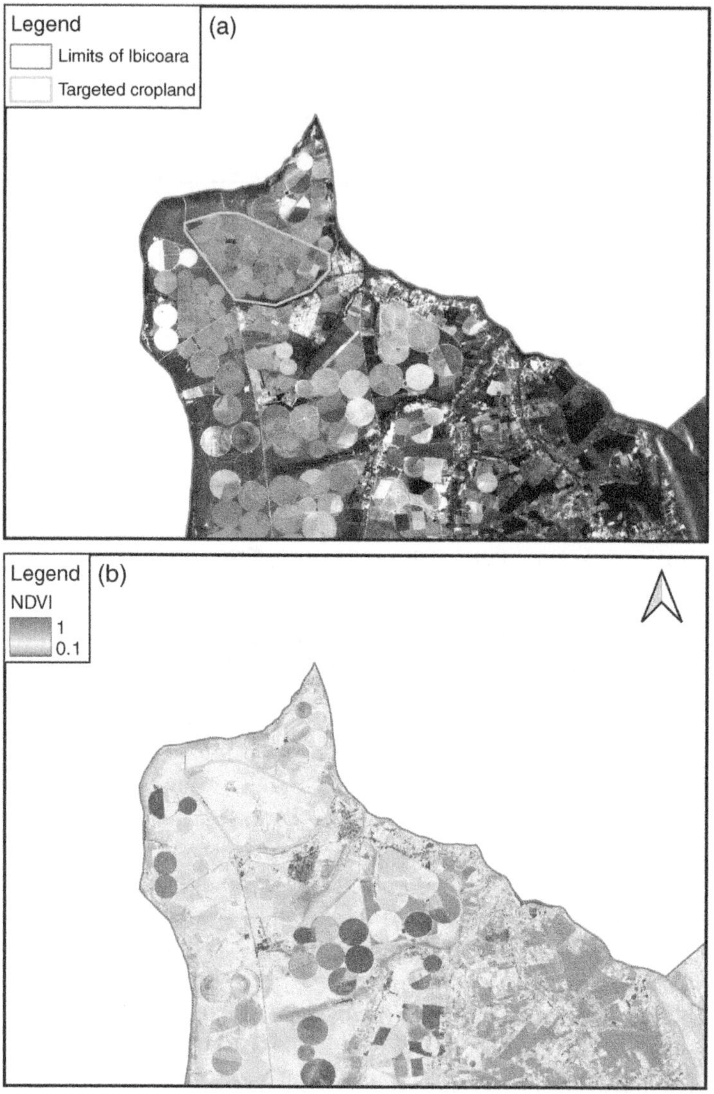

Figure 17.6 The targeted agricultural region shown in Figure 17.3c as: (a) a true color image; and (b) NDVI [dimensionless] both derived from the Sentinel-2A multi-spectral instrument (MSI) on 20 October 2020. For clarity, the region outside the domain of the municipality of Ibicoara was blanked.

17.6 Soil Moisture-Based Drought Indices

17.6.1 SMAP-Based Soil Surface Moisture Estimates

SMAP is an L-band passive microwave satellite launched in January 2015 that measures brightness temperatures using the interferometric radiometry method (1.4 GHz), from which soil moisture estimates for the topsoil are retrieved at a frequency of every of

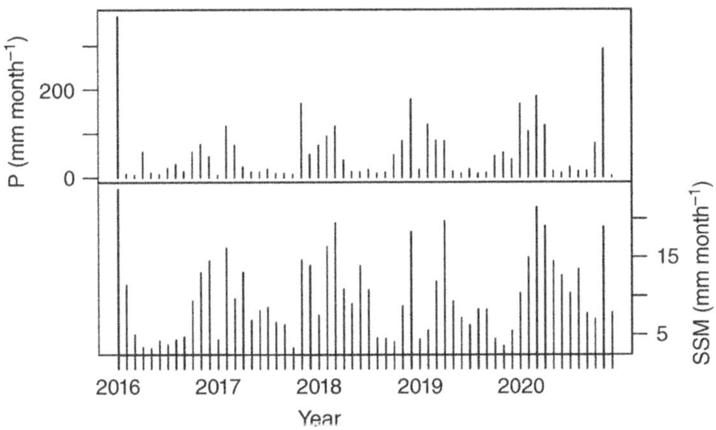

Figure 17.7 Same as Figure 17.3, but for SSM [mm/month] from 01/2016 to 12/2020. SSM was derived from the 10-km enhanced SMAP SM data set.

2–3 days at the equator (Entekhabi et al. 2010). The L-band band is not affected by the presence of clouds; therefore, it is appropriate for using SMAP to monitor SSM in agricultural regions under any climate condition (Martínez-Fernández et al. 2016).

Today, there are various operational SMAP-based SSM products available with different spatiotemporal resolutions and latencies. For example, the NASA-USDA Enhanced SMAP Global soil moisture data provides global data at a 10-km spatial resolution (including SSM in mm) from 04/2015 to present (termed as 10-km enhanced SMAP SM). This dataset is in the public domain and can be accessed at https://go.nasa.gov/3dA3mHl

In order to explore the relationship between the CHIRPS-based precipitation and SSM estimates derived from the 10-km enhanced SMAP SM data set, both were compared in the targeted region from 2016 to 2020. Figure 17.7 reveals a moderate linear correlation between these two variables (Pearson *r*: 0.673), meaning that the SMAP-based SSM estimates provide a better representation of accumulated precipitation than that from NDVI or EVI (Martínez-Fernández et al. 2016).

17.6.2 SMOS-Based Soil Surface Moisture Estimates

As SMAP, the Soil Moisture and Ocean Salinity (SMOS) is a passive microwave satellite. It was launched in November 2019 by the European Space Agency (ESA) and provides microwave L-band measurements (1.4 GHz, 21 cm), from which SSM and sea surface salinity estimates are retrieved (Kerr et al. 2001). Today, a wide variety of operational SMOS-based SSM products designed to monitor agricultural drought are easily accessible in most parts of the world (Kerr et al. 2010). Among them, the Level 3 (L3) soil moisture product with a coarse spatiotemporal resolution (25 km × 25 km and one day) was prepared by the SMOS Barcelona Expert Center (SMOS BEC) and available on its website (https://bit.ly/2NEFS9J) has shown a good performance when compared to soil moisture measurements in some regions of the NEB (Souza et al. 2018).

17.7 Drought Early Warning Systems

It should be noted that each satellite-based drought index has different strengths, weaknesses, and limitations. Therefore, agricultural drought early warning systems should be based on the combination of different satellite-based drought indices and other informational products related to this theme (e.g. crop type, growing period, crop water needs, drought resistance among others) to effectively identify water stress or other stressors that can affect crops at any stage of crop development. On the other hand, investments in new technologies and innovations such as mobile apps are also needed, as well as timely remote access in order to calculate the indicators and indices used by those systems. Beyond these recommendations, it is also necessary to integrate the in situ data generated by both the multiple federal and state monitoring networks in the NEB to validate the satellite data used and optimize the accuracy and reliability of drought forecasts, particularly as the impacts of climate change become more prevalent (Kumar et al. 2022).

As mentioned above, rain-fed crops have experienced marked declines in terms of yield during the multi-year drought in the entire NEB (see Figure 17.2). This situation could be exacerbated in the future by climate change and increased water demand for crops, increasing conflicts between different water use sectors. Although the agricultural drought early warning system constitutes a useful tool to help reduce the impacts generated by agricultural drought events, it is necessary to implement a wider holistic and integrated approach for enhancing the resilience of small farmers in the NEB, who often lose their harvests due to drought (Buriti et al. 2020).

17.8 Final Remarks

This chapter highlights the importance of satellite-based drought indices for monitoring agricultural drought in the NEB. The NEB is a drought-prone region with scarce observations and is considered a drought hotspot in future climate projections. The following key points can be drawn from this chapter:

- There are a plethora of satellite-based drought indices that can be used for agricultural drought monitoring in regions with scarce observations. Some of them, such as SPI, SPEI, NDVI, EVI, and SSM, were discussed in this chapter.
- The SPI, whose multi-scalar nature allows the evaluation of drought events at different time scales, is a drought index recommended by the WMO for drought monitoring. Similar to the SPI, the SPEI has a multi-scalar nature, while it also takes into account the effect of the reference evapotranspiration on drought occurrence.
- Regarding vegetation drought, several standardized indices have been proposed including the NDVI and EVI. However, these indices show a delayed response to droughts in crops. In this case, the SSM provides a better representation of the water stress of vegetation and surface soil moisture conditions.

Since each satellite-based drought index has different strengths, weaknesses, and limitations, the agricultural drought early warning systems should be based on the combination of an ensemble of satellite-based drought indices. Nonetheless, whenever possible, in situ data should be added to improve forecast accuracy and reliability.

17.9 Summary

According to the Intergovernmental Panel on Climate Change (IPCC), the incidence and severity of drought may increase in northeast Brazil (NEB) due to human-induced climate change. Most Brazilian agencies and research institutions are working on the development of operational applications to track agricultural drought and provide early warning of deteriorating crop moisture conditions. These applications are based on drought indices and biophysical parameters derived from observational networks, model output, and satellite estimates. The purpose of this chapter is to present a general overview with respect to the strengths and weaknesses of some satellite-based agricultural drought indices often used in NEB.

Acknowledgements

This work was supported by the National Council for Scientific and Technological Development (CNPq Universal 18/2021), Brazil, through "Satellite Desertification Monitoring Program in the Brazilian Semiarid Region", under the Grant/Award Number (403223/2021-0 to H.A.B). We thank Dra. Maria A. Moreno-Pizani and Dr. A. J. Farias for their valuable feedback on irrigated crops in the NEB; and Talmai Betancourt and Angel Lorenzo for their assistance during the data processing. We also thank the European Organisation for the Exploitation of Meteorological Satellites (EUMETSAT) for early access to the EUMETCast system within the user community of satellite users in South America (SAGEO).

References

Allen, R.G., Luis, S.P., Dirk, R., and Martin, S. (1998). *Crop Evapotranspiration-Guidelines for Computing Crop Water Requirements-FAO Irrigation and Drainage Paper 56*, vol. 9. FAO.

Andreoli, R.V. and Kayano, M.T. (2006). Tropical Pacific and South Atlantic effects on rainfall variability over Northeast Brazil. *International Journal of Climatology* 26: 1895–1912.

Barbosa, H.A. and Kumar, T.V.L. (2016). Influence of rainfall variability on the vegetation dynamics over Northeastern Brazil. *Journal of Arid Environments* 124: 377–387.

Barbosa, H.A., Alfredo, H., and Baethgen, W.E. (2006). A 20-year study of NDVI variability over the Northeast Region of Brazil. *Journal of Arid Environments* 67: 288–307.

Barbosa, H.A., Kumar, T.V.L., and Silva, L.R.M. (2015). Recent trends in vegetation dynamics in the South America and their relationship to rainfall. *Natural Hazards* 77: 883–899.

Barbosa, H.A., Kumar, T.V.L., Franklin, P. et al. (2019). Assessment of Caatinga response to drought using Meteosat-SEVIRI normalized difference vegetation index (2008–2016). *ISPRS Journal of Photogrammetry and Remote Sensing* 148: 235–252.

Brabo, J.M., Jacques, S., and Campos, J.N.B. (2009). Relationship between ocean climatic variability and rain-fed agriculture in Northeast Brazil. *Climate Research* 38: 225–236.

Brito, S.S., Cunha, A.P., Cunningham, C.C. et al. (2018). Frequency, duration and severity of drought in the Semiarid Northeast Brazil region. *International Journal of Climatology* 38: 517–529.

Buriti, C., Humberto, A.B., Franklin, P. et al. (2020). Un siglo de sequías: ¿por qué las políticas de agua no desarrollaron la región semiárida brasileña? *Revista Brasileira de Meteorologia* 35: 683–688.

Burney, J., Daniele, C., Jarrod, R. et al. (2014). Climate change adaptation strategies for smallholder farmers in the Brazilian Sertão. *Climatic Change* 126: 45–59.

Campos, J.N. (2015). Paradigms and public policies on drought in Northeast Brazil: a historical perspective. *Environmental Management* 55: 1052–1063.

Campos, J.N. and Carvalho, T.M. (2008). Drought and water policies in Northeast Brazil: backgrounds and rationale. *Water Policy* 10: 425–438.

Campos, J.N., Souza, F.A., and Lima, H.V. (2014). Risks and uncertainties in reservoir yield in highly variable intermittent rivers: case of the Castanhão Reservoir in Semi-Arid Brazil. *Hydrological Sciences Journal* 59: 1184–1195.

Cirilo, J.A. (2008). Public water resources policy for the semi-arid region. *Estudos Avançados* 22: 61–82.

Correia, W.L., José, O.J., Dimas, B.S. et al. (2019). Rainfall variability in the Brazilian Northeast Biomes and their interactions with meteorological systems and ENSO via CHELSA product. *Big Earth Data* 3: 315–337.

Entekhabi, D., Eni, G.N., Peggy, E.O. et al. (2010). The soil moisture active passive (SMAP) mission. *Proceedings of the IEEE* 98: 704–716.

Funk, C., Peterson, P., Landsfeld, M. et al. (2015). The climate hazards infrared precipitation with stations – a new environmental record for monitoring extremes. *Scientific Data* 2: 150066.

Giovannettone, J., Franklin, P., Humberto, B.A. et al. (2020). Characterization of links between hydro-climate indices and long-term precipitation in Brazil using correlation analysis. *International Journal of Climatology* 40: 5527–5541.

Hayes, M., Mark, S., Nicole, W., and Melissa, W. (2011). The Lincoln declaration on drought indices: universal meteorological drought index recommended. *Bulletin of the American Meteorological Society* 92: 485–488.

He, M., John, S.K., Yonghong, Y. et al. (2019). Satellite data-driven modeling of field scale evapotranspiration in croplands using the MOD16 algorithm framework. *Remote Sensing of Environment* 230: 111201.

Herwehe, L. and Scott, C.H. (2018). Drought adaptation and development: small-scale irrigated agriculture in Northeast Brazil. *Climate and Development* 10: 337–346.

Huete, A., Kamel, D., Tomoaki, M. et al. (2002). Overview of the radiometric and biophysical performance of the MODIS vegetation indices. *Remote Sensing of Environment* 83: 195–213.

Instituto Brasileiro de Geografia e Estatística (2009). Censo Agropecuário 2006. Viewed 30 January 2021, https://bit.ly/3hVQayD (accessed 18 May 2022).

Instituto Brasileiro de Geografia e Estatística (2010). 2010 Census (Censo 2010). Viewed 15 January 2021, https://bit.ly/3vnCi4p (accessed 18 May 2022)

Julien, Y. and Sobrino, J.A. (2010). Comparison of cloud-reconstruction methods for time series of composite NDVI data. *Remote Sensing of Environment* 114: 618–625.

Kerr, Y.H., Philippe, W., Wigneron, J.P. et al. (2001). Soil moisture retrieval from space: the soil moisture and ocean salinity (SMOS) mission. *IEEE Transactions on Geoscience and Remote Sensing* 39: 1729–1735.

Kerr, Y.H., Philippe, W., Jean-Pierre, W. et al. (2010). The SMOS mission: new tool for monitoring key elements ofthe global water cycle. *Proceedings of the IEEE* 98: 666–687.

Kogan, F.N. (1995). Droughts of the late 1980s in the United States as derived from NOAA polar-orbiting satellite data. *Bulletin of the American Meteorological Society* 76: 655–668.

Kumar, T.V.L., Durga, G.P., Aravindhavel, A. et al. (2022). Analysis of tropospheric warming and stratospheric cooling in the present and future climate from the suite of CMIP6 models. *Theor Appl Climatol*. https://doi.org/10.1007/s00704-022-04136-y

Liu, X., Xiufang, Z., Yaozhong, P. et al. (2016). Agricultural drought monitoring: progress, challenges, and prospects. *Journal of Geographical Sciences* 26: 750–767.

Marengo, J.A. and Bernasconi, M. (2015). Regional differences in aridity/drought conditions over Northeast Brazil: present state and future projections. *Climatic Change* 129: 103–115.

Marengo, J.A., Roger, R.T., and Lincoln, M.A. (2017). Drought in Northeast Brazil—past, present, and future. *Theoretical and Applied Climatology* 129: 1189–1200.

Marengo, J.A., Ana, M.A., Carlos, A.N. et al. (2020). Assessing drought in the drylands of Northeast Brazil under regional warming exceeding 4 °C. *Natural Hazards* 103: 2589–2611.

Martínez-Fernández, J.A., González-Zamora, N.S., Gumuzzio, A., and Herrero-Jiménez, C.M. (2016). Satellite soil moisture for agricultural drought monitoring: assessment of the SMOS derived soil water deficit index. *Remote Sensing of Environment* 177: 277–286.

McKee, T.B., Nolan, J.D., John, K. et al. (1993). The relationship of drought frequency and duration to time scales. *Proceedings of the 8th Conference on Applied Climatology* 17: 179–183.

Miranda, R.Q., Josiclêda, D.G., Magna, S.M. et al. (2017). Reliability of MODIS evapotranspiration products for heterogeneous dry forest: a study case of Caatinga. *Advances in Meteorology* 2017: 1–14.

Mishra, A.K. and Singh, V.P. (2010). A review of drought concepts. *Journal of Hydrology* 391: 202–216.

Mu, Q., Maosheng, Z., and Steven, W.R. (2013). MOD16A2 MODIS/terra net evapotranspiration 8-day L4 Global 500m SIN Grid V006 [Data Set]. Viewed 28 March 2021. https://lpdaac.usgs.gov/products/mod16a2v006/ (accessed 18 May 2022)

Nogueira, S., Maurício, M., and Margarete, V. (2018). Evaluating precipitation estimates from Eta, TRMM and CHRIPS data in the South-Southeast region of Minas Gerais State—Brazil. *Remote Sensing* 10: 313.

O'Neill, P.E., Steven, K.C., Eni, G.N., et al. (2016). SMAP enhanced L3 radiometer global daily 9 km EASE-grid soil moisture. Version 1, Viewed 10 January 2021, https://nsidc.org/data/SPL3SMP_E/versions/3 (accessed 18 May 2022).

Oliveira, P.T., Santos, C.M., and Lima, K.C. (2017). Climatology and trend analysis of extreme precipitation in subregions of Northeast Brazil. *Theoretical and Applied Climatology* 130: 77–90.

Paredes-Trejo, F., Humberto, B., and Kumar, T.V. (2017). Validating CHIRPS-based satellite precipitation estimates in Northeast Brazil. *Journal of Arid Environments* 139: 26–40.

Paredes-Trejo, F., Humberto, B., and Carlos, D.S. (2019). Evaluation of the performance of SM2RAIN-derived rainfall products over Brazil. *Remote Sensing* 11: 1113.

Pereira, P.S., Flavio, J., Ana, M.M. et al. (2014a). The influence of oceanic basins on drought and ecosystem dynamics in Northeast Brazil. *Environmental Research Letters* 9: 124013.

Pereira, V., Diego, G., Thiago, M. et al. (2014b). Exigências Agroclimáticas Para a Cultura Do Feijão (*Phaseolus Vulgaris L.*). *Revista Brasileira de Energias Renováveis* 3: 32–42.

Rossato, L., Regina, S.A., José, M. et al. (2017). Impact of soil moisture on crop yields over Brazilian semiarid. *Frontiers in Environmental Science* 5: 1–16.

Rozante, J., Daniel, V., Júlio, B.C. et al. (2018). Evaluation of TRMM/GPM blended daily products over Brazil. *Remote Sensing* 10: 882.

Sheffield, J. and Wood, E.F. (2012). *Drought. Drought: Past Problems and Future Scenarios.* Londres, UK: Routledge.

Silva, L.V., Thiago, D.V., Moacyr, A., and Pedro, T. (2018). Ocean–atmosphere feedback during extreme rainfall events in eastern Northeast Brazil. *Journal of Applied Meteorology and Climatology* 57: 1211–1229.

Solano, R., Kamel, D., Andree, J., and Alfredo, H. (2010). MODIS vegetation index user's guide (MOD13 Series). Viewed 15 February 2021, https://go.nasa.gov/3fFnc3i (accessed 18 May 2022).

Souza, E.B. and Ambrizzi, T. (2006). Modulation of the intraseasonal rainfall over tropical Brazil by the Madden–Julian oscillation. *International Journal of Climatology* 26: 1759–1776.

Souza, E.B., Kayano, M.T., and Tércio, A. (2005). Intraseasonal and submonthly variability over the eastern Amazon and Northeast Brazil during the autumn rainy season. *Theoretical and Applied Climatology* 81: 177–191.

Souza, A., Alfredo, N., Luciana, R. et al. (2018). Use of SMOS L3 soil moisture data: validation and drought assessment for Pernambuco state, Northeast Brazil. *Remote Sensing* 10: 1314.

Spatafora, L.R., Mercè, V., Adriano, C. et al. (2020). Validation of SMOS L3 AND L4 soil moisture products in the Remedhus (Spain) and CEMADEN (Brazil) networks. *Revista Brasileira de Geografia Física* 13: 691–712.

Stephens, J.C. and Stewart, E.H. (1963). *A Comparison of Procedures for Computing Evaporation and Evapotranspiration*, 123–133. Berkeley: *International Association of Scientific Hydrology*.

Sun, Y., Huazhong, R., Tianyuan, Z. et al. (2018). Crop leaf area index retrieval based on inverted difference vegetation index and NDVI. *IEEE Geoscience and Remote Sensing Letters* 15: 1662–1666.

Valadão, E.A., Leila, M.C., Paulo, S.L., and Rosane, C.H. (2017). Impacts of the Madden-Julian oscillation on intraseasonal precipitation over Northeast Brazil. *International Journal of Climatology* 37: 1859–1884.

Vicente-Serrano, S.M. (2006). Differences in spatial patterns of drought on different time scales: an analysis of the Iberian Peninsula. *Water Resources Management* 20: 37–60.

Vicente-Serrano, S.M., Santiago, B., and Juan, L.M. (2010). A multiscalar drought index sensitive to global warming: the standardized precipitation evapotranspiration index. *Journal of Climate* 23: 1696–1718.

Wang, Q., Jianjun, W., Xiaohan, L. et al. (2017). A comprehensively quantitative method of evaluating the impact of drought on crop yield using daily multi-scale SPEI and crop growth process model. *International Journal of Biometeorology* 61: 685–699.

Wilhite, D.A. and Pulwarty, R.S. (2017). *Drought and Water Crises*. In: *Integrated Drought Management Tools and Guidelines Series*, 2e (ed. W. Donald and S.P. Roger). Boca Raton, FL, USA: CRC Press.

Zargar, A., Rehan, S., Bahman, N., and Faisal, I.K. (2011). A review of drought indices. *Environmental Reviews* 19: 333–349.

18

Understanding Smallholder Farmers' Perceptions of and Adaptations to Climate Change: The Case of the Zambezi Region, Namibia

*Lee-Ann Steenkamp and Winnie Thebuho**

Stellenbosch Business School, University of Stellenbosch, Cape Town, South Africa

CONTENTS

18.1 Introduction

The Intergovernmental Panel on Climate Change (IPCC) views Africa as amongst the continents most susceptible to climate change, owing to various pressures, the scope of predicted climate repercussions, and its constrained potential for adaptation (Bernstein et al. 2008). Agricultural systems have become more exposed to variable weather patterns such as drought, elevated temperatures, harsh frosts and hailstorms, severe winds, and pests causing damage to growing crops (Khumalo 2021). Indeed, the IPCC warned in 2014 that climate change would exacerbate water scarcity and strain on agricultural systems, especially those in semi-arid environs (Pachauri et al. 2014).

* Lee-Ann Steenkamp and Winnie Thebuho conceptualized the study; Winnie Thebuho collected and analyzed the data under the supervision of Lee-Ann Steenkamp; Lee-Ann Steenkamp curated the data and wrote the manuscript, drawing on aspects of Winnie Thebuho's research assignment.

Climate Change and Agriculture: Perspectives, Sustainability and Resilience, First Edition. Edited by Noureddine Benkeblia.

In its 2019 Special Report on Climate Change and Land, the IPCC drew attention to the fact that smallholder farmers are particularly exposed to climate change risks – due to their primary source of income often being derived from agriculture – and also frequently suffer from chronic food insecurity (Shukla et al. 2019). Namibia is no exception. In fact, 17% of the Namibian population faces high levels of acute food insecurity and requires urgent humanitarian action to lessen food gaps, safeguard and restore livelihoods and avert acute malnutrition (Integrated Food Security Phase Classification (IPC) 2021). Several regions (including the Zambezi, which is the focus of this study) are classified as being 'in crisis' (Integrated Food Security Phase Classification (IPC) 2021).

Empirical studies have increasingly considered the connection between climate change adaptation strategies and agricultural productivity (Di Falco et al. 2012; Kurukulasuriya et al. 2006), as well as the influence of climate change on net farm income (Arshad et al. 2018; Ojo and Baiyegunhi 2020a; Seo and Mendelsohn 2008).

Some researchers investigated how crop yields were affected by climate change (Ajetomobi et al. 2010; Gbetibouo and Hassan 2005; Hossain et al. 2019; Huong et al. 2019; Kurukulasuriya and Ajwad 2007); while others examined the factors which influenced the implementation of climate change adaptation strategies (Ojo and Baiyegunhi 2020b; Tambo and Abdoulaye 2013; Thinda et al. 2020). Various studies sought to identify the determinants of climate change adaptation strategies (Adeagbo et al. 2021; Kalele et al. 2021; Mulwa et al. 2017).

Asrat and Simane (2018) suggested a two-step climate change adaptation process. First, farmers should perceive the impact of climate change on their productivity, and, secondly, they must respond to those changes through adaptation. Previous studies in sub-Saharan African countries concluded that most smallholder farmers perceive the climate to be varying, for example, by observing rainfall irregularity in Botswana and Malawi (Simelton et al. 2013); deforestation and soil degradation in Ethiopia (Tesfahunegn et al. 2016); erratic rainfall and droughts in Kenya (Ogalleh et al. 2013); fluctuating temperature in Nigeria (Tambo and Abdoulaye 2013); abnormal weather patterns in South Africa (Hitayezu et al. 2017); fewer days of rainfall in Tanzania (Mkonda et al. 2018); and worsening dry spells and heat in Zimbabwe (Makuvaro et al. 2018).

Understanding farmers' discernment of changes in the climate is crucial, as the perception of climate change is often a prerequisite for responding thereto (Dendir and Simane 2021). To that end, several recent studies in various African countries have evaluated farmers' perceptions and chosen adaptation strategies to lessen the detrimental effects of climate change, including in Ethiopia (Ayal and Leal Filho 2017; Dendir and Simane 2021; Hundera et al. 2019), Ghana (Asare-Nuamah and Botchway 2019; Assan et al. 2020; Martey and Kuwornu 2021), Kenya (Mairura et al. 2021; Nyang'au et al. 2021), Nigeria (Ayanlade et al. 2017; Ojo and Baiyegunhi 2021), South Africa (Lamega et al. 2021; Talanow et al. 2021), Tanzania (Mkonda et al. 2018), Uganda (Mubiru et al. 2018), and Zimbabwe (Makuvaro et al. 2018).

Overall, there is limited empirical data on Namibian farmers' climate change awareness and their adaptation approaches, specifically regarding the Zambezi region. Consequently, it is somewhat problematic to analyze varying perceptions and to formulate interventions aimed at specific agro-ecological zones. Thus, this study had as its main objective the examination of smallholder farmers' perceptions of and actual adaptation to climate change in the Zambezi region. First, we provide an overview of the Namibian economy, the country's emissions, and its climate profile. Next, we describe the materials and methods used in this study, followed by a discussion of the results. Finally, we conclude by proposing a number of recommendations as to how smallholder farmers' climate resiliency might be enhanced.

18.2 Background to Namibia

Namibia is an upper-middle-income country, which can be ascribed to its natural mineral wealth and comparatively small population (World Bank Group 2021a,b). The majority (about 51%) of the population is based in rural areas and is dependent upon agriculture (crop production, livestock rearing, and fishing) and the use of natural resources for livelihoods (World Bank Group 2021a,b).

The agricultural sector is a vital pillar of Namibia's economy, as illustrated by Figure 18.1, which shows that this sector contributes approximately 7 to 9% of the country's GDP. Indeed, agriculture indirectly affects the livelihoods of 70% of the population, which is mostly reliant on rain-fed crop production (World Bank Group 2021a,b). In addition, almost 48% of rural households are dependent on subsistence agriculture (World Bank Group 2021a,b).

The country's annual share of global CO_2 emissions is very low at 0.01% over the period 1991 to 2019, with its cumulative share at 0% (Ritchie et al. 2022). Nevertheless, as a

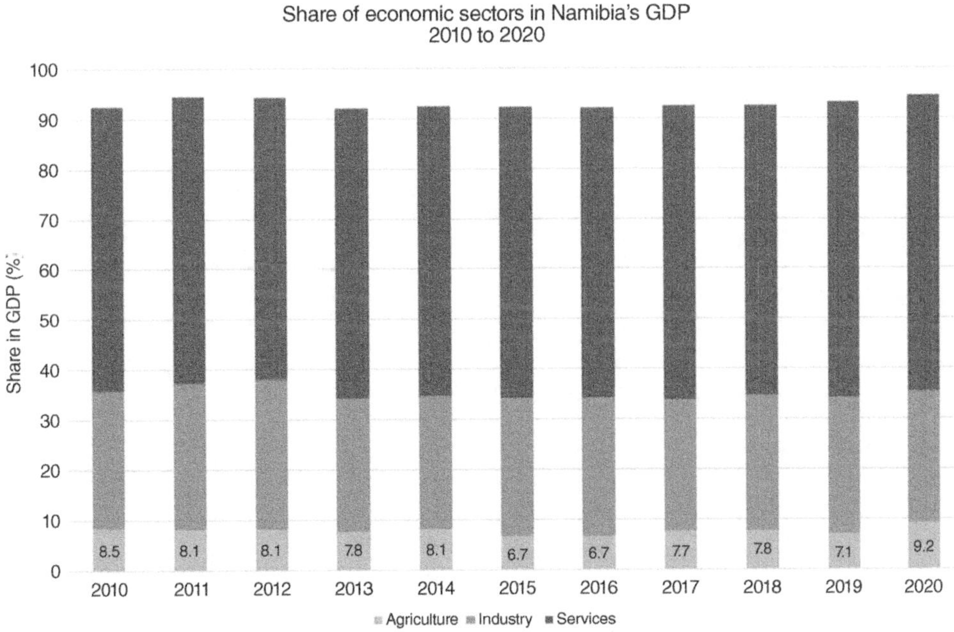

Figure 18.1 Share of economic sectors in Namibia's GDP from 2010 to 2020. *Source:* Authors' own, based on data from Statista (2021). Note: Figures have been rounded. The source does not provide any information regarding percentage points not adding up to or exceeding 100%. In 2020, the share of agriculture in Namibia's GDP was 9.15%, industry contributed approximately 26.32% and the services sector about 59.05%.

signatory to the Paris Agreement, Namibia submitted its Nationally Determined Contribution (NDC) in 2015, pledging to reduce its national emissions of greenhouse gases (GHGs) by 89% compared to business-as-usual by 2030. In 2021, Namibia submitted its second NDC, which provides for a more ambitious target (92%, compared to 89%) aimed at reaching net-zero emissions by 2050. It is evident from Figure 18.2 that land-use change and forestry, as well as agriculture, make up the largest components of the country's GHGs emissions.

Due to its dependence on rain-fed agricultural and livestock activities, Namibian farmers are more susceptible to climate change. In addition, poor households and communities have limited resources to manage climate risks and are consequently more vulnerable to climate-related shocks (World Bank Group 2021a,b). As a matter of fact, extreme weather events have become more frequent and severe of late and will probably worsen as a result of climate change (OECD/FAO 2021). Data from the Emergency Events Database (EM-DAT), as depicted in Figure 18.3, indicates that the country has endured various natural disasters over the past three decades, including droughts, floods, and epidemics.

The extreme rains and flooding that occurred in February/March 2009 across Namibia, Angola, and Namibia exacerbated the effects of the 2007/2008 drought (OECD/FAO 2021). This is due to heavy rainfall, which triggered riverine and flash floods. Subsistence farmers in the low-lying areas were particularly impacted and suffered damage and losses to crops and livestock (OECD/FAO 2021). Climate change will likely

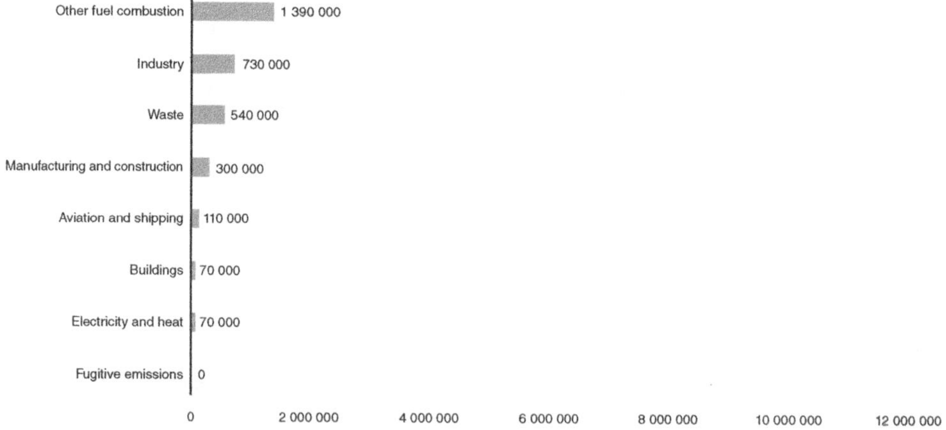

Figure 18.2 Namibia's GHG emissions by sector in 2019. *Source:* Authors' own, based on data from Ritchie et al. (2022).

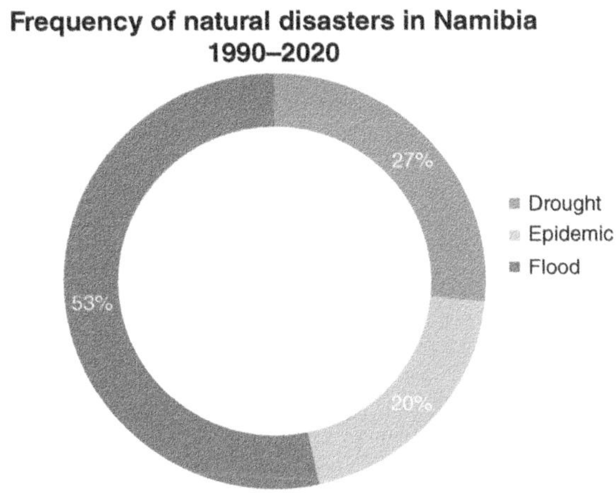

Figure 18.3 Frequency of natural disasters in Namibia from 1990 to 2020. *Source:* Authors' own, based on data from EM-DAT (2021). Note: epidemics include bacterial, parasitic and viral diseases; floods include flash and riverine floods. In sum, 27 natural disasters were recorded in this period.

have a manifold impact on agriculture – greater demands on water supply as a result of increasing temperatures; reduced harvests of rain-fed crops; greater occurrences of floods; and modified selections of the ideal geographical areas for cultivating arable crops and forests (OECD/FAO 2021, p. 117).

Ranking second in aridity after the Sahara Desert, Namibia is amongst the largest and driest countries in sub-Saharan Africa. The country experiences much climatic variability due to its changeable temperatures, erratic rainfall patterns, continual droughts, and water scarcity (World Bank Group 2021a,b). Figure 18.4 shows the average monthly rainfall and temperature from 1991 to 2020.

As mentioned in Section 18.1, few studies have been conducted to ascertain Namibian farmers' perceptions of and adaptation to climate change. These are briefly outlined in the following section.

18.3 Review of Empirical Literature Pertaining to Namibia

Newsham and Thomas (2011) focused on Ovambo farmers in North Central Namibia and considered the link between their agroecological knowledge and ability to adapt to climate change. The researchers conducted semi-structured interviews with farmers, agricultural extension staff, and policymakers, made use of focus groups, and conducted farm visits in the Omusati region during 2008.

Montle and Teweldemedhin (2014) conducted a case study on small-scale irrigation farmers of the Ndonga-Linena irrigation project in the Kavango region in north-eastern Namibia. Farm-level data were collected and interviews were conducted with all 30 farmers during February 2014 in an attempt to analyze their awareness of climate change and their adaptation strategies.

Mabuku et al. (2019) compared the flood adaptation strategies of rural households in the Mwandi District in Zambia and the eastern part of the Zambezi Region in Namibia. They held six focus group meetings and administered a questionnaire survey of 207 randomly sampled households in the flood-prone areas of the study during July 2015.

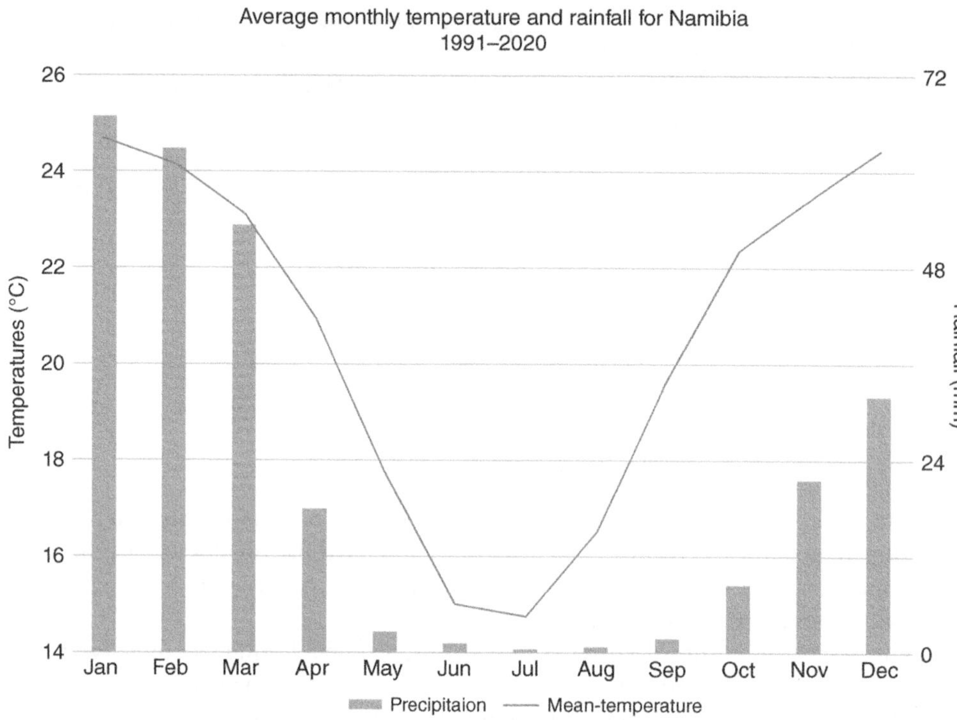

Figure 18.4 Average monthly temperature and rainfall for Namibia from 1991 to 2020. *Source:* Authors' own, based on data from the World Bank Group's Climate Change Knowledge Portal (World Bank Group 2021a). Note: the mean annual temperature for Namibia is 20.6 °C, and the mean annual precipitation is 269.2 mm.

During July 2017, Spear and Chappel (2018) conducted 31 interviews across three villages in the Onesi constituency in the Omusati region in north-central Namibia. They examined the agricultural practices used by smallholder crop farmers, their perception of changes in crop yield, their harvest prospects, and if they considered altering their agricultural practices.

Two studies were done on the isolated pastoral Himba community in the Epupa constituency in Kunene region, which occupies the north-west corner of Namibia. Shikangalah (2019) carried out interviews with 20 key informants from 19 villages during November and December 2016. The study drew on the perceptions of risks to livelihoods of the semi-nomadic pastoralists. Inman et al. (2020) conducted interviews with 30 pastoralists and 16 key informants from six villages between June and November 2018. The authors sought to discern the perceptions, experiences, and understanding of climate change and related impacts on the pastoralists' livelihoods.

Gitonga et al. (2020) located their study in three administrative regions in the semi-arid parts of northern Namibia, namely Oshana, Oshikoto, and Omusati. By analyzing representative data of 653 households from these areas, the researchers considered the importance of, and access to, climate prediction reports and how this information is integrated into farm production processes and used in adaptation strategies.

18.4 Study Area

Our research area is the Zambezi region (previously known as the Caprivi) in the north-eastern part of Namibia. As can be seen from the map in Figure 18.5, the Zambezi region connects Namibia to four neighbouring countries, namely Botswana, Zambia, Angola, and the tip of Zimbabwe. The 'arm of Namibia' is named after the Zambezi River and runs mostly parallel to the Caprivi Strip. The locus for this study is two of the Zambezi region's constituencies, namely Kabbe[1] (17.7196° S, 24.6336° E) in the northern lowlands and Linyanti (18.0735° S, 24.0220° E) in the southern highlands.

The Zambezi region is considered a tropical area, due to its heavy rainfall in the summer period of December to March, and its high temperatures (Zambezi Regional Council 2021). This area has the highest precipitation in the country, ranging from 600 mm in the west to 700 mm in the north-east (Integrated Environmental Consultants Namibia (IECN) 2011). The terrain mostly consists of wetlands, floodplains, marshes, and woodland (Zambezi Regional Council 2021). Moreover, it is also the region with the most perennial rivers, of which the Zambezi, Kavango, and Chobe are the most noteworthy. As a result, the area has an adequate water supply, evergreen vegetation, and abundant wildlife. It is also considered to have the biggest portion of arable land in Namibia (Integrated Environmental Consultants Namibia (IECN) 2011).

Great swathes of populated land lie in the seasonal flood plains. As a consequence, local populations and their agricultural production systems are affected by intensifying flooding events (Integrated Environmental Consultants Namibia (IECN) 2011). Conversely, bearing capacities for livestock and wildlife are under pressure during the dry season. This is due to the scarcity of surface water and nutrient-poor meadowlands during this time (Integrated Environmental Consultants Namibia (IECN) 2011). Temperatures range from an average maximum of 32 to 34 °C, and an average minimum of 8 to 10 °C (Integrated Environmental Consultants Namibia (IECN) 2011).

The Zambezi region has a population of 90 596, of which 14 518 live in Kabbe and 15 477 in Linyanti (Namibia Statistics Agency 2011). The region has a median age of 20 years, with females making up more than half (51%) of the population (Namibia Statistics Agency 2011). Notwithstanding that nearly 19% of the population aged six years and above in rural areas have never attended school, approximately 84% of the Zambezi region's population over 15 years is literate (Namibia Statistics Agency 2011).

Skilled agricultural/fishery workers make up the largest occupation group, at 42%, with wages and salaries (30%) being the largest contributor to household income (Namibia Statistics Agency 2011). Most residents are subsistence farmers, with local level fishing supporting food security and furthering economic activity (Integrated Environmental Consultants Namibia (IECN) 2011). Crop farming is the most common agricultural activity in the Zambezi region (41%), where more than half (52%) of the households are involved in crop farming (Namibia Statistics Agency 2011).

1 The Kabbe constituency was split into Kabbe South and Kabbe North in August 2013. However, for our research purposes, these two constituencies were treated as one district. Accordingly, all references to Kabbe include both these constituencies.

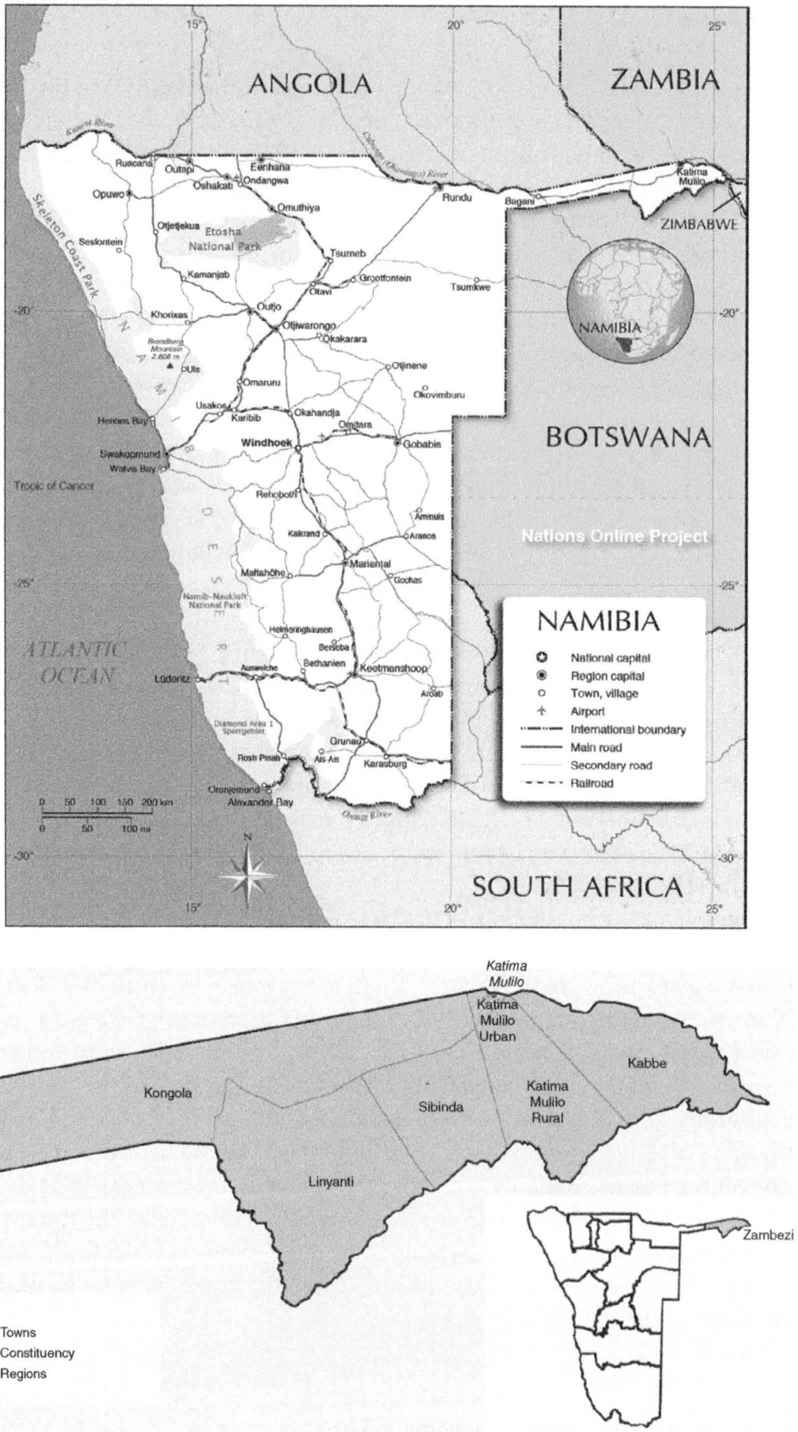

Figure 18.5 Map of the study area. *Source:* Nationsonline (2021) and Namibia Statistics Agency (2011).

18.5 Data Discussion

The study being primarily qualitative, we administered a semi-structured questionnaire to selected sample households. The questions were formulated in both open-ended and closed-ended forms and administered through face-to-face contact with the crop farmers. The survey explored smallholder farmers' discernment of climate change, their experience of its effects, and the ways in which they sought to adapt to these impacts.

As our goal was not to generalize regarding a population, but rather to obtain insights into a phenomenon (being smallholder farmers' perception of climate change), non-probability sampling was considered appropriate to maximize our understanding of this phenomenon. We made use of convenience and snowball sampling techniques. The research was conducted with the Stellenbosch University Research Ethics Committee approval (USB-2018-8240). Prospective respondents were made aware that they were not obliged to participate and that their responses would be anonymized. Individual informed consent was obtained from every participant. The inclusion criteria included age (only individuals aged 18 and older were allowed to participate), being a crop farmer in the Zambezi region, and willingness to partake in the study.

The survey was conducted during September and October 2018. The data collection was carried out by the second author, who spoke both English and the local language (Lozi) fluently, was born and raised in the study area, and was therefore familiar with the local dynamics. In the case of illiterate respondents, the second author read the questions to the respondents and recorded their responses.

Similar to the approach followed by Inman et al. (2020), due to dangerous road conditions, we selected villages that were reachable by a tarred road. We randomly selected one accessible village from each constituency, namely Ngoma in the Kabbe district and Mahachana in the Linyanti area. The snowball procedure was used to recruit respondents to great effect, as the news spread, and farmers wanted the opportunity to share their opinions. Although we had targeted a sample of 50 respondents from each village, we finally totalled 127 completed surveys – 64 from the Kabbe constituency and 63 from the Linyanti constituency.

18.5.1 Socio-Economic Characteristics of Respondents

Quantitative primary data were coded, entered, and analyzed in the Statistical Package for Social Sciences (SPSS). Descriptive statistics were calculated using frequencies and percentages. Table 18.1 summarises the socio-economic characteristics of the respondents, who were predominantly male (51%), within the age group of 60 years or above (35%), and who had lived in their particular constituency for over 20 years (75%).

About three-quarters of respondents (75%) had received some form of education, ranging from primary school (14%) and secondary school (33%), to tertiary level (28%). The remaining 24% of respondents had never received a formal education. These results are largely consistent with the census statistics reported in Section 18.4.

Table 18.1 Socio-economic characteristics of respondents.

Category	Total (n = 127)	Kabbe (n = 64)	Linyanti (n = 63)
Years of living in constituency			
0–5	1%	2%	0%
6–10	6%	8%	5%
11–20	18%	20%	16%
>20	75%	70%	79%
Gender			
Male	51%	61%	41%
Female	49%	39%	59%
Age categories			
18–25	5%	6%	3%
26–35	9%	9%	8%
36–49	20%	25%	14%
50–59	31%	19%	44%
60 and above	35%	41%	30%
Education			
Informal	24%	28%	21%
Primary	14%	14%	14%
Secondary	33%	38%	29%
Tertiary	28%	20%	37%
Household income (N$)			
0–1500	64%	67%	60%
1501–3000	24%	22%	25%
3001–5000	11%	11%	11%
5001 and above	2%	0%	3%
Livelihood[a]			
Crop farming activities	36%	36%	37%
Formal salary	25%	34%	16%
Social grant	65%	55%	76%
Other	4%	6%	2%

Note: Percentages are rounded to the nearest integer.
[a] Categories represent questions where respondents could select multiple options.

For the vast majority of respondents (65%), social grants were the primary source of livelihood, followed by farming income (36%) and formal (employment) income (25%). It is therefore not surprising that a similar proportion of respondents indicated a monthly household income of N$1500 or less (64%), as this corresponds with the monthly social grant received at the time.

18.5.2 Perceptions of Climate Change

Table 18.2 indicates various aspects of farmers' perceptions of climate change.

Although a significant portion of respondents (87%) indicated that they understood something about climate change (Figure 18.6a), only 64% attributed climate change to human activity (Figure 18.6b). The burning of fossil fuels was selected as the leading cause of climate change (91%), followed by deforestation (72%), natural events (47%), and carbon emissions from vehicles and large businesses (46%). Very few respondents (5%) thought that agricultural activities resulted in climate change (Figure 18.6c).

In respect of extreme weather events occurring in their particular area, 91% of respondents had experienced floods, 84% had faced drought, 58% suffered damaged crops and 32%

Table 18.2 Perceptions of climate change.

Category	Total (*n* = 127)	Kabbe (*n* = 64)	Linyanti (*n* = 63)
Do you understand anything about climate change?			
Yes	87%	81%	92%
No	13%	19%	8%
Are humans responsible for climate change?			
Yes	64%	56%	71%
No	36%	44%	29%
What causes climate change?[a]			
Other	2%	5%	0%
Agricultural activities	5%	6%	3%
Carbon emissions	46%	53%	38%
Natural events	47%	53%	41%
Deforestation	72%	66%	79%
Burning fossil fuels	91%	94%	87%
Have you experienced any of these extreme weather events in your area?[a]			
Washed over walls or waves of water	2%	2%	2%
Soil erosion	32%	38%	27%
Damaged crops	58%	53%	63%
Drought	84%	83%	86%
Floods	91%	94%	87%
What difficulties have you experienced in your areas as a result of climate change?[a]			
Other	18%	16%	21%
Lack of education	60%	63%	57%
Lack of funding	65%	64%	67%
Less harvesting	84%	83%	86%

Note: Percentages are rounded to the nearest integer.
[a] Categories represent questions where respondents could select multiple options.

had indicated soil erosion (Figure 18.6d). A small portion (2%) had experienced waves of water or walls being washed over, which is not surprising, given that this occurrence is most often associated with sea-level rise in coastal areas.

The Kabbe respondents indicated a higher frequency of floods (94%) and soil erosion (38%) compared to their Linyanti counterparts (87 and 27%, respectively), whereas the latter suffered more incidences of drought (86%, compared with Kabbe's 83%). These findings are in line with our expectations, as Kabbe is situated in the lowland floodplains and Linyanti in the highland, a more drought-prone area of the Zambezi region.

When confronted with the ways in which they experienced climate change (Figure 18.6e), farmers noted reduced harvesting as the largest climate change-induced impact (84%).

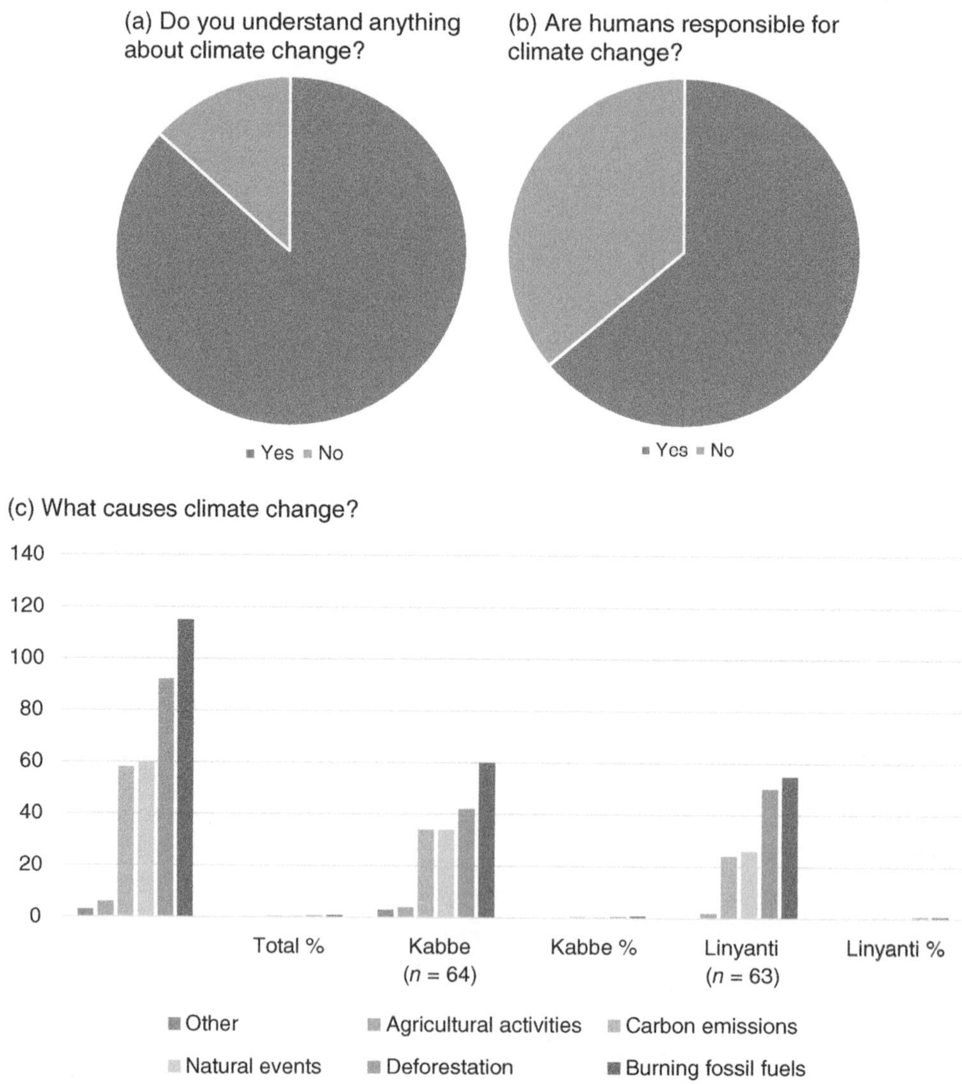

Figure 18.6 Farmers' climate change perceptions.

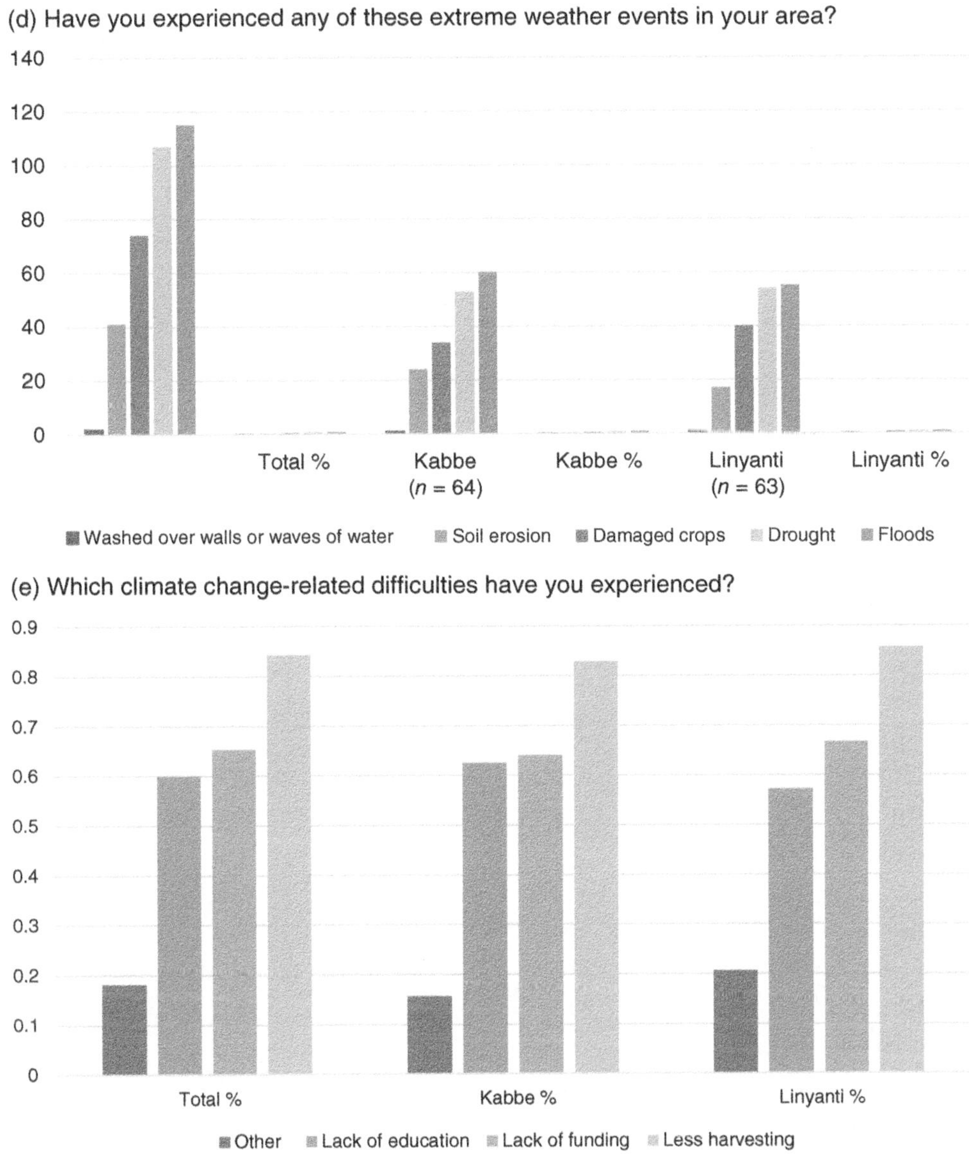

Figure 18.6 (Continued)

Other difficulties experienced include obtaining funding from financial institutions (65%) and access to climate change adaptation education (60%).

18.5.3 Adaptation Methods

Our study revealed that farmers employed various adaptation measures to protect their livelihood from climate change effects. All the respondents planted their crops during the rainy season, while 97% harvested during the dry season. The specific adaptation strategies are presented in Table 18.3 and are generally similar for the two constituencies.

Table 18.3 Adaptation strategies.

Category	Total (n = 127)	Kabbe (n = 64)	Linyanti (n = 63)
Adaptation strategies used[a]			
Minimum tillage	2%	2%	2%
No adaptation	5%	8%	2%
Soil cover	6%	6%	6%
Harvesting rainwater	9%	13%	5%
Early warning system	15%	17%	13%
Crop diversification	16%	23%	8%
Early harvesting	17%	22%	11%
Improved variety	17%	25%	10%
Crop rotation	21%	23%	19%
Planting trees	42%	34%	49%
Planting drought-resistant crops	45%	44%	46%
Moving to another location	67%	78%	56%
Cropping method influenced by climate change[a]			
Other	2%	2%	2%
Fertilizer application	43%	39%	46%
Planting date	47%	58%	37%
Varieties to grow	60%	61%	59%
Crop choice	72%	75%	70%
What assistance would be useful in your situation?[a]			
Relocation	28%	22%	33%
Food aid	31%	33%	30%
Credit/loan	66%	67%	65%
Crop insurance	69%	64%	73%
Awareness information	98%	98%	97%

Note: Percentages are rounded to the nearest integer.
[a] Categories represent questions where respondents could select multiple options.

It is interesting to observe (Figure 18.7a) that flood-prone Kabbe farmers indicated a much higher frequency of crop diversification (23%), early harvesting (22%), improved variety (25%), and moving to another location (78%). This is in contrast to the drought-susceptible Linyanti farmers, who have a comparatively greater preference for planting trees (49%) and drought-resistant crops (46%). Overall, minimum tillage was the least utilized strategy in both constituencies (2%), with relocation the most favoured adaptation response (67%).

Farmers were asked which cropping method they preferred as a means of adapting to climate change. The type (72%) and variety (60%) of crops to grow were the most favoured choices, followed by the planting date (47%) and fertilizer application (43%) (Figure 18.7b).

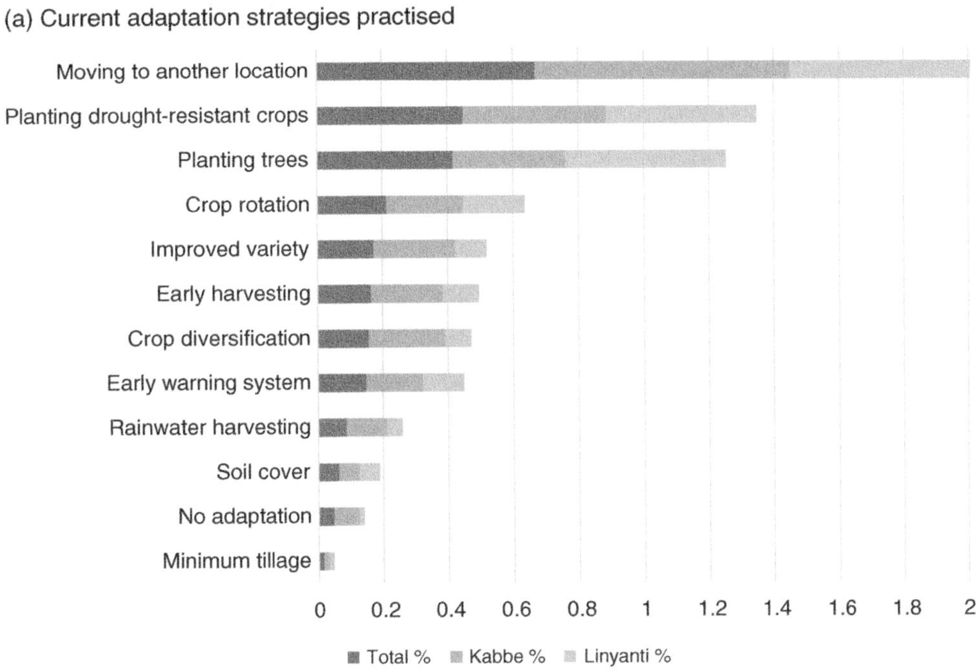

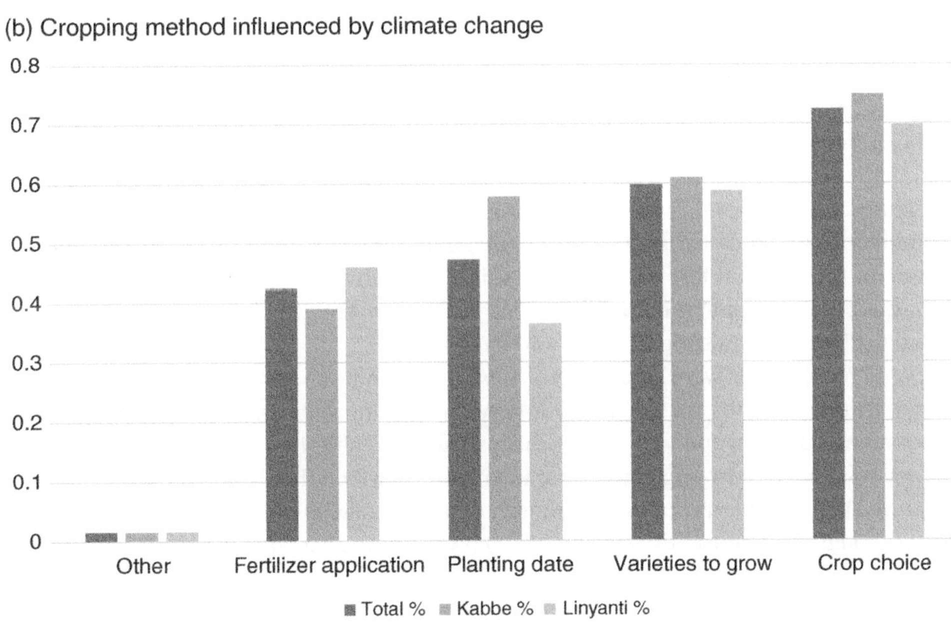

Figure 18.7 Farmers' climate change adaptation strategies.

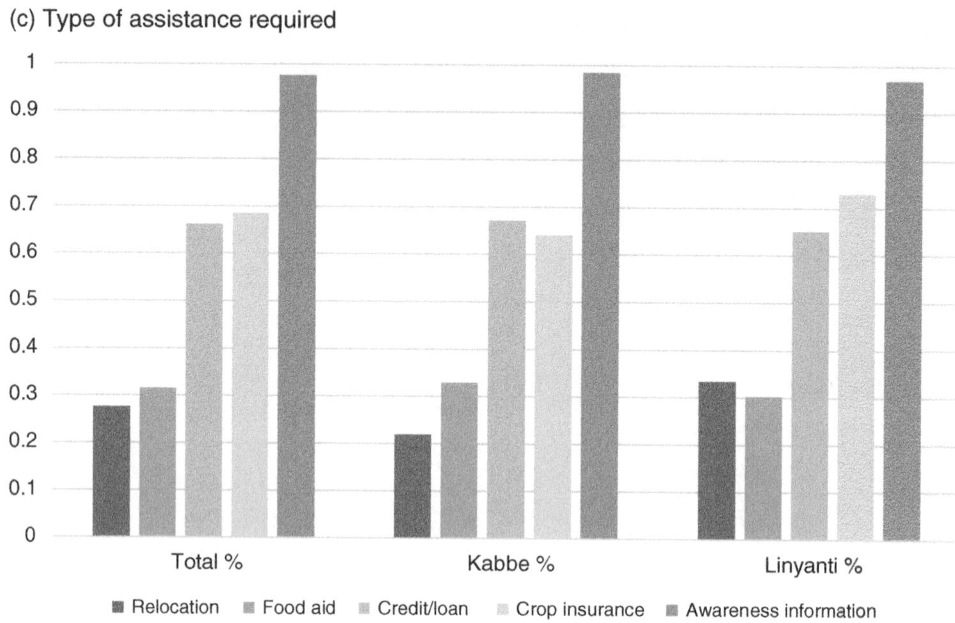

Figure 18.7 (Continued)

Kabbe farmers indicated a comparatively higher preference for the planting date (58%), compared with the Linyanti farmers (37%). This could likely be ascribed to the Kabbe region being more susceptible to flooding, which evidently influences the planting date to a greater extent.

The problem of limited information related to adaptation strategies is frequently cited as the main barrier to the adoption of new agricultural practices (Spear and Chappel 2018). Our study corroborates this theory, as almost all of the respondents (98%) expressed a need for enhanced climate change adaptation awareness information (Figure 18.7c). A larger portion of the Linyanti farmers required assistance with crop insurance (73%) and relocation (33%), as opposed to the Kabbe farmers. Both groups had fairly similar responses with regard to credit or loan support (66%) and food aid (31%).

Several respondents were able to identify existing climate change awareness programmes offered in their regions. These can be grouped into communications pertaining to deforestation, information, and weather, as encapsulated in Table 18.4.

Some respondents also suggested other interventions which they believed could help in the long run. These include increasing the number of tractors offered by local authorities, the provision of drought-resistant seeds, and advice on how to implement farm conservation methods.

18.6 Conclusion

Climate change adaptation is influenced by an awareness of climate change exposure and impending risks (Spear and Chappel 2018). On the one hand, if farmers are unaware of their vulnerability to climate change impacts, they will be less likely to respond; on the

Table 18.4 Identification of existing climate change awareness programmes.

Deforestation	• Not to cut down trees • Not to burn forests • To plant more trees
Information	• Awareness campaigns on adverse climate change impacts • Early warning programme • Use of drought-resistant crops, such as pearl millet • How to implement conservation agriculture
Weather	• Advice on best ploughing and planting time • Weather forecasts on TV and radio

other hand, if they recognize the looming effects of climate change, they will take steps to adapt (Spear and Chappel 2018). For example, in Kenya, farmers' acuities notably decided the uptake of climate-smart agro-technologies (Mairura et al. 2021); in Tanzania, after farmers noticed rising temperatures and declining precipitation, they switched to drought-resilient crops and enlarged the cultivation area to compensate for reduced harvests (Mongi et al. 2010); and in Ghana, farmers who experienced rainfall disturbances were apt to supplement mulch with residue as an adaptation strategy for evapotranspiration (Martey and Kuwornu 2021).

Consequently, this study has explored the perception of and adaptive measures to climate change by smallholder farmers in the Zambezi region in Namibia. The results of this study cannot be generalized to the entire population of smallholder farmers in Namibia. However, the insight from this study may guide policymakers in other developing countries (particularly sub-Saharan African countries) with similar agricultural and ecological environments to the present study area.

The Namibian climate is changing, with several droughts and floods recorded as natural disasters. Respondents from both the Kabbe and Linyanti constituencies had perceived changes in the climate and the occurrence of extreme events, particularly floods and droughts. In addition, farmers discerned the aggravating effects of climate change on their agricultural activities, for example, reduced harvests, damaged crops, and soil erosion.

Consequently, smallholder farmers adopted numerous adaptive measures in their agricultural endeavours. The most preferred adaptation response was relocation, followed by the planting of trees and drought-resistant crops. As in the case of Mabuku et al. (2019), we suggest that the Namibian government could provide flood victims with resettlement subsidies to support the relocation of households to higher ground during a flood disaster. In addition, policies for fair land redistribution to flood survivors will assist households in implementing some of these adaptation strategies (Mabuku et al. 2019).

Feedback from respondents revealed a clarion call for greater climate change adaptation awareness campaigns by the government. We therefore recommend that the government intensifies climate change education and mass awareness, especially in rural communities. Access to information could be enhanced by using all available media, including television, radio, extension agents, and community information centres (Asare-Nuamah and Botchway 2019). Capacity development programmes should be strengthened in rural areas and may include smart-agro-climate training (Ayanlade et al. 2017), on-farm demonstrations, and broader dissemination of adaptation strategy information (Adeagbo et al. 2021).

Although the provision of climate information could improve the adaptive capacity and food security of rural communities, it must also be integrated into farm production decisions (Gitonga et al. 2020). Other complementary forms of institutional support are also necessary, for example, extension services, reliability of network connectivity and communication infrastructure, and developing market value chains (Gitonga et al. 2020).

It should be recognized, though, that certain adaptation strategies can be capital-intensive, especially those requiring the acquisition of new or enhanced planting supplies and other technologies. Even if informed about climate change, without access to credit, farmers may find it difficult to undertake any adaptive measures due to the unaffordability of the necessary inputs (Ojo and Baiyegunhi 2021). Indeed, assistance with crop insurance (69%) and financial aid (66%) were high on the agenda of the respondents. Clearly, some form of government financial support is necessary to help smallholder farmers adapt to a changing climate.

Spear and Chappel (2018) posited that Namibians are prone to be overly reliant on the government to take action. This viewpoint, together with the fact that most of the respondents in our study are sustained by social grants, could diminish farmers' adaptative capacity, as it creates dependency, reduces the use of coping mechanisms, and inhibits innovation (Spear and Chappel 2018). Ultimately, care should be taken to create an enabling environment for smallholder farmers to adapt to and flourish in a changing climate, instead of developing an excessive dependence on government assistance. As one of the largest and most arid countries in sub-Saharan Africa, Namibia faces a supreme challenge in becoming climate-resilient.

References

Adeagbo, O.A., Ojo, T.O., and Adetoro, A.A. (2021). Understanding the determinants of climate change adaptation strategies among smallholder maize farmers in South-west, Nigeria. *Heliyon* 7 (2): e06231.

Ajetomobi, J.O., Abiodun, A., and Hassan, R. (2010). Economic impact of climate change on irrigated rice agriculture in Nigeria. Contributed Paper Presented at the Joint 3rd African Association of Agricultural Economists (AAAE) and 48th Agricultural Economists Association of South Africa (AEASA) Conference, Cape Town, South Africa, Sept, pp. 19–23.

Arshad, M., Amjath-Babu, T.S., Aravindakshan, S. et al. (2018). Climatic variability and thermal stress in Pakistan's rice and wheat systems: a stochastic frontier and quantile regression analysis of economic efficiency. *Ecological Indicators* 89: 496–506.

Asare-Nuamah, P. and Botchway, E. (2019). Comparing smallholder farmers' climate change perception with climate data: the case of Adansi North District of Ghana. *Heliyon* 5 (12): e03065.

Asrat, P. and Simane, B. (2018). Farmers' perception of climate change and adaptation strategies in the Dabus watershed, North-West Ethiopia. *Ecological Processes* 7 (1): 1–13.

Assan, E., Suvedi, M., Olabisi, L.S., and Bansah, K.J. (2020). Climate change perceptions and challenges to adaptation among smallholder farmers in semi-arid Ghana: a gender analysis. *Journal of Arid Environments* 182: 104247.

Ayal, D.Y. and Leal Filho, W. (2017). Farmers' perceptions of climate variability and its adverse impacts on crop and livestock production in Ethiopia. *Journal of Arid Environments* 140: 20–28.

Ayanlade, A., Radeny, M., and Morton, J.F. (2017). Comparing smallholder farmers' perception of climate change with meteorological data: a case study from southwestern Nigeria. *Weather and Climate Extremes* 15: 24–33.

Bernstein, L., Bosch, P., Canziani, O. et al. (2008). *Climate Change 2007 Synthesis Report.* Intergovernmental Panel on Climate Change.

Dendir, Z. and Simane, B. (2021). Farmers' perceptions about changes in climate variables: perceived risks and household responses in different agro-ecological communities, Southern Ethiopia. *Climate Services* 22: 100236.

Di Falco, S., Yesuf, M., Kohlin, G., and Ringler, C. (2012). Estimating the impact of climate change on agriculture in low-income countries: household level evidence from the Nile Basin, Ethiopia. *Environmental and Resource Economics* 52 (4): 457–478.

EM-DAT (2021). The Emergency Events Database – Universite catholique de Louvain (UCL) – CRED, D, Guha-Sapir, Brussels, Belgium. http://emdat.be/emdat_db (viewed 3 October 2021).

Gbetibouo, G.A. and Hassan, R.M. (2005). Measuring the economic impact of climate change on major South African field crops: a Ricardian approach. *Global and Planetary Change* 47 (2–4): 143–152.

Gitonga, Z.M., Visser, M., and Mulwa, C. (2020). Can climate information salvage livelihoods in arid and semiarid lands? An evaluation of access, use and impact in Namibia. *World Development Perspectives* 20: 100239.

Hitayezu, P., Wale, E., and Ortmann, G. (2017). Assessing farmers' perceptions about climate change: a double-hurdle approach. *Climate Risk Management* 17: 123–138.

Hossain, M.S., Arshad, M., Qian, L. et al. (2019). Economic impact of climate change on crop farming in Bangladesh: an application of Ricardian method. *Ecological Economics* 164: 106354.

Hundera, H., Mpandeli, S., and Bantider, A. (2019). Smallholder farmers' awareness and perceptions of climate change in Adama district, Central Rift Valley of Ethiopia. *Weather and Climate Extremes* 26: 100230.

Huong, N.T.L., Bo, Y.S., and Fahad, S. (2019). Economic impact of climate change on agriculture using Ricardian approach: a case of northwest Vietnam. *Journal of the Saudi Society of Agricultural Sciences* 18 (4): 449–457.

Inman, E.N., Hobbs, R.J., and Tsvuura, Z. (2020). No safety net in the face of climate change: the case of pastoralists in Kunene Region, Namibia. *PLoS One* 15 (9): e0238982.

Integrated Environmental Consultants Namibia (IECN) (2011). *Let's Act to Adapt*, 1–48. John Meinert Printing (Pty) Ltd.

Integrated Food Security Phase Classification (IPC) (2021). Namibia: Acute Food Insecurity Situation July - September 2020 and Projection for October 2020–March 2021. http://www.ipcinfo.org/ipc-country-analysis/details-map/en/c/1152879/?iso3=NAM (viewed 23 October 2021).

Kalele, D.N., Ogara, W.O., Oludhe, C., and Onono, J.O. (2021). Climate change impacts and relevance of smallholder farmers' response in arid and semi-arid lands in Kenya. *Scientific African* 12 (8): e00814.

Khumalo, T.F. (2021). Social issues related to climate change and food production (crops). In: *The Impacts of Climate Change* (ed. T.M. Letcher), 291–311. Elsevier.

Kurukulasuriya, P. and Ajwad, M.I. (2007). Application of the Ricardian technique to estimate the impact of climate change on smallholder farming in Sri Lanka. *Climatic Change* 81 (1): 39–59.

Kurukulasuriya, P., Mendelsohn, R., Hassan, R. et al. (2006). Will African agriculture survive climate change? *The World Bank Economic Review* 20 (3): 367–388.

Lamega, S.A., Komainda, M., Hoffmann, M.P. et al. (2021). It depends on the rain: smallholder farmers' perceptions on the seasonality of feed gaps and how it affects livestock in semi-arid and arid regions. *Climate Risk Management* 34: 100362.

Mabuku, M.P., Senzanje, A., Mudhara, M. et al. (2019). Strategies for coping and adapting to flooding and their determinants: a comparative study of cases from Namibia and Zambia. *Physics and Chemistry of the Earth, Parts A/B/C* 111: 20–34.

Mairura, F.S., Musafiri, C.M., Kiboi, M.N. et al. (2021). Determinants of farmers' perceptions of climate variability, mitigation, and adaptation strategies in the central highlands of Kenya. *Weather and Climate Extremes* 34: 100374.

Makuvaro, V., Walker, S., Masere, T.P., and Dimes, J. (2018). Smallholder farmer perceived effects of climate change on agricultural productivity and adaptation strategies. *Journal of Arid Environments* 152: 75–82.

Martey, E. and Kuwornu, J.K. (2021). Perceptions of climate variability and soil fertility management choices among smallholder farmers in Northern Ghana. *Ecological Economics* 180: 106870.

Mkonda, M.Y., He, X., and Festin, E.S. (2018). Comparing smallholder farmers' perception of climate change with meteorological data: experience from seven agroecological zones of Tanzania. *Weather, Climate, and Society* 10 (3): 435–452.

Mongi, H., Majule, A.E., and Lyimo, J.G. (2010). Vulnerability and adaptation of rain fed agriculture to climate change and variability in semi-arid Tanzania. *African Journal of Environmental Science and Technology* 4 (6): 371–381.

Montle, B.P. and Teweldemedhin, M.Y. (2014). Assessment of farmers' perceptions and the economic impact of climate change in Namibia: case study on small scale irrigation farmers (SSIFs) of Ndonga Linena Irrigation Project. *Journal of Development and Agricultural Economics* 6 (11): 443–454.

Mubiru, D.N., Radeny, M., Kyazze, F.B. et al. (2018). Climate trends, risks and coping strategies in smallholder farming systems in Uganda. *Climate Risk Management* 22: 4–21.

Mulwa, C., Marenya, P., and Kassie, M. (2017). Response to climate risks among smallholder farmers in Malawi: a multivariate probit assessment of the role of information, household demographics, and farm characteristics. *Climate Risk Management* 16: 208–221.

Namibia Statistics Agency (2011). *Zambezi Census Regional Profile*. Windhoek, Namibia: Namibia Statistics Agency.

Nationsonline (2021). Political map of Namibia. https://www.nationsonline.org/oneworld/map/namibia-political-map.htm (viewed 9 October 2021).

Newsham, A.J. and Thomas, D.S. (2011). Knowing, farming and climate change adaptation in North-Central Namibia. *Global Environmental Change* 21 (2): 761–770.

Nyang'au, J.O., Mohamed, J.H., Mango, N. et al. (2021). Smallholder farmers' perception of climate change and adoption of climate smart agriculture practices in Masaba South Sub-county, Kisii, Kenya. *Heliyon* 7 (4): e06789.

OECD/FAO (2021). *Building Agricultural Resilience to Natural Hazard-Induced Disasters: Insights from Country Case Studies*. Paris: OECD Publishing https://doi.org/10.1787/49eefdd7-en.

Ogalleh, S.A., Vogl, C., and Hauser, M. (2013). Reading from farmers' scripts: local perceptions of climate variability and adaptations in Laikipia, Rift Valley, Kenya. *Journal of Agriculture, Food Systems, and Community Development* 3 (2): 77–94.

Ojo, T.O. and Baiyegunhi, L.J.S. (2020a). Impact of climate change adaptation strategies on rice productivity in South-west, Nigeria: an endogeneity corrected stochastic frontier model. *Science of the Total Environment* 745: 141151.

Ojo, T.O. and Baiyegunhi, L.J.S. (2020b). Determinants of climate change adaptation strategies and its impact on the net farm income of rice farmers in south-west Nigeria. *Land Use Policy* 95: 103946.

Ojo, T.O. and Baiyegunhi, L.J.S. (2021). Climate change perception and its impact on net farm income of smallholder rice farmers in South-West, Nigeria. *Journal of Cleaner Production* 310: 127373.

Pachauri, R.K., Allen, M.R., Barros, V.R. et al. (2014). Climate Change 2014: Synthesis Report. Contribution of Working Groups I, II and III to the Fifth Assessment Report of the Intergovernmental Panel on Climate Change, IPCC, p. 151.

Ritchie, H., Roser, M. & Rosado, P. (2022). Namibia: CO_2 country profile. https://ourworldindata.org/co2/country/namibia

Seo, S.N. and Mendelsohn, R. (2008). Animal husbandry in Africa: climate change impacts and adaptations. *African Journal of Agricultural and Resource Economics* 2 (311-2016-5520): 65–82.

Shikangalah, R.N. (2019). Local perceptions of risks to semi-nomadic livelihoods in Namibia. *Journal for Studies in Humanities and Social Sciences* 8 (2): 96–116.

Shukla, P.R., Skea, J., Calvo Buendia, E. et al. (2019). IPCC, 2019: Climate Change and Land: An IPCC Special Report on Climate Change, Desertification, Land Degradation, Sustainable Land Management, Food Security, and Greenhouse Gas Fluxes in Terrestrial Ecosystems.

Simelton, E., Quinn, C.H., Batisani, N. et al. (2013). Is rainfall really changing? Farmers' perceptions, meteorological data, and policy implications. *Climate and Development* 5 (2): 123–138.

Spear, D. and Chappel, A. (2018). Livelihoods on the edge without a safety net: the case of smallholder crop farming in North-Central Namibia. *Land* 7 (3): 79.

Statista (2021). Namibia: Share of economic sectors in the gross domestic product (GDP) from 2010 to 2020. https://www.statista.com/statistics/510204/share-of-economic-sectors-in-the-gdp-in-namibia (viewed 3 October 2021).

Talanow, K., Topp, E.N., Loos, J., and Martín-López, B. (2021). Farmers' perceptions of climate change and adaptation strategies in South Africa's Western Cape. *Journal of Rural Studies* 81: 203–219.

Tambo, J.A. and Abdoulaye, T. (2013). Smallholder farmers' perceptions of and adaptations to climate change in the Nigerian Savanna. *Regional Environmental Change* 13 (2): 375–388.

Tesfahunegn, G.B., Mekonen, K., and Tekle, A. (2016). Farmers' perception on causes, indicators and determinants of climate change in Northern Ethiopia: implication for developing adaptation strategies. *Applied Geography* 73: 1–12.

Thinda, K.T., Ogundeji, A.A., Belle, J.A., and Ojo, T.O. (2020). Understanding the adoption of climate change adaptation strategies among smallholder farmers: evidence from land reform beneficiaries in South Africa. *Land Use Policy* 99: 104858.

World Bank Group (2021a). Climate Change Knowledge Portal. https://climateknowledgeportal.worldbank.org/country/namibia (viewed 4 October 2021).

World Bank Group (2021b). Climate Risk Profile: Namibia. https://climateknowledgeportal.worldbank.org/sites/default/files/2021-08/15931-WB_Namibia%20Country%20Profile-WEB.pdf (viewed 4 October 2021).

Zambezi Regional Council (2021). Overview. https://zambezirc.gov.na/regional-overview (accessed 3 October 2021).

Index

Climate Change and Agriculture: Perspectives, Sustainability and Resilience, First Edition. Edited by Noureddine Benkeblia.

C

f

g

n